SCIENCE PLUS

TECHNOLOGY AND SOCIETY

®

LEVEL BLUE

Project Directors

International: **Charles McFadden**
Professor of Science Education
The University of New Brunswick
Fredericton, New Brunswick

National: **Robert E. Yager**
Professor of Science Education
The University of Iowa
Iowa City, Iowa

Project Authors

Earl S. Morrison *(Author in Chief)*
Alan Moore *(Associate Author in Chief)*
Nan Armour
Allan Hammond
John Haysom
Elinor Nicoll
Muriel Smyth

This new United States edition has been adapted from
prior work by the Atlantic Science Curriculum Project,
an international project linking teaching, curriculum
development, and research in science education.

HOLT, RINEHART AND WINSTON
Harcourt Brace & Company
Austin • New York • Orlando • Atlanta • San Francisco • Boston • Dallas • Toronto • London

STAFF CREDITS

Editorial Director, Science and Social Studies
John Lawyer

Executive Editor
Robert W. Todd

Project Editors (Pupil's Edition)
Scott Snell
Robert Tuček (SourceBook)

Project Editor (Annotated Teacher's Edition)
R. Brent Lyles

Project Editor (Teaching Resources)
Jennifer Childers

Managing Editor, Science and Social Studies
Jim Eckel

Editorial Staff
Patrick Earvolino, Molly Gardner, Wendy Lym,
Gavin Mundy, Lance Wobus, Tanu'e White, Anne
Engelking, Christopher Hess, Meredith Phillips,
Daniel Chun, Janis Gadsden, Barbara Hofer,
Anne Geddes, Edward Connolly, Chris Parker,
Suzanne Lyons, John Gallo, Carolyn Biegert,
Rose Munsch

Copyediting
Steve Oelenberger, *Copyediting Supervisor*
Laurie Baker

Editorial Permissions
Lee Noble

Design
Richard Metzger, *Art Director*
Lisa Walston, Greg Geisler, Bob Prestwood,
Tonia Klingensmith, Jane Gilden, Jennifer Dix,
Stephen Sharpe, Sally Bess, Maria Lyle,
Heidi Haeuser, Alicia Sullivan, *Design Team*

Photo Research
Peggy Cooper, *Photo Research Manager*
Mavournea Hay, Cindy Bland Verheyden,
Sam Dudgeon, Victoria Smith, Diana Suthard,
Photo Research Team

Production
Gene Rumann, *Production Manager*
Amber Martin, Nancy Hargis, Belinda Barboza,
Shirley Cantrell, Jenine Street, *Production Team*

Electronic Publishing
Carol Martin, *Electronic Publishing Manager*
Kristy Sprott, Debra Schorn, Carla Beer,
JoAnn Davis, David Hernandez, Barbara
Hudgens, Heather Jernt, Mercedes Newman,
Rina Ouellette, Monica Shomos, Charlie
Taliaferro, *Electronic Publishing Team*

Morgan-Cain & Associates, *Cover Design*

The Quarasan Group, Inc., *Annotated Teacher's Edition Design and Production*

2000 Printing by Holt, Rinehart and Winston

7 032 00

ACKNOWLEDGMENTS

Project Advisors

Herbert Brunkhorst
Director, Institute for Science Education
California State University,
San Bernardino
San Bernardino, California

David L. Cross
Science Consultant
Lansing School District
Lansing, Michigan

Jerry Hayes
Associate Director, Science Outreach
Teacher's Academy,
Mathematics and Science
Chicago, Illinois

William C. Kyle, Jr.
*Director, School Mathematics
and Science Center*
Purdue University
West Lafayette, Indiana

Mozell Lang
Science Education Specialist
Michigan Department
of Education
Lansing, Michigan

Teacher Reviewers

Robert W. Avakian
Alamo Junior High
Midland, Texas

Paul Boyle
Perry Heights Middle School
Evansville, Indiana

Renae Cartwright
Cedar Park Middle School
Cedar Park, Texas

Kenneth Creese
White Mountain Junior High
Rock Springs, Wyoming

Harry Dierdorf
Educational Consultant
Carnegie Science Center
Pittsburgh, Pennsylvania

Pamela Jones
Birmingham Covington School
Birmingham, Michigan

Kevin Reel
Thacher School
Ojai, California

Larry Tackett
Andrew Jackson Junior High
Cross Lanes, West Virginia

Donald Yost
Cordova High School
Rancho Cordova, California

Academic Reviewers

David Armstrong, Ph.D.
Department of EPO Biology
University of Colorado
Boulder, Colorado

Bruce Briegleb
National Center for Atmospheric
Research
Boulder, Colorado

Kenneth Brown, Ph.D.
Professor of Chemistry
Northwestern Oklahoma State
University
Alva, Oklahoma

Linda K. Butler, Ph.D.
Division of Biological Science
University of Texas
Austin, Texas

Frederick R. Heck, Ph.D.
Associate Professor of Geology
Ferris State University
Big Rapids, Michigan

Arthur Huffman, Ph.D.
Department of Physics
University of California
Los Angeles, California

James Kaler, Ph.D.
Professor of Astronomy
University of Illinois
Urbana, Illinois

Gloria Langer, Ph.D.
University of Colorado
Boulder, Colorado

Doris I. Lewis, Ph.D.
Professor of Chemistry
Suffolk University
Boston, Massachusetts

R. Thomas Myers, Ph.D.
Professor of Chemistry Emeritus
Kent State University
Kent, Ohio

Alvin M. Saperstein, Ph.D.
Department of Physics
Wayne State University
Detroit, Michigan

Thomas Troland, Ph.D.
Associate Professor
Department of Physics and
Astronomy
University of Kentucky
Lexington, Kentucky

Blue-Ribbon Committee

The following teachers constituted
a special committee of existing
SciencePlus users who provided
information and insights on how to
improve the *SciencePlus* program.
Their input was invaluable.

**Patricia Barry, Carol
Bornhorst, Catherine
Carlson, Harry Dierdorf,
Sharon Effinger, Jeff Felber,
Barbara Francese, Ken
Horn, Roberta Jacobowitz,
Doug Leonard, Betsy Mabry,
Mary Beth McManus,
Debbie Melphi, Lynn
Roudabush-Novak, Donna
Robinson, Sandy Moerke-
Schaefer, Marvin Selnes,
Margaret Steinheimer,
Sandy Tauer, Joy Ward,
Gary Weaver, Nancy
Wesorick, Brenda West**

Project Associates

We wish to thank the thousands of
science educators, teachers, and
administrators from the scores of
universities, high schools, junior
high schools, and middle schools
who have contributed to the success
of *SciencePlus*.

CONTENTS

Unit 1 LIFE PROCESSES   2

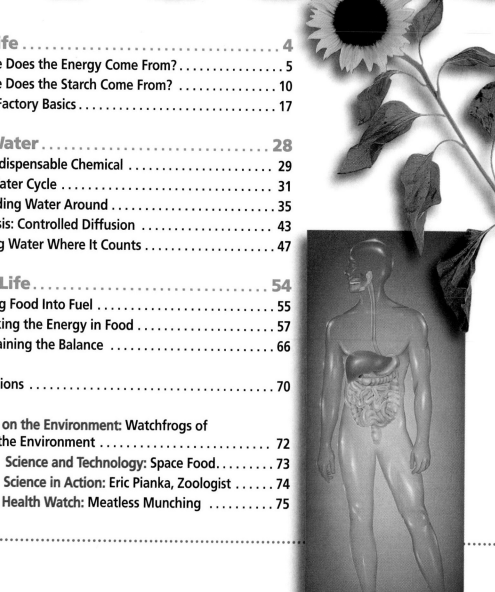

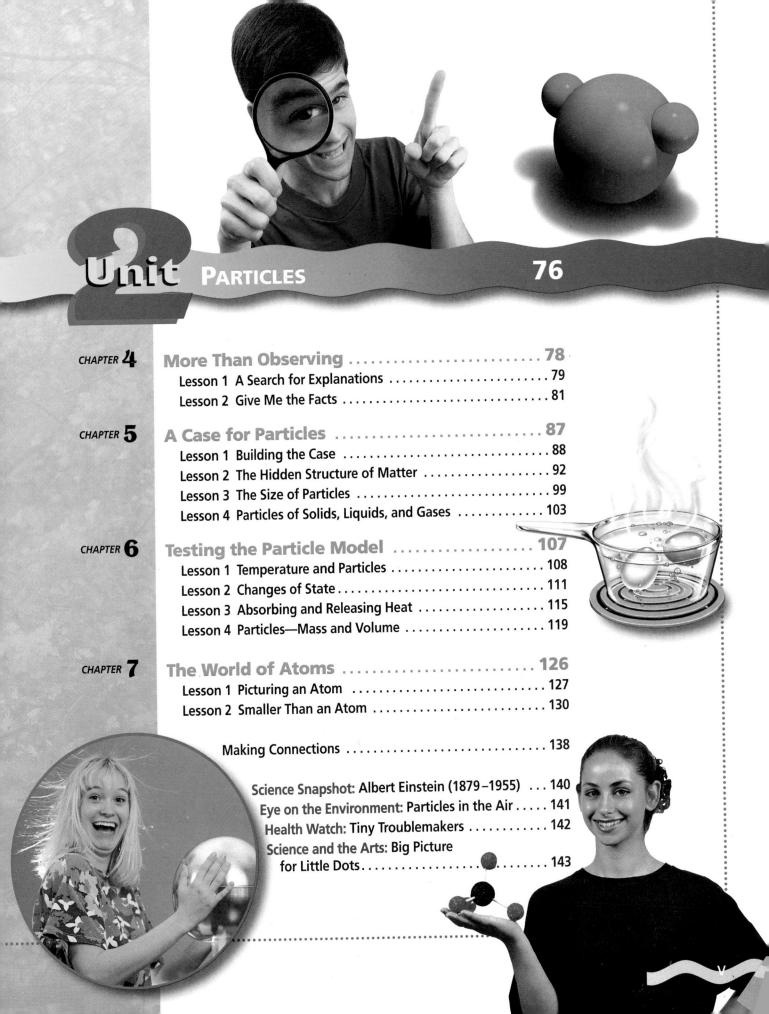

v

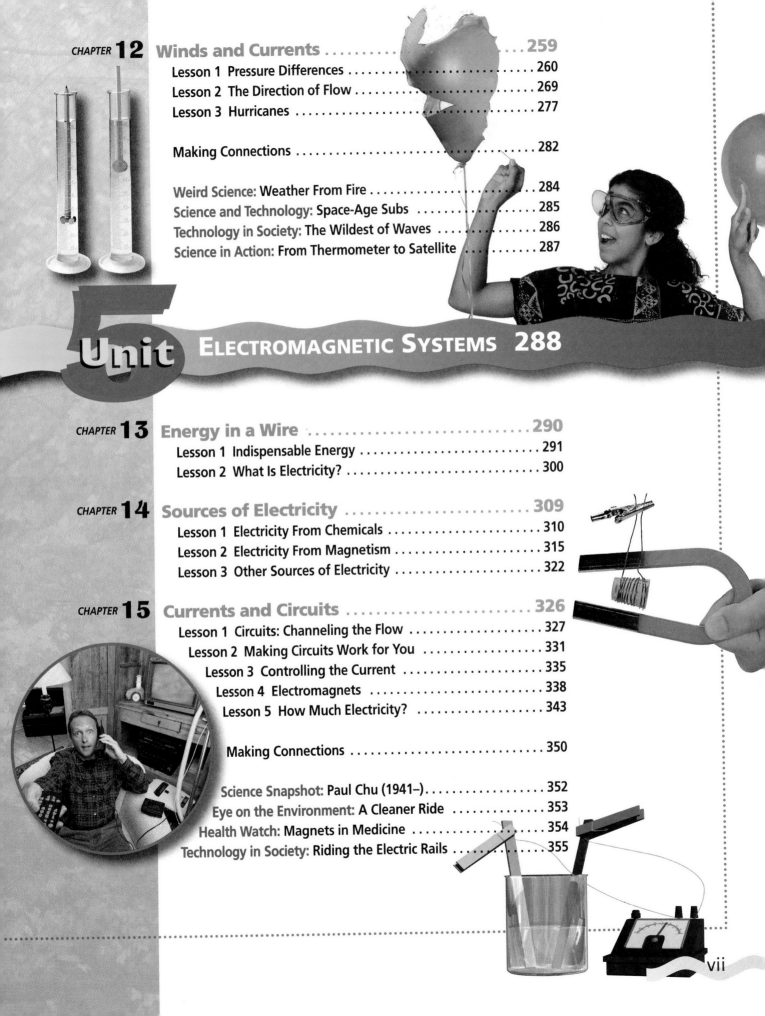

Unit 5
ELECTROMAGNETIC SYSTEMS   288

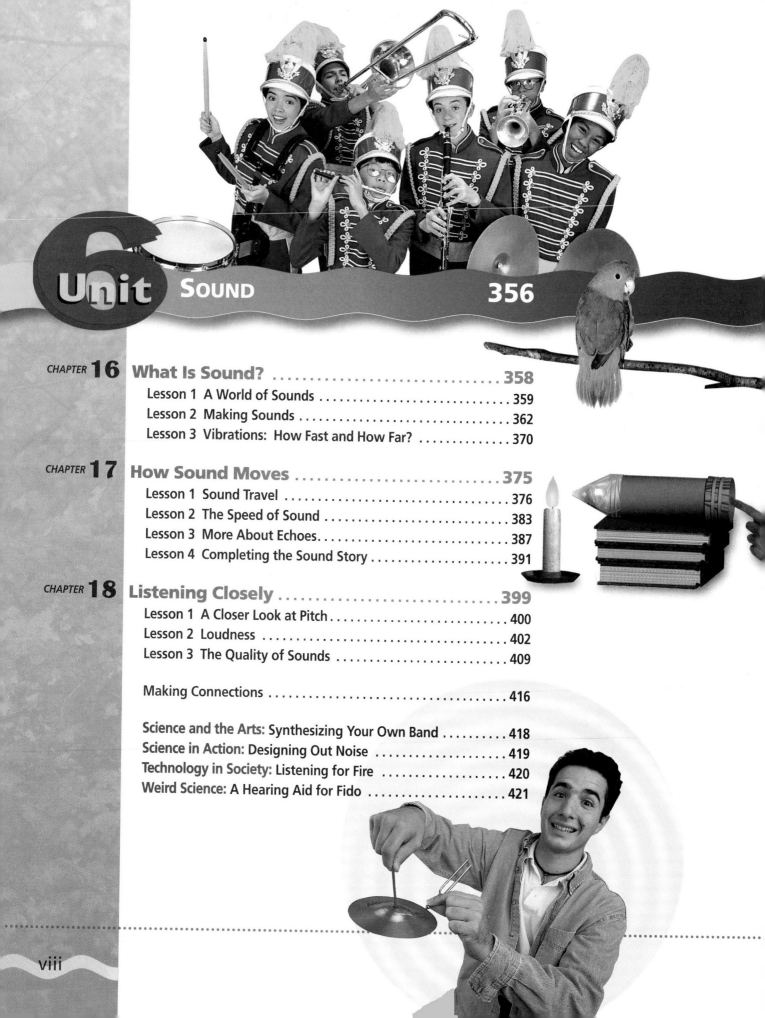

Unit 6 SOUND 356

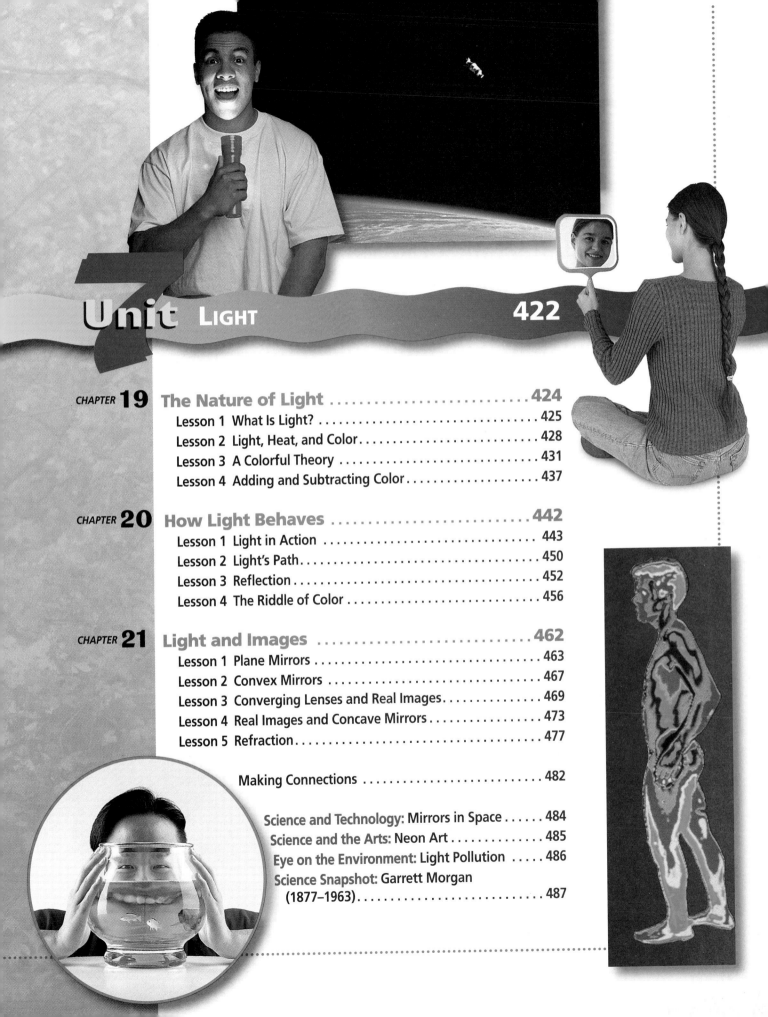

Unit 7 LIGHT 422

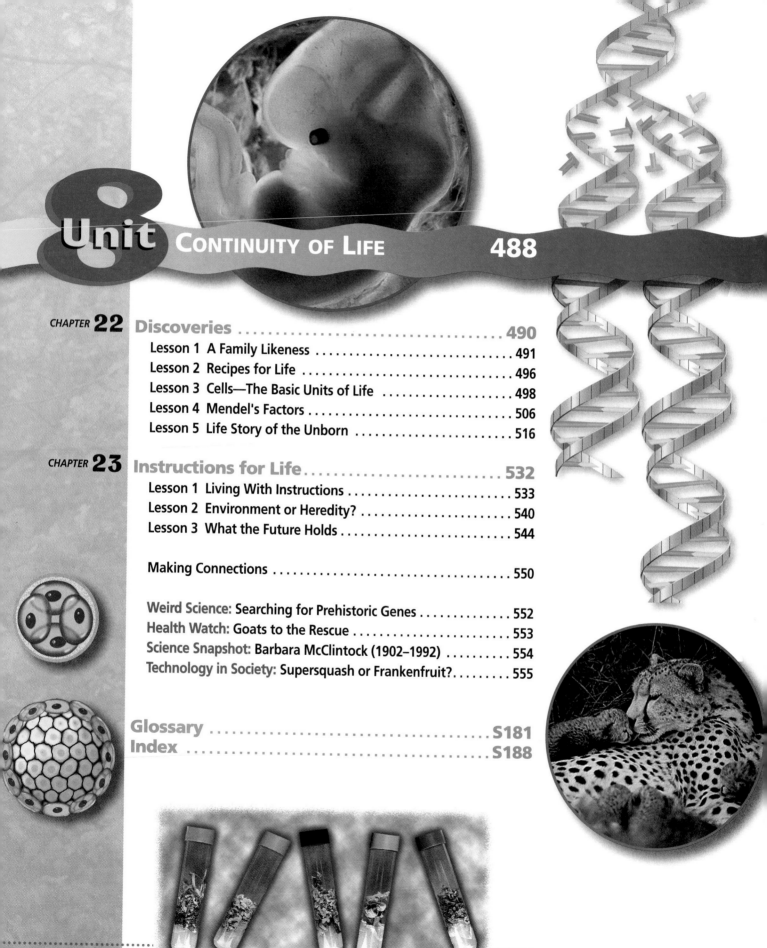

Unit 8 Continuity of Life 488

xi

TO THE STUDENT

This book was written with you in mind!

There are many things to try, to create, and to investigate—both in and out of class. There are stories to read, articles to think about, puzzles to solve, and even games to play.

GET INVOLVED!

The best way to learn is by doing. In the words of an old Chinese proverb:

Tell me—*I will forget*

Show me—*I may remember*

Involve me—*I will understand*

The activities in this book will allow you to make some basic and important scientific discoveries on your own. You will be acting much like the early investigators in science who, without expensive or complicated equipment, contributed so much to our knowledge.

What these early investigators had, and had in abundance, was curiosity and imagination. If you have these qualities, you are in good company! And if you develop sharp scientific skills, who knows?— you might make your own contributions to science someday.

Scientists are usually interested in understanding things that happen in nature. However, the discoveries that scientists make are often used by inventors and engineers. Using science in this way has resulted in our most sophisticated technology, including such things as computers, laser discs, nuclear reactors, and instant global communication.

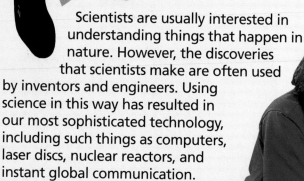

SCIENCE & TECHNOLOGY

There is an interaction between science and technology. Science makes technology possible. On the other hand, the products of technology are used to make further scientific discoveries. In fact, much of the scientific work that is done today has become so technically complicated and expensive that no one person can do it entirely alone. But make no mistake, the creative ideas for even the most highly technical and expensive scientific work still come from individuals.

A built-in reference section is located at the back of this book. It's called the SourceBook. **CHECK IT OUT!**

Keep a ScienceLog

A journal is an important tool in creative work. In this book, you will be asked to keep a type of journal, called a ScienceLog, to record your thoughts, observations, experiments, and conclusions. As you develop your ScienceLog, you will see your own ideas taking shape over time. This is often the way scientists arrive at new discoveries. You too may log some discoveries as you develop your own journal.

GO FOR IT!

Science is a process of discovery, a trek into the unknown. The skills you develop as you do the activities in this book—like observing, experimenting, and explaining observations and ideas—are the skills you will need in order to be a part of science in the future. There is a universe of scientific exploration and discovery awaiting those who take up the challenge.

SAFETY FIRST!

The study of science is challenging and fun, but it can also be dangerous. Don't take any chances! Follow the guidelines listed here, as well as safety information provided in the particular Exploration you are doing. Also, follow your teacher's instructions and don't take shortcuts—even when you think there is little or no danger.

Accidents can be avoided. The major causes of laboratory accidents are carelessness, lack of attention, and inappropriate behavior. These things reflect a person's attitude. By adopting a positive attitude and by following all safety guidelines, you can greatly reduce your chances of having an accident. Even a minor accident in a science laboratory can cause major injuries, so be very careful.

SAFETY GUIDELINES

GENERAL

Always get your teacher's permission before attempting any laboratory explorations. Read the procedures carefully, paying particular attention to safety information and caution statements. If you are unsure about what a safety symbol means, look it up here or ask your teacher. You cannot be too careful when it comes to safety! If an accident does occur, inform your teacher immediately, regardless of how minor you think the accident is.

EYE SAFETY

Wear safety goggles when working around chemicals, acids, bases, or any type of flame or heating device, and any other time when there is even the slightest chance that harm could come to your eyes. If any substance gets into your eyes, notify your teacher immediately. Treat any unknown chemical as if it were a dangerous chemical. Never look directly into the sun with an optical device, and never use direct sunlight as a light source for a microscope.

SAFETY EQUIPMENT

Know the location of and how to use the nearest fire alarms and any other safety equipment, such as fire blankets and eyewash fountains, as identified by your teacher.

NEATNESS

Keep your work area free of all unnecessary books and papers. Tie back long hair and secure loose sleeves or other loose articles of clothing such as ties and bows. Remove dangling jewelry. Don't wear open-toed shoes or sandals in laboratory situations. Never eat, drink, or apply cosmetics in a laboratory setting; food, drink, and cosmetics can easily become contaminated with dangerous materials.

SHARP/POINTED OBJECTS

Use knives and other sharp instruments with extreme care. Never cut objects while holding them in your hands. Place objects on a suitable work surface for cutting.

HEAT

Wear safety goggles when using a heating device or a flame. Whenever possible, use an electric hot plate instead of a flame as a heat source. When heating materials in a test tube, always slant the test tube away from yourself and others. Wear oven mitts, when instructed to do so, to avoid burns.

ELECTRICITY

Be careful with electrical wiring. When using a microscope with a lamp, do not place the cord where it could cause someone to trip. Do not let cords hang over a table edge in a way that could cause equipment to fall if the cord is accidentally pulled. Do not use equipment with damaged cords. Be sure your hands are dry and that the electrical equipment is in the "off" position before plugging it in. Turn off equipment when you are done.

Never taste, touch, or smell chemicals unless you are specifically directed to do so by your teacher.

If you are instructed to note the odor of a substance, wave the fumes toward your nose with your hand. Never put your nose close to the source. Never mix chemicals unless you are told to do so by your teacher.

CHEMICALS

Wear safety goggles when handling any potentially dangerous chemicals, acids, or bases. If a chemical is unknown, handle it as you would a dangerous chemical. Wear an apron and latex gloves when working with acids or bases or when told to do so in the Exploration or Activity. If a spill gets on your skin or clothing, rinse it off immediately with water for at least 5 minutes while calling your teacher.

ANIMAL SAFETY

Always obtain your teacher's permission before bringing any animal into the school building. Handle animals only as your teacher directs. Always treat animals carefully and with respect. Wash your hands thoroughly after handling any animal.

PLANT SAFETY

Do not eat any part of a plant or plant seed used in the laboratory. Wash hands thoroughly after handling any part of a plant.

GLASSWARE

Examine all glassware before using. Be sure that it is clean and free of chips and cracks. Report damaged glassware to your teacher. Glass containers used for heating should be made of heat-resistant glass.

CLEANUP

Before leaving, clean up your work area. Put away all equipment and supplies. Dispose of all chemicals and other materials as directed by your teacher. Make sure water, gas, burners, and electric hot plates are turned off. Hot plates and other electrical equipment should also be unplugged. Wash hands with soap and water after working in a laboratory situation.

concept mapping

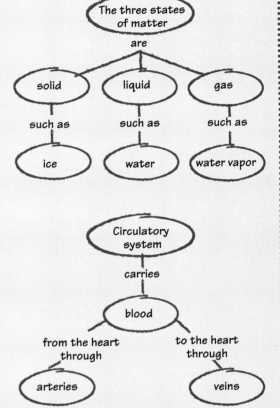

A WAY TO BRING IDEAS TOGETHER

What Is a Concept Map?

Have you ever tried to tell someone about a book or a chapter you've just read, and you find that you can remember only a few isolated words and ideas? Or maybe you've memorized facts for a test, and then weeks later you're not even sure what topic those facts are related to.

In both cases, you may have understood the ideas or concepts by themselves, but not in relation to one another. If you could somehow link the ideas together, you would probably understand them better and remember them longer. This is something a concept map can help you do. A concept map is a visual way of choosing how ideas or concepts fit together. It can help you see the "big picture."

How to Make a Concept Map

1. **Make a list of the main ideas or concepts.**

   It might help to write each concept on its own slip of paper. This will make it easier to rearrange the concepts as many times as you need to before you've made sense of how the concepts are connected. After you've made a few concept maps this way, you can go directly from writing your list to actually making the map.

2. **Spread out the slips on a sheet of paper, and arrange the concepts in order from the most general to the most specific.**

   Put the most general concept at the top and circle it. Ask yourself, "How does this concept relate to the remaining concepts?" As you see the relationships, arrange the concepts in order from general to specific.

3. **Connect the related concepts with lines.**

4. **On each line, write an action word or short phrase that shows how the concepts are related.**

Look at the concept maps on this page and then see if you can make one for the following terms: **plants, water, photosynthesis, carbon dioxide, and sun's energy.**

The answer is provided below, but don't look at it until you try the concept map yourself.

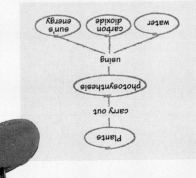

Unit 1

Life Processes

A mist-shrouded mountainside lies awash in lush tropical greenery. Every square meter, it seems, is occupied by riotous growth. Thanks to nearly optimal conditions, the plants in this picture grow at a furious pace, seeking the sunlight that makes their existence possible. To human eyes, the view is beautiful. But what function underlies such dazzling form?

The reader cannot help but notice how intensely green is this scene. Every conceivable shade of this color, it seems, is represented here. Obviously this is no accident. The "green" in greenery is clearly important, but how? The chemical that is the source of the green color actually enables plants to capture and store the energy of sunlight. This energy is transferred whenever a plant—or the animal that eats it—is consumed by another living thing.

Water also figures prominently in this photograph. Although not itself organic, water is perhaps the most important chemical of life. No living thing can exist without water; its unique chemical properties are critical to life processes.

When all of the ingredients of life—water, sunlight, and vital gases—are abundant, as they are here, life flourishes. But remove any one of these ingredients and life quickly ends—or never even begins.

Lush rain forest covers a mountain on the northeastern side of the island of Hawaii.

Energy for Life

1 Where do plants get their food?

2 Describe a typical plant.

What's inside a leaf?
3

Sciencelog

Think about these questions for a moment, and answer them in your ScienceLog. When you've finished this chapter, you'll have the opportunity to revise your answers based on what you've learned.

Where Does the Energy Come From?

Without plants, there would be no food.

The girl in the photograph is eating a popular meal—green leaves, slices of ripened plant ovaries, ground-up seeds, and cow muscle. Actually, most people call it a hamburger! You probably know that hamburgers and other foods provide us with the energy necessary to function. You need energy from food to go about your daily activities, such as walking, talking, thinking—and eating more food!

Now consider the following statement:

Do you think the statement above is true? To help you decide, think about the hamburger in the photograph. The lettuce, obviously, is a plant. The bun is made from wheat, so it is also plant matter. But how about the meat from the cow? Where did the cow get its food? Cows and other plant-eating animals eat plants directly. Humans and other meat-eating creatures eat the animals that ate plants. Somewhere in the food chain there is a connection to plants.

Without plants, there would be no food. How can that be true? If we didn't have plants to eat, we would still have meat, right?

Right, but who wants to eat just meat?

Besides that, where does the meat come from? That animal has to eat something.

Yeah, and most of the animals we use for food eat plants, as far as I know.

Okay, but how does the plant get food?

Here is your challenge: Figure out what should go where the question mark is. By thinking and doing the Explorations in this chapter, you will figure out the answer to the question, "How do plants get food?" Get started by forming a group with several other students, and do the following:

1. Read and discuss possible answers to Question 1 on page 4. In talking, you may come up with ideas you hadn't thought of before.

2. Work out an answer that your group considers reasonable.

3. Make a poster that clearly illustrates your group's answer.

4. Decide what you could do to test your answer. On the back of your poster, illustrate your plan with labeled diagrams.

5. Share your ideas with the other groups, explaining both sides of your poster to them.

As you do the Explorations in this chapter, see how closely they resemble the plan you suggested on the back of your poster.

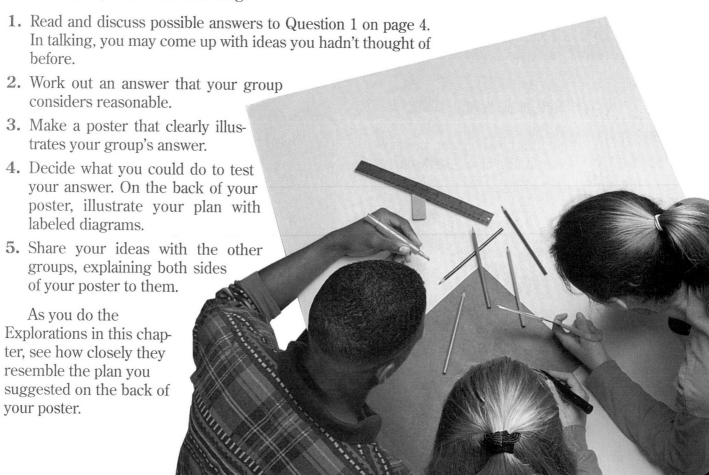

Some Food for Thought

Before you get serious about figuring out how plants get food, you have to know what kind of food is in plants. The food in plants is **starch**. The starch in plants provides almost 70 percent of the world's food supply.

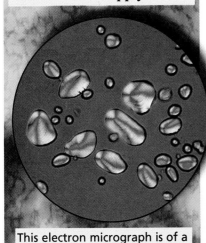

This electron micrograph is of a starch-storage cell in a potato. The sphere-shaped objects are granules of pure starch.

If you look at the photograph above, you will see some starch granules inside a potato plant. The photograph was taken using an electron microscope. (An electron microscope is a powerful instrument that uses a beam of electrons to form an enlarged image.) You can't use an electron microscope in class, but there is another way to tell whether something contains starch, which you will learn in Exploration 1. You will also use the technique later in the unit as you answer the question, "Where do plants get food?"

The Search for Starch

PART 1

A Simple Test

You Will Need

- 2 containers
- cornstarch
- an eyedropper
- a stainless steel scoop
- a glass stirring rod
- iodine solution
- sugar
- water

Add iodine to each container to test for starch.

What to Do

1. Mix some cornstarch in water, and then add two drops of iodine solution. What happens?

   Be Careful: Iodine is poisonous and stains clothing and skin. Wear a lab apron.

2. Repeat the test with plain water. Then add a little sugar to the water and repeat the test. How do the results of these tests differ from the result of the test with cornstarch?

3. When testing for a substance, a *positive* result signals that the substance is present; a *negative* result means it is not present. What signaled the presence of starch? For each of your tests for starch, was the result positive or negative?

Be sure to save iodine solution for later Explorations.

Exploration 1 continued ▶

PART 2

Starch in Green Leaves

You Will Need

- oven mitts
- water
- a geranium plant
- a large beaker
- a small beaker
- methanol
- a hot plate
- tincture of iodine
- an eyedropper
- white paper
- latex gloves

What to Do

1. Remove two leaves from a healthy plant that has been growing in a sunny spot.

2. Remove the green color from the leaves. To do this, first heat water in a large beaker until it boils.

3. Turn off the heat source. While wearing oven mitts, move the beaker to a place away from any heat sources.

4. Put the leaves into a small beaker and cover them with methanol.

   Be Careful: Methanol is poisonous and highly flammable. Keep methanol away from open flames and other heat sources. Wear a lab apron, latex gloves, and goggles.

5. Put the small beaker into the large beaker containing hot water, as shown in the diagram. The methanol will become warm enough to boil.

6. Soak the leaves in the warm methanol for 5 to 10 minutes. The methanol should become green as it dissolves the coloring of the leaves. Simultaneously, the leaves should become pale.

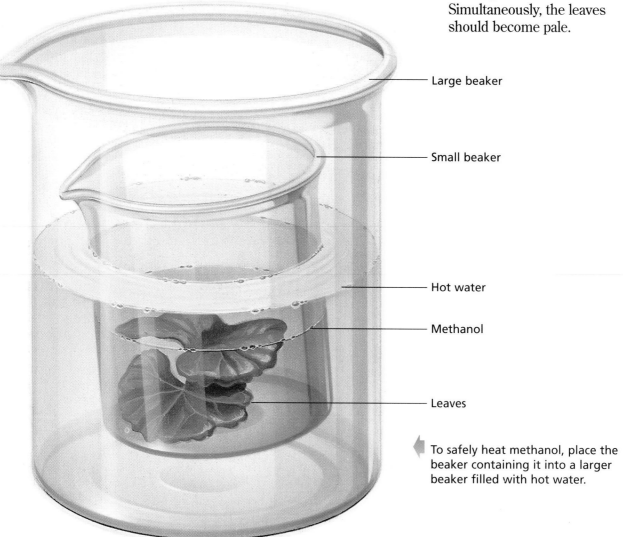

— Large beaker

— Small beaker

— Hot water

— Methanol

— Leaves

To safely heat methanol, place the beaker containing it into a larger beaker filled with hot water.

7. Remove the leaves from the methanol and rinse them with tap water. Spread them face up on a piece of white paper.

8. Treat one leaf with three or four drops of the iodine solution. Then place the leaf in water to *fix* the color (make it permanent). Compare the treated leaf with the untreated leaf. Look for any slight change in color.

Questions

1. Sketch the two leaves in your ScienceLog, noting any changes observed.

2. What evidence shows you that starch is present in a plant leaf?

3. Why was it necessary to heat the methanol that you used to soak the leaves?

4. What was the purpose of using two leaves if only one was treated with iodine?

Starch—Essential for Life

The starch in plants is essential to other living things. Even organisms that do not feed on plants depend on them indirectly for food. Carnivorous (meat-eating) animals, for example, prey upon other animals. But the preyed-upon animals either ate plants or ate animals that ate plants. Consider the statement, "All meat is ultimately grass." What does it mean? Write your explanation in your ScienceLog.

Starch is a carbohydrate—a complex organic compound. Learn more about organic compounds on pages S2–S6 of the SourceBook.

Where Does the Starch Come From?

In Explorations 2, 3, and 4 you will be figuring out where the starch in plants comes from. As you do the Explorations, think about what light, water, and carbon dioxide could have to do with the starch in plants.

EXPLORATION 2

Light and Starch

The leaf you tested for starch in Exploration 1 was from a plant that had received plenty of light. If a plant receives no light, will it also have starch in its leaves?

You Will Need

- a geranium plant (grown outdoors)
- thin cardboard
- scissors
- straight pins
- materials and equipment to test leaves for starch

What to Do

1. Put a geranium plant in a dark but warm place for 4 days. The soil should be moist but not too wet.

2. After the 4 days have passed, test some of the plant's leaves for starch, as in Exploration 1. Record your findings in your ScienceLog.

3. Now cover one leaf with cardboard, as shown in the illustrations. (Important—the leaf must still be attached to the plant!) Affix a letter or a number to the top surface of the leaf, using the method shown in the illustration.

4. Put the plant in bright light, such as under a grow lamp, for 24 hours. Then remove the cardboard pieces and test the covered and uncovered parts of the leaf for starch as you did in Part 2 of Exploration 1. Record your findings.

Questions

1. What effect did darkness have on the amount of starch in the plant?

2. Which part of this Exploration, steps 1 and 2 or steps 3 and 4, suggests that light has an important role in determining the amount of starch in green plants? Why?

3. Could you perform steps 3 and 4 without first doing steps 1 and 2? Explain your answer.

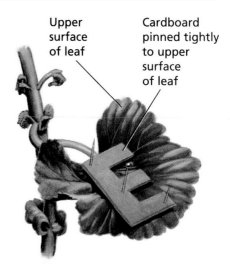

Upper surface of leaf

Cardboard pinned tightly to upper surface of leaf

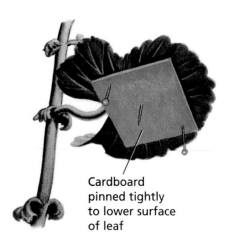

Cardboard pinned tightly to lower surface of leaf

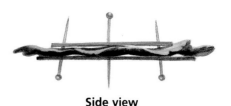

Side view

Water—How Essential Is It?

When you started this chapter, you may have thought that plants get their food only from the soil. If you did, you are in good company. That's what the great philosopher Aristotle thought too! But a Belgian scientist of the seventeenth century, Jan Baptista van Helmont, questioned this belief and decided to look into the matter further. He decided to find out what role water had in plant growth.

What happens when a plant is deprived of water? You have probably seen what happens when you or somebody else forgets to water a houseplant. Plants that don't get enough water wilt. If they continue to be deprived of water, they eventually die.

Van Helmont performed an experiment that convinced him that water was so important that it was completely responsible for the great change in mass that occurs in growing plants. In other words, he concluded that water was completely responsible for growth in plants.

What do you think? Before you decide, let van Helmont tell you about his experiment, translated from his own words. Don't let the odd wording and spelling throw you. People spelled and spoke quite differently in van Helmont's day. (The masses have been changed to metric units.)

The corn plant on the left is healthy because it received adequate rainfall. The corn on the right is stunted because it did not receive enough water.

Journal

I have learned from this handicraft operation that all vegetables do immediately and materially proceed out of the element of water only. For I took an earthen vessel in which I put ninety kilograms of earth that had been dried in a furnace, which I moystened with rainwater, and I implanted therein the trunk, or stem, of a Willow tree, weighing two kilograms and about two-hundred and fifty grams. At length, five years being finished, the tree sprung from thence did weigh seventy-six kilograms. But I moystened the earth vessel with rain-water, or distilled water (always when there was need) . . . and lest the dust should be co-mingled with the earth, I covered the lip or mouth of the vessel with an iron plate covered with tin, and easily passable with many holes. I computed not the leaves that fell off, in the four autumns. At length, I again dried the earth of the vessel, and there were found the same ninety kilograms, wanting a few grams. Therefore almost seventy-four kilograms of wood, bark, and roots arose out of water only.

Jan Baptista van Helmont

Here is a summary of van Helmont's experiment. Fill in the missing data in your ScienceLog.

Mass of dried earth = ⁇

Mass of willow tree at the beginning of experiment = ⁇

Van Helmont's experiment indicated that water was an important raw material for plants. Somehow, plants used water to build new tissues and structures. It was left to others, however, to find out exactly how the conversion of water into plant matter takes place.

Van Helmont was a careful experimenter. Can you give some examples that show this? But even though he was careful, van Helmont made a major mistake when he concluded that water alone was completely responsible for the increase in mass in plants. What do you think was wrong with his conclusion?

In the next lesson you will perform Explorations involving gases that van Helmont knew nothing about. You will also be getting more information that will allow you to answer the question, "Where does the starch in plants come from?"

Mass of dried earth at the end of the experiment = ⁇

Mass of grown tree = ⁇

What About Carbon Dioxide?

Scientists know that starch is composed of three substances: carbon, hydrogen, and oxygen. You cannot see hydrogen or oxygen (they are colorless gases), but you can see carbon. If you cook carrots too long, they eventually turn black. What happens when bread gets stuck in a toaster? It, too, turns black. In fact, any food that is cooked too long will turn black because food is made from living things and all living things contain carbon.

Where do plants get their carbon? The answer is: from the air. As you will learn in a later unit, air is made up of different gases. One of these gases is carbon dioxide, which is produced whenever carbon-containing substances are burned. A burning candle produces carbon dioxide, for example. You can test for the presence of carbon dioxide with the chemical *limewater,* which consists of calcium hydroxide dissolved in water. Limewater turns milky when it comes into contact with carbon dioxide.

Testing for Carbon Dioxide

Try the following activity. Take two small jars with lids and put a small warming candle in each one. Light only one candle. Then put the lids on both containers. After the candle flame goes out, pour limewater into both jars. Remove the candles. Gently swirl the liquid, keeping the lids on the jars. What happens to the limewater? Why is a jar with an unlit candle also used?

1

2

3

EXPLORATION 3

Carbon Dioxide and Starch

You Will Need

- sodium hydroxide solution (4%)
- baking soda (sodium bicarbonate) solution (5%)
- a lamp with 100 W bulb
- materials and equipment to test leaves for starch
- 2 large jars with lids
- 2 test tubes
- a wax pencil
- 2 leaves

What to Do

Arrange the two jars as shown in the illustration below. Jar 1 contains a 4% solution of sodium hydroxide, which absorbs carbon dioxide, thus removing it from the air.

Be Careful: Handle sodium hydroxide solution carefully—it can burn your skin!

The second jar holds a 5% baking soda solution. To make a 5% solution, add 1 part baking soda to 19 parts water. Baking soda contains carbon; its chemical name is *sodium bicarbonate*. If all of the carbon dioxide in the air is used up by the plant, the baking soda releases more carbon dioxide into the air.

Label the jars with a wax pencil. Put the jars in strong light for 3 to 5 days. Then test each leaf for starch. Do you observe differing amounts of starch? What do you conclude about the relationship between carbon dioxide and starch in plants? How could you have tested for the presence of carbon dioxide in the two jars?

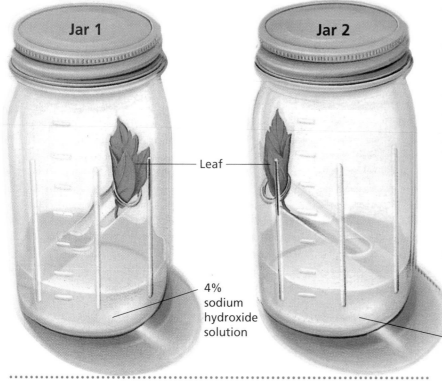

Jar 1

Jar 2

Leaf

4% sodium hydroxide solution

5% sodium bicarbonate (baking soda) solution

Put It All Together!

So now what do you think about how plants get food? If you decided, as a result of doing these Explorations, that plants use light, water, and carbon dioxide to make starch, you are right. Plants actually *make* their own food in their leaves. To do this, they need energy from the sun plus two raw materials: water and carbon dioxide. The process by which plants make food is called **photosynthesis**. Like many scientific terms, this one is a combination of two Greek words: *photo,* which means "light," and *synthesis,* which means "putting together." Given what you learned in doing the Explorations, why do you think the process is called *photosynthesis*?

Ultimately, photosynthesis is responsible for feeding practically all the organisms on Earth! Plants use photosynthesis to make their own food. Animals feed directly on plants, and other animals feed on those animals. At your next meal, think about the fact that none of the food on your plate could exist without photosynthesis.

Now, here is your next challenge:

What do plants give off during photosynthesis?

Guess Which Gas!

You Will Need

- a lamp with 100 W bulb
- 2 test tubes
- a wooden splint
- baking soda
- 2 clear plastic containers
- 2 funnels
- an elodea sprig
- water
- matches

What to Do

1. Arrange the materials as shown in the illustrations.

2. Leave each apparatus in bright light for at least 24 hours. What happens?

3. If a test tube has collected gas, remove it from the container, making sure to hold your finger over the end of the tube so that the gas does not escape.

4. Insert a glowing (not burning) wooden splint into the test tube. If the gas is carbon dioxide, the splint will go out immediately. If the gas is oxygen, the splint will burst into flames.

Questions

1. What discoveries did you make?

2. Why is a setup without a plant sprig also used?

3. Notice that baking soda dissolved in water was used again in this Exploration. What purpose did this serve?

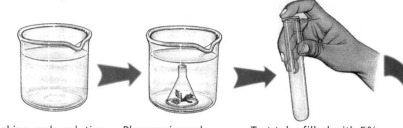

5% baking soda solution | Place sprig and funnel in container. | Test tube filled with 5% baking soda solution

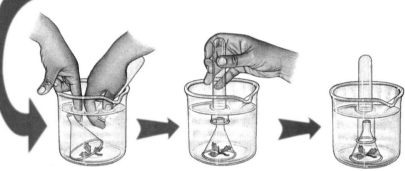

Invert the test tube while holding your finger over its mouth. | Lower the filled test tube over the funnel. | **Wait 24 hours**

Setup With Control

Baking soda in water

Sprig of elodea

Testing for gas

Insert the glowing splint quickly.

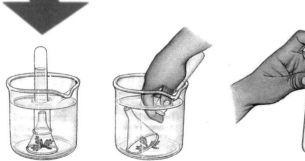

A Formula for Food

Photosynthesis can be likened to the processes that occur in a factory: raw materials come in, energy is used, a finished product results, and wastes are produced. Now, check what you have learned about leaves as "food factories."

1. What important product results from photosynthesis?
2. What is one essential raw material for photosynthesis?
3. What other raw material is required?
4. What is the source of energy for photosynthesis?
5. What is a waste product of photosynthesis? Why is this waste product important to other living things?

Now, write a description of photosynthesis based on what you have learned so far in this unit. Use the following terms in your description: light, oxygen, waste, starch, water, raw material, carbon dioxide, food, manufacture, and product. Your description of photosynthesis could turn out to be quite long. A much shorter description, which still conveys a great deal of information, is the following simple equation:

Copy this equation into your ScienceLog. Then replace each question mark with the number of the question above that relates to it. What does the arrow mean?

A Final Word on Photosynthesis

You have learned that photosynthesis produces a food called starch. However, when it is first made, the food in the leaf is actually in a form that is chemically less complex than starch—a simple sugar. Many sugar particles join together in a specific way to form the more complex starch particle. The sugar is changed into starch when the plant stores food for future use. When the plant needs food, it changes some of its stored starch back into sugar. If this is so, why didn't you test for sugar? Mostly because it is easier to test for starch than for sugar, and the presence of starch is evidence that food is being manufactured by the leaf.

You can more accurately rewrite the photosynthesis word equation as follows:

Food Factory Basics

Plant Structure

A well-designed structure is a thing of beauty. Architects and engineers strive to design buildings, bridges, machines, and other structures that are elegant as well as efficient and functional. Structures are also found in nature. Take plants, for example. How do they rate in terms of their efficiency? their functionality? their "ingenuity" of design?

Quickly draw a plant. What parts did you show—leaves? stem? roots? What other plant parts might you include? How about flowers, fruit, and seeds? When we say the word *plant,* we usually think first of a large, important group that we call the seed plants. This group includes many familiar plants. Look at the drawing of a typical seed plant.

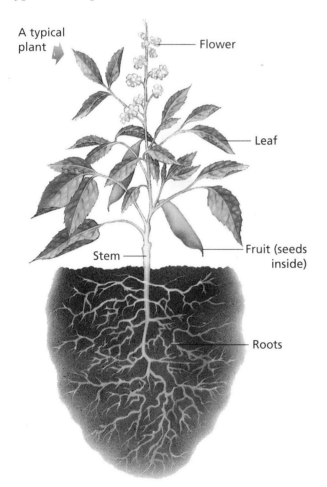

A typical plant

Flower

Leaf

Fruit (seeds inside)

Stem

Roots

As you know, parts of your body that have special functions, such as your heart or your stomach, are called **organs**. Plant parts may also be called organs because they perform special functions for a plant. What are the special functions of each of the plant organs shown in the drawing at lower left?

Plant Organ	Functions
Roots	
Stem	
Leaves	
Flowers	
Fruit	
Seeds	

Listed below are a number of functions (tasks) that plant parts perform. Which plant organ carries out which function? Copy the names of the plant organs into your ScienceLog and then match the letters indicating functions with the appropriate plant organs.

Functions of Plant Organs

More than one organ may perform certain functions, so you may want to use a letter more than once.

a. Make food by photosynthesis
b. Hold plant firmly in the soil
c. Produce the leaves
d. Exchange gases with the air
e. Allow the plant to reproduce
f. Store food for later use
g. Hold leaves up to the light
h. Produce a new plant without using seeds
i. Absorb water and minerals from the soil
j. Hold the developing seeds
k. Let water vapor escape into the air
l. Conduct water and minerals upward to the leaves

Out of the Ordinary

You have just identified functions of stems and roots. You also know where to look to find these parts on a plant. Some stems and roots, however, grow in unusual locations on a plant, where they have special functions to perform.

Examine the plants shown at right. Refer to the exercise you have just completed to help you

- identify what is unusual about the location of one part of each plant.

- explain what special function that part might perform because of its location.

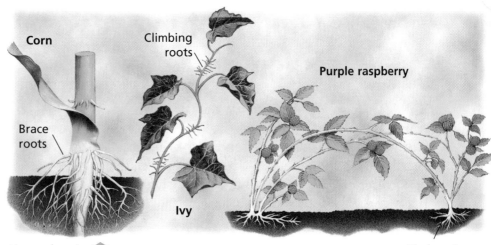

Corn
Climbing roots
Purple raspberry
Brace roots
Ivy
Unusual roots
Tip layering

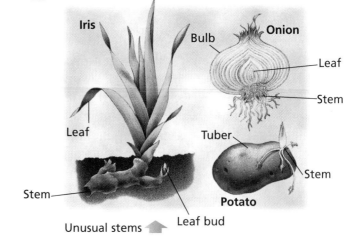

Iris
Bulb
Onion
Leaf
Stem
Leaf
Tuber
Stem
Stem
Potato
Unusual stems
Leaf bud

Edge
Blade
Vein
Petiole

Leaves

A leaf is specially structured for its main task—the production of food by photosynthesis. The thin green part is called the **blade**. Each kind of plant has a leaf blade that distinguishes it, by size or shape, from other kinds of plants.

A network of **veins** gives strength to the leaf. The pattern of veins is different for each kind of plant. Materials move into and out of the leaf through the veins.

The leaf is fastened to the stem, usually by means of a stalk called the **petiole**. The leaves of some plants are fastened directly to the stem, without petioles.

Sycamore

Note the different vein patterns in these leaves. How might you describe them?

Elm

Lily

Grass

Beech

Carrot

How might you describe the leaf edges shown here?

Maple

Mimosa

Oak

Simple or compound? How can you tell?

Although leaves show great variation, they all perform the same basic functions for plants. In the next Exploration you will closely examine the outside surfaces of leaves and their internal structure.

EXPLORATION 5

A Good Look at Leaves

ACTIVITY 1

An Outside View

You Will Need

- a magnifying glass or low-power binocular microscope
- a variety of fresh leaves

What to Do

Choose a leaf. Look carefully at both its upper and lower surfaces. Which surface has more features? Look for patterns of veins, hairlike projections, and other structures. Compare the leaf to leaves from several other plants.

Choose one type of leaf and write descriptions of both of its surfaces. Let a classmate read your description and try to identify the leaf you described.

To get a closer look at a leaf, you will need to use a microscope. You will also need to prepare a slide of leaf tissue that is thin enough for light to pass through. You will do this in the next Activity.

Exploration 5 continued ▶

A Close-Up View

ACTIVITY 2

You Will Need

- a microscope
- 2 lettuce leaves
- a microscope slide and coverslip
- a clear plastic ruler
- an eyedropper
- water

What to Do

You are going to make a wet-mount slide of the cells of the lower surface of a lettuce leaf, and then you are going to make a wet-mount slide of the upper surface. For the lower-surface slide, use your fingernail to gently scrape away the upper part of the leaf so that only the thin lower layer remains. Mount the thin leaf section on a slide, as shown in the illustrations at right.

Look for the kinds of cells shown below. Do you see any slit-like openings?

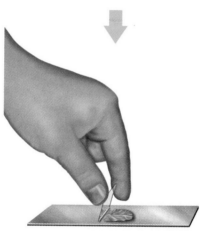

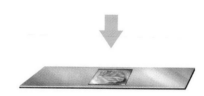

These openings are called *stomata* (singular *stoma*). Study them closely. Then draw and label one stoma and the cells around it. You should see two kinds of cells: guard cells and regular epidermal (surface) cells.

Now follow the same process for the upper surface of a leaf.

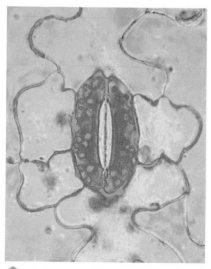

⬆ A stoma (with *guard cells* stained purple)

Interpreting Your Observations

Working with a classmate, find the answers to the following questions:

1. On which leaf surface (top or bottom) did you observe more stomata?

2. Estimate the number of stomata on one square centimeter of the lower surface of the leaf. (Hint: How much magnification are you getting with the microscope? What do you estimate to be the area of the field of view in the microscope?)

3. What purpose do you think the stomata might serve?

4. How might the guard cells open and close the stomata?

5. Do you think the stomata close at night? Why or why not?

6. What do you predict would happen if grease were smeared on the lower surface of the leaves of a healthy plant and left there for several days? Why would this happen?

In case you still have questions about the purpose that stomata serve, their function will be explained further, beginning on the next page.

Stomata Data

As you saw in Exploration 5, the stomata are located mainly on the underside of a leaf. Stomata allow water vapor and other gases to pass into and out of the leaf. Despite their important function, stomata are extremely tiny—the width of a stoma is about $\frac{1}{10}$ the thickness of this page! Each stoma is flanked by a pair of guard cells, which control the size of the opening. When the guard cells absorb water and swell, the stoma is opened. The guard cells relax when water leaves them—this closes the stoma. Did you guess that this is how stomata work? Usually stomata open during the day and close at night. Considering what you know about photosynthesis, why does this make sense?

An Inside View

The drawing below shows a cross section of a typical leaf. Imagine that you could shrink to the size of a water or carbon dioxide particle. You could journey into a leaf and make some interesting discoveries about this "food factory." Study the diagram of the leaf, and then answer the questions below.

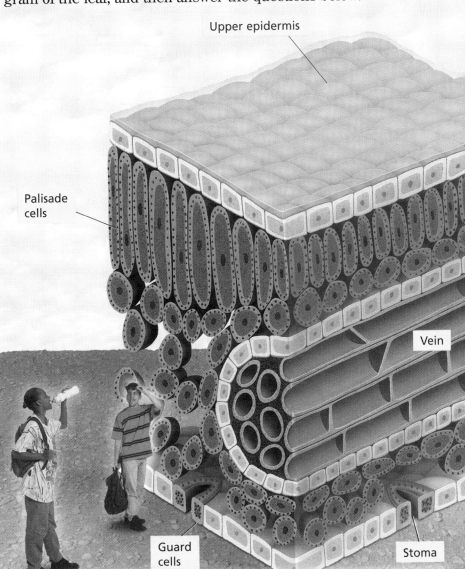

Upper epidermis

Palisade cells

Vein

Guard cells

Stoma

1. In your miniaturized state, what are two ways you could enter the leaf?

2. If you were a carbon dioxide particle, where would you wait, along with large numbers of other gas particles, until you could be used?

3. Where would you find huge, bulging "gates"?

4. Where would you exit the leaf? Would a water particle and a carbon dioxide particle exit at the same place? Why?

5. Where would you find a pipeline system to carry liquids such as water?

6. Can you find the elongated cells near the top of the leaf? Would these cells be well lit? Might these be good places to make food?

7. Food is also made in the loosely packed, somewhat rounded cells farther inside the leaf. Would these be as well lit as the elongated cells? Could as much food be made there?

Making a Good Microscope Drawing

You don't have to be an artist to make a good microscope drawing. Below is a drawing of onion-skin cells viewed under a microscope. It is good because

- a sharp, hard pencil was used on plain unlined paper; thick markers, soft pencils, and colored pencils were not used.
- details are shown, and only one cell is shown with all of the details.
- labels are outside the drawing and neatly printed.
- labeling lines end at the feature being identified.
- there is no blurring, shading, or added color.
- the magnification is provided.
- the drawing was done neatly and carefully.

In the next Exploration, you will try your hand at making a microscope drawing.

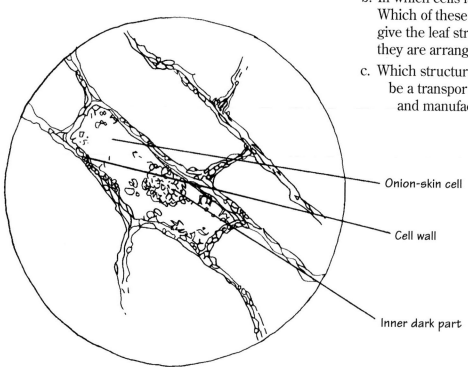

Onion-skin cell, magnification = 300x

EXPLORATION 6 ·

A Microscopic Activity

You Will Need

- a prepared slide of a leaf cross section
- a microscope
- a sharp, hard pencil
- unlined paper

What to Do

In this activity, you will use a microscope to view a cross section of a leaf, and make drawings of what you see.

1. Examine the cross section under a microscope. Use the cross-sectional view of a leaf on page 21 to help you identify the kinds of cells present.
2. Draw a detailed diagram of each type of cell you see. Label the various kinds of cells.
3. Answer each of the following questions, even if you did not see all of the cell types in your cross section.

   a. Which cells protect the leaf, preventing the inside from drying up?

   b. In which cells is food manufactured? Which of these cells can also function to give the leaf strength because of the way they are arranged?

   c. Which structure looks like it might be a transportation system for water and manufactured food?

   d. Which cells control the flow of gases into and out of the leaf?

Why Green?

You have been investigating the process by which plants produce food. You now know about the raw materials plants use, their source of energy, and the products of photosynthesis. You have no doubt noticed that most living plants are green. Obviously this color must be important, but why?

First of all, this green color is due to a substance called *chlorophyll*. When you tested for starch in Exploration 1, you dissolved chlorophyll and removed it from the leaf so that you could see the results of the test for starch. What function do you think chlorophyll performs? The next Exploration will give you a good idea.

What Does Chlorophyll Do?

To figure out the function of chlorophyll, you need leaves, such as those from a coleus, that have some white areas (indicating a lack of chlorophyll).

You Will Need

• materials and equipment to test leaves for starch

What to Do

Give the plant a good chance to make food (you know the conditions), and then remove a couple of leaves. Test for starch as you did previously. (Remember to heat the methanol safely!)

Make a brief report in your ScienceLog explaining your idea about the function of chlorophyll. Include a diagram showing the areas with and without starch.

Exploration 7 continued ▶

A Crash Course in Colors

Light from the sun seems to have no particular color. But, in fact, sunlight is a mixture of many different colors blended together. We can see the colors that make up sunlight by passing it through a prism, or by looking at a rainbow. If sunlight is white, then why do the things around us appear to have color?

Suppose that a substance absorbed all of the colors of sunlight except red, which it reflected. What color would you perceive the substance to be? Red, of course. Substances appear to have color because they absorb some colors and reflect others.

A *pigment* is a substance that absorbs light. Some pigments absorb all the colors of light, and some absorb certain colors but reflect the rest. In this way pigments give substances their color. Chlorophyll is a pigment that reflects one color of light. What color is this?

As you know, sunlight has energy. Chlorophyll absorbs certain colors of light, so chlorophyll absorbs energy. How do you think plants put this energy to work?

The process of photosynthesis puts the energy absorbed by chlorophyll to work, chemically combining carbon dioxide and water to form sugars. Further reactions change these sugars into starches. Now that you know this, you can take the following mini-quiz.

Crash-Course Quiz

1. Write a word equation for photosynthesis, placing *chlorophyll* in the equation.

2. Study the diagram of a leaf cell shown below. Which of the cell bodies contain chlorophyll (and make the plant's food)? Find out what these chlorophyll-containing bodies are called.

3. In which leaf cells is chlorophyll found?

4. Which type of leaf cell would you expect to be the chief food producer? Support your inference with at least two observations you have made.

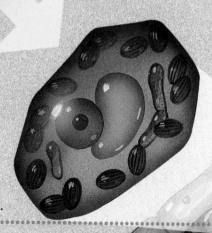

Sunlight

Energy from the sun

Chlorophyll in cell absorbs solar energy to carry out photosynthesis.

Topics to Discuss

Form small groups and discuss any two of the following topics. You may need to do some research first. Then share your conclusions with the whole class.

Topic 1

A tree releases water through its stomata. How, and from where, did the tree get this water? How do you think the water reaches the leaves?

Topic 2

a. Why are leaves often broader at the bottom of a tree?

b. How do plants without green leaves (such as the purple velvet vine below) make food?

c. Carbon dioxide is fairly scarce, making up only about 0.035 percent of the atmosphere. How do plant leaves get enough carbon dioxide to carry out photosynthesis?

d. How do the leaves of evergreen trees survive cold winters?

Topic 3

How do plants in closed terrariums manage to get raw materials for photosynthesis?

Topic 4

Discuss the design of a specific leaf, taking into consideration the following questions:

a. In what ways is the leaf well suited for its particular environment?

b. Can you find any "design flaws"? If so, identify them and suggest improvements.

Topic 5

Coniferous trees, such as pines, have leaves that are radically different in structure from those of plants such as maple trees, sycamore trees, or orange trees.

a. Identify some of the differences.

b. How might these structural differences reflect differences in each plant's way of life?

Pine   Orange   Maple

Sycamore

CHALLENGE YOUR THINKING

1. How Is a Plant Like a Windmill?

You have been asked to respond to the issue raised in the cartoon below. How would you go about providing evidence to support your response?

3 I don't think so. Plants need the sun's energy to make food. So the sun is the ultimate source of all energy.

4 I don't agree, Sam. What about windpower? The sun doesn't cause windmills to turn, does it?

2 So does that mean that the source of all energy is plants?

5 Yeah, and what about hydroelectric dams? Flowing water is what causes their turbines to rotate, right?

1 Okay, we know that food ultimately comes from plants. And we know that plants make their own food.

2. It's Not Easy Being Green

The diagram at left shows the *absorption spectrum* (pattern of light absorption) for a typical plant. Light that is not absorbed is reflected. Study the diagram and then answer the following questions. Remember that white light (such as that from the sun) is a mixture of all the different colors of light.

a. What causes plants to appear green?

b. If a plant received only green light, how might it grow compared to a plant grown in normal light?

c. Would a plant grown in red light do well? blue light?

Absorption — More / Less

Red   Yellow   Green   Blue   Violet

3. Back to the Drawing Board

This chapter began with the question, "How do plants get their food?" Has your thinking changed since then? Here are some suggestions to help you organize your ideas.

a. Begin by writing a brief report about how your thinking has changed.

b. Look at the poster your group made. Discuss and write comments on the poster.

c. Produce a new poster and compare it with the first one.

d. Prepare a way to explain to some fifth-grade students how plants get food.

4. Tony's Hypothesis

Your friend Tony (who isn't taking this class) has a hypothesis about the starch in plants. "The starch in plants comes either from the soil they grow in or the water they use," Tony said. "Plants just won't grow where the soil or water isn't starchy." How could you test Tony's hypothesis?

5. The Day the Earth Stood Still

Imagine that you hear the following report on the radio:

Scientists have noticed that plants all over the world are behaving strangely. The leaves of plants fold up when they are in sunlight and they open up in the dark. The stomata of some plants remain shut all day, and then they open at night . . .

Continue this story and describe other behaviors that would be strange for plants. How would other living things be affected by such behaviors?

ScienceLog

Review your responses to the ScienceLog questions on page 4. Then revise your original ideas so that they reflect what you've learned.

A World of Water

1 How much of you is water?

2 How does water get from the roots to the top of this tree?

3 How does water move within plants?

4 Plants don't use all the water they get. Where does the extra water go?

ScienceLog

Think about these questions for a moment, and answer them in your ScienceLog. When you've finished this chapter, you'll have the opportunity to revise your answers based on what you've learned.

The Indispensable Chemical

If you looked at the Earth from space, what would impress you more than anything else—its smallness in the vast emptiness of space? the lack of national boundaries? How about the colors? Blues and whites totally dominate the view. What is the significance of these colors? They signal the presence of liquid water. Look around you. Water is everywhere. One of the remarkable things about our planet is that conditions are just right for liquid water to exist. If our planet were a little colder or a little warmer, or had a weaker gravitational field, this would not be the case. And without an abundant supply of liquid water, life as we know it could not exist.

Water Facts

What is this stuff we call water, and why is it so important? Here are just a few facts about this remarkable chemical.

- About 70 percent of the Earth's surface is covered by water; that's a total of 1.4 billion km^3. Three percent of this water is fresh (and 75 percent of that is ice).
- All living things are made mostly of water. A chicken is about 75 percent water, an earthworm is about 80 percent water, and a tomato is about 95 percent water. You are about 65 percent water.
- Water can dissolve more substances than any other substance. It is the closest thing there is to a universal solvent.
- Water has a higher *heat capacity* (absorbs more heat without increasing in temperature) than almost any other substance.
- Water is the only substance on Earth that is naturally present in solid, liquid, and gaseous states.
- About 265,000 km^3 of rain or snow fall every year (enough to cover Arizona to a depth of 1 km).
- A person will drink about 60,000 L of water in an average lifetime.

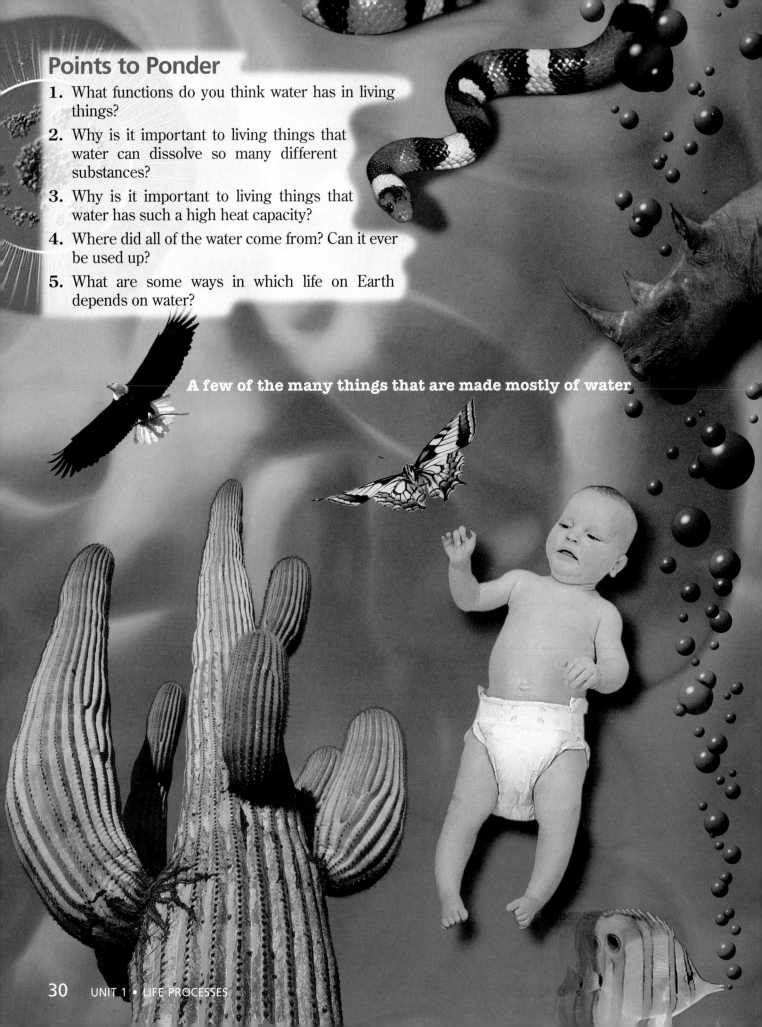

Points to Ponder

1. What functions do you think water has in living things?

2. Why is it important to living things that water can dissolve so many different substances?

3. Why is it important to living things that water has such a high heat capacity?

4. Where did all of the water come from? Can it ever be used up?

5. What are some ways in which life on Earth depends on water?

A few of the many things that are made mostly of water

LESSON

2

The Water Cycle

So far in this unit, you have been looking at various aspects of the food-making process in leaves, which is called photosynthesis. Photosynthesis is not the only process that occurs in leaves, as the following demonstration will show.

Transpiration: A Demonstration

Use a geranium or some other hardy plant. Give the plant some water, and then cover the leaves with a plastic bag and seal the bag carefully around the stem. Observe the setup over a period of 24 hours. Record any changes you see. What do you think causes the change(s)?

The process you observed in the previous demonstration, in which water evaporates from the plant's leaves, is called **transpiration.** Transpiration occurs continually. From which part of the leaf do you think the water exits?

How does water evaporate? Think about how a clothes dryer works. The clothes dryer produces heat. The heat is absorbed by water in the clothes, causing the water to change into a gas—water vapor. This is evaporation.

A similar process causes some of the water in leaves to evaporate as well. You might be surprised to learn that a mature tree gives off as much as 15 L of water daily through its stomata. Imagine how much water an entire forest gives off! Imagine how much heat is absorbed by all this water as it evaporates!

The Brazilian rain forest transpires so much water that rain clouds form almost every day.

Where Does the Water Go?

What happens to the water given off by plants? Evidence of transpiration can be seen in the photograph of the rain forest on the facing page. What is the evidence? The clouds you see contain a lot of moisture. How do clouds form? In the transpiration demonstration, you saw water droplets on the inner surface of the plastic covering around the plant. How did the water from the plant get back into the liquid state again? Is there a connection between this phenomenon and cloud formation?

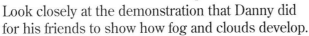

EXPLORATION 1

From a Gas to a Liquid

Look closely at the demonstration that Danny did for his friends to show how fog and clouds develop.

1. Danny poured some water into a large juice bottle, rinsed it around, and poured the water out.
2. He held a match at the mouth of the bottle, lit it, and inserted it inside, holding it there for several seconds before removing it and putting it out.
3. Immediately, Danny put his mouth over the opening and blew into the bottle several times.
4. Then he held the bottle up to show the swirling, whitish (foglike) substance that appeared inside it.

Try this for yourself. Then try it again, but leave out step 2.

"This is really mysterious," Hannah said. "I guess the air in the bottle represents the atmosphere, but how did the fog form?"

Raquel wondered, "There are no plants in the bottle to transpire, but I suppose some water vapor forms from the water poured into the bottle."

"How does the match help?" Ian asked. "I saw some smoke enter the bottle. Is that important?"

Perhaps you also have questions about Danny's demonstration. Write down any questions that occur to you. If you have no questions, write your explanation for the formation of the swirling, cloudlike substance in the bottle.

Exploration 1 continued ▶

Danny produced a copy of a page from a reference book. "I thought you might find that demonstration interesting," he said. "Here's an explanation for cloud formation in the atmosphere. Read it and see if you can pick out the details that correspond to the things I did and the observations you made."

Check to see whether your questions have been answered. If you already wrote an explanation, check it and change it if you find that improvements are needed.

What About the Heat?

What happens to the heat when water changes from the liquid state to the gaseous state, that is, when it evaporates? What happens to the heat when water condenses, as it does during cloud formation?

Summarize your thoughts about these questions. Include the following words: evaporation, condensation, transpiration, water, heat, energy, plants, stomata, atmosphere, and cloud.

The processes of transpiration, evaporation, and condensation are constantly recurring.

When water droplets in the clouds reach a certain mass, they fall, and *precipitation*—rain, sleet, or snow—begins. Precipitation returns water to the Earth, where it is used by plants and animals. Each of the changes you've learned about in this lesson is part of the **water cycle.**

The drawing below illustrates the water cycle.

1. What process takes place at each labeled area?
2. For each process, is heat being absorbed or released?

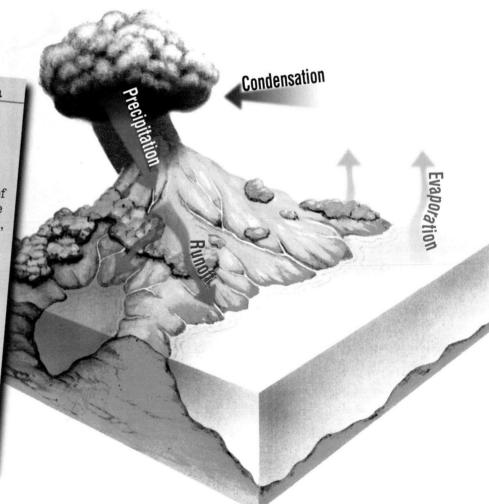

23 How Clouds Form

Heat evaporates millions of tons of water into the air daily. Not only does water evaporate from lakes, rivers, and oceans, but an amazing amount of water transpires from the leaves of plants. As warm, moist air rises, there is less pressure on it from above, so it expands. This causes it to fall in temperature. As the air cools, some of the moisture it carries begins to condense. The tiny droplets of water that form produce clouds. This process of condensation in the atmosphere is aided by dust or smoke particles, on which the water droplets can form.

3 Spreading Water Around

Inside the leaf

Stomata

Vein

Underside of leaf

How does a plant obtain the water it uses? As you may already know, this water enters the plant through its roots and travels upward. The water contains dissolved minerals that combine with sugar to make other kinds of materials used by the plant. The plant uses these materials to grow and to maintain itself.

A seedling needs a generous supply of water. Its *root hairs* ensure that this need is met. Just look at the many hairs on a sprouting radish seed. You can sprout radish seeds between two damp pieces of paper towel. Keep the paper towels slightly moist at all times. Once the seeds have sprouted, estimate the number of root hairs on a single seedling. But don't just guess. Use some kind of systematic method. You might want to use a microscope.

Now look at the close-up drawing of a section of a root. Note that each root hair is a projection of one of the cells that make up the outside of the root tip.

Root hairs grow among the tiny particles of soil. Moisture often fills the spaces between soil particles, and this is where plants get most of their water.

Inside the stem

Inside the root

Closely examine some radish seedlings to see root hairs.

How do the root hairs provide a plant with its water and mineral requirements? The answer is, through the process of **diffusion.** Under normal conditions, water *diffuses* inward from the soil, through the walls of the root hair, and into the other cells. Now you know *what* happens. However, you still do not know *how* it happens. What causes water and minerals to enter root hairs? To understand this you will need to look more closely at the process of diffusion.

Defining Diffusion

Here are some other situations in which diffusion occurs. Thinking through what happens in each situation will help you form a definition for *diffusion.*

a. A sugar cube is left in a beaker of water for a while. What happens?

b. Several drops of food coloring are placed into a beaker of water at room temperature. What happens? The experiment is then repeated using very hot water. What do you predict will happen?

c. Invisible fumes of ammonia gas rise from a container full of concentrated ammonium hydroxide with the stopper removed. A person stands about 3 m away. How soon will she smell the ammonia gas?

How is diffusion at work here? ⬇

In situation (a), the sugar diffuses throughout the water and seems to disappear, yet you can taste the sugar in every part of the water. In situation (b), the food coloring diffuses throughout the water until the mixture is the same color throughout. The process occurs much more quickly in the hot water. In situation (c), the ammonia gas diffuses through the air, eventually reaching your nose. Soon it can be smelled in every part of the room. Now write a definition of *diffusion*. Can you suggest other examples of diffusion?

You will find additional information to help you define *diffusion* more precisely later in this lesson.

A Theory of Diffusion

As you probably already know, scientists have learned that all matter is composed of incredibly tiny particles. (You will take this up in more detail in Unit 2.) These particles are in constant motion. The hotter the temperature, the faster the particles move. These particles are far too small to be seen, and therefore you cannot directly observe their motion. But you can see indirect evidence of their motion. Diffusion, for example, is evidence of particle motion.

A diagram may help you understand the connection between particles and diffusion. Here is a drawing of situation (a) from the previous page. This is only a model of what happens. You can't really see the sugar and water particles.

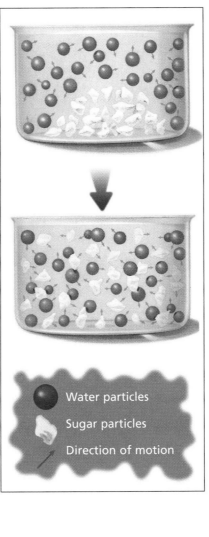

● Water particles
◇ Sugar particles
↗ Direction of motion

Refining the Definition of Diffusion

Once you have seen the connection between particles and diffusion in situation (a), you can work in small groups to try to explain how diffusion occurs in each of the other two situations. Make use of the ideas suggested by the theory of particles. After you have finished, flex your creative muscles a little bit with the following exercise: Write a short account of the experiences of a particle undergoing diffusion.

A Law of Diffusion

The most important law governing the process of diffusion states:

"The particles of a substance always diffuse from a region of high concentration to a region of low concentration."

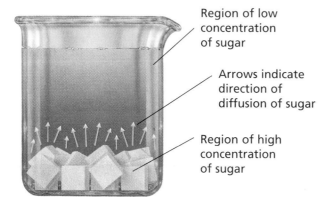

Region of low concentration of sugar

Arrows indicate direction of diffusion of sugar

Region of high concentration of sugar

Do you understand why the diffusion process results in solvent and solute becoming uniformly mixed? This is because diffusion continues until the particles of the substances involved are completely and randomly scattered.

Turn back to the situations discussed earlier. In your ScienceLog, identify for situation (b) the areas of high and low concentration of food coloring and its direction of diffusion. For situation (c), identify the regions of high and low concentration of ammonia gas and its direction of diffusion.

A Further Refinement

Here is a quick summary of what you have learned so far about diffusion.

- One substance intermingles with another.
- The intermingling seems to be uniform.
- The two substances that intermingle do not need to be mixed mechanically; mixing seems to occur naturally.

Using these observations, we can formulate a good working definition of the process of diffusion:

Diffusion is the uniform intermingling of the particles of one substance with those of another substance. This occurs naturally because of the motion of both types of particles.

Compare this definition to your own. Do you need to refine your definition in any way?

The Root of the Matter

Look at the illustration of a plant's root hair. Where is there an area of higher concentration of water particles? an area of lower concentration? In what direction will diffusion take place? If water is to diffuse from the soil into the root-hair cells, the water must pass through both the cell wall and the cell membrane, which enclose the contents of each root-hair cell. Water does, in fact, pass easily through both the cell wall and cell membrane. After this it must go from cell A to cell B, again through the membrane and wall of each cell. Diffusion through a membrane occurs in living things. In the Exploration that follows, you will study this special kind of diffusion.

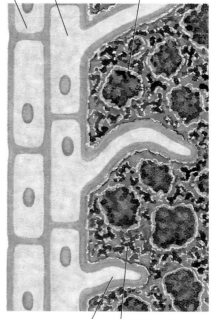

Cell B Cell A

Water around soil particles

Low water concentration

High water concentration

Root hair

Water and air in spaces between soil particles

Diffusion Through a Membrane

Here are two models of diffusion through root hairs.

ACTIVITY 1

You Will Need

- a support stand and clamp
- an egg
- vinegar
- a wide-mouthed bottle
- a needle
- a thin glass tube (10 cm)
- a candle or silicone sealant
- matches
- a Petri dish
- water

An egg is a single cell—a very large one. This cell, like all cells, is protected from its surroundings by membranes.

What to Do

1. Find a way to support the egg so that only the small end rests in the vinegar. The shell will become very thin in about 24 hours.

2. Fill the bottle with water and place the egg (thinned shell down) on the mouth of the bottle as shown. Use the needle to poke a hole through the top of the egg large enough to accommodate the glass tube.

3. Insert the glass tube. It should just penetrate the inner membrane, which is right under the shell. Seal the glass tube in place with candle wax or silicone.

4. Leave the egg in place for several hours.

5. Record your observations.

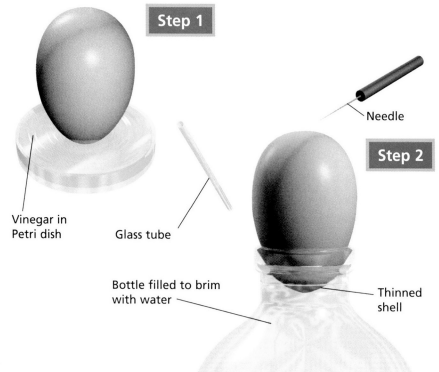

Step 1

Step 2

Needle

Vinegar in Petri dish

Glass tube

Bottle filled to brim with water

Thinned shell

Step 3

Step 4

Support for glass tube

Exploration 2 continued ▶

ACTIVITY 2

You Will Need

- a support stand with clamp
- a thin glass tube (20 cm)
- an animal membrane
- sugar solution (5%)
- a rubber band
- a large beaker or similar container
- water

What to Do

Arrange the equipment as shown in the illustration, and then leave it for several hours.

Questions

1. What evidence is there in Activities 1 and 2 that water has passed through the membranes?

2. For each of Activities 1 and 2, identify in your ScienceLog the regions of high and low concentrations of water. Identify the direction of diffusion of the water.

3. What can you say about the concentration of the egg contents in Activity 1? the sugar in Activity 2? Did these substances also diffuse through the membranes? How could you find out?

4. Develop an explanation for the results of this Exploration.

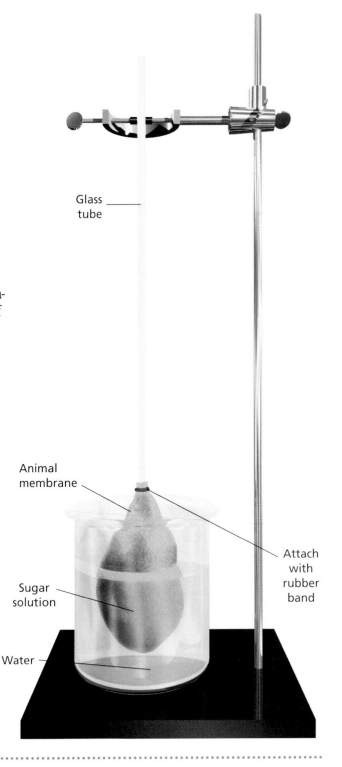

Glass tube

Animal membrane

Attach with rubber band

Sugar solution

Water

A Model of Cell Membranes

Exploration 2 demonstrated that not all particles can pass through every type of cell membrane. To better understand the selection process, consider what happens in the following situations, which involve common substances.

- Does water, salt, or sand pass through glass?

- Does water, salt, or sand pass through a fine-mesh screen?

- Does water, a solution of salt and water, or sand pass through filter paper?

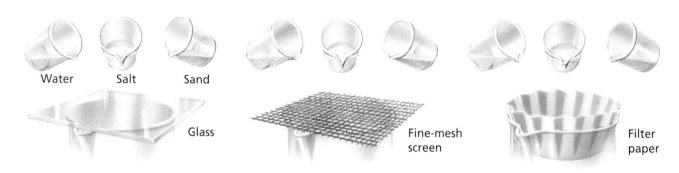

Water   Salt   Sand

Glass

Fine-mesh screen

Filter paper

The word *permeable* means "open to passage," while *impermeable* means "closed to passage." Of the glass, screen, and filter paper, which is permeable to all three substances—water, salt, and sand? Which is impermeable to all three substances? Which is permeable to some substances but impermeable to others? What might you call a material with such a property?

The screen is a permeable material, while the glass is an impermeable material. The filter paper is a *semipermeable* material. In general, what do you think determines whether materials are permeable, semipermeable, or impermeable? (Think about what you have learned so far in this unit.)

Semipermeable membranes allow some particles to pass through them, depending on the size and shape of the particles. In general, semipermeable membranes prevent larger particles from passing through them.

Suppose, for a moment, that the egg and the animal membrane in Exploration 2 are accurate models for the way diffusion takes place in cells. What kind of membranes do cells have: permeable, semipermeable, or impermeable? What does this model suggest about the way a cell interacts with the surrounding environment?

The action of the cell membrane is similar to that of the filter paper, which allowed passage of salt and water particles, but prevented passage of the sand. In other words, the membranes allow some particles to pass through, while acting as a barrier to other, larger particles. Cell membranes are, therefore, semipermeable.

The next Exploration will reveal more about semipermeable membranes.

Semipermeable Membranes

You Will Need

- two 14 cm to 16 cm lengths of dialysis tubing
- three 250 mL beakers
- 20 mL of cornstarch and water mixture (5%)
- 20 mL of 5% sugar solution (use corn syrup and water)
- iodine solution
- Benedict's solution
- 2 paper clips
- a hot plate
- a graduated cylinder
- water
- a test tube
- an eyedropper
- thread
- test-tube tongs
- latex gloves

Iodine

Testing for starch

What to Do

Arrange the materials as shown in the illustrations below.

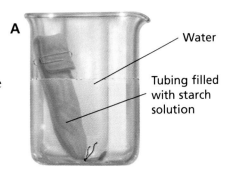

A

Water

Tubing filled with starch solution

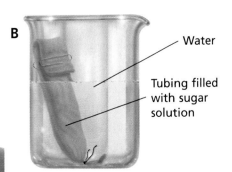

B

Water

Tubing filled with sugar solution

Check the liquids in the beakers after about 20 minutes and then again 48 hours later. To determine whether starch particles have passed through the tubing in *A* and whether sugar particles have diffused in *B,* you will have to apply the chemical test for each of these substances to the liquid in the beakers. You already know how to test for starch, shown at left. The test for sugar is shown at right. Report your results in your ScienceLog, and then answer the questions at the end of the Exploration.

Caution: Benedict's solution can irritate the skin. Handle with care.

A Test for Sugar

Place 2 mL of the solution to be tested for sugar in a test tube. Add an equal amount of Benedict's solution. Put the test tube in boiling water as shown below. Watch carefully for 5 minutes.

When heated, Benedict's solution by itself is blue. Benedict's solution heated with a concentrated sugar solution is brick-red. Benedict's solution heated with a weak sugar solution is greenish yellow.

Questions

1. Is the dialysis tubing selective in what it allows to pass? Which substance(s) does it permit to pass through?

2. What conclusions would you draw about the relative sizes of starch particles and sugar particles?

3. What property do cell membranes and dialysis tubing have in common?

Testing for sugar

Osmosis: Controlled Diffusion

Diffusion of water (or another liquid) through membranes is called *osmosis,* from the Greek word *osmos,* meaning "push." Osmosis takes place continuously in living things. The cells of all plants and animals are surrounded by membranes that allow smaller particles to pass through but stop larger ones. Even when two different kinds of particles are both small enough to pass through a cell membrane, the smaller ones are likely to get through more quickly. You can see this for yourself in Exploration 4.

Observing Osmosis

Potatoes have a very high concentration of water, which is why they are used in this Exploration. You are going to study the effect of putting potato slices into different solutions.

You Will Need

- a potato
- 4 beakers
- salt water
- a potato peeler
- a knife

What to Do

Peel a potato and cut four slices that are the same size. Place a slice of potato into each of four beakers that contain either air alone, tap water, a dilute salt solution (20 g/L), or a concentrated salt solution (150 g/L). After about 30 minutes, remove the slices and examine them. Record all observations in your ScienceLog.

Follow-Up

1. Describe the size and rigidity of each of the four slices before and after being placed in a beaker.
2. What seems to be the effect of tap water on a slice?
3. What effect does the concentrated salt solution have on a slice?

Exploration 4 continued

4. Compare the concentration of water in each potato slice and in each of the beakers. Remembering the rule for diffusion, state the direction of water flow through the cell membranes (i.e., into or out of the cells). Does this explain your results?

5. Plant cells become less rigid (more spongy) when water leaves them. How does this principle explain what happens to plants when they are deprived of water? (They wilt.)

6. Did any slice seem to become more rigid than the others? Which liquid was it in? Explain what happened to this slice. Which slice seems to have lost some water? a lot of water? How do you explain this?

Osmosis Explained

In the beakers containing salt solutions, water particles move out of the cells and into the beaker because initially, the water concentration is greater inside the cells than outside. A kind of pressure forces the water toward the outside until the solution outside has the same concentration as the solution inside. Diffusion always occurs in both directions, from the inside of the cell outward and from the outside of the cell inward. Diffusion occurs fastest, though, where the difference in concentration is greatest. So even though some water diffuses from a salty solution into a cell, it is more than offset by the water diffusing outward from the cell into the salt water. The reverse situation occurs when cells are placed in tap water.

EXPLORATION 5

Osmosis: How Fast?

Purpose: to obtain information about how fast osmosis can occur

You Will Need

- a carrot
- a coring knife
- a metric ruler
- a 20 cm long glass tube fitted with a rubber stopper
- molasses
- candle wax or silicone sealant
- matches
- materials and equipment to test for sugar (see Exploration 3)
- a support stand and clamp
- a beaker
- water
- graph paper
- a grease pencil

What to Do

1. With a coring knife, cut a hole in the carrot about 5 or 6 cm of the way down its length. The hole should be just large enough so that a one-holed stopper will fit tightly.

2. Fill the hole in the carrot with molasses.

3. Push the stopper and glass tube into the hole in the carrot. Seal any openings between the stopper and the carrot with candle wax or silicone sealant.

4. Place the carrot in a beaker filled with water, as shown in the illustration.

5. Mark the initial level of the molasses in the glass tube. This is the level at time zero.

6. Record the level of the molasses in the glass tube several times over the next 3 days, at the same time each day. Before school, during lunch hour, and after school would be good times.

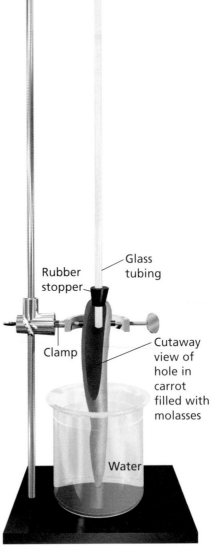

Rubber stopper

Glass tubing

Clamp

Cutaway view of hole in carrot filled with molasses

Water

Graph your findings. Interpret your results using the following questions:

Questions

1. What was the rate of rise of the molasses in millimeters per hour (a) during day 1, (b) during day 2, and (c) during day 3?

2. Use your graph to indicate the level of the molasses at the following points after time zero:
   a. 12 hours
   b. 18 hours
   c. 30 hours
   d. 60 hours

3. Use the sugar test to determine whether any molasses particles went through the carrot into the water.

4. Which do you think is made up of smaller particles, water or molasses? Justify your answer.

5. What happened to the molasses concentration inside the carrot during osmosis?

6. What causes the process of osmosis to slow down? Under what conditions do you think it would stop?

Applied Osmosis

Six examples of applied osmosis are presented here. Can you figure out how osmosis explains each of these common occurrences?

1. When old, spongy potatoes are being prepared for cooking, which would it be better to soak them in, tap water or salt water? Explain.

2. Why is salt sometimes used to kill weeds?

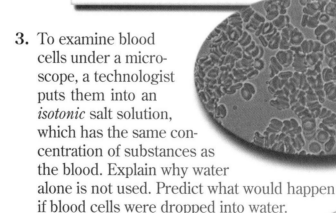

3. To examine blood cells under a microscope, a technologist puts them into an *isotonic* salt solution, which has the same concentration of substances as the blood. Explain why water alone is not used. Predict what would happen if blood cells were dropped into water.

4. Placing sugar on a grapefruit makes the grapefruit become moist very quickly; a sweet syrup forms over its surface. Explain why this happens.

5. The plants below suffered "fertilizer burn." How does osmosis help explain what happened?

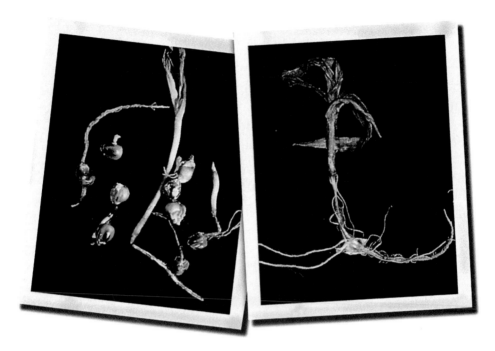

6. Celery becomes limp when it dries out. What might you do to restore its original crispness?

Putting Water Where It Counts

As you know, water enters the roots of plants by diffusion. But what happens then? How does the water get from the root hairs to the leaves, where it is needed?

The concentration of water around the root hairs is normally higher than the concentration of water inside the root hairs. This difference in concentration results in the diffusion of water into the plant, causing a phenomenon known as *root pressure*. Root pressure forces a steady stream of water into the plant. At the same time, the membranes surrounding the root-hair cells prevent the vital contents of the cells from leaking out. The root hairs pass the water they absorb from cell to cell. Eventually the water reaches special cells that function very much like pipes. These cells carry water to the leaves.

In a living plant there is an unbroken "pipeline" of water from roots to leaves. This becomes pretty impressive when you realize that some trees are over 100 m tall. Can root pressure alone push the water that high, or are there other forces at work? What do you think? Come up with as many ways as you can to explain how plants are able to move water to such great heights.

Bark

These long cells, which carry food up and down, are living.

Woody core of the stem

These long cells, which carry water and dissolved minerals up the stem, are hollow, dead cells.

Cells in this region multiply in number, causing the stem to increase in width.

Cutaway view of a woody stem

How does water get all the way from the roots to the tops of these trees?

Study the ideas that are presented on these two pages. At one time or another, each has been suggested as the mechanism responsible for delivering water to the tops of the highest trees. Which, if any, seem logical to you? Which, if any, of the following examples do not seem logical?

Example A

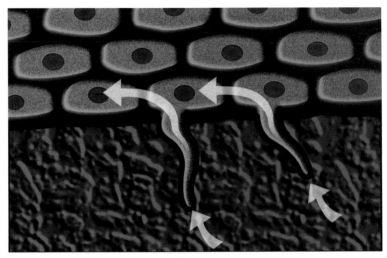

Root pressure caused by osmosis

Example B

Water rises through the process of *capillary action*. Why does it do so? Can you think of other situations in which this occurs? In each case, water rises through tiny passages.

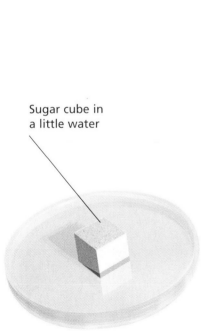

Sugar cube in a little water

Strip of paper towel in water

Glass tube with very small (hair-sized) opening

Example C

Here, atmospheric pressure forces the water up the glass tube of the dropper. The partial vacuum created by water evaporating from the leaves creates a similar situation.

Eyedropper

Squeeze the bulb of the eyedropper.

Place the dropper with the squeezed bulb in water.

Release the bulb, and the water rises.

Example D

Particles of water have a strong attractive force for one another. Notice how droplets of water cling to each other. (This is called *cohesion.*)

Tap turned on slightly

Example E

The strong attraction of water to itself causes water particles to pull each other along. As each water droplet evaporates or drips from the leaf, it "tugs" on the droplet next to it. This tug is transferred back through the unbroken chain of water droplets to the root.

Stem with a continuous column of water particles, starting at the roots

Theories of Water Movement

In looking at the examples on pages 48 and 49, you discovered four kinds of forces that could account for the rise of water in the stems of plants. Here are four explanations in more detail:

- **Osmosis**   Water particles enter the root cells more quickly than they leave the cell, so pressure in the cells increases, pushing water upward. This is called root pressure.

- **Capillary Action**   The narrow passages inside the plant stem cause the water particles to rise by capillary action. Capillary action results from *adhesion* (the attraction between the water and the walls of the passages) and cohesion.

- **Transpiration**   Water constantly evaporates through the stomata of the leaves (transpiration). This lowers the pressure in the cells at the top of the plant, allowing atmospheric pressure to push more water upward.

- **Cohesion**   The force of cohesion keeps the water column intact. Between the pull of water caused by transpiration and the push caused by root pressure, the water column is able to move upward in a plant.

Which explanation do you think scientists accept today? Draw a diagram of a tree, and label it to show the explanation you find most convincing.

I don't think water can get to the top of the tallest tree by just osmosis. I mean, how far can root pressure push water upward?

I agree. And it doesn't seem as if capillary action or transpiration could do it either—at least not by themselves.

But what about cohesion? Maybe water droplets could pull each other along, all the way up the tree.

Evaporation at the top of the tree—transpiration, I mean—would leave spaces so the water would keep rising . . .

Yeah, as long as there was still a complete column and root pressure. I think cohesion must be the key force involved.

As it turns out, all four phenomena assist the upward movement of water in plants. However, cohesion is most important in moving water to the tops of the tallest trees. Root pressure can only move water upward a few meters. Capillary action is slow and cannot push water very high at all. Transpiration creates only a slight vacuum, so atmospheric pressure would not be able to push the water very far.

Defying Gravity

Each of the illustrations below shows liquids overcoming gravity. Match each picture with one of the explanations for the upward movement given below.

- attraction between water particles (cohesion)
- atmospheric pressure
- capillary action
- pressure caused by osmosis
- some other force

To find out more about the characteristics of water, see pages S6–S7 in the SourceBook.

A few gravity-defying acts

Starch solution

Water

Wick

Water bulges over top of container without overflowing

CHALLENGE YOUR THINKING

1. Cell Show and Tell

Casey showed the class the three diagrams below. She said that they show a cell under three different conditions: hypotonic, isotonic, and hypertonic.

Hypotonic solution

Isotonic solution

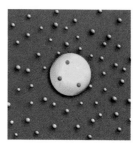
Hypertonic solution

Then she challenged the other students to answer the following questions:

 Cell

 Dissolved particle

a. What do *hypotonic, isotonic,* and *hypertonic* mean?

b. In which direction are water particles moving in each example?

c. What might eventually happen to each of the cells?

If you were in Casey's class, how would you answer the questions?

2. See You Later, Honey

Honey is basically an extremely concentrated solution of sugar and water. It is also remarkably durable; it has been known to last for centuries without spoiling. How does the principle of osmosis help to explain honey's ability to resist spoilage?

3. Diffusion Confusion

Your 9-year-old brother has just read about diffusion and he has a few questions. Help him answer them.

a. "How does stuff know when to start diffusing?"

b. "How does it know when to stop diffusing?"

c. "When stuff diffuses, why does it always go from a concentrated area to a less-concentrated area? Why doesn't it ever go in the other direction?"

4. How to Kill a Plant Without Really Trying

How does the principle of osmosis help explain why plants wilt when they don't get enough water?

5. Water, Water, Everywhere, But . . .

Why will drinking salt water make you thirstier than not drinking any water at all?

6. How Is a Leaf Like a Wet Towel?

A leaf has been compared to a wet towel on a clothesline.

a. What conditions help a towel to dry?

b. Suggest what you could do to find out whether these conditions would also increase the transpiration rate of a plant.

c. What beneficial effect might transpiration have on plants on a hot day?

7. Keep It Simple

Design a poster to show how a plant gets water and how the water travels through the plant. The design should be large, attractive, and simple enough for a fifth-grade student to understand. Show it to someone. Is the poster clear enough? What changes would help?

ScienceLog

Review your responses to the ScienceLog questions on page 28. Then revise your original ideas so that they reflect what you've learned.

3

Maintaining Life

Animals need plants. Do plants need animals in any way?

1

2 **How is the energy in your food released in you?**

What keeps heat near the Earth's surface?

3

ScienceLog

Think about these questions for a moment, and answer them in your ScienceLog. When you've finished this chapter, you'll have the opportunity to revise your answers based on what you've learned.

Turning Food Into Fuel

LESSON 1

You know that the energy needed by living things to carry out life functions comes from food. But exactly how do living things—people, for example—get energy from food?

Food in its raw form is not usable by our bodies. Before our bodies can put food to work it must be converted into a form suitable for absorption by individual cells. It's not just a matter of breaking down the food into tiny pieces, though. It must also be broken down chemically into simpler water-soluble compounds that can pass by diffusion through cell membranes.

The process of breaking down food into substances usable by the body is called **digestion**. Even a simple chemical compound such as starch is too complex to pass through the membranes of individual cells. The starch must be converted into a simpler substance. What do you think that simpler substance is?

The Mouth—Where It All Starts

Put an unsalted soda cracker into your mouth and chew it slowly, letting it soften thoroughly. Hold the food in your mouth and describe any change in taste. Does it taste sweeter or less sweet than when you first put it in your mouth?

Monica did the following demonstration to show how the mouth breaks down starch into a simpler substance.

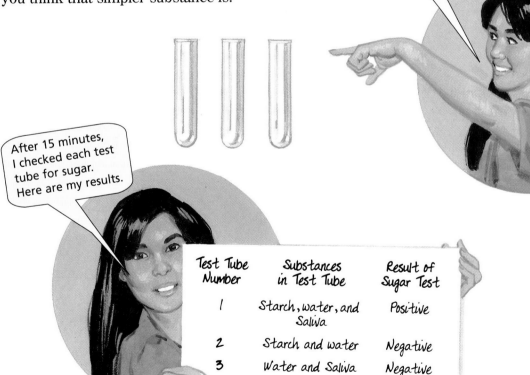

In one test tube, I put starch, water, and some saliva from my mouth. In the second test tube, I put starch and water. And in the third test tube, I put water and saliva.

After 15 minutes, I checked each test tube for sugar. Here are my results.

Test Tube Number	Substances in Test Tube	Result of Sugar Test
1	Starch, water, and Saliva	Positive
2	Starch and water	Negative
3	Water and Saliva	Negative

Why is it necessary for the starch to be broken down into sugar? What substance breaks down the starch into sugar?

CHAPTER 3 • MAINTAINING LIFE **55**

The Rest of the Story

Digestion breaks down starch into a simple sugar called *glucose,* which cells can absorb. Once food is completely digested, it is able to diffuse through cell membranes. In the small intestine, most of the digested food diffuses into the blood vessels. The bloodstream then carries the food to all parts of the body. The small particles of digested food are able to pass from the blood vessels into the body cells by means of diffusion through thin-walled vessels called capillaries. (How does diffusion determine the direction that the food particles move?) Once the digested food is in the body cells, it is ready for the next process.

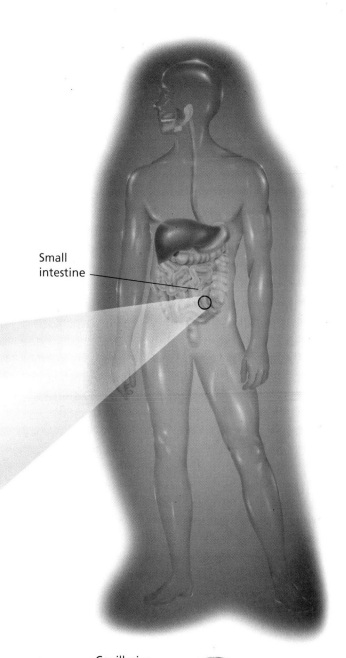

Small intestine

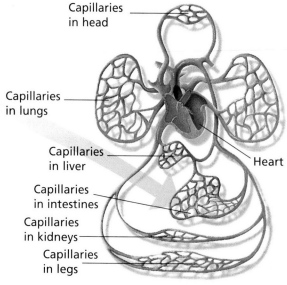

Capillaries in head

Capillaries in lungs

Capillaries in liver

Heart

Capillaries in intestines

Capillaries in kidneys

Capillaries in legs

Unlocking the Energy in Food

A Matter of Energy

Perhaps in earlier studies you burned a peanut (or some other food) to find out how much energy it contained. Where did the energy in the peanut come from? Where does it go when you eat the peanut? How is the energy released? The series of photographs below might give you some ideas. Write a brief caption of about two or three sentences for the series of photographs.

Energy from the sun

Energy stored as food

Energy released and used to activate muscles

How is the energy in gasoline released in an engine? How is the energy in candle wax released as heat and light energy? The energy within digested food in your body is released in a somewhat similar way. Look at the following diagrams. In what ways is the release of energy similar in each case? How does it differ?

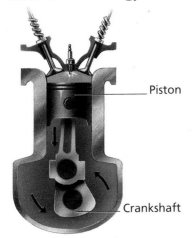

Piston

Crankshaft

A mixture of air and gasoline is ignited. The exploding mixture pushes the piston down, turning the crankshaft.

Air current

Oxygen combines with candle wax, releasing heat energy.

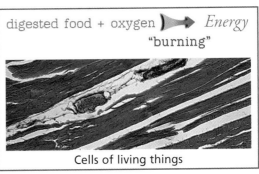

digested food + oxygen ➤ *Energy* "burning"

Cells of living things

You have probably noticed that living cells release energy differently from the candle or engine. The release of energy that occurs in the cells of living things—both plants and animals—is called **respiration**.

Human

Air sac cluster

Oxygen-poor blood

Oxygen-rich blood

Close-up of air sac

Carbon dioxide and oxygen exchanged here

The Function of Breathing

Air, which contains oxygen, is necessary for combustion (burning), such as the explosive combustion of gasoline that takes place in a car engine. Likewise, oxygen is essential for respiration in the cells of living things. But how does oxygen reach the cells? In animals, the process begins with breathing.

While you were reading the last few sentences, were you conscious of the fact that you were breathing? How many times a minute do you breathe? Do you have to think about it for it to happen? Unless something goes wrong with your "breathing system," such as a stuffed nose, you are normally not even aware of the flow of air into and out of your lungs. In winter, why are you more likely to be aware of your breathing?

Fish

External flap covering gill

View of gill filament (under flap)

Water carrying dissolved oxygen

Oxygen-rich blood

Oxygen-poor blood

Expanded view inside gill filament

How do different animals get the oxygen their cells need? Study the examples on these pages to find out.

Grasshopper

Tracheae

Spiracle

Outside

Oxygen

Carbon dioxide

Skin

Blood vessel

Earthworm

How does a tree get its oxygen? It may surprise you, but plants also "breathe." Their cells must carry on respiration, just as do the cells of animals.

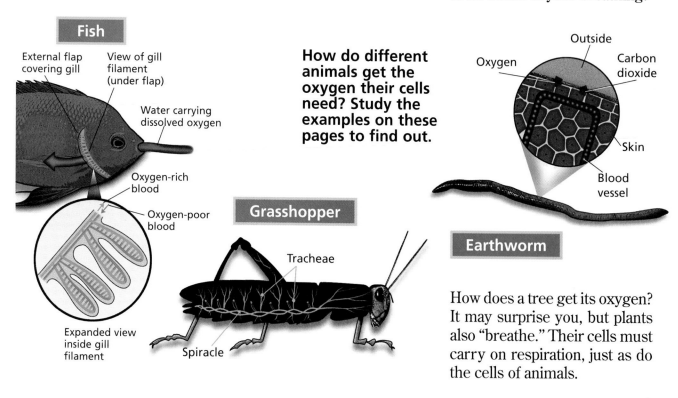

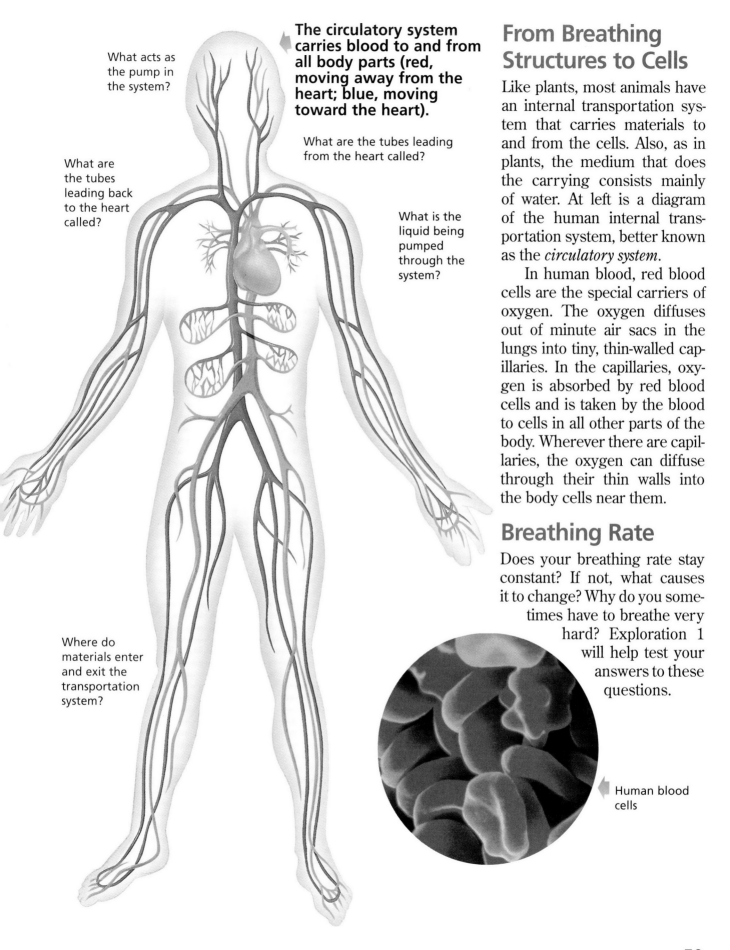

What acts as the pump in the system?

What are the tubes leading back to the heart called?

The circulatory system carries blood to and from all body parts (red, moving away from the heart; blue, moving toward the heart).

What are the tubes leading from the heart called?

What is the liquid being pumped through the system?

Where do materials enter and exit the transportation system?

From Breathing Structures to Cells

Like plants, most animals have an internal transportation system that carries materials to and from the cells. Also, as in plants, the medium that does the carrying consists mainly of water. At left is a diagram of the human internal transportation system, better known as the *circulatory system.*

In human blood, red blood cells are the special carriers of oxygen. The oxygen diffuses out of minute air sacs in the lungs into tiny, thin-walled capillaries. In the capillaries, oxygen is absorbed by red blood cells and is taken by the blood to cells in all other parts of the body. Wherever there are capillaries, the oxygen can diffuse through their thin walls into the body cells near them.

Breathing Rate

Does your breathing rate stay constant? If not, what causes it to change? Why do you sometimes have to breathe very hard? Exploration 1 will help test your answers to these questions.

Human blood cells

Testing Your Breathing and Pulse Rates

You will need a partner to help you count and measure time. (Use a stopwatch or a watch with a second hand to measure the time accurately.)

First find out your breathing rate while sitting quietly. Have a partner count the number of breaths you take in 1 minute. Try not to think about your breathing. Repeat the procedure five times and average the results. Then take your pulse using the following methods. Count the number of beats in 30 seconds and multiply by 2 to get your resting pulse rate. Repeat the procedure five times and average the results.

Find out your breathing rate while walking around for 2 minutes. Again, average the results of six readings. Calculate your pulse immediately following this activity using the method outlined above.

Find out how many times you breathe in 1 minute while running. (You may also run in place.) Again, calculate your pulse rate immediately after the activity.

Record all results in your ScienceLog in a data table of your own design.

Interpreting Your Data

1. What effect does exercise have on your breathing rate? on your pulse rate?

2. Is there any relationship between your breathing and pulse rates? If so, what is it? Why do you think this relationship exists?

3. Why did you repeat your measurements to find average breathing and pulse rates?

4. How are breathing and blood circulation related to respiration in the cells?

5. What part of the body do you think might regulate the rate at which respiration occurs?

6. Did you find that you got hot as you exercised? What does this suggest about the relationship between respiration and burning?

The Ins and Outs of Respiration

The process of respiration uses oxygen and releases energy, as does the combustion process in a car engine or the burning of a candle. Each of these processes produces wastes. In Exploration 2 you will learn more about the waste products of respiration.

Be Careful: Do not draw the limewater into your mouth. Blow gently so that the limewater does not splash out of the test tube.

EXPLORATION 2

Products of Respiration

TEST 1

You Will Need

- a drinking straw
- limewater
- a test tube

What to Do

Blow *gently* through a straw placed into clear limewater in a test tube. What happens? Is there carbon dioxide in the air you breathe out? Might carbon dioxide be a product of the respiration process that takes place in the cells of living things?

TEST 2

You Will Need

- a mirror or a shiny piece of metal
- cobalt chloride paper

What to Do

Exhale so that your breath strikes the surface of the mirror or piece of metal. Examine the shiny surface closely for any change in appearance. Repeat this process several times. Wipe the surface with a piece of cobalt chloride paper. This paper is blue when dry but changes to pink when wet. This color change signals the presence of water.

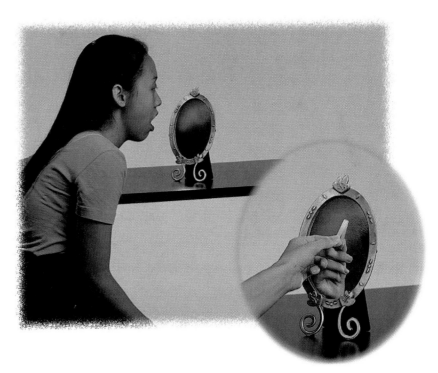

Exploration 2 continued

Questions

1. What was the appearance of the mirror after you exhaled on it?

2. What color was the cobalt chloride paper before you wiped the mirror? after you wiped the mirror?

3. What was the substance on the mirror?

4. Where did it come from?

5. What conclusion can you draw regarding another product of respiration?

Respiration is part of metabolism. Learn more on pages S16–S17 of the SourceBook.

Water and carbon dioxide are produced by the burning of most fuels, for example, gasoline or candle wax. Living things, in the process of respiration, give off these same waste products. Carbon dioxide diffuses into the blood from the cells and is carried back to the lungs to be breathed out. (This is the reverse of the way that oxygen reaches the cells.) What do you think happens to water, the other product of respiration?

It might have occurred to you that the water and carbon dioxide you detected in your breath might have been present *before* you breathed in. How might you determine this experimentally?

Now you are ready to devise a definition of *respiration* that takes into consideration the substances and the energy required, the products, and the location at which the process takes place. When you have written down your definition, complete this word equation describing respiration:

digested foods + ? → energy + ? + ?

Now look at the following simple chemical equation:

fuel (gasoline, coal, wood, etc.) + oxygen → carbon dioxide + water + energy

What does this equation tell you about the process of respiration?

Plants Also Respire

Plants must use some of the food they make to carry out their own life processes. Respiration is, therefore, just as essential to their existence as is photosynthesis. However, because photosynthesis is taking place, it is difficult to actually observe a green plant taking in oxygen and giving out carbon dioxide. To observe respiration in plants, it is easier to use germinating (sprouting) seeds, because they have not yet begun to make their own food through photosynthesis. Instead, they are still using food stored inside them. You will observe plant respiration in the next Exploration.

Plant Respiration

You Will Need

- 20 seeds (such as radish or bean seeds)
- 4 test tubes
- 4 stoppers
- a wooden splint
- limewater
- a graduated cylinder

What to Do

1. Put stoppers in two test tubes and set them aside. These test tubes will contain only air.

2. Place 20 moist seeds into the remaining 2 test tubes. (Water helps seeds carry on their life processes.)

Moist seeds

3. Stopper these test tubes tightly and leave them alone for 2 days.

4. After 2 days, remove the stopper from one test tube that contains seeds and quickly insert a lit wooden splint. Did the splint go out? What does this suggest that the seeds might have done?

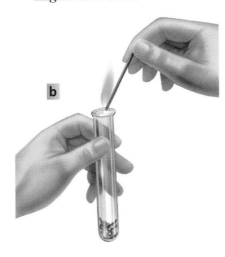

5. Unstopper the second test tube that contains seeds and quickly add 5 mL of limewater.

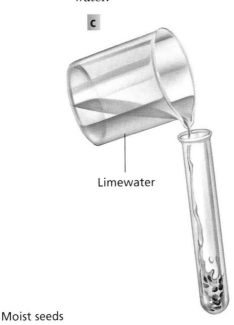

Limewater

6. Stopper the tube again, and gently swirl it for a minute. Write your observations in your ScienceLog.

7. Perform the same tests on the test tubes that contain only air, and compare the results with those obtained using the test tubes that contain seeds. How can you explain your observations?

   What do the results of this Exploration suggest concerning respiration in seeds? Write a summary of your results, along with your conclusions, in your ScienceLog. From your studies so far in this unit, make a list of differences between plant and animal processes.

Swirl

Hot Plants?

What do you think? How could you investigate Kimberly's question? Devise an experiment using seeds.

Controlling Respiration

While reading about respiration, did you wonder how your body knows where energy is needed, how much energy is needed, and so on? The answer lies in the glands, the "process controllers" of the body. The glands produce chemicals that regulate or control different functions in the body.

One of the most important glands, the thyroid gland, regulates the rate of respiration in body cells; that is, it controls how quickly food is "burned" in the cells. It does this by means of a chemical substance, *thyroxin*, that diffuses from the thyroid gland into the blood, which carries it to other body cells. The respiration process is monitored and regulated by the nervous system, which consists of the brain and nerves.

Another vital gland is the pancreas. It is tucked between the stomach and the first part of the small intestine. The pancreas makes *insulin*, a chemical that enables sugar to be used as "fuel" for respiration.

These two glands, the thyroid and the pancreas, are directly involved in the process of respiration. There are many other glands in the body as well, not all of which regulate respiration. You will study these in later science courses.

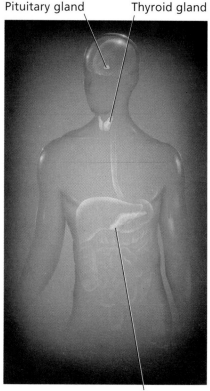

Pituitary gland

Thyroid gland

Pancreas

Respiration: A Summary

Imagine an individual cell in your body carrying on respiration.

1. What does the cell need? How do these materials reach the cell?

2. What does the cell give off? How do these materials leave, first from the cell and then from the body?

3. How do substances get into and out of the cell?

4. Why does the cell carry on respiration? Why is this process so important?

5. Many parts of the body are either directly or indirectly involved with respiration. How do each of the following parts of the body contribute to respiration at the level of individual cells: teeth, heart, skin, kidneys, lungs, salivary glands, thyroid gland, brain, capillaries, pancreas, and stomach?

Body Systems

Question 5 in the summary above named parts of many systems in the body. For example, the teeth and stomach are part of the **digestive system**. The kidneys and skin are two organs in the **excretory system**. The heart and capillaries are part of the **circulatory system**. The lungs are part of the **respiratory system**. The brain is part of the **nervous system**. All of these systems in your body work together to perform respiration and digestion, as well as other life processes.

Look on pages S16–S18 of the SourceBook for more information about body systems.

Maintaining the Balance

So far in this unit, we have talked about life processes carried out on a small scale, within a single plant or animal. Let us now turn our attention to life processes carried out on a much larger scale, that of the entire planet.

Our Earth is a very special place. It is the only world we know of where life exists. No other planet in our solar system is even close to being habitable. Life is able to exist on Earth for many reasons, not the least of which is the delicate balance that exists between plants and animals. Animals take in oxygen and give off carbon dioxide, and plants take in carbon dioxide and give off oxygen. Plants and animals need each other.

Consider a sealed terrarium, in which plants and animals live in balance with each other. Once the terrarium is sealed, no additional materials enter and no materials leave. What kinds of plants and animals would you put into a terrarium to make it self-sustaining? What additional materials from outside the terrarium are required? How could the terrarium environment be damaged so that everything in it died?

The Earth is much like a sealed terrarium, only on a much larger scale, of course. Balance must be maintained in order for life to survive. A balance between plants and animals, and between oxygen and carbon dioxide, has existed on Earth for many millions of years. However, it seems now that the actions of human beings can upset this balance.

Work with several others to explore the situation and to write a brief report about it. Your report will consist of six paragraphs. The first sentence of each paragraph is given on the facing page. Complete each paragraph by developing the idea in the opening sentence. Some research will be needed to gather information. Provide a title for the report. (Note: The numbers used to indicate the separate paragraphs should not be used in your report.)

1. Carbon dioxide is absolutely necessary for plants and animals.

2. At present, it appears that green plants cannot remove the carbon dioxide from the air as fast as humans produce it.

3. An excess of carbon dioxide is a threat.

4. The greenhouse effect is beneficial up to a certain level.

5. It is difficult to predict all the possible effects of a higher concentration of carbon dioxide in the atmosphere.

6. There are ways to lessen the potential problem of an excess of carbon dioxide in the air.

Points for Discussion

Choose one or more of the following topics and discuss them in small groups. Record your findings for presentation to the class.

1. Some people have suggested that in the future, human energy needs could be met by *biomass* alone—that is, energy would come solely from organic materials, such as plants or animal wastes. What are the possible advantages and disadvantages of this approach? Would this be a viable way to meet energy demands? Why or why not?

2. As cities expand, trees often have to be cut down to permit the widening of streets and the building of houses and businesses. Do we need to conserve trees in our cities and industrial areas? Consider these facts: One large tree can absorb 2300 g of carbon dioxide in 1 hour. This is the amount given off by 10 single-family houses. During the same hour, that tree can give out about 1700 g of oxygen. How might large trees help control the level of carbon dioxide in the air around your town or city?

3. So far, you have learned that carbon dioxide gets into the air through cellular respiration and the burning of fuels. Are there other ways? If so, what are some of them? Do a little research. What are some sources of carbon dioxide emission? How many of these carbon dioxide sources might you find in a typical home?

How is the Earth's atmosphere affected by pollutants from automobiles?

by burning forests?

1. The Magic Square

Match the number of each structure in the illustrations with the letter of the best description, and place the number in the appropriate box. If your choices are correct, the sum of the numbers in each row, column, and diagonal will be the same. The mathematical term for such a box is *magic square*. (Some questions refer to earlier chapters.)

a where diffusion of digested food into the blood primarily occurs

B gland that controls the rate of respiration in body cells

C water pipes in the food factory

d protective waterproof covering for the factory

e the food factory

F air space in the factory

g air pipe with strong supporting rings

H where insulin is made

i center of blood circulation

J where digestion of starch begins

K the regulators for opening and closing the food-factory doors

L cell layer where most of the factory doors are located

m site of the actual manufacture of food

n the service doors of the factory

o capillaries, where diffusion of digested food from the blood into the body cells occurs

P where the blood exchanges gases with the air

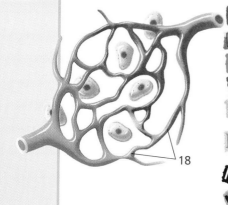

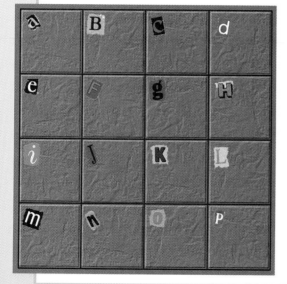

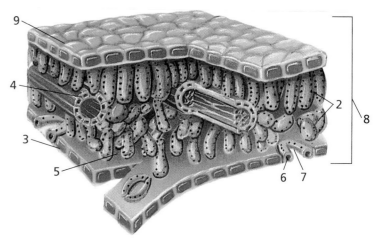

2. How Useful?

Write a report on the usefulness of plants. Use all available resources. Be sure to write the report in your own words. Prepare a list of all the resources you used.

3. Lunch to Go

Suppose your lunch consisted of a peanut butter sandwich, an apple, and a tall glass of milk. You would get energy from the meal. The energy would help you to move and to grow. Explain clearly how this energy is linked to the energy from the sun. Then trace the path of this food as it is used as energy by your body.

4. Lost in Space

In a space biosphere, a plant's water supply would have to be maintained. How would you retrieve the water lost by plants through transpiration?

ScienceLog

Review your responses to the ScienceLog questions on page 54. Then revise your original ideas so that they reflect what you've learned.

Making Connections

Unit

The Big Ideas

In your ScienceLog, write a summary of this unit, using the following questions as a guide:

1. Where does food come from?
2. Why are leaves like food factories?
3. Why is water important to living things?
4. What role does osmosis play in living things?
5. How do living things get energy from food?
6. Why do we breathe?
7. What is respiration?
8. How is respiration similar to burning?
9. How do plants and animals depend on each other?
10. How are humans changing the natural environment?

Checking Your Understanding

1. Below are a number of statements about topics you studied in this unit. Suggest ways in which each statement could be scientifically verified.

    a. Plants give off oxygen only during daylight.
    b. Carbon dioxide is a raw material used in the manufacture of food.
    c. Plants exchange gases through their stomata.
    d. Water has an attraction to itself.
    e. Osmosis affects living cells.
    f. Carbon dioxide is a waste product of respiration.
    g. All living things require water.
    h. Temperature affects the rate of respiration in plants.
    i. Plants need sunlight to produce food.
    j. Osmosis creates pressure.
    k. The light of the sun, not its warmth, is what makes life on Earth possible.

2. Read the following statement:

Diffusion begins when you add a substance to water, and it ends when the substance is completely mixed in the water.

Does the statement really tell the whole story, or is there more to it? Explain.

3. Plants are sometimes afflicted with an inherited abnormality (called *albinism*) that results in the plants having no color whatsoever. What kind of problems do you think this might cause for a plant? What would the chances be for this plant to survive and reproduce?

4. Look at the illustrations below. Each tells a story. It is your job to figure out what that story is.

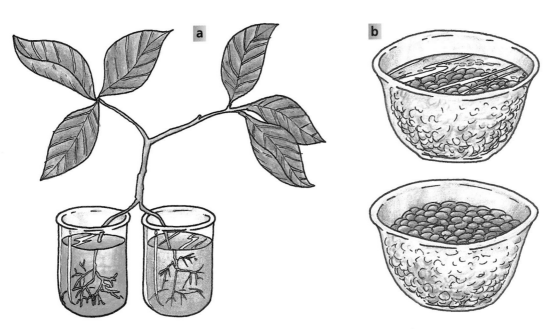

  a. After a few days the leaves were mottled with blue and red patches.
  b. After a few hours, the water seemed to disappear. What happened?

5. (concept map) Make a concept map using the following terms: carbon dioxide, breathing, water, respiration, oxygen, and energy.

6. How is it that the same water is used over and over again in the natural world?

Watchfrogs of the Environment

In 1990 scientists made a troubling discovery. Populations of frogs, toads, salamanders, and other amphibians were decreasing all over the world. Equally troubling were the questions this discovery raised. Why were they dying? Was there a problem with their environment? If there was a problem, could it harm people too?

A Featherless Bird?

Amphibians can be compared to the canaries that miners once carried down mine shafts. If a canary died, the miners knew that poisonous gases were in the air and that the miners were in danger. Similarly, amphibians may be warning us of dangers in our own environment.

Amphibians are particularly sensitive to their surroundings. They live most of their lives underwater or in damp soil. As they breathe, they absorb oxygen through their thin skin. Any pollutants, such as acid rain, industrial chemicals, or pesticides, can easily pass through their skin. Thus, pollutants can drastically affect an amphibian's health. Similarly, only a jellylike coating protects their eggs from the outside world, making the eggs vulnerable to pollutants as well.

No Trees Means No Tree Frogs

What is causing the shrinking of the amphibian population? The likeliest reason is the destruction of their natural habitats as forests are cut down and wetlands are filled in. But there may be another reason: the decrease of *ozone* in the atmosphere. Ozone is an oxygen molecule that absorbs much of the sun's harmful ultraviolet rays before they reach the Earth's surface. Exposure to these rays can damage amphibian eggs, resulting in mutations or even death.

Cleaning Up Our Planet

Perhaps the best way we can help our amphibian friends is to reduce our impact on the environment. For instance, we can recycle paper so that fewer forests need to be cut down. We can buy natural, organic detergents so that when we wash clothes, we do not add pollutants to our water. By carpooling or using public transportation, we can reduce air pollution, too. Little things can add up!

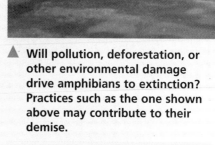

▲ Will pollution, deforestation, or other environmental damage drive amphibians to extinction? Practices such as the one shown above may contribute to their demise.

Take a Breather

Both plants and animals carry out respiration in order to live, but different organisms respire in different ways. Find out how a tree, a fish, an amphibian, and a human respire. How are the processes different for each organism? How are the processes alike?

Space Food

*T*hey say that dining in space leaves a lot to be desired. The menu is limited, the service is nonexistent, and fresh food is out of the question. But don't blame the chef; preparing food for space travel is a difficult task.

Eating Out—Way Out!

Before the first spaceflight, scientists did not know whether eating food in space would be possible without gravity to aid in swallowing. So one of the first things John Glenn did on an American spaceflight in 1962 was try to eat. As Glenn soon discovered, swallowing food is not a problem. However, making space food both nutritious and good to eat is another story.

Food used for space travel must meet several requirements. There are no refrigerators on space shuttles, so the food must be nonperishable and must remain free of harmful bacteria. The stress of space travel is hard on the body, so the food must be rich in minerals and high in calories. Also, limited storage requires dense food and compact packaging. Finally, food must be packaged in special containers that will keep it from floating away.

What's for Dinner?

Early attempts to make space food resulted in small, dried food cubes and tubes of food paste. As you can imagine, these

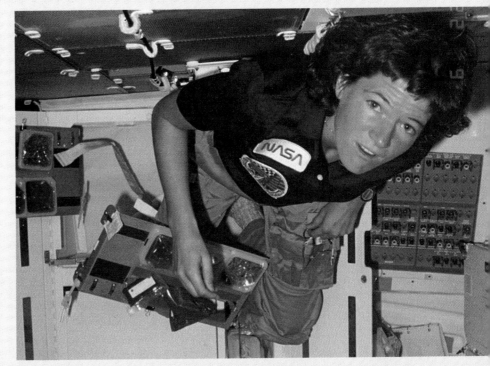

▲ Astronaut Sally Ride preparing to have a meal on a space shuttle. The food tray is specially packaged to keep the food from floating away.

foods made unappetizing meals. Over time, however, nutritionists developed better methods of packaging tasty, conventional food so that it would meet the astronauts' needs.

Thanks to these advances, a typical meal on today's space shuttle might consist of smoked turkey, mixed Italian vegetables, mushroom soup, strawberries, butterscotch pudding, and tropical fruit punch. The food is canned or specially packaged in foil to solve the problems of storage and weightlessness. Most spices would float away in space, but astronauts can still flavor their food with liquid pepper and other flavorings that will stick to the food. Some of the food is freeze-dried for

easier storage. Before eating this food, astronauts mix it with water that is produced as a byproduct of the shuttle's fuel cells. It's a complicated process, but despite all of the hassle, most astronauts would repeat their dining experiences. After all, the food may not be fancy, but you can't beat the view.

Plan It Yourself

How is planning a backpacking trip like planning for spaceflight? Plan a menu for a three-day backpacking trip. How much water would you need and how would you get it? What other concerns will you need to think about?

Eric Pianka, Zoologist

Eric Pianka became involved in the study of lizards when he was four or five years old. "On a trip across the country with my family, I saw a big, green lizard at a roadside park," Eric explains. "I tried to catch it, but all I got was the tail. At that moment, I knew I had to find out everything I could about the kind of life it led." After years of study, Eric is now a world-famous professor of zoology at the University of Texas.

The Ecology of Desert Lizards

In his research as a zoologist, Eric has focused on the ecology of desert lizards. He goes to a desert, collects lizards, and examines and classifies them. Then he compiles data and interprets it in books or papers. As Eric puts it, "I try to answer questions like, Why are there more lizards in one place than

in another? How do they interact with each other and with other species? How have they adapted to their environment?"

Learning From Wildlife

Eric believes that research on lizards and other animals may help protect our environment. "Everyone always asks, 'Why lizards?' I turn that around and say, 'Why you?' The general attitude is that everything on Earth has to somehow serve humans. By looking at how other species have lived and died and changed over millions of years, we can gain a better understanding of the world we live in."

Exploring the Deserts of the World

One of the things Eric likes best about his job is being in the wilderness and seeing things

▲ The collared lizard lives in rocky regions of the southwestern United States.

that few people have ever seen before. "I've been almost everywhere! I spent a lot of time studying deserts in the western United States. I've been to deserts in southern Africa, India, and Chile. My most current (and oldest) interest is in the deserts of Australia. I haven't had a chance to study the Brazilian Amazon yet, but that's my goal for the future!"

A Project Idea

Select a common animal that lives in your area and that can be easily observed. Spend a couple of hours watching what it eats, what it does, and where it goes. What other animals does it interact with? What specific features of this animal make it well suited to its environment? Carefully document everything you observe. Did you discover anything you didn't already know?

▼ Eric Pianka with a perentie in Australia's Great Victoria Desert. Perenties are some of the largest lizards in the world, ranging up to 2.4 m in size.

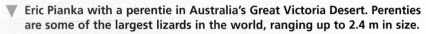

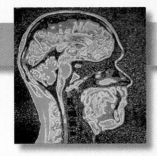

Meatless Munching

What'll it be today, the hamburger special, chicken surprise, or the garden-fresh veggie platter? More and more people are opting for the veggie platter. In fact, research indicates that over 12 million Americans are now eating vegetarian, and this number appears to be growing.

It's Not Just Salads

When you think of a vegetarian diet, you might think only of vegetables. Of course, vegetables are the mainstay of a vegetarian diet, but not all vegetarian diets are alike. Many vegetarian diets include dairy products, eggs, fruits, and nuts. Some vegetarian diets are based solely on plant products, while semivegetarian diets may also include some fish and poultry.

Why the Trend?

Different people have different reasons for going vegetarian. Some people believe that a vegetarian diet is healthy because a decrease in consumption of animal products can lower their intake of saturated fat and cholesterol. Many people also have ecological reasons for eating a vegetarian diet. For example, the grains that are fed to livestock can supply more nutritional value per volume than the meat we get from the livestock. And producing a serving of meat requires more land, water, and chemicals than producing a serving of grain. Finally, many vegetarians believe that it is unethical to kill animals for meat when plant substitutes are available.

▼ **All of the foods shown here are the products of plants and are commonly used in vegetarian diets.**

Benefits and Risks

A vegetarian diet can be a healthy choice. Recent statistics suggest that a vegetarian diet may reduce the risk of heart disease, adult-onset diabetes, and some forms of cancer. However, this type of diet takes some careful thought. It is not simply a matter of eliminating meat. People may replace the missing meat with too many dairy products and eggs, which are higher in fat than most meats. Others may substitute high-carbohydrate foods (such as pasta) and junk food (such as french fries) for their meat choices instead of increasing their fruit and vegetable intake. This could lead to nutritional deficiencies. The key to consuming the recommended amount of calories and nutrients for a healthy diet is to eat a wide variety of nutritious, low-fat foods. Actually, this is good advice whether you want to decrease your meat intake or not.

Prepare a Healthy Menu

Choose one of the nutrients that meats provide, and do some research to find a vegetable substitute. What difficulties did you encounter in your search? Would you eat the substitute you found?

Unit 2

Particles

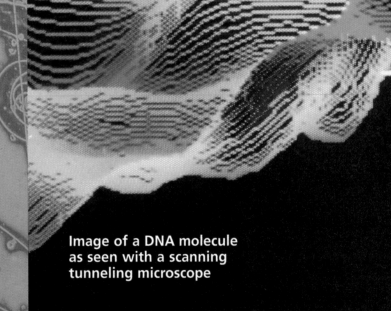

Image of a DNA molecule
as seen with a scanning
tunneling microscope

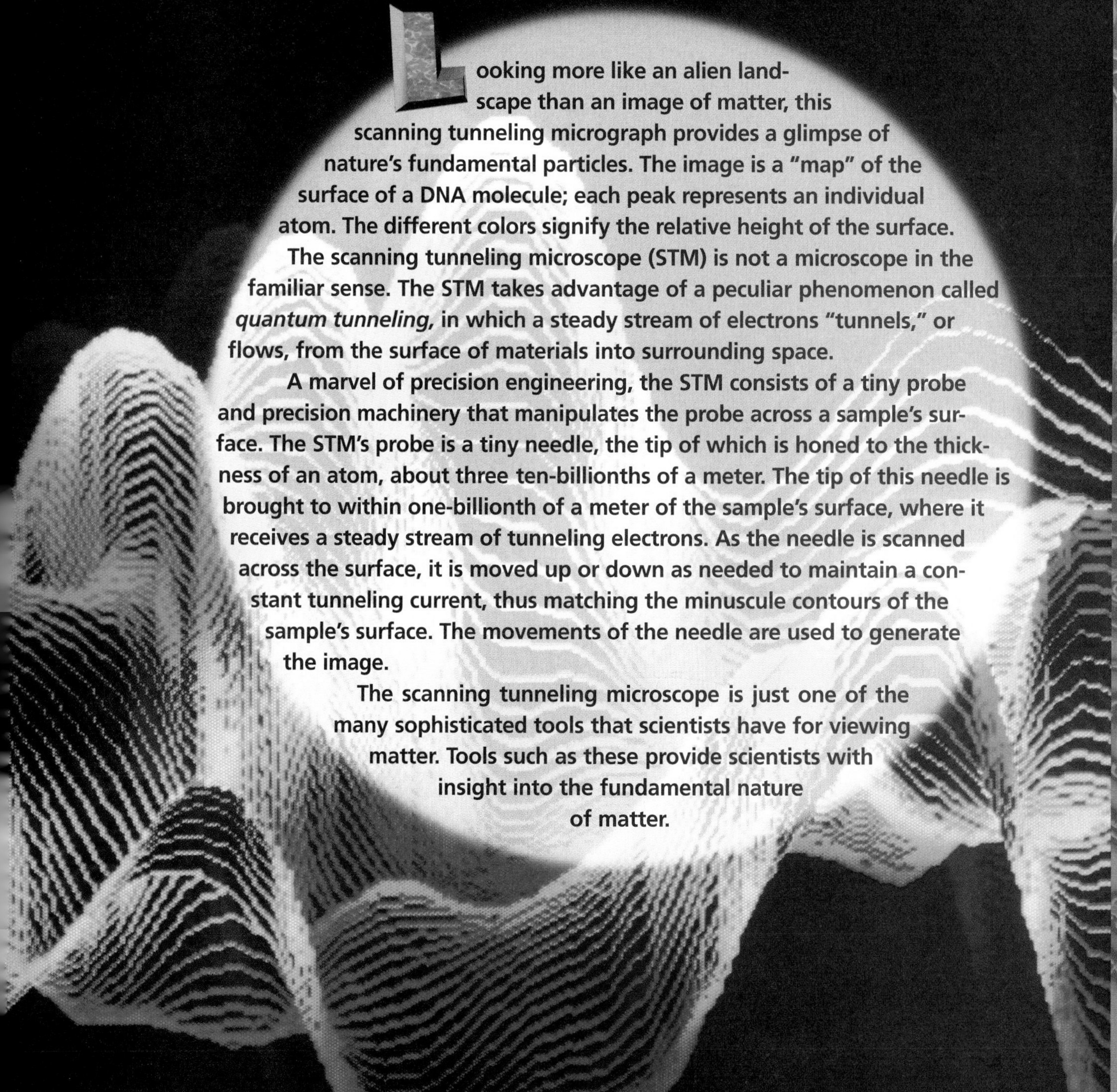

Looking more like an alien landscape than an image of matter, this scanning tunneling micrograph provides a glimpse of nature's fundamental particles. The image is a "map" of the surface of a DNA molecule; each peak represents an individual atom. The different colors signify the relative height of the surface.

The scanning tunneling microscope (STM) is not a microscope in the familiar sense. The STM takes advantage of a peculiar phenomenon called *quantum tunneling*, in which a steady stream of electrons "tunnels," or flows, from the surface of materials into surrounding space.

A marvel of precision engineering, the STM consists of a tiny probe and precision machinery that manipulates the probe across a sample's surface. The STM's probe is a tiny needle, the tip of which is honed to the thickness of an atom, about three ten-billionths of a meter. The tip of this needle is brought to within one-billionth of a meter of the sample's surface, where it receives a steady stream of tunneling electrons. As the needle is scanned across the surface, it is moved up or down as needed to maintain a constant tunneling current, thus matching the minuscule contours of the sample's surface. The movements of the needle are used to generate the image.

The scanning tunneling microscope is just one of the many sophisticated tools that scientists have for viewing matter. Tools such as these provide scientists with insight into the fundamental nature of matter.

More Than Observing

1 What observations might this scientist be making? To what inferences might these observations lead?

2 If you could see individual air particles, what might they look like?

3 Some people collect models of cars or trains. How do such models differ from scientific models?

ScienceLog

Think about these questions for a moment, and answer them in your ScienceLog. When you've finished this chapter, you'll have the opportunity to revise your answers based on what you've learned.

A Search for Explanations

What Is the Smallest Object You Can See?

Angelina's class went to the planetarium. As they sat back in their seats and looked up at the domed ceiling, the lights went out. Stars appeared on the dark ceiling. Angelina felt as if she were in outer space. Then a voice began to speak:

"Imagine that you have journeyed far, far away to another galaxy. You are now on your way home to Earth. You are approaching our galaxy, the Milky Way. The entire Milky Way appears as a mere point of light in space because of the vast distance separating the two galaxies. As you come closer, the Milky Way begins to resolve into billions of stars, one of which is our sun. Move even closer and the planets can be seen shining in the reflected light of the sun. Come closer still and Earth is seen as a sphere of land, ice, and water. When you land on Earth, you find yourself surrounded by a myriad of objects of every shape, size, and color imaginable. With a sharp eye, you can detect tiny particles of dust in the air of a sunlit room. And with the aid of optical instruments, you can view even smaller particles."

What is the limit for detecting smaller and smaller bits of matter? Is there no limit to the size of objects that can be detected? The answers to these questions are important steps in the search for explanations about why matter behaves the way it does. This unit will lead you on a search for these explanations.

The Sizes of Things

Some objects and distances are very large. The diameter of the solar system, for example, is about 10,000,000,000,000 m. Other objects and distances are very small. The size of one cell in your body is about 0.000001 m. But it can be very awkward to work with such large or small numbers. One useful way to refer to the sizes of very large and very small things is to use **exponents**. Exponents are a kind of shorthand. Count the number of zeros in the figure that shows the diameter of the solar system. When it is expressed using an exponent, the figure becomes 10^{13} m. The size of a body cell expressed with an exponent is 10^{-6} m.

Now examine the chart below. Explain the rule for using exponents to express both very large and very small numbers.

The list at right gives more examples of the sizes of different objects, using exponents. The measurements are approximate, not exact.

Where would you place the following in the list?

- width of a pencil
- diameter of a hair
- distance from your home to school
- diameter of Venus's orbit

Suggest other objects that could be added to the list.

Sample exponent values
$10^{10} = ?$
$10^4 = 10,000$
$10^3 = 1000$
$10^2 = 100$
$10^1 = 10$
$10^0 = 1^*$
$10^{-1} = 0.1$
$10^{-2} = 0.01$
$10^{-3} = 0.001$
$10^{-4} = 0.0001$
$10^{-10} = ?$

Meters	
10^{26}	Diameter of the known universe
10^{25}	
10^{24}	
10^{23}	
10^{22}	
10^{21}	Diameter of the Milky Way
10^{13}	Diameter of the solar system
10^{12}	Diameter of Jupiter's orbit
10^{11}	Diameter of Earth's orbit
10^{10}	
10^9	Diameter of the moon's orbit
10^8	
10^7	Diameter of the Earth
10^6	Distance across the continental United States
10^5	
10^4	Deepest part of the Pacific Ocean
10^3	
10^2	Height of a 25-story building
10^1	Length of your classroom
10^{0*}	Height of a young child
10^{-1}	Width of a hand
10^{-2}	Width of a finger
10^{-3}	Diameter of a thread
10^{-4}	Diameter of a fine sand particle
10^{-5}	
10^{-6}	Diameter of a body cell
10^{-7}	
10^{-8}	
10^{-9}	
10^{-10}	Diameter of an atom

Any number to the power of 0 is always 1.

Give Me the Facts

A Question of Proof

You have probably heard people say, "I'll believe it when I see it!" With average vision you can see objects as small as 10^{-4} m. With a microscope, objects as small as 10^{-7} m can be detected. What are the facts about even smaller objects? You cannot see them, so how do you know they exist?

You must rely on *circumstantial evidence*. In scientific terms, that means you must make observations and use those observations (facts) to make inferences.

Observation vs. Inference

How does an inference differ from an observation? The article below offers some clues. While reading the story, try to find at least three inferences made by the defense lawyer and the prosecuting attorney. What facts (observational evidence) support each inference? Are all the inferences true? In your ScienceLog, record your answers in a table similar to the one shown at right.

The AIDS virus (orange dots) is approximately 1.2×10^{-7} m across.

Inference	Supporting observation (evidence)
1.	
2.	
3.	

Jury Still Out

Phoenix, AZ

After a full day of deliberation, the jury has yet to reach a verdict in the trial of Ike Swipe, who is accused of robbing the corner store. Yesterday, District Attorney Ivana Burnham summed up the prosecution's case. She reminded the jury of the facts brought out by the prosecution:

- The accused party was seen in the area of the robbery.
- His blood type was found on the doorknob. He had a cut on his right hand.
- His fingerprint was found on the countertop.
- He was observed spending more than the usual amount of money at the horse races the next day.

- The defendant has a past record of robbery.

Burnham stated that since Swipe was in the area of the crime and has a past criminal record, he must have committed the crime. The blood type found on the doorknob also matches Swipe's, so Swipe cut himself while breaking the store's window. Finally, the prosecutor declared that the money the defendant spent at the races was the money taken during the robbery.

On the other hand, defense attorney Sibyl Quibbler claimed that many people were seen in the area of the robbery and that Swipe is no more

a suspect than any of them. The cut was the result of a dish-washing accident and had nothing to do with the broken window at the store. As for the blood stain on the doorknob, Swipe has a very common blood type. The fingerprint was probably left on the countertop when the defendant bought a paper at the store a few hours earlier. "The evidence is entirely circumstantial," Quibbler noted. Judge Hugo Furst reminded the jury that in order to reach a verdict of guilty, they must find that the evidence shows—beyond a reasonable doubt—that the accused committed the crime.

Max Reads From His ScienceLog

After reading the article titled "Jury Still Out," I think the jury should give a verdict of "not guilty." No one saw the accused at the scene of the crime, and there is no real proof that he did it. I just don't think the circumstantial evidence is good enough to let anyone make an inference that will send that man to prison.

Do you agree or disagree with Max's opinion? Explain why. Are all inferences based on circumstantial evidence? Describe the similarities and differences between the inferences that scientists make and those that the lawyers and Max made.

Observation or Inference?

You live in an ocean of air, but you cannot see it. Why not? The answer is based on circumstantial evidence. Such an explanation depends on knowledge about the unseen structure of matter. For instance, it is easy for a scientist to observe how liquids or solids behave, but it is more difficult to explain *why* they behave as they do. A scientist must draw inferences based on observations. In this sense, a scientist does exactly what a jury does!

Copy the chart below into your ScienceLog. Working in small groups, examine the statements about air. Do you agree with each one? Place a check mark next to the statements you agree with. Are your decisions based on observations, or are you making inferences? What evidence supports each decision? Discuss your decisions with your classmates. As a group, do you agree or disagree with each statement?

Statements about air	Observation	Inference	Evidence supporting your decision
1. Air can be squeezed into a smaller space.	✔		I can pump air into my bicycle tire.
2. Air is invisible.			
3. Air has volume.			
4. Air has mass.			
5. Air moves.			
6. If we could see air, we would see many particles.			
7. Air behaves like a sponge.			
8. There is water in air.			
9. Create your own statement about air.			

Using Models

Based on your observations and inferences, you can come to some conclusions about air. Do you think air is made of particles, as suggested in statement 6 on page 82? Or is air more like a sponge that can be compressed and expanded, as suggested in statement 7?

These kinds of questions help you develop a mental picture or idea about the structure of air. They help you form a *model* for the concept you are thinking about. So far, you have considered two types of models. In statement 6, the model is an idea: air is made up of particles. You cannot actually see air—you can only observe how it behaves. Therefore, you form an idea about what air is really like based on your observations of its behavior. In statement 7, you compare air to an actual object—a sponge. In what ways is air like a sponge? In what ways is it different? Do you think the particle model is better or worse than the sponge model? Why? How can the models help you understand air and its behavior?

More About Models

Francine created the model below to represent the workings of the inner ear and how sounds are transmitted across the eardrum. Find a diagram of the inner ear in a reference book, and then figure out what each feature of the model represents. Do you find that Francine's model helps explain how the ear works? Do you think it is a good model? Why or why not?

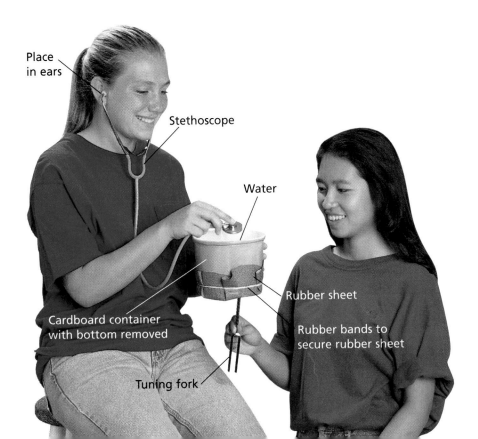

Place in ears

Stethoscope

Water

Cardboard container with bottom removed

Rubber sheet

Rubber bands to secure rubber sheet

Tuning fork

Making Ramón's Model

Ramón agreed with statement 6 on page 82, so he devised the following model. Now it's your turn to build Ramón's model.

You Will Need

- a balloon
- 5 small plastic beads

What to Do

Place the plastic beads inside the balloon. Then blow up the balloon—but not too much! Tie the balloon closed. The balloon now contains not only real air but also the plastic beads, which represent air particles. Gently shake the balloon until the beads rattle around inside. Can you see the model air particles in motion? Can you feel them in motion? What observation about air does this model explain?

Visualizing, Explaining, and Predicting

Were the models made by Francine and Ramón useful? In your ScienceLog, describe how models help us to do the following things:

- Visualize a complex idea or structure.
- Explain observations and make inferences.
- Make predictions that can be tested through further observations and experiments.

What word or phrase best describes a model? Suggest another example of a model.

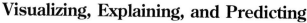

"Nobody noticed me until I bought my Shred Jeans™. Now that I've got 'em, I'm the head cheerleader, I'm an A+ student, and I look GREAT! Get your Shred Jeans™ today . . . they'll change your life. They sure changed mine."

1. These Jeans Changed My Life!

Obviously, the advertisement above is making some pretty ridiculous claims. It gives several examples of how people draw incorrect inferences from available observations. With another student, make a list of five ridiculous claims that you can support without actually lying. Identify the circumstantial evidence you used to make your claim. Then explain how you can make better inferences based on your observations.

2. This Model Is All Wet

Design a model for water that illustrates the observations you have made about it. What happens when liquid water turns into ice? into steam? Can your model represent these changes? What observations does your model have trouble explaining? After you create your design, summarize its strengths and weaknesses as a model.

3. It's Like This, Kid . . .

Choose a scientific concept or process that you have studied either in this course or in a previous science course. Describe a model that could help a fifth-grade student better understand the concept or process.

4. Road Trip

Choose one of the pictures shown here. Make a list of as many observations as you can about the photograph. Then make as many inferences as you can about the photo, and write them on a separate list.

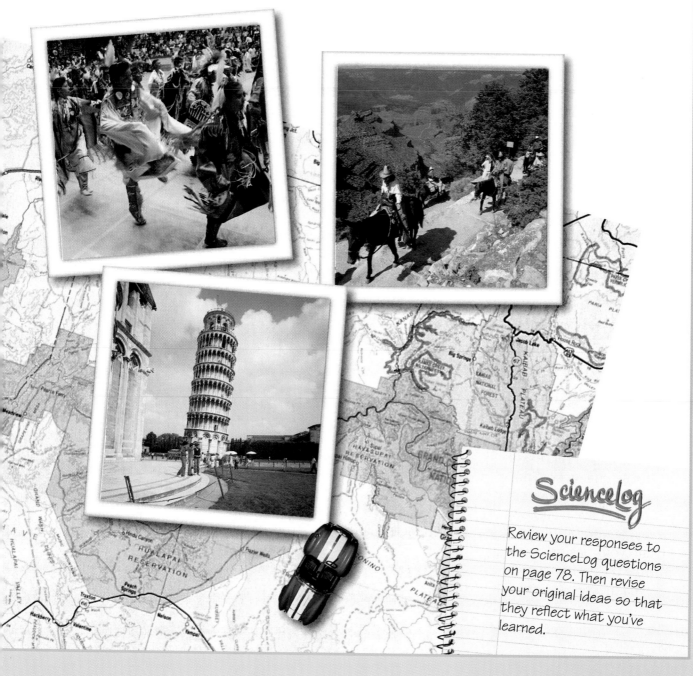

Science Log

Review your responses to the ScienceLog questions on page 78. Then revise your original ideas so that they reflect what you've learned.

A Case for Particles

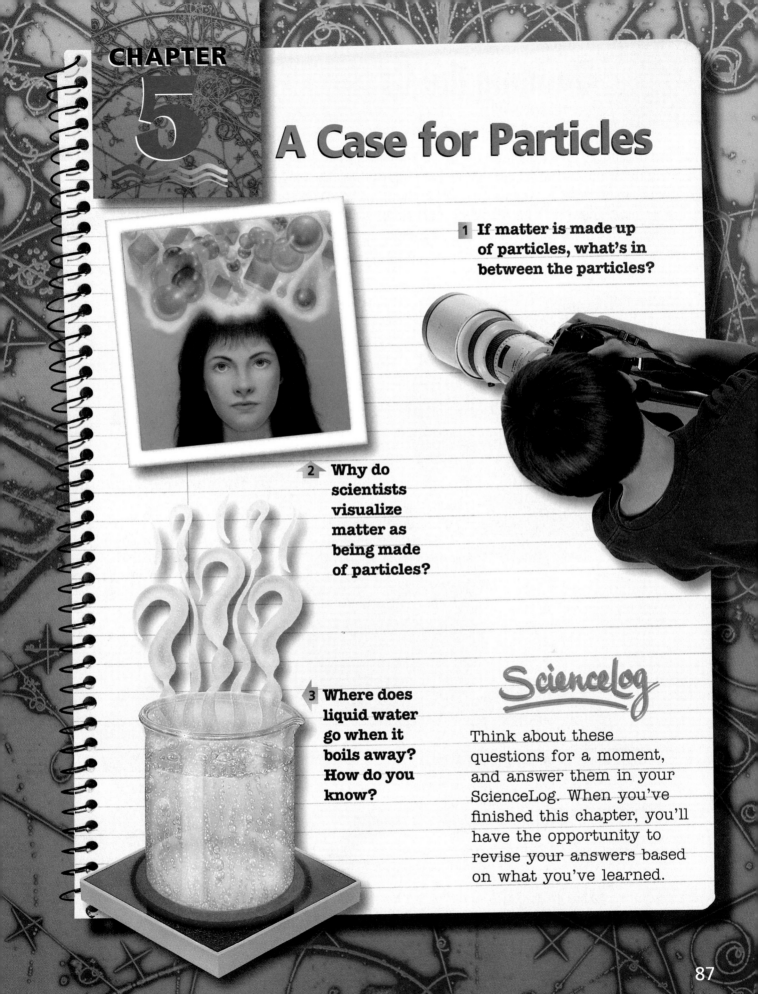

1 If matter is made up of particles, what's in between the particles?

2 Why do scientists visualize matter as being made of particles?

3 Where does liquid water go when it boils away? How do you know?

ScienceLog

Think about these questions for a moment, and answer them in your ScienceLog. When you've finished this chapter, you'll have the opportunity to revise your answers based on what you've learned.

Building the Case

You are the judge, jury, and attorney in a landmark case—a case that will determine whether all matter is composed of **particles**. This case may raise as many questions about matter and its behavior as it answers. The following experiments will provide the observations and information you will need to make some important inferences as you prepare your case in favor of the particle theory of matter—or against it.

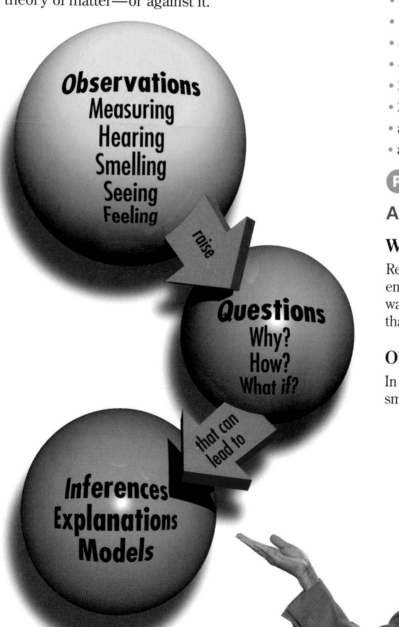

Observations
Measuring
Hearing
Smelling
Seeing
Feeling

raise

Questions
Why?
How?
What if?

that can lead to

Inferences
Explanations
Models

EXPLORATION 1

Making the Case

You Will Need

- red food coloring
- an eyedropper
- a stirring rod
- a 100 mL beaker
- 500 mL of sand
- water
- 500 mL of dried peas or beans
- 40 mL of rubbing alcohol
- 4 large containers
- 25 mL of salt
- 2 graduated cylinders
- a funnel
- a stopwatch or clock

PART 1

A Thought Experiment

What to Do

Read the observation and inferences about liquid and frozen water. Then answer the questions that follow.

Observation

In the freezer, ice cubes become smaller over time.

Questions

- Where does the ice go?
- How does it disappear?
- Can ice be prevented from disappearing?

Inference and Possible Explanation

- Perhaps ice (water) is made up of particles.
- Maybe some of these particles escaped from the solid state to form a gas, which floated away.

Follow-Up

1. Name another substance that changes directly from a solid into a gas.
2. Could a gas change directly into a solid? If so, think of some examples.
3. Do these observations and explanations support the idea that water is made up of particles? Why or why not?

PART 2

Seeing Red

What is the largest amount of water in which you could dissolve a drop of red food coloring and still detect its color? Here is a way to find out.

What to Do

Thoroughly dissolve a drop of food coloring in 50 mL of water. Now divide this solution into two equal parts. Wash 25 mL down the sink, and add 25 mL of water to what remains. Once again the total volume of the solution is 50 mL. Is the solution still colored red?

The concentration of the food coloring has been diluted to one-half of the original amount. Repeat the dilution process once more. Can you still see the red coloring in the water? Your beaker now contains one-quarter of the original drop of food coloring. Repeat the procedure—keeping accurate records—until you no longer see the red color.

Before going on to Part 3, discuss the following questions with a partner:

1. Is the color spread evenly throughout the solution, or are bits of food coloring clumped together?
2. Do you think there may be some food coloring left in the solution at the end, even though you cannot see any? How much of the food coloring do you have in the beaker of water at the end of the experiment? How do you know?

3. If matter is made up of particles, what can you infer about the size of the food-coloring particles?
4. Does the experiment support the particle theory of matter? Why or why not?

PART 3

Pour Judgment

What to Do

Fill three large containers with the substances listed below. Do not mix the substances.

- dried peas or beans
- sand
- water

Now pour each substance into an empty container. Did either of the first two substances resemble water in the way they poured? What might you infer about matter from this experiment?

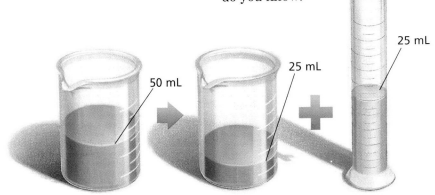

Exploration 1 continued ▶

PART 4

When 1 + 1 ≠ 2

What to Do

Carry out the following three activities. After making careful observations, use them to develop inferences about the unseen structure of matter.

1. Pour 50 mL of sand into a 100 mL graduated cylinder. Then pour 50 mL of water into another 100 mL graduated cylinder. Carefully pour the water into the sand. Record the volume of the mixture. Suggest an explanation for why the combined volume is not 100 mL.

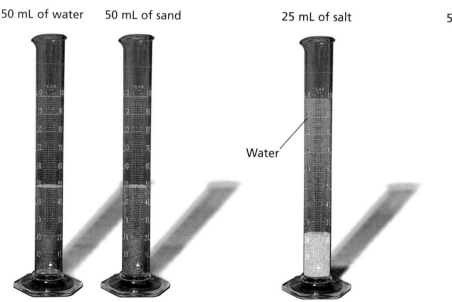

50 mL of water   50 mL of sand   25 mL of salt   50 mL of water   40 mL of alcohol

Water

2. Put 25 mL of salt into a graduated cylinder. Add enough water to bring the combined volume of salt and water to 100 mL. Without spilling the contents, gently shake the cylinder for a minute or two. Record the volume after shaking. How do you explain the final volume of salt and water?

To read volume, locate the curve at the top of the liquid. Read at eye level the lowest point of the curve.

3. Pour 50 mL of water into a graduated cylinder. Then pour 40 mL of alcohol into a second cylinder. Pour the alcohol into the water and stir. Is the volume of the two combined liquids 90 mL? Explain.

Drawing Conclusions

Do your observations support the idea that matter consists of particles? Why or why not? In your ScienceLog, summarize your case for or against the particle theory of matter.

Particles on Trial

A ninth-grade class was asked to build a case for or against the particle model. Here are a few of their replies. Imagine you are the teacher—what comments would you write on each report? Has each student built a good case for the existence of particles? How could they improve their case?

Marco's Case

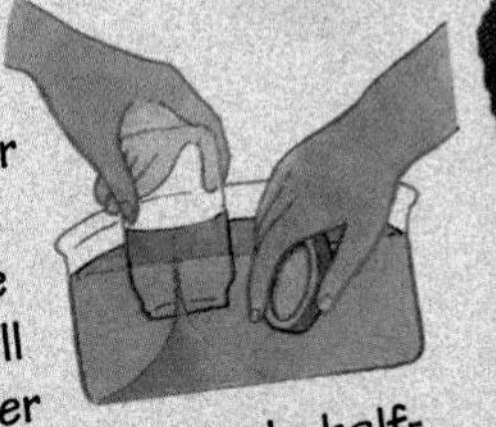

In sixth grade we did an experiment in which a jar full of sand was turned upside down in a container of water. Then we took the top off the container. We thought that the water would fill the jar when the sand came out, but the water only went halfway up. This showed that the sand had only half-filled the jar. Air had filled the other half by fitting into the spaces between the sand particles. Otherwise, the water would have completely filled the jar. When salt dissolves in water, I think the same sort of thing happens. The salt fills in the spaces between the water particles.

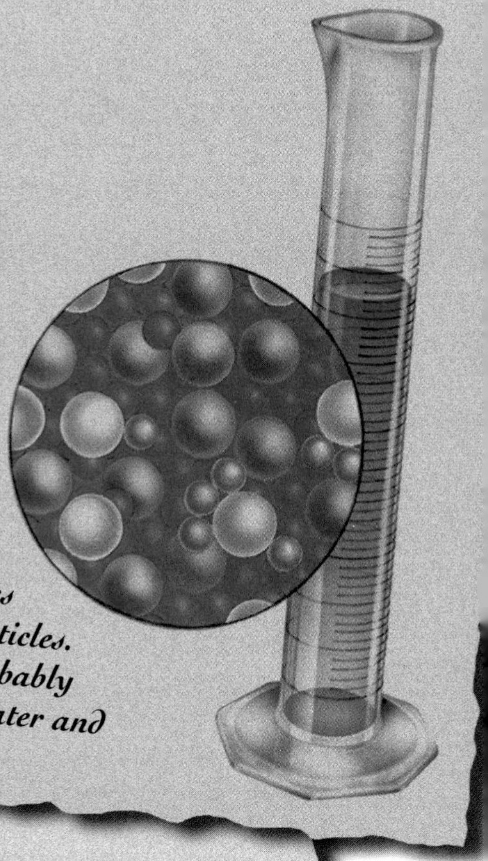

Bob's Case

Matter must be made of particles. It says so right in the book.

Nikki's Case

I think the particle model is correct. In Part 4 of the Exploration, water filled the spaces between the sand particles. This is also what probably happened when the water and alcohol were mixed.

Your Case

In a few paragraphs, revise your own case for or against the particle model. Share it with a few classmates, and ask for their opinions. You should use your summary and evidence from the Exploration, as well as any other evidence you have observed. Now test the strength of your case. Does it explain why you cannot see air? Does it explain why you see sugar when it is in the sugar bowl, but not when it is dissolved in water? How does your case for or against the existence of particles answer these questions?

The Hidden Structure of Matter

Democritus and John Dalton were born more than 2000 years apart in two different countries, Greece and England. But while thinking about the nature of matter, they both separately arrived at the same conclusion: Matter is made up of particles—what Democritus called **atoms**. During the two millennia separating the lives of these men, this idea was largely ignored. How did Dalton arrive at a conclusion that had been neglected for 2000 years? Read Constructing a Particle Theory to find out.

> According to convention, there is a sweet and a bitter, a hot and a cold, and according to convention there is color. In truth there are atoms and a void.

> I should apprehend there are a considerable number of what may properly be called elementary particles, which can never be metamorphosed [changed] one into another.

Constructing a Particle Theory

Recall your study of chemicals from earlier science courses. You found out that all pure substances on Earth can be classified as either *elements* or *compounds*. All matter is made of elements. During chemical changes, elements combine to form compounds, or compounds decompose into elements. Compounds can also change into other compounds.

By Dalton's time, much had been discovered about chemical changes and the elements involved in these changes. Antoine Lavoisier had discovered the role that oxygen plays in combustion. Water had been decomposed into the elements hydrogen and oxygen by passing an electric current through it. Metals could be obtained from ores by chemical changes, and at least 40 different elements had been identified.

Time Out for Facts

- About 2000 years ago, people had already discovered the elements gold, silver, iron, lead, tin, mercury, sulfur, and carbon without understanding that they were elements.

- By 1735, alchemists had added zinc, arsenic, and phosphorus to the list of known elements.

- Hydrogen, oxygen, and nitrogen were discovered between 1765 and 1775. Why do you think elements like gold were discovered long before elements such as oxygen and hydrogen?

John Dalton began his scientific investigations at a very early age. When he was 12 years old, he organized a school in an old barn and became a schoolmaster.

As an observant and curious scholar, Dalton explored many scientific questions. One of these explorations led to some rewarding conclusions. Trace Dalton's path of investigation by completing the flowchart below. You'll find the information you need in the next paragraph.

Dalton knew that by passing an electric current through water, the water could be separated into the elements hydrogen and oxygen. He also observed that by identifying the amount of one element in a quantity of water, he could predict the amount of the other element. For example, if 16 g of oxygen were formed by decomposing water, then 2 g of hydrogen would always be formed as well. In other words, the mass of the oxygen was always eight times greater than the mass of the hydrogen. Dalton asked himself, "From this information, what can I infer about the structure of matter?"

Perhaps this activity will help you understand the answer Dalton found. Measure 16 g of modeling clay, and form it into a ball. This ball represents an oxygen atom. Next, form two 1 g balls of clay. These represent two hydrogen atoms.

Attach the model atoms so that they resemble the illustration below. You have just simulated a chemical change between oxygen and hydrogen atoms that forms a new substance—water.

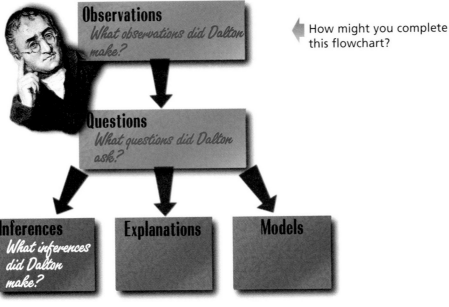

How might you complete this flowchart?

Time Out for Discovery

1. How many times heavier is your model oxygen atom than your model hydrogen atom? Does your answer support Dalton's observation?

2. Add your water-particle model to those formed by your classmates. How does the total mass of model oxygen atoms compare with the total mass of model hydrogen atoms?

3. Consider a quantity of water. If the water consists of oxygen and hydrogen atoms, how will the total mass of oxygen atoms compare with the total mass of hydrogen atoms?

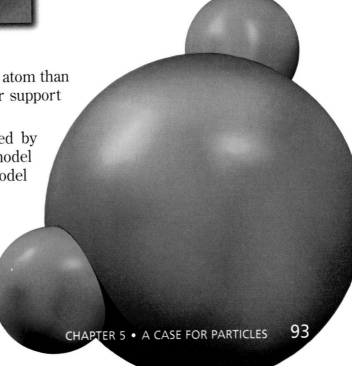

In 1803 Dalton wrote the following entry in his journal: "An enquiry into the relative weights of the ultimate particles is, as far as I know, entirely new. I have lately been prosecuting this enquiry with remarkable success."

 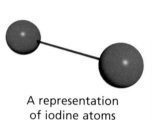

A representation of iodine atoms

 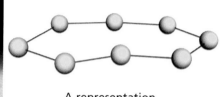

A representation of sulfur atoms

Iodine (left) and sulfur (right) are examples of elements. What do the representations show?

Of course, the "ultimate particles" Dalton spoke of were much like Democritus's atoms. Although he could not see atoms, Dalton made the following inferences:

1. An atom of one element has a mass that is different from the mass of an atom of any other element. (How much heavier was your model oxygen atom than your model hydrogen atom?)

2. An atom of one element is identical to any other atom of the same element. (Did you make your two model hydrogen atoms the same or different?)

3. An atom of one element cannot be changed into an atom of another element. (Did you alter or destroy your model oxygen and hydrogen atoms by making your model water particle?)

Because Dalton made careful observations about the masses of different elements and asked the right questions, evidence emerged to support a new model for the structure of matter:

Matter is composed of atoms.

Time Out for Analysis

1. What inferences were made in designing the model? Consider the examples below. Not all of these inferences are necessarily true.

   a. Atoms are round.

   b. The two atoms of hydrogen are identical.

   c. Atoms are different colors.

   d. In forming water, two hydrogen atoms combine with one oxygen atom.

   e. An atom of oxygen is 16 times more massive than an atom of hydrogen.

   Which of these inferences are similar to Dalton's? Which of these inferences can you support? Which can't you support?

2. Shown at right are the symbols Dalton used to represent some atoms.

Dalton inferred that in the methane molecule there were four hydrogen atoms for each carbon atom.

Dalton knew that the total mass of the carbon atoms in methane was three times greater than the total mass of the hydrogen atoms. How much heavier is each carbon atom than each hydrogen atom?

Try this: Form a 12 g ball of clay and four 1 g balls of clay. Construct a model of a methane molecule. How does the mass of the model carbon atom compare with the total mass of the four model hydrogen atoms?

This lesson started with quotations from Democritus and Dalton. Rewrite each quotation in your own words.

Dalton's symbols for some atoms and molecules

Elements and Compounds

Both Dalton and Democritus concluded that all matter is composed of particles called atoms. It follows that pure substances, such as gold, silver, and sulfur, consist of one kind of atom. Such substances are called **elements**. Scientists have identified 91 naturally occurring elements on Earth and therefore 91 kinds of atoms. When you simulated a chemical change between oxygen and hydrogen atoms, you simulated the formation of a new substance— water. Water particles are called molecules. **Molecules** are particles that are a combination of two or more atoms. Water, sugar, and carbon dioxide consist of molecules made of two or more *different* kinds of atoms. These substances are called **compounds**. The number of existing compounds is practically limitless.

Water and sugar are compounds. Each is made up of more than one type of element.

Summing It Up

- Elements consist of atoms; each element is made of one kind of atom.

- Elements cannot be divided into simpler substances by chemical changes.

- There are 91 naturally occurring elements on Earth.

- Elements combine with each other to form compounds.

- A molecule of an element consists of two or more of the same kind of atoms. A molecule of a compound consists of two or more different kinds of atoms. A molecule is the smallest particle of a compound that still has the characteristics of the compound.

- Compounds can be broken down into elements by chemical changes.

- The number of compounds that exist or that can be made through chemical changes is essentially unlimited.

I know that atoms make up elements, but don't they make up molecules too?

Molecules can be made up of different kinds of atoms—like carbon, oxygen, and hydrogen. An element has only one kind of atom in it.

If you'd like to know more about molecules and atoms, see pages S22–S26 of the SourceBook.

Testing Your Understanding

1. 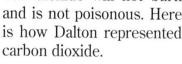 Use the ideas in the summary to draw a concept map showing the relationship between atoms, elements, molecules, and compounds.

2. Both carbon dioxide and carbon monoxide consist of carbon atoms and oxygen atoms. Yet these substances have different properties. Carbon monoxide will burn and is poisonous, while carbon dioxide will not burn and is not poisonous. Here is how Dalton represented carbon dioxide.

a. How would he have represented carbon monoxide?

b. How are the masses of these two compounds different?

3. Earlier you made models of water molecules using clay balls to represent oxygen and hydrogen atoms. The model oxygen atom had a mass of 16 g. Each model hydrogen atom had a mass of 1 g. Why do you think these masses were chosen?

Even More Models

Here are more models of molecules that you can make. Use clay balls of different colors to represent each element. In order to make your representation fit Dalton's findings, your model atoms should have the masses listed below. (You could also use a fraction of the suggested masses. For example, you can divide all of the masses by two, and your models will still show how much more massive one type of atom is than another.)

For a model carbon (C) atom, use 12 g of clay.
For a model hydrogen (H) atom, use 1 g of clay.
For a model nitrogen (N) atom, use 14 g of clay.
For a model oxygen (O) atom, use 16 g of clay.
For a model sulfur (S) atom, use 32 g of clay.

Compound	Elements Forming Each Molecule	Atoms in the Molecule
water	hydrogen (H), oxygen (O)	H—O—H
hydrogen sulfide	hydrogen (H), sulfur (S)	H—S—H
carbon dioxide	carbon (C), oxygen (O)	O=C=O
methane	carbon (C), hydrogen (H)	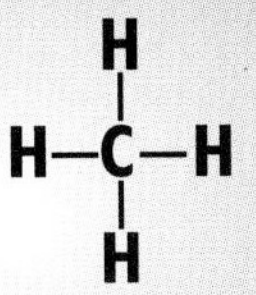 H—C—H with H above and below
butane	carbon (C), hydrogen (H)	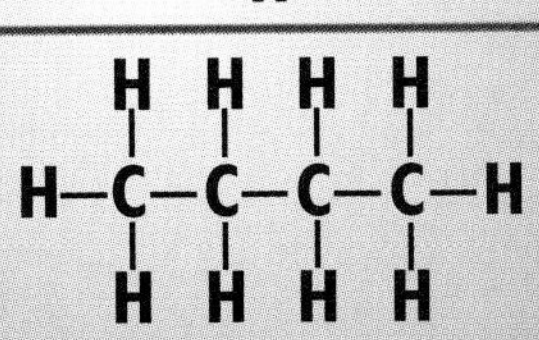 H—C—C—C—C—H with H above and below each C
ammonia	nitrogen (N), hydrogen (H)	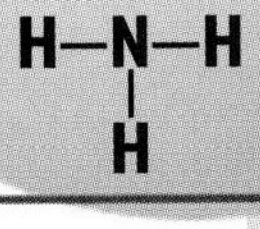 H—N—H with H below
glucose (sugar in honey)	carbon (C), hydrogen (H), oxygen (O)	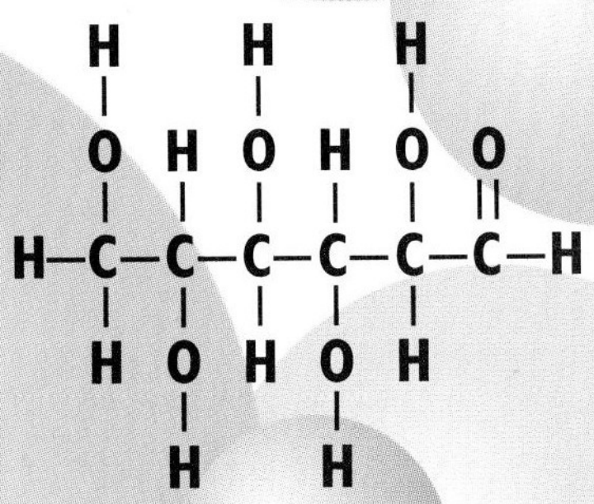 H—C—C—C—C—C—C—H with O and H groups
alcohol (from fermentation)	carbon (C), hydrogen (H), oxygen (O)	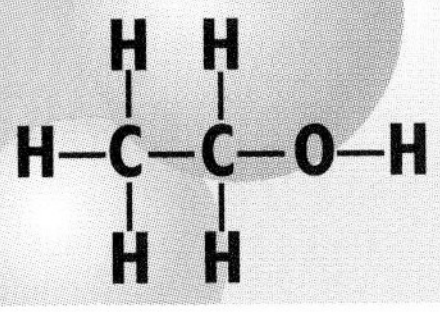 H—C—C—O—H with H above and below

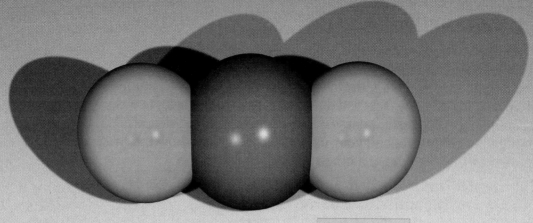

Carbon dioxide

In Three Dimensions

Shown here are three-dimensional models of some fairly common molecules so tiny that even the most powerful microscopes cannot clearly capture their form. Carefully study each model, and answer the questions that follow.

Methane

Water

Questions

1. How are these molecules alike? How are they different?

2. How many atoms are there in each molecule? How many kinds of elements are shown?

3. In forming a molecule of a compound, were the individual atoms of the elements destroyed?

LESSON 3

The Size of Particles

Sugar and Starch Molecules

Starch and sugar are two compounds that consist of the same elements, just arranged differently. The molecules of starch and sugar are made up of carbon, hydrogen, and oxygen atoms.

After performing the following experiment, name three differences in the properties of sugar and starch.

You Will Need

- a graduated cylinder
- 5 mL of cornstarch
- 5 mL of dextrose
- a jar with a lid
- a stirring rod
- a large beaker
- 100 mL of hot water
- an egg
- a straight pin
- iodine solution
- Benedict's solution
- a watch or clock
- a hot plate
- a hot-water bath
- 2 test tubes
- an oven mitt or test-tube tongs
- an eyedropper
- latex gloves

What to Do

1. Mix 5 mL of cornstarch with 5 mL of dextrose. (Dextrose is a sugar.) Add this mixture to 100 mL of hot water in a beaker. Stir.

2. Crack an egg in half, and save the larger end of the shell, which contains the air sac.

3. Using a straight pin, carefully remove part of the large end of the shell to expose the air sac. Be careful not to puncture the air-sac membrane. (You will, however, need to break the membrane that lies flush with the eggshell.)

4. Pour 5–10 mL of water into the shell, and float the shell in the sugar-cornstarch-water mixture.

5. After 15 minutes, pour half of the liquid in the eggshell into a test tube. Pour the remaining half into another test tube.

6. Test the liquid in one test tube with a few drops of iodine solution. A blue color indicates the presence of starch. Did starch molecules move through the air-sac membrane into the liquid in the shell?

7. Test the remaining liquid by adding eight drops of Benedict's solution to the second test tube.

Caution: Benedict's solution can irritate the skin. Handle with care.

Heat the liquid *gently* in a hot-water bath to avoid splattering. Use an oven mitt or test-tube tongs to handle the hot test tube. A red or yellow color indicates the presence of sugar. Did sugar molecules pass through the air-sac membrane?

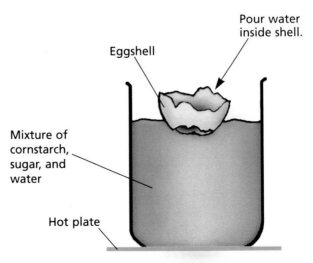

Pour water inside shell.

Eggshell

Mixture of cornstarch, sugar, and water

Hot plate

Something to Think About

1. As a conclusion to the experiment, which of the following statements do you think is most correct? Which statement is an inference?

   a. Sugar molecules have properties that are different from those of starch molecules.

   b. The experiment showed that some molecules pass through an egg membrane.

   c. Since sugar molecules passed through the membranes and starch molecules did not, sugar molecules may be smaller in size.

   d. Iodine solution is a test for starch, while Benedict's solution is a test for sugar.

2. What is the function of the air-sac membrane? What molecules do you think pass through the membrane as the egg develops into a chicken, and why?

Flashback!

You observed the passing of water through a membrane in Chapter 2. As you may recall, this process is called osmosis. Osmosis is an important process in every living organism. Water is transferred by osmosis from cell to cell. Review and explain the observations you made in Exploration 2 of Chapter 2, using the particle model of matter in your explanation. Can you define *osmosis* using the particle model?

A Mini-Experiment— Locking in the Smell

Wrap a clove of garlic in a piece of plastic wrap. Now crush the garlic between your thumb and forefinger. Can you still smell the garlic? How many layers of plastic wrap are needed to seal in the smell? What can you conclude from this mini-experiment?

Huge Numbers of Tiny Particles

If some particles are so tiny that they can pass through openings invisible to the human eye, then atoms and molecules must be extremely small. Because these particles are so small, a great number of them are needed to make up even a tiny bit of matter. The images on this page illustrate just how small atoms and molecules are.

A single drop of water contains approximately 3×10^{21} (3,000,000,000,000,000,000,000) water molecules. If you started to count these water molecules at the rate of one per second, it would take you . . . Well, you do the calculation. How many minutes? How many hours?

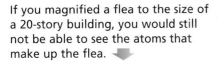

If you magnified a flea to the size of a 20-story building, you would still not be able to see the atoms that make up the flea.

More Large Numbers!

You know that atoms are small. In fact, each of the aluminum atoms in a piece of aluminum foil has an approximate diameter of 0.00000003 cm (3×10^{-8} cm). How many atoms thick is the aluminum foil?

Here is the equation that you will need to get the answer:

$$\text{Number of atoms} = \frac{\text{thickness of foil in centimeters}}{\text{thickness of one atom in centimeters}}$$

You still have a problem, though. You need to find the thickness of the aluminum foil in centimeters, so try the following exercise:

1. Cut a piece of foil that measures exactly 10 cm $\times$ 10 cm.

2. Find its mass in grams.

3. Find its volume in cubic centimeters (cm^3) by dividing the mass by 2.7 g/cm^3—the *density* of aluminum. (Density is the mass of a substance divided by the volume of the substance. You can learn more about density in Unit 4.)

4. Find its area in square centimeters (cm^2):

   area = length $\times$ width.

5. Divide volume (length $\times$ width $\times$ thickness) by area (length $\times$ width) to find the foil's thickness:

   V/A = thickness (cm).

6. Now use the first equation to calculate how many atoms thick a piece of aluminum foil is. Are you surprised?

Modern Miracle Fabrics

Is your jacket water-repellent? If it is made of plastic, it is. Some fabrics, such as nylon, are actually formed of plastic threads and can withstand some rain. Other fabrics are bonded synthetic materials. Gore-Tex® is one such fabric. It contains many tiny pores that are large enough to let individual water molecules through but that prevent the passage of water droplets. As you perspire, the water in your perspiration evaporates and its particles pass through the openings, from inside to outside. This keeps you dry. Can you think of some situations in which you would need such a fabric?

Particles of Solids, Liquids, and Gases

From circumstantial evidence, you have developed a model of the structure of matter. So far, your model includes the following ideas:

1. All matter consists of particles.

2. Particles are different sizes, although all are small.

3. Elements are made up of particles called *atoms*, and compounds are made up of particles called *molecules*.

In the Explorations that follow, you will discover more ideas about the structure of matter to add to these three.

Compressed air in this rocket forces water through the opening at the bottom. As the water is forced downward, the rocket is propelled upward. Why do you think both air and water are needed to make this rocket work?

EXPLORATION 5

Compressing Gases, Liquids, and Solids

Fill a plastic syringe (without the needle) with air. Then, with your finger over the end, push down on the plunger as hard as you can. What does this experiment tell you about the particles that make up a gas?

Now fill the syringe with water, and again try to push down on the piston. What is the difference between the particles that make up a gas and those that make up a liquid?

Finally, consider this: If you could get a solid (such as a piece of chalk) into the syringe, could you compress it? What differences are there between the particles of a solid and those of a liquid or a gas?

To conclude this Exploration, add a fourth statement to the particle model of matter. (See the list of ideas in the first column on this page.)

EXPLORATION 6

Particles on the Move

Your model of matter is becoming more and more useful because it can explain more observations. Now you will make a few more observations of the behavior of matter. In each instance, explain your observations in terms of what the particles in the solid, liquid, or gas are doing.

You Will Need

- food coloring
- an eyedropper
- ice water
- hot water
- a balloon
- a plastic soft-drink bottle
- an ice chest with ice
- rubbing alcohol
- 2 microscope slides
- matches
- test-tube tongs
- a beaker
- cotton balls
- a metal lid from a jar
- perfume
- a candle

STATION 1

Place a drop of food coloring into very cold water and another drop into very hot water. Explain the difference in behavior.

Exploration 6 continued

STATION 2

Place a balloon over the mouth of a 2 L or 3 L plastic soft-drink bottle. Place the bottle into a container of hot water for a few minutes. Now quickly place it into a container of ice water. Use the particle model to explain what happens.

STATION 3

Heat a microscope slide with a match. Then, after extinguishing the flame, place one drop of alcohol on the heated slide and one drop on an unheated slide. Using the particle model, explain the differences you observe.

STATION 4

Pour ice water into a beaker. Now breathe on the side of the beaker. What do you observe? Explain this observation in terms of what you think the water molecules in your breath are doing.

STATION 5

Place a cotton ball on a metal lid. Add a few drops of perfume to the cotton. From how far away can you smell the perfume? What do you think the liquid particles that make up the perfume are doing?

STATION 6

Observe a burning candle. What forms at the top of the candle (not the top of the flame)? What happens after the candle is blown out? Explain these observations in terms of what the particles of wax are doing.

Analysis, Please!

1. Now add at least one more statement to the three given on page 103.

2. Here are six words that help describe the processes you observed in Stations 1–6: condensation, expansion, diffusion, evaporation, melting, and solidification. Which word(s) would you associate with each station?

Expanding the Model

Mr. Chin's class expanded the particle model of matter, as described on page 103, by adding more ideas. If you agree with the statements they added, suggest at least one observation to support each statement.

More Ideas

1. Particles in gases are far apart.

2. Particles that make up liquids and solids must be as close together as possible.

3. Particles move.

4. Particles in a hot substance move faster than particles in a cold substance.

5. The faster gas particles move, the more pressure they exert on the sides of a balloon.

6. Liquid particles can become gas particles, and gas particles can become liquid particles.

1. Invisible Aerobics

Make a table like the one shown here. The first column lists some words that describe the ways particles may move. Which state of matter—solid, liquid, or gas—is most likely to exhibit each kind of movement? Suggest an everyday event that is similar to the way particles move. One has been done for you.

Word	State of matter	Your analogy
Wriggling	Solid	Like students wriggling while sitting in their seats
Vibrating	?	?
Tumbling	?	?
Bouncing	?	?
Flying	?	?
Shaking	?	?
Whirling	?	?
Sliding	?	?

2. Changes in Behavior

The following pictures illustrate the behavior of particles in solids, liquids, and gases. Write a sentence or two that would explain to a fifth-grader what is happening in each picture.

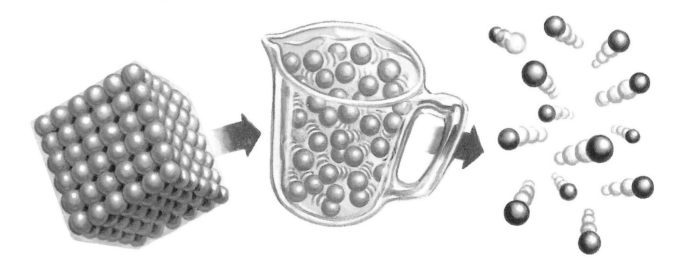

3. What's the Matter?

Here are some students' descriptions of solids, liquids, and gases. To which particular characteristic of the particles is each student referring?

Solid—made up of the staying-at-home type of particles

Liquid—consists of particles slip-sliding away

Gas—particles with claustrophobia

Now it's your turn to create an unusual definition of each state of matter. Share it with a friend, and see whether he or she can discover which term you are describing.

4. Air Apparent

In Chapter 4, Ramón used plastic beads in a balloon to simulate particles of air. Explain how each of Ramón's observations (listed below) describes the behavior of air and therefore supports the particle model of matter.

a. "When the balloon was shaken gently, the plastic beads rattled around. I could feel and see them hitting the sides of the balloon."

b. "Shaking the balloon harder caused more frequent and harder collisions."

c. "When I doubled the number of plastic beads in the balloon, the number of collisions with the sides of the balloon increased."

d. "When I stopped shaking the balloon, the plastic beads formed an orderly pattern in the bottom of the balloon."

ScienceLog

Review your responses to the ScienceLog questions on page 87. Then revise your original ideas so that they reflect what you've learned.

Testing the Particle Model

Why does the thermometer **1** read no higher than 100°C even though the stove continues to supply heat to the water?

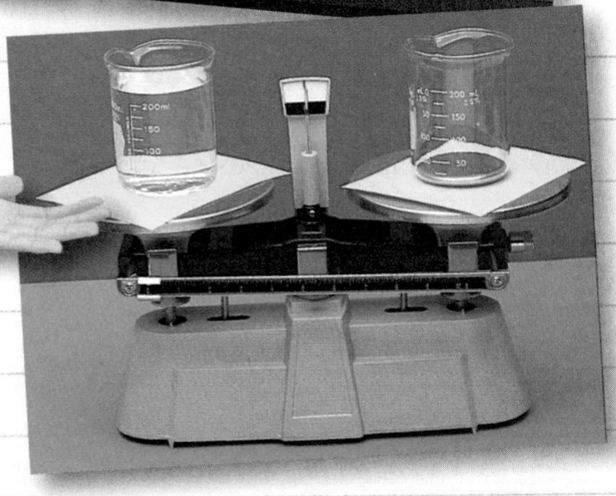

100°C

2 How might you explain this situation?

ScienceLog

Think about these questions for a moment, and answer them in your ScienceLog. When you've finished this chapter, you'll have the opportunity to revise your answers based on what you've learned.

3

Why might a person shiver on a hot day?

Temperature and Particles

You know that matter is made up of molecules that are in constant motion. The particle model of matter suggests that when heat is added to matter (solid, liquid, or gas), the molecules move faster and faster and farther and farther apart. Similarly, as matter cools, the molecules slow down and move closer together. The particle model suggests that this is true for all states of matter—solids, liquids, and gases. The following Exploration shows the way in which two students chose to test these inferences.

EXPLORATION 1

Slowing Down and Speeding Up Particles

Examine the notes jotted down by Sedrick and Leilani on the following two experiments. Then do the following:

1. For each experiment, devise a good title in the form of a question.
2. Draw an appropriate conclusion for each experiment.

Sedrick's Experiment

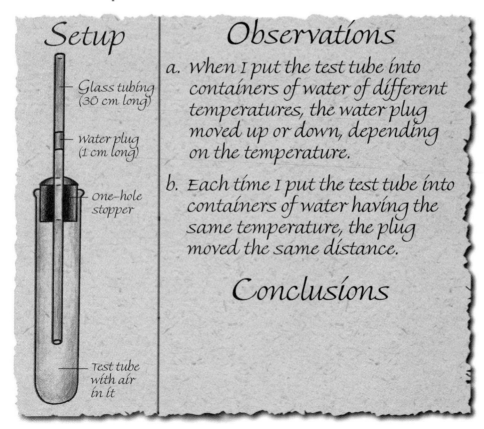

Setup

- Glass tubing (30 cm long)
- water plug (1 cm long)
- One-hole stopper
- Test tube with air in it

Observations

a. When I put the test tube into containers of water of different temperatures, the water plug moved up or down, depending on the temperature.

b. Each time I put the test tube into containers of water having the same temperature, the plug moved the same distance.

Conclusions

Leilani's Experiment

Setup

In this experiment, I filled two test tubes with two different liquids, glycerin and water.

Here are my two test tubes.

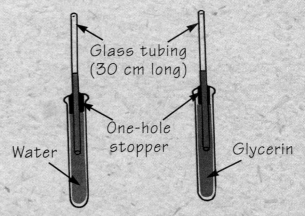

Glass tubing (30 cm long)

One-hole stopper

Water

Glycerin

Procedure

When I put the stoppers in, I had to make sure that the liquids rose about one-third of the way up the glass tub-
ing from the stopper. I also had to make sure there was no air left in the test tubes. Then I put both test tubes in a beaker of ice and water. The water level had to almost cover the test tubes. After the liquids in the glass tubing stopped moving, I marked the level on each piece of glass tubing as "O cm"—my baseline. The baseline for each tube did not have to be at the same level. I then put my test tubes into four more water baths of different temperatures. I measured how far from the baseline the liquid moved up the glass tube each time.

Data

Place mark here. This is "zero level."

Ice cubes

Ice water

Temperature	Height of liquid at different temperatures	
	Glycerin	Water
5°C	0 cm	0 cm
24°C	0.9 cm	0.4 cm
45°C	3.8 cm	2.6 cm
50°C	4.2 cm	3.1 cm
63°C	5.5 cm	4.1 cm

Conclusions

Aisha's Secret

Use the particle model to explain why Aisha's technique works.

An Eggsperiment

Why does an egg sometimes crack when you boil it? To find an answer, try this activity at home.

Start with a pot of cold water, and heat an egg in it. Watch the large end. What do you observe? How does your observation explain why eggs sometimes crack?

Repeat the experiment with a new egg, only this time pierce the large end with a thumbtack. Start over again with cold water and heat the egg. Again, what do you observe?

Why do some eggs crack when boiled, while others don't? Can you suggest why it is not a good idea to put a raw egg in boiling water, even if you have pierced the large end?

Ask someone in your family how he or she boils eggs without cracking them.

LESSON 2 Changes of State

Does heating a material always cause a temperature change within the material?

"Of course," said Derrick. "When you heat water, it gets hotter, and the pot it's in gets hotter too."

"And," added Wendy, "because the particles making up the water and the pot are vibrating at a faster rate than they were before heating, the particles take up more space, causing the materials to expand."

That explanation seems to make good sense. Now examine what happens when heat is *removed* from a substance. Believe it or not, this will help you answer the question above.

EXPLORATION 2

The Temperature Connection

In this Exploration, you will start with stearic acid at 75–80°C and observe how its temperature changes as it cools for 30 minutes.

You Will Need

- stearic acid
- a scoop
- a beaker of hot water (75–80°C)
- a hot plate
- a thermometer
- a test tube
- test-tube tongs
- a test-tube rack
- a watch with a second hand
- wire-loop stirring device

Be Careful! Do NOT try to pull the thermometer out of the solid; this could break the thermometer.

What to Do

1. Fill a test tube halfway with stearic acid and place it in the hot water.

2. Once the stearic acid has melted, place the test tube in the test-tube rack and measure the temperature of the stearic acid.

3. Slip the wire-loop stirring device over the thermometer. Record the temperature of the molten liquid every 30 seconds. Mix the liquid before each reading by moving the stirring device up and down. Record your data in a table in your ScienceLog.

4. Record the temperature until *all* of the liquid has become solid.

5. Remove the thermometer and stirring device from the frozen stearic acid by melting the solid again, as in step 1. Clean the thermometer if necessary.

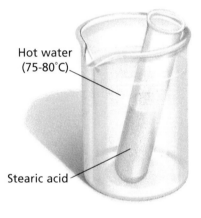

How to melt stearic acid

Hot water (75-80°C)

Stearic acid

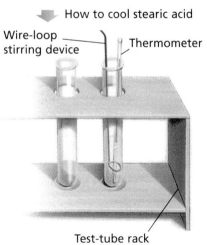

How to cool stearic acid

Wire-loop stirring device

Thermometer

Test-tube rack

Analysis, Please!

1. In your ScienceLog, prepare a graph of your results. Place *time* on the horizontal axis and *temperature* on the vertical axis.

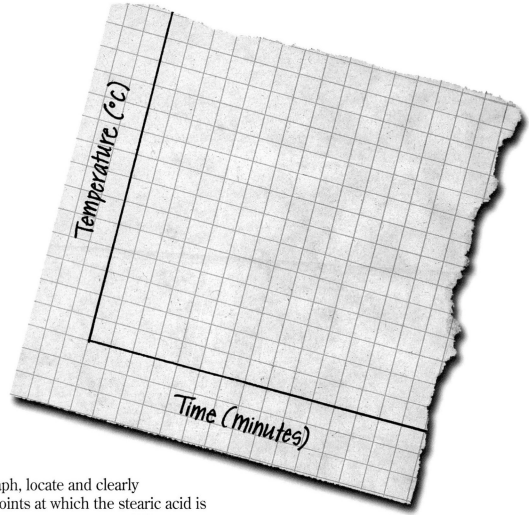

2. On the graph, locate and clearly label the points at which the stearic acid is
   a. liquid only.
   b. solid only.
   c. a mixture of solid and liquid.

3. How does your graph resemble those of other students in the class? How does it differ?

4. What is (a) the melting point and (b) the freezing point of stearic acid? Compare your answer with those of your classmates.

5. The graph should reveal that, at one point, the temperature does not drop as you might expect. Explain this observation.

6. Return to the question that began this lesson: Does adding heat to a material always cause a temperature change within the material? Have you changed your answer?

Derrick's Logbook

I was surprised by the results of the experiment, although I shouldn't have been. After all, I already knew that you can't cool liquid water below 0°C using only ice, and I knew that you can boil water all day without its temperature going over 100°C.

Here is what I learned by doing this experiment. I drew two graphs to illustrate. The first graph shows what I now predict will happen as stearic acid is heated. The second is what actually happened as the stearic acid cooled.

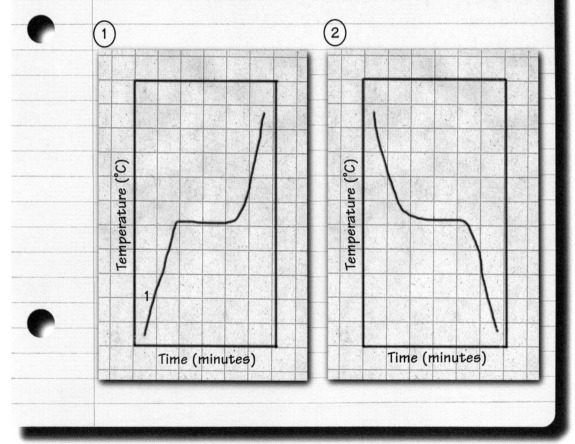

Derrick intends to label the parts of his graphs. The labels are listed below in no particular order. First, copy his graphs (on a larger scale) into your ScienceLog. Then, with a classmate, decide where each label belongs. Write the number of each label in the appropriate place on the graphs. Derrick has already placed one label.

1. Heat energy is added to solid stearic acid.
2. Stearic acid starts to form a solid.
3. The particles of solid stearic acid are moving more slowly.
4. The solid stearic acid returns to room temperature.
5. The particles of solid stearic acid are vibrating at a greater rate.
6. The temperature of liquid stearic acid goes up.
7. Melting—no change in temperature
8. The temperature of liquid stearic acid goes down.
9. The stearic acid melts.
10. The particles of stearic acid release heat energy.
11. Freezing—the temperature remains constant.
12. The freezing point of stearic acid

Follow-Up

Ice melts at 0°C. Stearic acid melts at 70°C. In which substance does the force of attraction between molecules appear to be greater? Explain the reason for your choice.

LESSON 3
Absorbing and Releasing Heat

When energy is added to or released by a substance, one of two things may happen: the temperature of the substance may change, or the material may undergo a change of state. In your investigations with stearic acid, at what point was there a change in temperature? a change of state?

When enough heat energy is added to a solid such as stearic acid or ice, the solid increases in temperature and then starts to melt. But during the melting process, the temperature of the solid-liquid mixture no longer increases. This change in state causes the plateau you saw on your graph. For stearic acid, the plateau is at 70°C. For water, the plateau is at 0°C. These are the melting points of stearic acid and ice. During melting, heat is *absorbed*, or taken *in,* by the substance being heated. *Endo* means "in." Melting is therefore an example of an **endothermic** change.

When a liquid solidifies, heat is released. During the solidification process, the temperature of the liquid-solid mixture remains constant. This time period of constant temperature corresponds to the plateau on your graph; it is the freezing point for the substance. For stearic acid the plateau occurs at 70°C. For ice the plateau occurs at 0°C. *Exo* means "outside." Because heat energy is released, or given off, freezing is called an **exothermic** change.

Freezing—heat energy being released

Melting—heat energy being absorbed

In the remainder of this lesson, you will be exploring these ideas in more detail.

EXPLORATION 3

Salol and Changes of State

This Exploration involves changes of state, this time for a substance called *salol* (phenyl salicylate). As you follow the procedure, identify the points when

- melting is occurring.
- freezing is occurring.
- energy is being absorbed.
- energy is being released.
- an exothermic change occurs.
- an endothermic change occurs.

You Will Need

- a glass microscope slide
- a test-tube clamp
- a stainless steel scoop
- a pair of tweezers
- a match
- salol
- a magnifying glass or low-power microscope

What to Do

1. Using the scoop, add a small amount of salol to a glass slide.

Be Careful! Salol may cause irritation to the skin. Don't handle it with your bare hands.

Exploration 3 continued ▶

2. Heat the slide with a match until the salol melts, and then immediately remove the match.

3. Extinguish the match, let the slide cool, and add a small crystal of salol to the slide. This is called a "seed crystal."

4. Using a magnifying glass or low-power microscope, observe what happens.

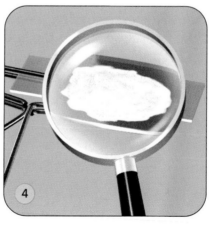

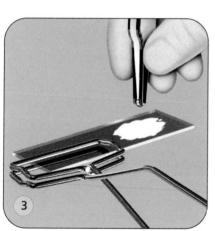

Hot Dogs, Cold Sweats, and Other Phenomena

Panting dogs, a sweating person, a refrigerator, and even the Earth itself have something in common. They all use a similar process to exchange heat. Read on to find out how this works.

You've probably experienced the coolness that results when you dip your finger in alcohol, when you sweat, or when you wet your finger to test the wind direction. Why does this occur? What's happening involves a process called *evaporative cooling*. In each example above, a change of state occurs: a liquid evaporates to form a gas. When a liquid evaporates or a solid melts, heat energy is absorbed from the surroundings. As sweat on your skin absorbs nearby available heat (including heat from your body), the sweat changes into water vapor. Since heat has been removed from the skin during this process, you feel cooler. How does this process work for a panting dog or explain how a refrigerator cools our food? Perhaps the following Exploration will help you with these questions.

Cooling Through Changes in State

You Will Need

- 3 thermometers
- 3 pieces of hollow shoelace (at least 5 cm long)
- a fan
- rubbing alcohol
- water

What to Do

1. Record the room temperature.
2. Prepare three *wet-bulb thermometers* by inserting each thermometer bulb into one end of a piece of shoelace. Secure each bulb with masking tape.
3. Wet one bulb in alcohol and the other two in water.
4. Place one of the water-dipped thermometers in a quiet spot on a table top.
5. Place the other water-dipped thermometer and the alcohol-dipped thermometer in front of a fan.
6. Every 30 seconds, record the temperature of each thermometer.

Making Sense of It

1. How did the temperatures change in the three thermometers? Can you explain why?
2. Why did the thermometer that was dipped in alcohol read a lower temperature than the one that was dipped in water?
3. How does this experiment help explain why you feel cool when getting out of the shower? How does it help explain why dogs pant?
4. If this experiment were done on another day, the readings on the thermometers might be higher or lower. Why would a wet-bulb thermometer read lower temperatures on some days than on others?
5. Many of the main points of the particle model are listed on pages 103 and 104. This model is also called the kinetic theory of matter. *Kinetic* means "motion." Why is this a good name for the particle model we have developed?

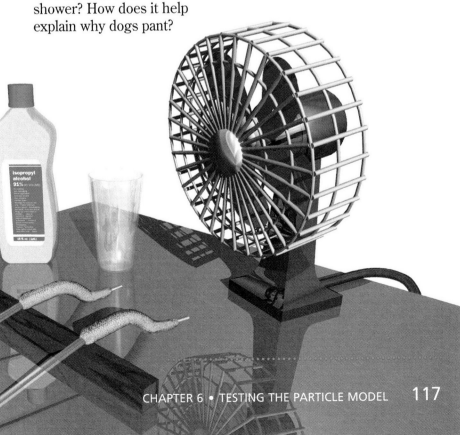

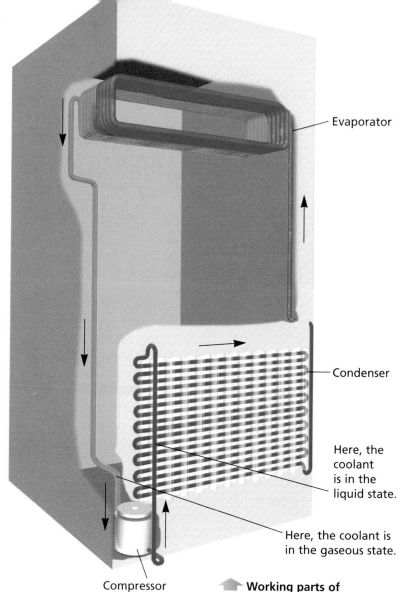

Evaporator

Condenser

Here, the coolant is in the liquid state.

Here, the coolant is in the gaseous state.

Compressor

Working parts of a refrigerator

Explanations, Please!

1. **a.** Did you know that changes of state are involved in maintaining a colder temperature inside your refrigerator than outside? Examine the diagram of a refrigerator. Discuss with a friend why it is colder inside the refrigerator than it is outside.

b. Also discuss the following with a friend: You could warm a room by turning on an oven and opening the oven door. But can you cool a room by opening the door of a refrigerator? (Notice that the back of the refrigerator has pipes containing a coolant. The coolant circulates in these pipes.) If not, why not?

2. The water cycle is illustrated in the diagram at left. Use the diagram to explain how heat is transferred over large distances by the processes of evaporation and condensation. Use the words *endothermic* and *exothermic* in your explanation.

Condensation

Evaporation

Particles—Mass and Volume

Identifying Substances

Many everyday observations and phenomena can be explained by the particle model. Perhaps you can help Jon with a question that he brought to class.

Jon found two cubes of the same volume, but their masses were quite different.

He wondered how the particle model explains how two objects with the same volume could have different masses.

If they have the same volume, why not the same mass?

A Comparison of Two Objects Having the Same Volume

Object	Mass (g)	Volume (cm³)
Cube 1	15.2	2.1
Cube 2	3.4	2.1

Write your own explanation based on the particle model. Then try the following Explorations, which will explore this question in more detail.

EXPLORATION 5

Measuring Volumes

Here are three methods for determining volume.

METHOD 1

Using an Overflow Can

You Will Need

- an overflow can
- a graduated cylinder
- a small block of wood
- a straight pin

What to Do

1. Fill the can with water so that the water flows out of the spout. Catch and discard the overflow.
2. Stick the straight pin firmly into the block of wood.
3. Holding the block by the pin, carefully submerge it. Avoid putting your fingers into the water. (Why?)
4. Collect the water that overflows in a graduated cylinder. What is the volume of the water? the block of wood?

Exploration 5 continued

METHOD 2

Using a Ruler

You Will Need

- the block of wood from Method 1
- a metric ruler

What to Do

The volume of objects with a box-like shape (such as your block) can be determined using the following procedure:

1. Measure the length, width, and height of your block of wood in centimeters.

2. Calculate its volume in cubic centimeters (cm^3) using this formula:

Volume = length × width × height

⬆ What is the volume of this block in cubic centimeters?

3. How does the measured volume in milliliters compare with the calculated volume in cubic centimeters? Which do you think provides the better answer?

METHOD 3

Using a Graduated Cylinder

You Will Need

- a graduated cylinder
- an iron bolt

What to Do

A more accurate method for finding the volume of small objects is to use a graduated cylinder. Follow the sequence in these diagrams to determine the volume of an iron bolt.

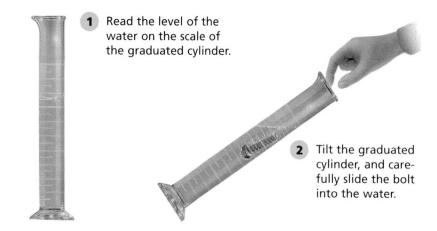

1 Read the level of the water on the scale of the graduated cylinder.

2 Tilt the graduated cylinder, and carefully slide the bolt into the water.

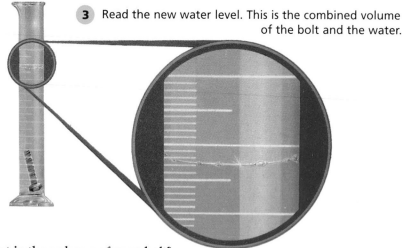

3 Read the new water level. This is the combined volume of the bolt and the water.

What is the volume of your bolt?
Remember to read the volume as you learned to on page 90.

Mass and Volume of Different Materials

You Will Need

- a graduated cylinder
- a balance
- a small iron object
- a small aluminum object

Note: You may use materials other than iron and aluminum as long as everyone else uses the same materials.

Iron		Aluminum	
Mass (g)	Volume (mL)	Mass (g)	Volume (mL)

What to Do

1. Find the mass and volume of your iron object. Your classmates will likely be using iron objects that have different volumes. In your ScienceLog, record your results in a class data table like the one shown above. Since you will be sharing your results with others in your class, do the measurements carefully.

2. Now find the mass and volume of your aluminum object. Add these values, as well as those of your classmates, to your data table.

3. Prepare a graph by placing *mass* on the vertical axis of the graph and *volume* on the horizontal axis. Choose scales to include the largest of your mass and volume values.

4. Draw a straight line (or line of best fit) through the points that represent the iron objects and another line through the points that represent the aluminum objects. The lines should go through the origin (0, 0). Why?

5. Is it possible to identify a material by knowing only its mass and volume? Could the materials that Jon brought to class be iron or aluminum? How do you know?

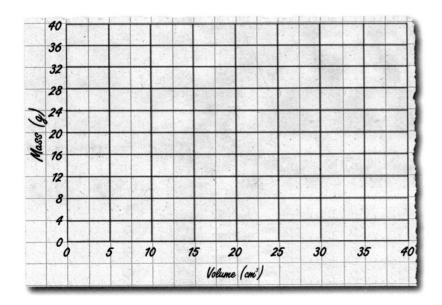

Jon's Class Project

Each student in Jon's ninth-grade class collected mass and volume values for a variety of materials. Here is the graph they prepared from the data.

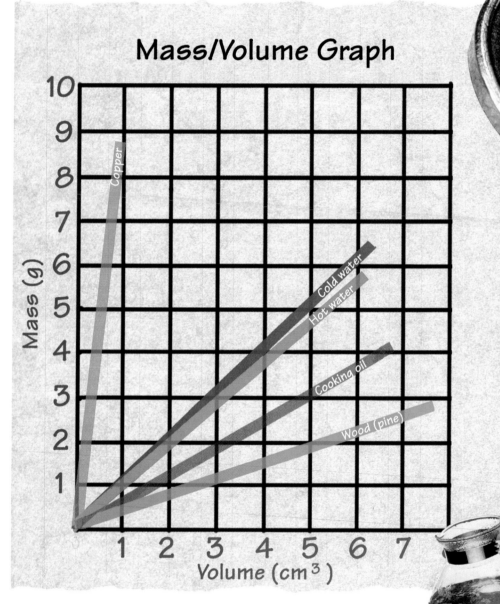

Mass/Volume Graph

Mass (g) — 1 through 10

Copper
Cold water
Hot water
Cooking oil
Wood (pine)

Volume (cm³) — 1 2 3 4 5 6 7

Try to interpret their data in terms of your understanding of the particle model of matter.

1. Why does 1 cm³ of copper have a mass of 8.9 g, and why would 2 cm³ of copper have a mass of 17.8 g?

2. Why would the mass of 5 cm³ of water be less than the mass of the same volume of copper?

3. Why would $5\,cm^3$ of cold water have a greater mass than the same volume of hot water? Be sure to explain this observation based on what you discovered earlier about the behavior of particles.

4. From your experiment and the class's, determine the mass of $1\ cm^3$ (1 mL) of each of the following materials: iron, aluminum, copper, cold water, oil, and wood. By determining the mass of each of these materials per unit volume, you have found the *density* of each substance. (You will learn more about density in Unit 4.)

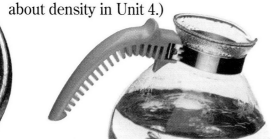

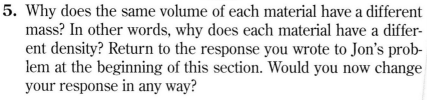

5. Why does the same volume of each material have a different mass? In other words, why does each material have a different density? Return to the response you wrote to Jon's problem at the beginning of this section. Would you now change your response in any way?

6. From your experience, which of the materials in the graph will float in cold water? Which ones will sink? Suggest how you could use the class's graph to predict which materials will float and which will sink.

1. A Matter of Particles

Can the particle model of matter explain the phenomena below? If so, how?

a. When a sunbeam shines into a room, you can see dust particles dancing about in the path of light.

b. If you could trap a little smoke inside a transparent box and put it under a microscope, you would see the particles of smoke moving in a haphazard pattern.

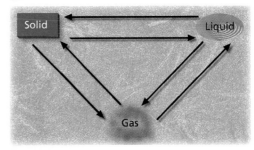

2. Name This Diagram!

Copy the diagram at right into your ScienceLog. Then complete it by following the instructions below.

a. Come up with a title for the diagram that summarizes the diagram and is accurate, descriptive, and catchy!

b. Place one of the following words on each arrow:
- melting
- freezing
- condensing
- vaporizing (or evaporating)
- subliming (changing directly from solid to gas or vice versa)

c. Label each arrow as signifying either an endothermic or exothermic change.

3. Solid Evidence

Three solids were heated, and their temperatures were plotted against the heating time (shown at left).

a. Which substance(s) melted when heated? How do you know?

b. Which substance has the highest melting point?

c. Which substance would seem to have the strongest forces of attraction between its particles?

4. Heavy Subject

The density of an object is the ratio of its mass to its volume $(D = m/v)$. Study the diagram and chart below. What kind of wood is the block probably made of?

Wood	Density
balsa	0.13 g/mL
birch	0.64 g/mL
pine	0.42 g/mL

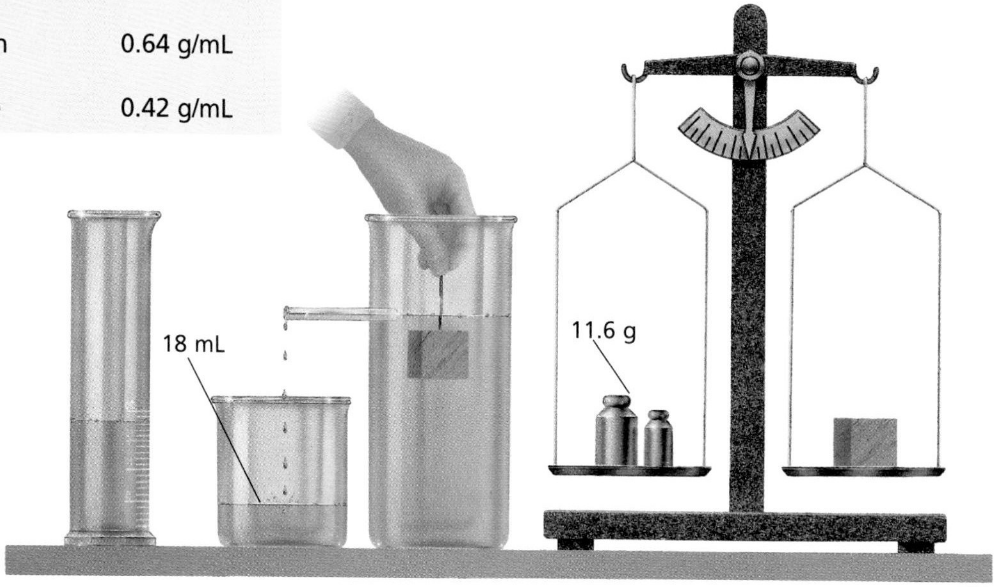

18 mL

11.6 g

5. Hammering It Home

Janet concluded that according to the particle model of matter, a hot iron nail should have less mass than a cold iron nail. Is she correct in her thinking? Why?

6. Food for Thought

Michael said his data proved that a can of soup would sink in water. Was he right?

284 mL

323 G

Andy's Vegetable 284mL SOUP

ScienceLog

Review your responses to the ScienceLog questions on page 107. Then revise your original ideas so that they reflect what you've learned.

CHAPTER 7

The World of Atoms

1 Do you think there is anything smaller than an atom?

2 How do we know what atoms look like?

3 How small do you think an atom is?

ScienceLog

Think about these questions for a moment, and answer them in your ScienceLog. When you've finished this chapter, you'll have the opportunity to revise your answers based on what you've learned.

Picturing an Atom

Viewing the Invisible

The particle model explains many observations about the behavior of matter. According to the model, all matter is made up of particles. Dalton called these particles atoms. As you found out in Chapter 6, everything on Earth is made up of 91 different elements. Since there are 91 different elements that exist naturally on Earth, there must be 91 different kinds of atoms.

Imagine one of these atoms magnified large enough for you to see. What would it look like? How does one kind of atom differ from another? Why is an atom of gold different from an atom of oxygen? In this section you will follow in the footsteps of others who, through their insight and creativity, provided us with a mental picture of the unseen structure of matter and in the process answered these and many other questions.

Before seeing how others pictured an atom, draw one yourself. In your diagram, include your thoughts about the nature of the atom. Does it have a definite shape? Is it made up of identifiable parts, or is it solid like a ball bearing? Adding labels to your diagram will help you to communicate your ideas and thoughts to others. Then read on as scientists from the past give their views on the nature of the atom.

So what do you think atoms are like?

And are atoms all the same size?

These are difficult questions because atoms are quite tiny—so small, possibly, that we will never be able to see them.

Even so, I picture atoms as indestructible spheres. My work indicates that the masses of atoms vary by kind. Each kind of atom has its own characteristic mass.

But atoms interact with each other, right?

Most certainly, yes. But the atoms are not consumed. They merely join together in some impermanent way—always in precise ratios.

For example, one can devolve water into hydrogen and oxygen, using an electric current. The ratio of gases produced indicates that oxygen atoms are 16 times as massive as hydrogen atoms. And when these gases are blended and ignited, water forms.

Reflecting

1. What are the main characteristics of atoms as described by John Dalton?

2. How is his picture of an atom similar to yours?

J. J. Thomson's View of the Atom

The year is 1897. You are listening to a lecture given by Joseph John Thomson, professor of physics at the famous Cavendish Laboratories of Cambridge University in England. He has just shown the audience what he thinks an atom looks like. How is his model different from the one suggested by John Dalton? How does it compare with your diagram of an atom?

I see the atom as being made up of electricity. I call my picture the "plum pudding" model of an atom. Note that in this model negatively charged particles called electrons (the plums) are embedded in a sphere of positive charge (the pudding). A more familiar analogy may be a muffin with raisins. The raisins represent the electrons, while the rest of the muffin represents the positive charge. Read on to find out how my plum pudding model led to important new discoveries about the atom.

Smaller Than an Atom

Rethinking the Plum Pudding Model—Ernest Rutherford's Atom

As a scientist, Ernest Rutherford was very familiar with the "plum pudding" model of the atom, having studied under J. J. Thomson at Cambridge University. However, Rutherford's research led him to another view of the structure of the atom.

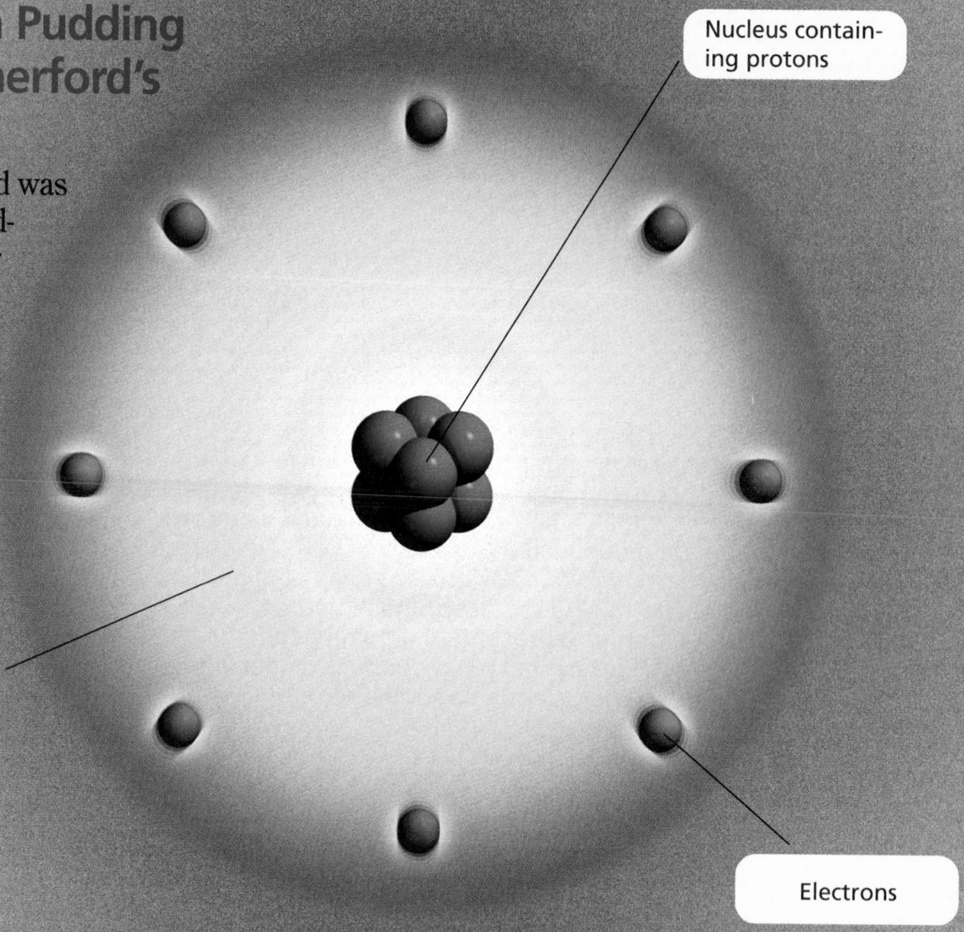

Nucleus containing protons

Empty space

Electrons

Listen in as he describes his atomic model.

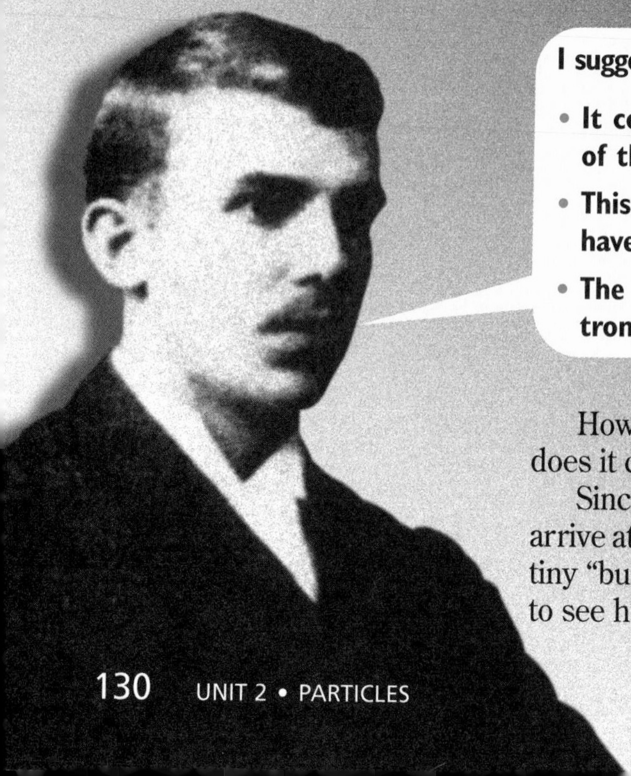

I suggest that the atom has the following characteristics:

- **It consists of a small core, or nucleus, that contains most of the mass of the atom.**

- **This nucleus is made up of particles called protons, which have a positive charge.**

- **The protons are surrounded by negatively charged electrons, but most of the atom is actually empty space.**

How is Rutherford's diagram different from Thomson's? How does it compare with yours?

Since the atom is far too small to see, how did Rutherford arrive at these surprising conclusions? The answer is that he shot tiny "bullets" at atoms to probe their internal structure. Read on to see how this was done.

Probing the Atom

SIMULATION 1

The "Plum Pudding" Atom

A simulation uses a model to test predictions about the behavior of real objects, processes, and systems. In this simulation, "bullets" will be shot at a model of an atom to find out more about its internal structure. An empty box, open at both ends, will represent the atom as conceptualized by J. J. Thomson. Pieces of popcorn will represent the electrons embedded in a sea of positive charge. The overall atom is neutral.

Predict

What would happen if table-tennis balls were shot at this "atom"? Would they pass straight through, or would they be deflected in different directions?

Observe

Here is what happens.

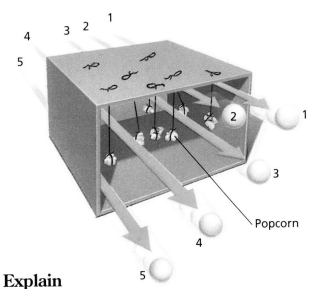

Popcorn

Explain

Why were the balls not deflected by this model of the atom?

SIMULATION 2

Probing Another Structure for the Atom

Predict

This time you will not be able to see inside the box. Can anything be inferred about what is inside the box from the following simulation? Again, balls are shot at the "atom." The diagram below shows what happened.

Observe

Ball 1 passes straight through.
Ball 2 deflects to the right.
Ball 3 bounces back.
Ball 4 passes straight through.
Ball 5 passes straight through.

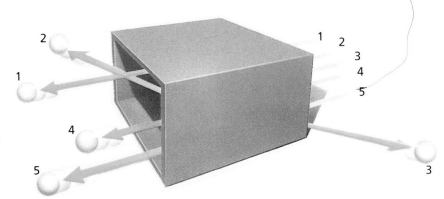

Explain

How must the arrangement inside this model atom differ from that in Simulation 1? Draw a diagram to show what you infer the inside of this "atom" might look like.

The Actual Experiment: Rutherford's Gold Foil Experiment

Rutherford and his assistants performed similar experiments by shooting atomic "bullets" at real atoms. Instead of table-tennis balls, Rutherford used helium nuclei, each with a positive charge. These charged atoms are called alpha particles. The alpha particles were shot at the gold foil. The gold foil had been made extremely thin so that the alpha particles did not have to pass through great numbers of atoms. What do you think Rutherford predicted before the experiment was performed?

What Rutherford Predicted

Since alpha particles are much more massive than electrons, Rutherford predicted results similar to those in Simulation 1.

What Rutherford Observed

Rutherford was extremely surprised by the results; he compared the experiment to firing a cannon at a piece of tissue paper and having the shells bounce back. Examine the diagram below for his observations.

1. Which simulation produced results most similar to the gold foil experiment?
2. What was the purpose of the fluorescent screens in the gold foil experiment?
3. Which alpha particles passed straight through the gold atoms?
4. Which particles probably made a direct hit on something more massive than themselves?
5. Which particles may have approached, but not directly hit, something with a positive charge?

The Setup for Rutherford's Famous Gold Foil Experiment

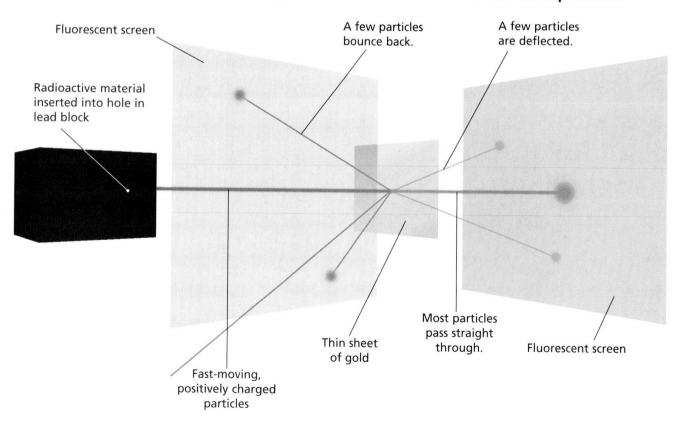

Fluorescent screen

A few particles bounce back.

A few particles are deflected.

Radioactive material inserted into hole in lead block

Fast-moving, positively charged particles

Thin sheet of gold

Most particles pass straight through.

Fluorescent screen

Rutherford's Nuclear Atom

Based on his observations, Rutherford arrived at a new model of the atom. The main features were described earlier but are repeated here. Respond to the question that follows each feature.

Atoms consist of a core, or nucleus, that contains most of the mass of the atom. (What evidence from the experiment suggested this?)

The nucleus has a positive charge. (Why did Rutherford infer that the nucleus has a positive charge rather than a negative charge or no charge at all? Hint: What would happen to a positive alpha particle that approached but did not hit a positive nucleus? What would happen if the nucleus had a negative charge?)

Most of the atom is empty space. (On what observation did he base this inference?)

The Bohr Model

In 1913, Niels Bohr, a Danish scientist who worked with Rutherford, developed a theory that electrons travel around the nucleus in orbits like those of the planets around the sun. This theory is called the *Bohr model*. Just as Dalton's theory had to be changed in light of Rutherford's discoveries, Bohr's model has been modified in response to more recent discoveries.

I propose that the electrons of an atom travel around the nucleus in specific orbits. Take a look at my model of a hydrogen atom, shown below. A circular path represents the fast-moving electron, which orbits a nucleus consisting of a single proton.

▲ Even though the Bohr model has been replaced by more accurate models, it is still widely used to represent atoms today.

Exploring the Size of the Atom

PART 1

Modeling the Mass of Electrons and Protons

You Will Need
• a graduated cylinder
• long-grain rice

What to Do

1. Set one grain of rice aside. This represents the mass of one electron.

2. Measure 125 mL of rice. This will contain approximately 2000 grains of rice and represents the mass of one proton.

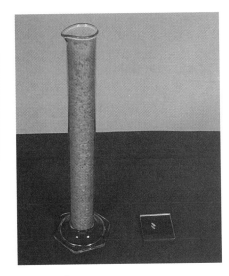

Questions

1. Compare the mass of 125 mL of rice with the mass of one grain of rice. According to this activity, how much more massive are protons than electrons?

2. How does this model demonstrate Rutherford's inferences about the structure of the atom?

Exploration 2 continued

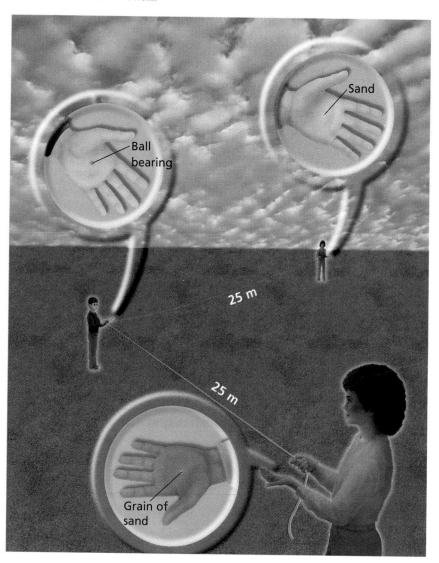

PART 2

Modeling the Space Within Atoms

You Will Need

(for each group of three students)

- two 25 m lengths of string
- a small ball bearing or BB
- 2 grains of sand

What to Do

(This activity is best done outdoors.)

1. Have one person hold the ball bearing. (This ball bearing represents the nucleus of an atom.)

2. The person holding the "nucleus" will also hold one end of each string while the other two students, each holding a grain of sand and the other end of a string, pace out 25 m in different directions. (The grains of sand represent the electrons in an atom.) The position of each student represents the distance of the closest electrons to the nucleus in an atom.

Questions

1. How does this simulation demonstrate Bohr's theory of the structure of the atom?

2. Why did most of the alpha particles in Rutherford's gold foil experiment pass directly through the gold foil, while only a few were deflected back?

Sand

Ball bearing

25 m

25 m

Grain of sand

More Particles in the Nucleus?

James Chadwick, while working with Ernest Rutherford in 1932, discovered that another particle, in addition to the proton, is found in the nucleus of the atom. This particle has no charge and is about the same mass as the proton. It received the name *neutron*.

James
Chadwick

To read more about the neutron and other particles smaller than the atom, see pages S35—S38 of the SourceBook.

Particle Summary

Particle	Charge	Mass (compared to electron)	Location
proton	positive (+)	almost 2000 times as massive	in the nucleus
neutron	neutral	almost 2000 times as massive	in the nucleus
electron	negative (–)		outside the nucleus

1. Where is most of the mass of the atom concentrated? What contributes the most to its mass?

2. If the mass of the proton is said to be one atomic mass unit (1 amu), what would be the approximate mass of an atom of lithium, which has a nucleus made up of three protons and four neutrons? (Ignore the comparatively tiny mass of the electrons.)

3. Below are Bohr models for atoms of helium, lithium, and beryllium. Each red orbit represents the path of a single electron.

   a. What do all atoms have in common?

   b. How do atoms of different elements differ?

   c. Draw diagrams for the following atoms:

      • nitrogen (7 protons and 7 neutrons in its nucleus)

      • oxygen (8 protons and 8 neutrons)

   • argon (18 protons and 21 neutrons)

4. By insightful experiments and reasoning, John Dalton discovered that each kind of atom has its own distinctive mass. (For example, oxygen atoms are 16 times more massive than hydrogen atoms.) Of course, Dalton didn't know anything about electrons, protons, and neutrons. How would you explain to him the reasons for his conclusion?

• Protons
• Neutrons
/ Electron path

He

Li

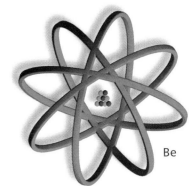

Be

CHALLENGE YOUR THINKING

1. Neutral Neighbors

Examine the Bohr model of an atom of the element carbon.

a. How many of each kind of particle make up this atom?

b. Is there a charge on this atom, or is it neutral? Why?

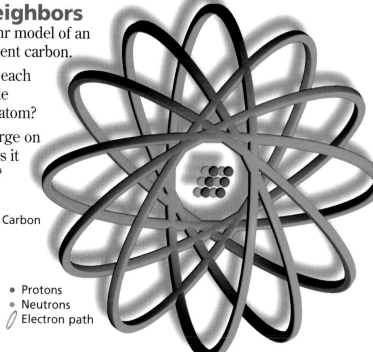

Carbon

- Protons
- Neutrons
- Electron path

2. Same but Different

Here are Bohr models of atoms representing different forms of hydrogen.

a. How are they similar? How are they different?

b. What identifies an atom as hydrogen and not as some other atom, such as helium?

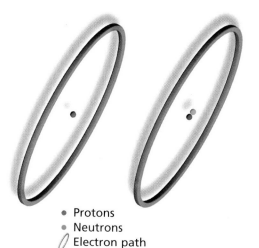

- Protons
- Neutrons
- Electron path

3. Timely Discoveries

Draw a time line that describes the discoveries made about the nature of the atom. Include the people encountered in this section and their discoveries. Then add two more names to the time line: Henri Becquerel and Marie Curie. What were their contributions? How might you find out?

1800   1900

4. Ballpark Figure

Here is another analogy for the atom: If the nucleus of an atom were as large as a grape, then the atom itself would be as large as Yankee Stadium. Use this analogy to explain the results of Rutherford's gold foil experiment.

5. Hair-Raising Experience

Here's an activity you can try:

When an object is rubbed against something else, electrons (negatively charged particles) are transferred from one object to another.

a. Describe two or three occasions when this has happened to you.

b. If electrons are added to a plastic strip when it is rubbed with silk, what is the charge on the plastic strip?

c. What happens when this strip is brought close to small pieces of paper? (Try this yourself, using a comb that you have run through your hair.)

6. Piecing It Together

Choose whatever materials you wish, and construct a model of an atom for an element that you have not encountered in this unit. The model must show the number of electrons and protons. Your model could be rated in the following way:

- creative use of materials

- numbers of protons and electrons accurately portrayed

- bonus values if neutrons are included in your model

ScienceLog

Review your responses to the ScienceLog questions on page 126. Then revise your original ideas so that they reflect what you've learned.

Unit 2

Particles

The Big Ideas

In your ScienceLog, write a summary of this unit, using the following questions as a guide:

1. How do observations differ from inferences?
2. What are models, and why are they useful?
3. What evidence is there that matter is made up of particles?
4. What everyday observations can be explained by the particle model of matter?
5. Who was John Dalton, and what did he conclude about the makeup of matter?
6. How does the particle model explain the properties of solids, liquids, and gases?
7. What is the effect of temperature changes on the particles making up matter?
8. What are some examples of endothermic changes? examples of exothermic changes?
9. Why do many objects with the same volume have different masses?
10. What is the general structure of the atom, and who helped discover this structure?

Checking Your Understanding

1. Why do we feel hotter on hot, humid days than on hot, dry days? Use the particle model of matter to explain your reasoning.

2. The following story contains at least five observations that can be explained with the particle model. List them and give explanations for each.

"One more dive and then we gotta go. We can't be late for dinner again. Mom'll get mad." Ben and Josh each dove off the cliff, neatly splitting the water. "It shouldn't take too long to dry in this sun," said Ben. "Oh great," groaned Josh. "My front tire's flat. Guess I should've filled it before we left. We'll have to walk it to the gas station." With the tire pumped up, the brothers raced home to make up for lost time. The breeze felt cool on their damp skin and hair. When they got home, their mother said, "Put your wet things in the dryer and come eat. I want one of you to mow the grass before it gets dark, while it's still dry. There'll be too much dew to mow in the morning." "No prob, Mom," said Josh, "Ben'll do it. Say, dinner smells great."

3. A balloon initially had a mass of 6.2 g and a volume of 2.3 L. What might have been done to the balloon to bring about the following changes?

	Mass	Volume
a.	6.2 g	3.2 L
b.	6.5 g	2.9 L
c.	6.0 g	2.5 L

4. Air fresheners are often placed in different areas of a home, such as kitchens, bathrooms, and basements. Over a period of weeks, the fragrant part of the air freshener gradually disappears. What happens to it? Use the particle model to explain.

5. **concept map** Copy the concept map at right into your ScienceLog. Then complete the concept map using the following words: electrons, elements, atoms, nuclei, molecules, matter, neutrons, negative charge, protons, no charge, positive charge, and compounds.

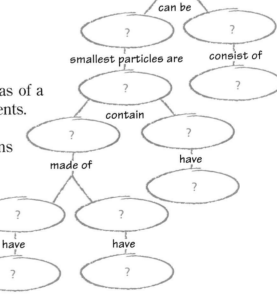

Science Snapshot

Albert Einstein (1879–1955)

The year 1905 was an extraordinary time in the development of modern physics. In that year, a 26-year-old German patent clerk living in Switzerland published four papers that would forever change our understanding of the physical nature of matter and energy. This young patent clerk's name was Albert Einstein, considered today to be one of the greatest scientists to have ever lived.

The Existence of the Atom

Even though we have only recently developed the technology to see individual atoms, scientists have been collecting evidence for their existence for centuries. In 1827, for example, an English botanist by the name of Robert Brown placed tiny pollen grains on the surface of completely still water. He noticed that even though the water was perfectly still, the pollen grains moved around erratically. Scientists at the time thought that this effect, which is now called Brownian motion, might be caused in some way by living organisms. However, when it was demonstrated that even nonliving particles suspended in a fluid would undergo this motion, scientists

▶ **Albert Einstein overturned our view of reality when he was only 26. He won the Nobel Prize for physics when he was 42.**

began to consider other explanations. Some accepted the view that a fluid must be composed of particles that are in constant motion and that these particles periodically bang into the suspended particles, causing them to move.

Enter Einstein

Many scientists tried to develop theoretical explanations to support Brownian motion, but it was not until Einstein's paper on the subject that Brownian motion was placed upon a firm foundation. Einstein derived the mathematical equations that govern Brownian motion and used these equations to determine the size of the particles.

Einstein's creative mind and incredible mathematical skills

▶ **The zigzag path of a particle executing Brownian motion**

put him in the spotlight as a world-renowned physicist, but Einstein was more than just a scientist. He was also very active in human affairs, and his actions showed a deep concern for the politically and economically oppressed. He also loved classical music and even played the violin.

Einstein moved to the United States in 1933 and became a U.S. citizen in 1940. He died at his home in Princeton, New Jersey, in 1955. On his deathbed, he spoke his last words in German to a nurse who did not understand the language. We will never know what he said.

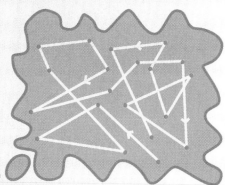

$E = mc^2$

Albert Einstein pushed the boundaries of physics with both philosophical and mathematical reasoning. He is perhaps best known for his theory of relativity and his famous equation, $E = mc^2$. Do some research to learn more about Albert Einstein. And in the process, discover the meaning of his famous equation.

Particles in the Air

Take a deep breath. You have probably just inhaled thousands of tiny specks of dust, pollen, and other particles. These particles, called particulates, are harmless under normal conditions. But if concentrations get too high or if they consist of harmful materials, they are considered to be a type of air pollution.

Where There's Smoke . . .

Unfortunately, dust and pollen are not the only forms of particulates. Many of the particulates in the air come from the burning of various materials. For example, when wood is burned, it releases particles of smoke, soot, and ash into the air. Some of these are so small that they can float in the air for days. The burning of fuels such as coal, oil, and gasoline also creates particulates. The particulates from these sources can be very dangerous in high concentrations. That's why particulate concen-

trations are one measure of air pollution. Because they are visible in large concentrations, they can also be responsible for the dirty look of polluted air. But don't be fooled—even clean-looking air can be polluted.

Eruptions of Particulates

Volcanoes can be the source of incredible amounts of particulates. For example, when Mount St. Helens blew its top in 1980, it launched thousands of tons of ash into the surrounding air. The air was so thick with ash that the area became as dark as night. For several hours, the

ash completely blocked the light from the sun. When the ash finally settled from the air, it covered the surrounding landscape like a thick blanket of snow. This layer of ash killed both plants and livestock for several kilometers around the volcano.

One theory about the extinction of dinosaurs is that a gargantuan meteorite hit the Earth with such velocity that the resulting impact created enough dust to block out the sun for years. During this dark period, plants were unable to grow to support the normal food chains. Consequently, the dinosaurs died out.

It's a Matter of Health

Because many particulates are so small, our bodies' natural filters, such as nasal hairs and mucous membranes, cannot filter all of them out. When inhaled, particulates in the lungs can cause irritation. Over time, this irritation can lead to diseases such as bronchitis, asthma, and emphysema. The danger increases as the level of particulates in the air increases.

◀ When the ash from Mount St. Helens settled from the air, it created scenes like this one.

Cigarette Smoke

Since the burning of most substances creates particulates, there must be particulates in cigarette smoke. Do some research to find out if the filters on cigarettes are effective at preventing particulates from entering the smoker's body. Your findings may surprise you!

◀ The burning of most fuels adds harmful particulates to the air.

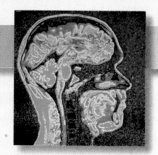
Tiny Troublemakers

You're minding your own business when suddenly your body is infiltrated by invaders so tiny that millions of them could fit on the point of a sharpened pencil. They immediately start using the chemicals and energy inside your cells to make more invaders. Soon thousands of your cells are infected, your nose begins to run, and you start to sneeze. You've just been invaded by a common cold virus!

Thank goodness it was only a cold virus—it could have been worse. Scientists have identified more than 1400 different viruses. Some cause problems such as cold sores and warts. Others cause serious diseases such as polio, measles, chickenpox, influenza, and acquired immune deficiency syndrome (AIDS). The Ebola virus, for example, can cause severe internal bleeding that can lead to death within three days of being infected.

Living or Nonliving?

Because of their makeup, viruses are hard to classify. Unlike living organisms, viruses are not made up of cells. Instead, a typical virus has only two

► Some viruses look more like machines than living organisms. But once inside a cell, viruses become part of the organism's living system.

basic components—a core of nucleic acid and an outer coat of protein. On its own, a virus is a lifeless particle that is totally inert. It cannot perform any of the functions of a living organism. It can exist in the environment for weeks, months, or even years without any negative effects or changes.

Once inside a living cell, however, a virus takes on very different characteristics. The virus is no longer inert. It takes over the cell nucleus and redirects the cell to produce proteins that

◄ Viruses range in size from about 0.01 to 0.03 micron (µm) in diameter. (A micron is 0.001 millimeter!) In contrast, the smallest bacteria are about 0.4 µm in diameter. Shown here are Ebola viruses.

are necessary for the production of more viruses. The new viruses then burst out of the cell and go on to infect other cells. One virus inside a single cell can produce thousands of new viruses!

Once inside a cell, a virus has many of the characteristics of a living thing—most obviously, it can reproduce. A virus is much like a parasite, using a cell as its host. In fact, some scientists speculate that those viruses which kill their hosts have not fully adapted to their environment. By killing the host, the virus is, in effect, sabotaging own ability to function and reproduce. This is why Ebola epidemics have been short lived.

A Deadly Virus

HIV—the virus that causes AIDS—doesn't actually kill its host directly. Instead, the virus weakens the body's immune system to the point that the body cannot protect itself from other infections. Do some research to find out about how HIV causes AIDS. Also, find out about the role of the drug AZT in the fight against AIDS.

Big Picture for Little Dots

*T*ake a look at the painting on this page. What do you see? Probably a holiday crowd on an island, right? Take a closer look and you'll see something else—the whole image is created with tiny, uniform dots of single colors! This method of painting is called *pointillism*, and it is as scientific as it is artistic.

Seurat's Dots

Pointillism was developed by the French artist Georges Seurat in the 1880s. Seurat noticed that when two small dots of different colors are placed next to each other, the two colors seem to form a new color when seen from a distance. Seurat then began to develop ways in which different combinations of colored dots could be used to create subtle changes in form and color.

It's Not Connect-the-Dots

Seurat immersed himself in the study of how white light and color interact. As you might know, white light is actually made up of seven colors—red, orange, yellow, green, blue, indigo, and violet—known as spectrum colors. Paints contain tiny grains of colored substances called pigments. When light hits the paint, the pigments absorb some of the colors of the light and reflect the rest. The color you see is actually the color of light that is reflected by the pig-

ments. Mixing different-colored pigments changes which colors of light are absorbed. This type of mixing is called subtractive color mixing because the resulting mixture absorbs, or subtracts, colors from white light. Most of the colors you see work this way. For example, a red shirt appears red because it absorbs all colors of light except red—only red light is reflected. Mixing many pigment colors together will make a dark brown or black paint because together the pigments will absorb almost all light.

Adding It Up

Mixing light is quite different from mixing pigments. Mixing colors of light is called additive color mixing. For example, overlapping a red and a green spotlight will produce a yellow light. Mixing all of the colors of light produces white light because all of the different parts of white light are added together.

Seurat used this information as he painted. He used only pure spectrum colors in his paintings. He placed dots next to each other in such a way that the light reflected by one dot would combine with the light reflected by a second dot to create the color he wanted. It's almost as if

▲ *Sunday Afternoon on the Island of La Grande Jatte* **by George Seurat, at the Art Institute of Chicago. Look closely at the inset to see the dots of color.**

the light reflected from each dot becomes its own little spotlight of color. In other words, Seurat produced the *effect* of additive light using subtractive pigments. Amazing!

A color television set also uses the principles of pointillism. Look at the screen with a magnifying glass to see for yourself.

Test It Out!

Create a small picture using pointillism. Using only the seven spectrum colors, experiment with different dot combinations to see what other colors you can make.

Unit 5

MACHINES, WORK & ENERGY

This amazing contraption, known as the Featherstone Kite, dazzles the eye with its complexity and wild form. Even though it is a whimsical device and works only in the imagination, the Featherstone Kite contains many working mechanical systems. Whirling fins, clanking chains, and spinning flywheels all transfer and transform energy.

It is often said that machines "save work." Actually, machines harness energy to do more work. It would be inconceivable to build a skyscraper or a highway without the aid of machines. Even tasks like digging a hole or shoveling snow would be difficult without the aid of a simple machine like a shovel.

Machines are often the product of necessity, solutions to practical problems. What practical problem do you think the inventor of the first wheel might have been trying to solve? Sometimes, the invention of a machine sets off a technological revolution. The invention of the steam engine set off the Industrial Revolution. And the world changed forever with the invention of the automobile.

Although most inventors have practical goals in mind when devising their machines, many have mixed engineering with a sense of humor to create whimsical, delightful machines like the one shown here.

The Featherstone Kite, a moving sculpture on display at the Mid-America Museum, Hot Springs, Arkansas

Putting Energy to Work

1 What is work? How might it be measured?

2 How is energy converted from one form to another by a roller coaster?

3 What is energy? How is work related to energy?

ScienceLog

Think about these questions for a moment, and answer them in your ScienceLog. When you've finished this chapter, you'll have the opportunity to revise your answers based on what you've learned.

Machines for Work and Play

Can you think of a time when people did not use machines of any sort? If you went back in time 5500 years, you would discover the first plow. How was this invention beneficial? How did it change people's lives?

There were also great construction projects long ago. During the construction of Egypt's Great Pyramids more than 4000 years ago, huge earthen ramps were constructed to help raise enormously heavy stone blocks into position. A ramp is a kind of machine. A pulley is another kind of machine. Pulleys have been in use for over 3000 years. Archimedes is said to have been the first to use a combination of pulleys. These pulleys were used to haul ships ashore.

A sampling of machines

Try to place the following machines in chronological order. In other words, begin the list with the machine that was invented first, followed by the machine that was invented second, and so on. Also try to determine the approximate century in which each machine was invented. Write your answers in your ScienceLog.

- motorcycle
- internal-combustion engine
- jet aircraft
- steam locomotive
- automatic clothes-washing machine
- CD player
- electric motor
- weighing scales
- gears
- mechanical clocks
- blender
- gasoline-powered lawn mower

Machines in Your Life

Now compare your answers with the actual dates of invention shown below.

- weighing scales—3500 B.C.
- gears—100 B.C.
- mechanical clocks—1300
- steam locomotive—1804
- electric motor—1830
- internal-combustion engine—1860
- motorcycle—1885
- gasoline-powered lawn mower—1902
- blender—1923
- automatic clothes-washing machine—1937
- jet aircraft—1939
- CD player—1979

Were you surprised by some of the early dates? Notice that several of the machines were invented in the 1800s. At that time, in the wake of the Industrial Revolution, machines were developed quickly and in large numbers.

Take a look at the machines shown on this and the facing page. What does each machine do? How does each use energy?

Now take a moment to consider the following questions:

- How would you define *machine*?
- How many machines have you used so far today?
- Do machines require energy? If so, where do they get it?
- What are some hand-operated machines?
- Do machines do work for you?
- Do machines save work or make it easier?
- What is work?
- What is energy?

You can answer some of these questions now, but you probably cannot answer all of them. This unit will help you answer these remaining questions and, possibly, help you to become an inventor, too!

Machines—Simple . . .

When many people think of machines, they think of complex devices such as cars, photocopiers, or computers. But many machines are quite simple. Just look at the devices at left and below. What does each one do?

. . . And Not So Simple

The two machines pictured here are not simple. They are complex **mechanical systems**. Each of these complicated machines is composed of many simpler machines—*subsystems*—all working together to perform some overall function. What does each of these mechanical systems do?

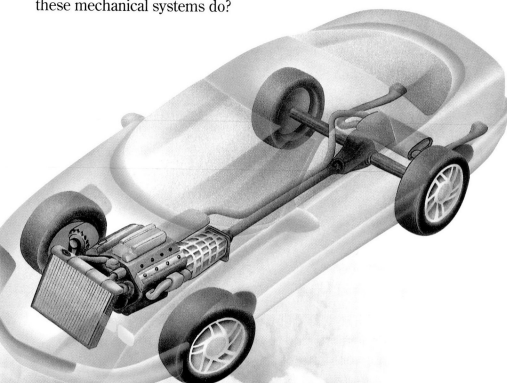

Familiar Mechanical Systems

You use mechanical systems all the time. Some, like a bicycle, are complex. Others, like a ballpoint pen, are simpler. But even the pen has a surprising number of parts and subsystems, each of which performs functions that contribute to the pen's overall function. The subsystems of a bicycle are more obvious. The braking subsystem is just one example. How many more subsystems of a bicycle can you identify? What does each do? What parts make up each subsystem?

You are able to move faster on a bicycle than you can move by walking. How does the bicycle make possible this increase in speed? As you progress through this unit, you will be investigating many mechanical systems, their subsystems, and their functions. You will also discover how they can increase either force or speed.

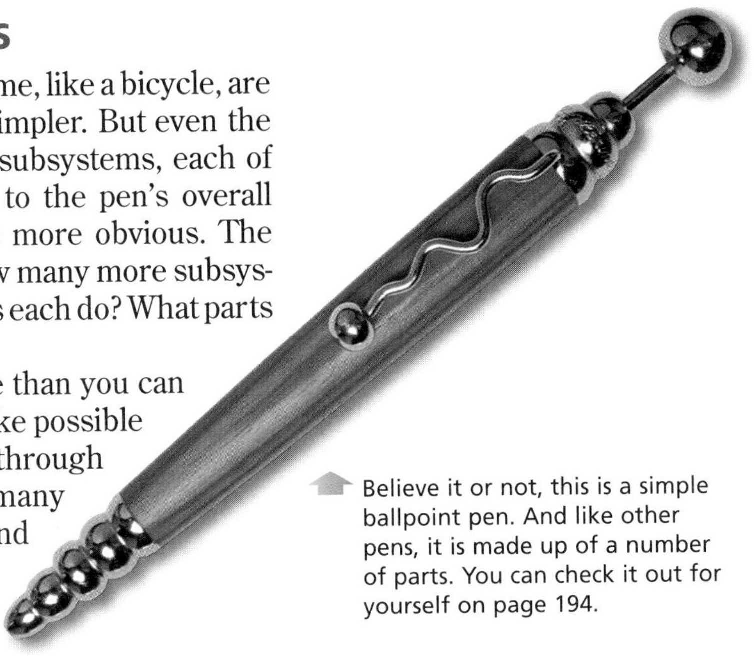

Believe it or not, this is a simple ballpoint pen. And like other pens, it is made up of a number of parts. You can check it out for yourself on page 194.

Work and Energy

Some machines are hand-powered—you have to do some work to use them. A shovel is a very simple kind of machine that requires you to do all the work. If you have to dig a large hole, a shovel helps to make your work easier. Imagine clearing away the dirt without it—using only your hands! Of course, it's even easier to dig a large hole using a backhoe. In this case, though, you don't do the real work—the machine does. The energy to dig up the dirt and to move the backhoe forward comes from the gasoline burned by the machine's engine. You do just a little work in operating the controls.

Notice the words *work* and *energy* in the preceding paragraph. What do they mean? You probably have a fairly clear idea of what work and energy are, whether you realize it or not. The lessons that follow will help you define both terms in a scientific sense.

LESSON 2

The Idea of Work

What Is Work?

Ken and Monica are having a conversation in which the word *work* is used in several different ways. How many different meanings can you find?

"Hey Ken, guess what—I'm going to *work* part time after school."

"That's great, Monica! What kind of *work* will you do?"

"I'm going to *work* in the stockroom at the Minit Mart."

"That doesn't sound like too much *work*."

"I don't know. I hope it *works* out."

"It's probably not as much *work* as mowing grass."

"No kidding—pushing a mower is really hard *work*."

"It's not as hard as home*work*."

"Yeah, thinking is pretty hard *work*!"

"By the way, what are you *working* on in science class?"

"We've started this really interesting unit called Machines, *Work*, and Energy."

"Sounds interesting. Speaking of science, I've been *working* on an idea for a science fair project."

"Speaking of *work*, we'd better get back to it or we'll never finish."

As you can see from the conversation on the previous page, the word *work* can be used in many different ways with very different meanings. Look at the statements again. Now, in your ScienceLog, group the statements by their meanings. Which statements suggest that work is

- a vocation or a job?
- anything that occupies our time?
- the opposite of leisure or recreation?
- a way to earn money?
- connected with forces?
- connected with motion?
- an assignment?

Some statements fit into more than one category.

In science, *work* has a very precise meaning. For example, thinking or preparing for an exam would not be considered work. **Work, in the scientific sense, requires that a force be applied to an object and that the object move.** Try to group the sentences on the previous page and the pictures on these two pages into examples of work in a scientific

sense and work in a nonscientific sense.

Another word that is often used with work is *energy*. No doubt you often use this word, but do you really know what energy is? *Energy*, too, can have different meanings. Construct some sentences using the different meanings of *energy*. One way to start would be to look up *energy* in a dictionary.

Work and energy—how are they related? In this unit, you will explore both concepts and learn how one relates to the other.

The Scientific Idea of Work

Work in Progress

It's moving day in the library. The new shelves have arrived and the librarian has recruited 10 volunteers to help reshelve the books.

"Okay, here's your job," the librarian explains. "The books are stacked on the floor by the shelves. You need to place the books back on the shelves according to the call number found on each book."

Each volunteer has a different task. Look at the illustrations on this page showing the work of five of the volunteers at one specific moment. Then answer these questions:

- Who is doing the least amount of work?
- Who is doing the greatest amount of work?
- Are some volunteers doing the same amount of work?
- Can you place the volunteers in order according to the amount of work each does?

Assume that each book has the same weight.

Lifting a book 2 m obviously requires twice the work needed to lift a book 1 m. Therefore, Kyle must do twice as much work as Marie. What about Sandra? Did you conclude that the amount of work depends on the distance through which the force is exerted?

To lift two books 1 m requires twice the work needed to lift one book 1 m. So Roberta does twice as much work as Marie. When you lift two books, you exert twice the force needed to lift one book. Therefore, the amount of work also depends on the size of the force exerted.

Do you see why Gene does more work than Roberta and Kyle? The order of the volunteers, from least to most work done, is Marie, Kyle/Roberta, Sandra, and Gene.

On the next page is a second set of illustrations to study. Answer the same questions as before. Try to figure out a way to calculate the relative amount of work each student is doing.

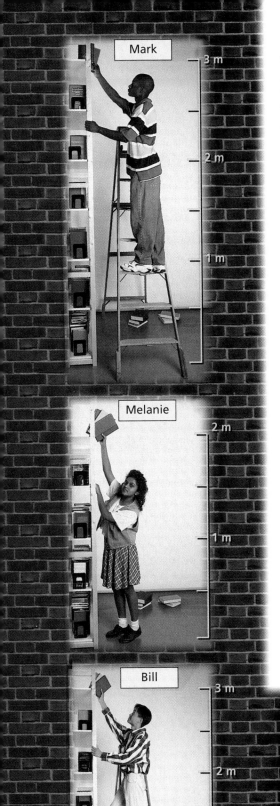

Calculating Work

Did you find a good method for calculating how much work each volunteer does? If you multiply the force exerted by the distance through which the force is exerted, you obtain a quantity that indicates the amount of work done:

work done = force × distance

Suppose that each book weighs 10 newtons (N); a newton is a unit of force. How much work does Marie do when she lifts one book up to the second shelf? She exerts an upward force of 10 N to lift the book, and she exerts this force through a distance of 1 m.

work = 10 N × 1 m = 10 Nm = 10 J

The unit for work is the *joule* (J). It is named after the English scientist James Prescott Joule. A joule is actually a fairly small amount of work, so most work is measured by the *kilojoule* (kJ). How many joules are in a kilojoule?

How much work must Bill do to lift the books to the top shelf?

Now that you know the scientific formula for work, calculate the work being done by each set of volunteers. Do your results verify your earlier rankings of the volunteers in terms of how much work they are doing?

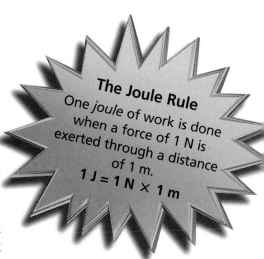

The Joule Rule
One joule of work is done when a force of 1 N is exerted through a distance of 1 m.
1 J = 1 N × 1 m

What Work Feels Like

Here are a few tasks that will give you an idea of what a joule of work feels like.

TASK 1

A Work-Experience Display

You Will Need

- 1 sheet of paper
- scissors
- measuring tape or a meter stick
- 100 g, 1 kg, 5 kg masses
- masking or transparent tape

What to Do

1. Make a set of labels like the ones below.

1 J (100 g)	2 J (100 g)	3 J (100 g)
10 J (1 kg)	20 J (1 kg)	30 J (1 kg)
50 J (5 kg)		100 J (5 kg)

2. Choose one label. To what height above the floor must you raise the indicated mass to experience the amount of work shown on the label?

3. Lift the mass to this height and tape the label onto the wall.

4. Repeat for each of the labels.

   You are also lifting part of your body when you do this work. How might you calculate how many extra joules of work are needed?

TASK 2

You're Pushing It!

You Will Need

- a chair
- a bathroom scale

What to Do

1. Hold the bathroom scale up against the back of the chair as shown in the illustration. As smoothly as you can, push the chair with a person seated in it for a distance of 30 cm. Does it matter where you place the bathroom scale as you push?

2. Determine the force needed to move the chair the measured distance. Calculate the work done.

3. Would you do the same amount of work if you lifted the chair and the person to a height of 30 cm? Explain.

4. Will the amount of work done change if a different person sits in the same chair? Explain.

Figuring Force
If the scale measures in kilograms, multiply by 10 to get the force in newtons; if it measures in pounds, multiply by 4.5.

TASK 3

Comparing Different Examples of Work

You Will Need

- a force meter
- various objects

What to Do

Try these activities:

a. Open a door.
b. Pull out a drawer.
c. Lift this book from the floor to a position directly over your head.
d. Pull a 1 kg mass along the floor for 2 m.
e. Pull down a screen or blind.

   Figure out how many joules of work are required to do each task. First design a way to use a spring scale or force meter to determine the size of the force needed for each task. Then measure the distance through which the force is applied.

TASK 4

Work While You Exercise

You Will Need

- a meter stick
- a pull-up bar

What to Do

1. Calculate your weight in newtons.
2. Do a pull-up.
3. Have someone measure how high you lift your weight.
4. Calculate the work you do in one pull-up and in five pull-ups. What muscles are doing this work?
5. Repeat this task, this time doing a vertical jump. At the highest point of your jump, have someone measure the distance from your heels to the floor with your legs straight.

TASK 5

Machines Also Work

Answer the following questions about the diagram below:

a. How much upward force must the pulley machine put on 1 kg of gravel to lift it?

b. If the gravel is raised 0.5 m, how much work does the machine do?

You Will Need

- 1 kg bag of gravel (or other 1 kg mass)
- 2 pulleys, with cord
- a force meter or spring scale

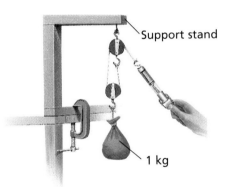

Support stand

1 kg

What to Do

1. Arrange the setup as shown.
2. Measure with a force meter the force needed to operate the machine.
3. Measure how far you need to pull the force meter to lift the gravel 0.5 m.
4. Calculate the work you put on the machine. Compare your work with the work the machine puts on the gravel.

   Why is it helpful to use the machine?

A Powerful Idea

By now you should have a good feel for the scientific meaning of *work*. But what is the scientific meaning of *power*, an energy-related word we often use? Take a moment to write down your own definition of *power*. Using your definition, which is more powerful, a horse or a car? a diesel locomotive or a jet air-craft? How powerful are you? How would you find out?

What to Do

Find a staircase or a steep hill that isn't too tall. Perhaps your school's gym has a rope climb. Measure the vertical distance to the top. If necessary, ask your teacher how to do this. Find out how much work you would do in climbing to the top. As fast as you can, climb the rope or run to the top of a hill or staircase (don't stumble!). Time how long it takes. To find your power in joules per second, or *watts*, divide the amount of work you did by the time it took to do it. How does the definition of *power* you wrote ear-lier compare with the defini-tion of *power* that you just learned?

To find out more about power, see pages S44–S45 of the SourceBook.

LESSON 3

Work and Energy

What Do You Need to Do Work?

Look at the illustration shown here. A lot is going on. Write a caption to explain what's happening in terms of work and energy. Where do you think all this energy came from?

The Work/Energy Story that follows will give you many new ideas that relate work to energy. As you read it, see how many blanks you can complete without looking at the answers.

The Work/Energy Story

Energy enables us to do (1). If you lift a 10 N block to a height of 2 m, you will have done (2) J of work on it. If you now let the block fall on a nail, it hits the nail with a (3) and the nail is pushed a short distance into the wood. In other words, the block does (4) on the nail. Since energy enables work to be done, the raised block must have (5). How did the block get it? The block got the energy by the (6) you put on the block to raise it.

The energy of raised objects is called potential (stored) energy. Since you did 20 J of work on the block, the potential energy of the raised block must be (7) J. You, of course, got your energy from the (8) you ate. If you had felt the top of the nail, it would have been warm. As the block did work on the nail, some of the kinetic energy was changed into (9) energy. Note that in every case when work was done, (10) was transferred.

Answers: 1. work, 2. 20, 3. force,
4. work, 5. energy, 6. work, 7. 20,
8. food, 9. heat, 10. energy

158   UNIT 3 • MACHINES, WORK, AND ENERGY

Important Forms of Energy

Potential Energy

Alexander has a lot of *potential* as an archer. What does *potential* mean? It doesn't mean that he's a good archer yet. But he does have the ability to become one.

A raised object has **potential energy**. This does not mean that the raised object is doing work in that position. Rather, it means that the object can perform work on something else if it is released.

When you lift the object, you do work to overcome the gravitational force on the object. The object now has potential energy. When the object is released, gravitational force acts on it, causing it to fall. Because of its motion, the falling object is able to do work on other objects.

A raised object is only one example of potential energy. Potential energy also takes other forms. Work can be done on objects by overcoming other forces, and in so doing, the objects gain potential energy. For example, if you pull back on a bowstring, you are overcoming opposing elastic forces—the tensions of the bow and the bowstring. The work you do against the opposing forces gives potential energy to the bow and bowstring. When the bowstring is released, the elastic forces of the bow and bowstring do work on the arrow, sending it flying.

Potential-Energy Study

The illustrations on these pages show objects that have potential energy. Working with a partner, answer the following questions for each situation:

1. Which objects or substances have potential energy?
2. How is the potential energy stored?
3. What opposing force is overcome to store the potential energy?
4. How will the potential energy be put to work?

Now consider the answers to these questions for the bow-and-arrow situation on page 159. The bent bow and stretched bowstring gain potential energy (1) from the archer's work in stretching the bow (2). This work overcomes the elastic forces exerted by the bow and string (3). The stored energy enables the string to do work on the arrow (4).

Note that the energy represented by (d) and (e) is stored in fuel and food, respectively. This energy is released by chemical changes. For this reason, the potential energy in food and fuel is often called **chemical energy**.

Another Form of Energy: An Investigation

In this Exploration, you will investigate the following questions through a series of real and imagined experiences.

• Do moving objects do work?
• Do moving objects have energy?
• If moving objects have energy, where do they get it?
• What factors affect the amount of energy that moving objects might have?

The first two experiences are thought experiments; only the last one is a laboratory experience. Put on your thinking cap!

EXPERIENCE 1

Does the moving bowling ball in the illustration below do work when it hits the pins? If the bowling ball does work, what must the moving ball have? Where does this energy come from?

Rodney puts the bowling ball in motion by applying force to it. He does this by swinging the ball forward. If Rodney puts 100 N of force on the ball through a swing of 1 m, how much work does he do? How much energy does the bowling ball have? How much work will the bowling ball be able to do when it strikes the pins?

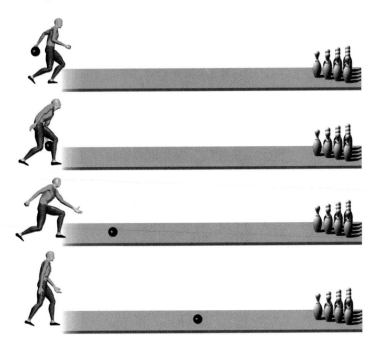

A simple calculation shows that about 100 J of work was done to the ball. Therefore, the ball was given 100 J of energy, which enabled it to do 100 J worth of work to the bowling pins. This form of energy—the energy of moving bodies—has a name. It is called **kinetic energy**. *Kinetic* comes from an ancient Greek word meaning "motion."

EXPERIENCE 2

A 2 kg brick is raised to a height of 3 m. How much potential energy does it have? When the brick is released, it accelerates downward. What kind of energy does it have now? About how much kinetic energy will it have as it just begins to strike the nail? Where does this energy come from? Does the moving brick do work on the nail? How much work does it do?

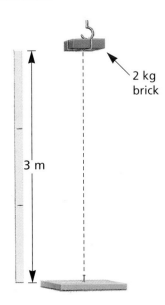

2 kg brick

3 m

• In Experience 1, the source of the kinetic energy was Rodney. Where did Rodney get the energy to do this work?

• In Experience 2, the source of the brick's kinetic energy was the potential energy it had before it was dropped. Whenever energy is transformed from one form into another, work is done. Therefore, work must have been done as the potential energy of the brick was changed into kinetic energy. What did the work on the brick?

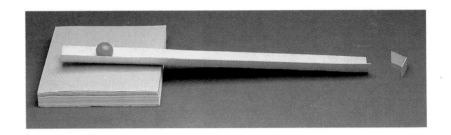

What factors affect the amount of kinetic energy a moving body has?

You Will Need

- a trough (made of metal or cardboard approximately 60 cm long)
- 3 identical books
- 3 balls, one twice the mass of the lightest, and the other 3 times the mass of the lightest
- a piece of cardboard (approximately 2 cm × 15 cm)

What to Do

1. Raise the trough by placing a book under it. Let the lightest ball roll down the trough to strike a piece of cardboard folded as shown. What happens? Is work done by the ball? What kind of energy does the ball have before it starts moving? after it starts rolling? The distance the cardboard moves is an indication of the amount of kinetic energy the rolling ball has. Allow the ball to roll again, and measure the distance the cardboard moves from the starting point.

2. Raise the trough by adding another book. Allow the ball to roll from the same spot on the trough. How does the speed of this ball compare with that of the ball in step 1? Measure and record the distance the cardboard moves. How does the kinetic energy compare in each case? What is one characteristic of the ball that affects how much kinetic energy it has?

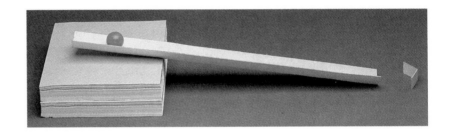

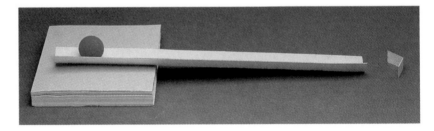

3. Try using three books. Record the distance that the piece of cardboard moves.

4. Place the trough on one book again. Allow the ball that has twice the mass as the lightest ball to roll down the trough. Compare the distance the cardboard moves with the distance it moved in step 1. What causes the difference? What is another characteristic that affects the amount of kinetic energy?

5. Check your conclusion from step 4 by using the ball that is 3 times the mass of the lightest ball.

Questions

1. Where does the kinetic energy of the balls come from? How does the potential energy of the ball (before rolling) compare in steps 1, 2, and 3? How does its kinetic energy (just before striking the cardboard) compare in steps 1, 2, and 3?

2. Which factors (variables) do you think affect the amount of kinetic energy a moving body has—color, speed, shape, mass, composition, location, others not listed? Make a list in your ScienceLog.

Energy Changes

How many different types of energy have you seen so far? Can one type of energy change into another type of energy? Look over the last few pages and identify some examples of where this happens.

The following Exploration features a number of investigations. Some are thought investigations that can be performed best by talking about them; others you may actually perform.

EXPLORATION 4

An Energetic Discussion

Working in groups, discuss question 1, below, and any two others that interest you. Your group should be prepared to discuss the results with the class.

1. Refer to the Potential-Energy Study on pages 160 and 161 to answer the following questions:

   a. How is potential energy changed to kinetic energy in each situation in the Potential-Energy Study?

   b. When energy is transferred, work is done. What agent is doing the work as potential energy changes into kinetic energy?

   c. How does kinetic energy do work in image (h)? Any moving part of a machine has kinetic energy, which is sometimes called **mechanical energy**. Can you think of other examples?

2. Hold a marble at the rim of a large bowl. Release the marble. What happens? You know that potential energy can change into kinetic energy. Can kinetic energy change into potential energy? When does the marble have the most potential energy? the most kinetic energy?

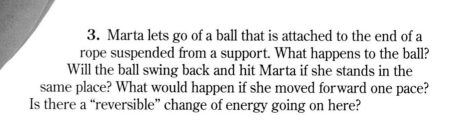

3. Marta lets go of a ball that is attached to the end of a rope suspended from a support. What happens to the ball? Will the ball swing back and hit Marta if she stands in the same place? What would happen if she moved forward one pace? Is there a "reversible" change of energy going on here?

4. Consider the acrobatic act pictured at right. One acrobat is standing on a raised platform. A second acrobat is standing on one end of a seesaw. Their initial positions are both labeled *1*. The illustration shows two other positions for each acrobat after the one on the platform jumps off (positions *2* to *5*). What happens? Describe the potential and kinetic energy of each acrobat in each position. How do you explain the fact that the second acrobat ends up on a platform higher than the platform used by the first acrobat?

5. Diagram A represents the lifting of a brick through a vertical distance of 6 m, in two steps. The brick has a mass of 2 kg. Diagram B represents the release of the same brick.

   In your ScienceLog, calculate the amount of potential and kinetic energy that you think the brick has in each position in the two diagrams.

6. Think about dragging a table across the floor. What kind of energy does the table have as it moves? When you stop applying a force to the table, it stops moving. Where does the energy go?

7. Touch a nickel to your chin. Place this nickel flat on a piece of paper. While pressing down hard on it with your fingers, move it back and forth briskly 10 to 20 times. Again touch the nickel to your chin. What do you observe? Explain any energy changes.

8. Look at the stop-action photo below. It shows a bouncing golf ball. When is the potential energy the smallest? Why does the ball stop bouncing? What other type of energy might be present?

Diagram A		Potential energy	Kinetic energy
6 m	Step 2	?	?
3 m	Step 1	60 J	0 J
0 m		0 J	0 J

Diagram B		Potential energy	Kinetic energy
6 m		?	?
3 m		?	?
0 m		?	?

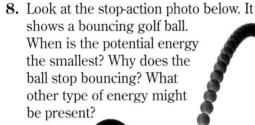

Discovering New Forms of Energy

You have already seen how different types of energy can change from one form to another. Now identify the energy changes in each of the **energy converters** pictured on this page. Look not only for the familiar energy forms you have studied, but also for new ones such as electrical, light, heat, and sound energy. In your ScienceLog, make a table similar to the one shown below, and complete it for energy converters (a) through (d).

Example	Energy converter	Energy change
(e)	Battery	Chemical to electrical

CHALLENGE YOUR THINKING

1. Pushing the Issue

Suppose your mass is 50 kg. Assume that your hands and arms support half your weight and that in doing one push-up you elevate this weight through a total distance of 40 cm. How many push-ups would you need to do to expend 1 kJ of work?

2. Power Play

a. Assume it took you 0.5 seconds to do a push-up in question 1. Use this to calculate the power you exerted in doing a single push-up, in both watts (W) and kilowatts (kW).

b. A 40 W light bulb uses 40 J of electrical energy each second to produce light and heat. How many such light bulbs would have to burn to equal the rate of work of one of your push-ups?

3. Climbing Potential

a. Mount Everest is 8848 m high. How much work would you have to do to get to the top? How much potential energy would you have at the top?

b. A glass of milk provides 500 kJ of energy, one-quarter of which is available for work. How many glasses of milk would provide you with enough energy to scale Mount Everest? Where does the rest of the energy go?

4. Stop! Look! Listen!

"The one thing constant in this world is change."

For the next 5 minutes, think about this statement and then about your surroundings. Make a list of all the energy converters and energy changes that you recognize. Try to identify where work is being done during each energy change.

5. The Race Is On

Excitement was in the air! It was the day of the big Overnight Bike Race. Vic felt confident. He had trained well by riding his bicycle; doing daily push-ups, sit-ups, and pull-ups; jogging; and lifting weights at the gym. Now, 20 minutes into the journey, he reached the top of a steep hill. In the valley below he could see the dam and power station. Water gushed down a sluiceway, making generators hum and electricity surge through the power lines. As Vic sped down the hill, he caught occasional glimpses of evergreens amid colorful maples. "Nature's power stations," he mused.

Vic braked as he approached the first rest station at the bottom of a steep hill. He was the fifth to arrive. He glanced at his watch—he had been racing for 56 minutes.

a. How many energy changes can you identify in the story so far?

b. Name the forms of energy involved in each change.

c. Where possible, name the energy converter.

d. Continue the story. Hide some more energy changes in the narrative, and then have a classmate try to identify them.

ScienceLog

Review your responses to the ScienceLog questions on page 146. Then revise your original ideas so that they reflect what you've learned.

Harnessing Energy

2 **How does the simple machine shown above help you do work?**

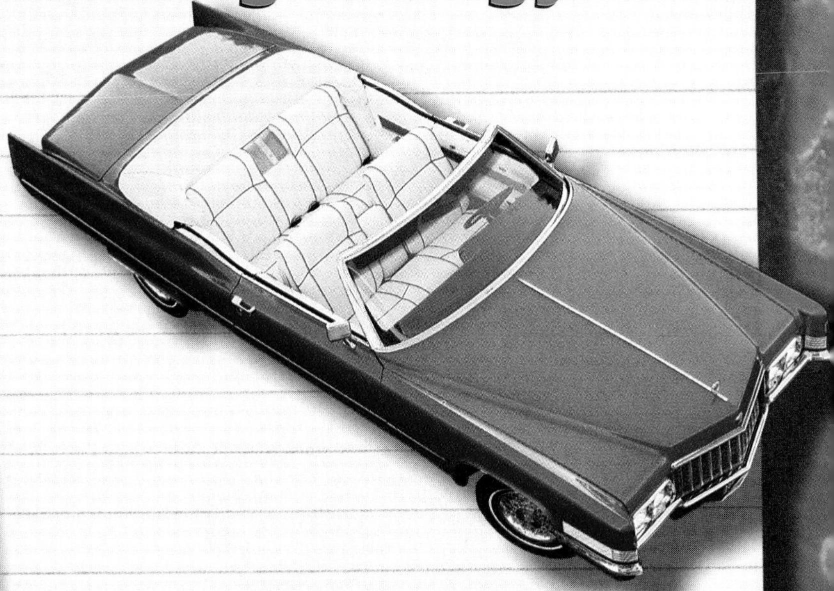

1 **What makes one machine more *efficient* than another?**

What are some of the ways in which machines transfer energy?

3

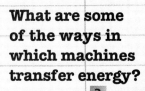

ScienceLog

Think about these questions for a moment, and answer them in your ScienceLog. When you've finished this chapter, you'll have the opportunity to revise your answers based on what you've learned.

169

Lightening the Load

Tony's Problem

Tony is the supervisor on a construction site. He needs to move a 100 kg load to a platform that is 1 m high. Tony knows that each member of his crew can safely exert a lifting force of about 250 N. How many of his crew should Tony get to lift the load? (Remember, the weight of a 1 kg mass is about 10 N.) However, Tony is the only crew member who isn't busy. But he can't lift the load by himself. What type of device could he put together so that one person could lift the load onto the platform?

Helping Tony

Here's your chance to design a device that might help Tony.

1. Examine the diagrams on the next page.
2. Design and construct a device for lifting the mass with as little effort as possible. Your device doesn't actually have to be able to lift 100 kg! Instead, design your machine to scale. For instance, it might lift a 1 kg mass with an effort of 2 N.

Here is some of the information you will need in order to test your machine:

- the size of the force (in newtons) required to lift the load using your machine
- the distance through which you must apply that force to raise the load to the desired height
- the amount of work you do in moving the load
- the comparison of the amount of work you do with your machine to the amount of work you would do without the machine

What might you do to improve your machine? Can you lift the load with an even smaller force? If you are able to do this, is there a trade-off with something else? Does your machine actually *save* work?

Some Possible Devices to Help Tony

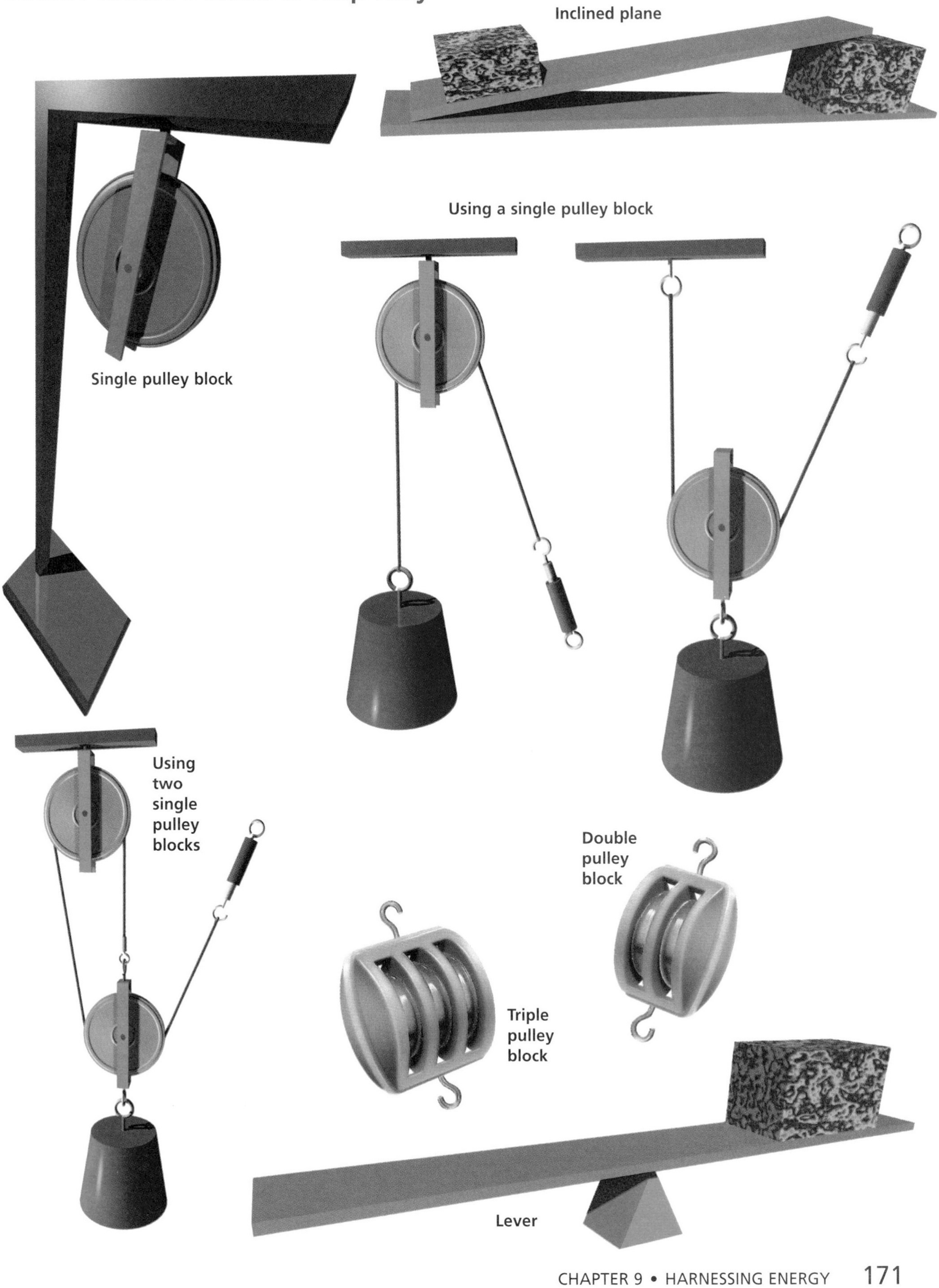

Inclined plane

Single pulley block

Using a single pulley block

Using
two
single
pulley
blocks

Double
pulley
block

Triple
pulley
block

Lever

A Look at Two Solutions

Leona's Machine

Leona used an arrangement of pulleys to solve Tony's problem. Below is Leona's report about the machine she made. From the information in her report, solve the problems on the facing page.

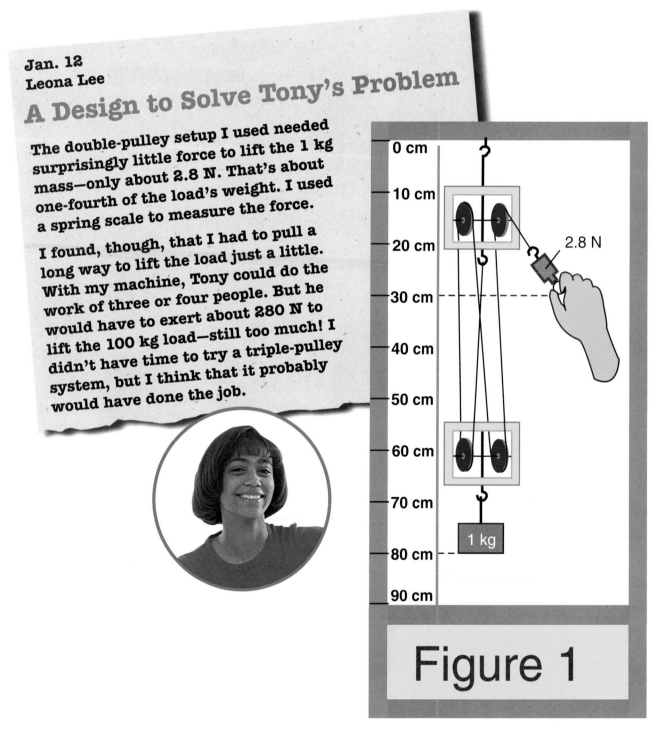

Jan. 12
Leona Lee

A Design to Solve Tony's Problem

The double-pulley setup I used needed surprisingly little force to lift the 1 kg mass—only about 2.8 N. That's about one-fourth of the load's weight. I used a spring scale to measure the force.

I found, though, that I had to pull a long way to lift the load just a little. With my machine, Tony could do the work of three or four people. But he would have to exert about 280 N to lift the 100 kg load—still too much! I didn't have time to try a triple-pulley system, but I think that it probably would have done the job.

2.8 N

Figure 1

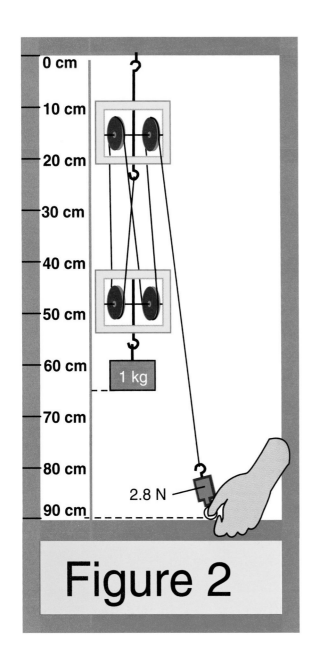

Figure 2

Problem 1

a. How much force did Leona put on the string?

b. How far did she move the string?

c. How many joules of work did Leona do in using the machine?

Machines are commonly used to make work easier. The work put into or done on a machine is called **work input**.

Problem 2

a. What upward force does the machine (pulley and strings) place on the load? (This is the same as the force you would need to lift the load without a machine.)

b. How far is the load raised by the machine?

c. How much work was done on the load by the machine?

The work done by a machine on a load is commonly called **work output**, which is the work you get out of the machine, regardless of the work put into it.

Problem 3

How does the work output in Problem 2 (c) compare with the work input in Problem 1(c)? What do you think accounts for these results? From your own experience, how would you modify the following principle?

The work you put into a machine is (greater than, less than, the same as) the work you get out of it.

Pam's Machine

Pam had a different solution to Tony's problem. She tried three different versions of her idea, which involved using a slanted board. A slanted board is a type of simple machine called an *inclined plane*. Pam tried three boards of different length.

Complete her table in your ScienceLog, and then write some conclusions that she might draw from her solution. The questions below will help you.

Questions

1. Does the length of the board affect the force needed to pull the load? If so, how?

2. Compare the *work output* (the amount of work actually done by the machine) with the *work input* (the amount of work put into the machine) for each of the three boards. What do you notice? Would a long inclined plane and a short inclined plane be equally good at converting work input to work output? How and why might they differ?

3. Compare the work input and work output of Leona's machine with those of Pam's machine. Which of the two machines is better at converting work input to work output?

4. Which machine do you think Tony should use and why?

A SLANTED BOARD is a type of machine called an "inclined plane." I used a mass of 1 kg. I tried boards of different length.

FORCE to pull the mass along slanted board	
DISTANCE that the mass is moved	
WORK INPUT	
WEIGHT	
VERTICAL (HEIGHT) distance the mass is raised	
WORK OUTPUT	

Board Machine

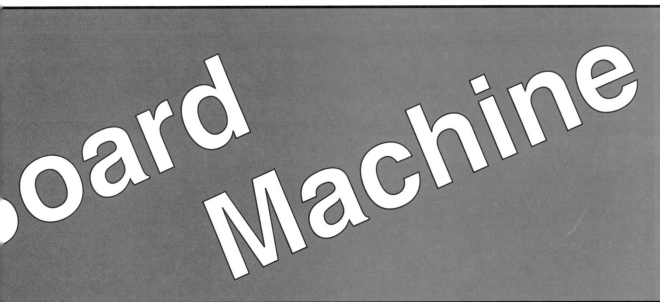

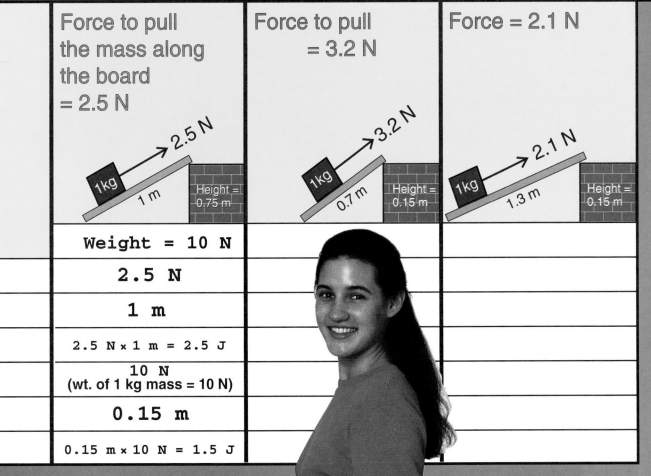

	Force to pull the mass along the board = 2.5 N	Force to pull = 3.2 N	Force = 2.1 N
	1kg →2.5 N 1 m Height = 0.75 m	1kg →3.2 N 0.7 m Height = 0.15 m	1kg →2.1 N 1.3 m Height = 0.15 m
Weight = 10 N			
2.5 N			
1 m			
2.5 N × 1 m = 2.5 J			
10 N (wt. of 1 kg mass = 10 N)			
0.15 m			
0.15 m × 10 N = 1.5 J			

Some Machines Are Better Than Others

Which of the following machines would you prefer to use for a job?

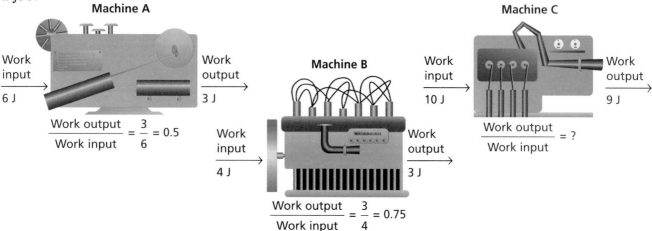

Machine A

Work input
6 J

Work output
3 J

$$\frac{\text{Work output}}{\text{Work input}} = \frac{3}{6} = 0.5$$

Machine B

Work input
4 J

Work output
3 J

$$\frac{\text{Work output}}{\text{Work input}} = \frac{3}{4} = 0.75$$

Machine C

Work input
10 J

Work output
9 J

$$\frac{\text{Work output}}{\text{Work input}} = ?$$

If you chose Machine C, you probably did so because it is better at converting work input to work output. Machine C is more *efficient.* There is a way to measure how efficient a machine is. The fraction

> **work output**
> **work input**

tells you how much of the work put into a machine becomes useful work done by the machine. This fraction expresses the efficiency of the machine. Efficiency can be expressed as a percentage by multiplying the efficiency value by 100. For instance, Machine A is 0.5 × 100 = 50 percent efficient.

Machine B is 75 percent efficient. This means that 75 percent of the work put into Machine B emerges as useful work done by the machine. What is the efficiency of Machine C? What is the efficiency of Leona's machine? How efficient are each of the inclined planes of Pam's machine? Do you observe any differences? Can you explain them?

Now consider the machine you designed to solve Tony's problem. Compute its efficiency. Compare your machine's efficiency with that of each of your classmates' designs. Which machine has the highest efficiency? How do you think your machine might be made more efficient?

What Do You Think?

Look at Machine D at right. Do you think that such a machine is possible? Why or why not? How can Pam's and Leona's results help you reach a conclusion?

In fact, no machine can be 100 percent efficient. Some energy is always "wasted" and does not go into work output. What do you think happens to this energy?

Machine D

Work input = 100 J

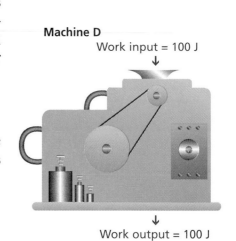

Work output = 100 J

The Advantage of Machines

If machines do not *save* work, is there any advantage in using them? Indeed there is! With a machine, one person can do the work of many. This is because a small force can be turned into a larger force. Leona, using her machine, can lift the load of 10 N with a force of only 2.8 N. She increases her force almost fourfold! The number of times a force exerted on a machine is increased by the machine is called the **mechanical advantage** of the machine. For Leona's machine, the mechanical advantage is calculated as follows:

$$\text{mechanical advantage} = \frac{\text{force exerted by the machine}}{\text{force exerted on the machine}} = \frac{10 \text{ N}}{2.8 \text{ N}} = 3.6$$

Of course, you never get something for nothing. The smaller force put on the machine must be moved through a longer distance, while the larger force that the machine exerts moves a shorter distance.

A Trade-Off: Force for Distance

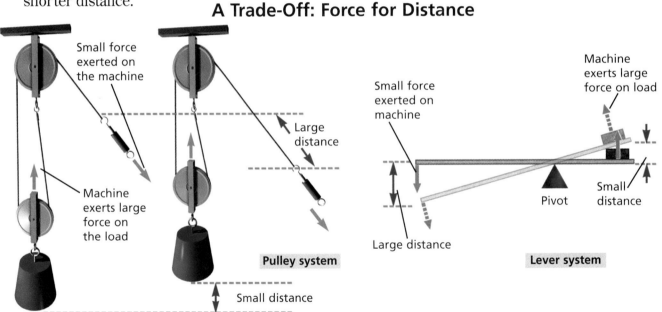

Small force exerted on the machine

Machine exerts large force on the load

Large distance

Small distance

Pulley system

Small force exerted on machine

Machine exerts large force on load

Large distance

Pivot

Small distance

Lever system

Some machines work differently. They multiply distance at the expense of the force. You will see some of these on pages 178 and 179.

1. What is the mechanical advantage of each of Pam's inclined planes? of Leona's machine? of the machine you constructed?

2. Suppose that Leona used a triple-pulley setup and found that a 2 N force could lift the 1 kg mass. What would be the mechanical advantage?

3. Now suppose that Tony used Leona's triple-pulley system to raise his 100 kg load. How much force, in newtons, would be needed to raise the load? How many people would be needed for the job?

Which Machines?

Some familiar machines are shown on these pages. They are all simple machines, although in one or two cases you might not believe it! *Simple* machines make up *complex* machines such as bicycles and backhoes. Your task is to analyze each of the simple machines shown.

1. Classify each machine as one of the following:
   - lever
   - inclined plane
   - pulley
   - some other type of machine

2. Study the machines. Which ones do you think
   - multiply (increase) the force put into them?
   - multiply (increase) the distance put into them?
   - change the direction of the force put into them?

3. Where do you exert a force on each machine?

4. Where does the machine exert a force on something else?

A sketch showing the forces might be the best way of answering questions 3 and 4. Make sure your sketch shows the forces involved.

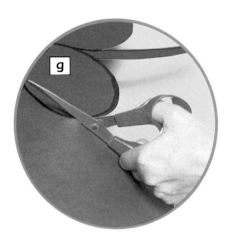

Machines and Energy

You have seen that machines can convert energy from one form to another. You have also seen that machines can multiply force or distance, but never both at the same time. An increase in force always comes at the expense of distance, and an increase in distance comes at the expense of force. What accounts for this?

Efficiency is another matter altogether. As you know, no machine is 100 percent efficient, but where does the "lost" energy go? Does it simply disappear, or is it converted into some unseen form? In this lesson you will tackle these questions and others. You will also learn a bit more about the ways in which machines convert and convey energy.

EXPLORATION 1

Miguel's Machine

Miguel devised a machine that would lift loads easily. With his machine, he found that he could lift a load of 10 N with a force of only 2 N.

Miguel realized, however, that he had to turn the handle a long way to lift the load just a little. In fact, when he turned the handle all the way around once, the load came up by just the distance around the spool.

Evaluate Miguel's machine by completing his evaluation report in your ScienceLog. Could Miguel's machine be used to solve Tony's problem on page 170?

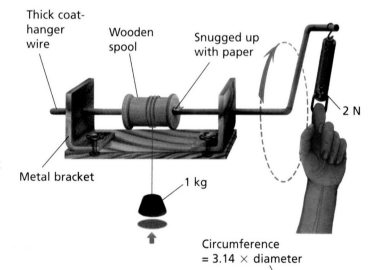

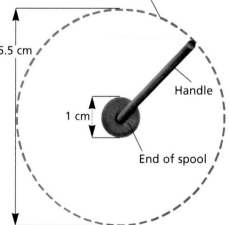

Circumference = 3.14 × diameter

My Machine: An Evaluation	
Distance handle moves in one turn = _____?_____	Work input = _____?_____
Distance load rises when handle makes one turn = _____?_____	Machine's work output = _____?_____
Force put on handle = _____?_____	Efficiency = _____?_____
Upward force put on the load by the machine = _____?_____	
Mechanical advantage = _____?_____	
Conclusion and Recommendation: This machine would have an even greater mechanical advantage if . . .	

A Closer Look at Efficiency

Count Rumford

James Joule

Miguel put 0.34 J of work into his machine. How much energy did he transfer to the machine? His machine did 0.31 J of work on the load. How much energy did the machine transfer to the load? Was some energy lost? It seems so. Where did the lost 0.03 J of energy go?

Situations like the one above puzzled scientists such as Count Rumford (1753–1814) and James Joule (1818–1889). Through many observations and experiments, the two scientists independently came to the conclusion that the energy apparently lost in machines was actually only diverted; the energy was turned into heat by friction. Their research established that heat was a form of energy and not a form of matter, as was commonly believed at that time. Rumford and Joule reasoned that the amount of work actually done by a machine is equal to the amount of energy put into the machine minus the energy lost to friction. Each scientist concluded that energy is *conserved* in nature. In other words: **Energy can be transferred and transformed, but the total amount of energy in a system always stays the same.**

How would you apply Rumford's and Joule's findings to Miguel's machine?

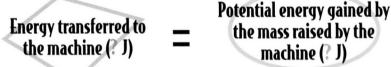

Energy transferred to the machine (? J) = Potential energy gained by the mass raised by the machine (? J) + Heat energy caused by friction (? J)

Compute the energy values for one turn of the handle. What would the energy values be for 10 turns of the handle? If the machine had less friction, how would these values change? How would the value for the efficiency change? In your ScienceLog, complete the following principles.

- A machine is a device for transferring or converting __?__ .
- Some of the work (energy) input to a machine is used to overcome friction. This produces __?__ .
- Energy input is equal to energy __?__ plus other forms of energy (such as heat) that are lost to use.

To find out more about how friction reduces the efficiency of machines, turn to pages S55–S56 in the SourceBook.

Do you now understand why the energy you put into a machine (work input) is always greater than the energy you get out of the machine (work output)? Why is this so?

How much heat energy was produced as the load was lifted 15 cm by Leona's machine (pages 172–173)? by each of Pam's inclined planes (pages 174–175)? How could both of these machines be made more efficient? What is the relationship between a machine's efficiency and its loss of useful energy?

Wheels and Axles

Miguel's machine is commonly called a **wheel and axle.** Why? Did Miguel apply his force to the *wheel* or to the *axle* of the machine? Where does the machine apply its force on the load?

Wheels and axles are normally found together—neither is very useful without the other. Now turn back to the pictures on pages 178 and 179. Which ones show a wheel and axle? For each example, determine where the force is applied and where the machine exerts a force on something else.

Solving a Technological Problem

You have already encountered a wheel-and-axle machine. Miguel designed such a machine. What steps might he have followed in designing and building it? Try to approach the problem as he might have.

As you follow the steps on the next page, you will be tracing out the steps commonly used in solving technological problems.

You Will Need

Choose from the following readily available materials and tools:

- spools of various sizes
- mailing tubes
- film canisters with lids
- wire coat hangers
- knitting needles
- craft sticks
- milk cartons
- shoe boxes
- plastic-foam or paper cups
- cardboard
- plywood
- a 1 kg mass
- a hand saw
- white glue
- cans
- wire
- straws
- string
- pliers
- corks
- scissors
- a force meter

What to Do

1. First review Miguel's problem. He wanted to use simple materials to make a hand-powered wheel and axle strong enough to lift a load of 10 N with a much smaller force.

2. Observe Miguel's machine again. Identify each of its working parts.

3. Make decisions regarding the following:
   a. What will I choose for the spool part of the axle?
   b. What will I use for the wire part of the axle?
   c. How will I construct the wheel?
   d. How can I attach the spool to the wire so that both turn together?
   e. What size would be best for each of these parts?
   f. What will I use for a supporting frame?
   g. How strong should each part be?

4. Make a simple diagram showing your design solution. Name all the materials you plan to use.

5. Construct a model of the wheel and axle according to your design.

6. Test its operation. Are some improvements necessary to make it operate better? If so, make them.

7. Determine your machine's mechanical advantage. Also determine its efficiency and the amount of energy it changes into heat due to friction with every turn.

8. Think of other ways you might critically judge your machine, for example, appearance, strength, and maximum load it can carry.

9. Compare your wheel and axle with those of your classmates. Identify strengths and weaknesses in their wheels and axles as well as in your own.

10. List suggestions to make your machine better.

Congratulations! You have a workable hand-powered wheel and axle of sufficient mechanical advantage to do the job you wanted. You did this by "technological problem solving." The diagram at right shows the steps. Match what you did in this Exploration (steps 1–10) with the steps in the diagram (A, B, C, and D). More than one Exploration step may match each diagram step.

Exploration 3 extends the problem. Your task will be to adapt your machine so that it can be powered by some source of energy other than your hand.

Technological Problem Solving

A Understanding the Problem

B Developing a Plan
- Identifying alternative design solutions

C Carrying Out the Plan
- Constructing a model and troubleshooting

D Evaluating
- Evaluating your design
- Proposing improvements

Powering Your Machine

Can you suggest another way to power your wheel-and-axle machine, other than by hand? You might consider using any of the following sources of energy:

- water
- steam
- wind
- electricity
- a rubber band
- a raised weight

How would you connect your energy source to your machine?

PART 1

What to Do

1. Choose an energy source.
2. Decide on a possible way to connect the source to your machine. If you need to modify the machine, determine how you will do it.
3. Check your design. Ask yourself whether you are using simple and readily available materials and equipment. Modify your design if necessary.
4. Draw your design.
5. Show your design to others, and ask for their opinions and advice.

PART 2

What to Do

1. Obtain the necessary materials and equipment, and construct a model of the powered wheel and axle according to your design.
2. Connect the machine to the source of energy you have chosen.
3. Evaluate your model. Does it work? If not, why does it not work? If it works, what size load will it lift? How does it operate compared with the models of others? Propose some improvements to your model.
4. How good a problem solver were you? Exploration 2 showed you a model for solving technological problems. Draw a chart in your ScienceLog like the one on page 183, and complete it by adding details of what you did at each of steps *A–D*.

Rubber-band energy

Water energy

Steam energy

Electric energy

Gravitational energy

Other Wheels and Axles

Let's look at some more examples of wheel-and-axle machines. As you will see, the following machines are different in a key way from those you have seen so far.

Look carefully at the images below. How are these wheels and axles different from Miguel's machine? In each diagram, visualize the circle that is traced out by the handle or pedal as it rotates the machine's axle. How does the circumference of this circle compare with the circumference of the machine's wheel?

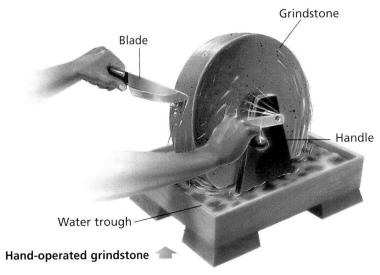

Grindstone

Blade

Handle

Water trough

Hand-operated grindstone

In using the grindstone, a person applies a force to the handle, which turns through a smaller circumference than does the rim of the grindstone. How does this wheel and axle affect the speed, distance, and size of the output force?

Analyze the operation of the unicycle shown at right in the same way. Which of these statements is true for these two examples of a wheel and axle?

- Force is multiplied at the expense of distance.
- Distance is multiplied at the expense of force.

Suppose you push the pedals of the unicycle through one complete turn.

1. Through what distance does each pedal move?
2. Through what distance would the wheel move along the road?
3. How much has the pedal distance been multiplied?
4. Suppose you move the pedals through one full turn in 1 second. What is the speed of the pedals in centimeters per second? What is the speed of the unicycle?
5. The speed of the pedals has been multiplied. How did this happen? What's the trade-off?

Diameter of wheel = 60 cm

Diameter of pedal swing = 35 cm

LESSON 3

Transferring Energy

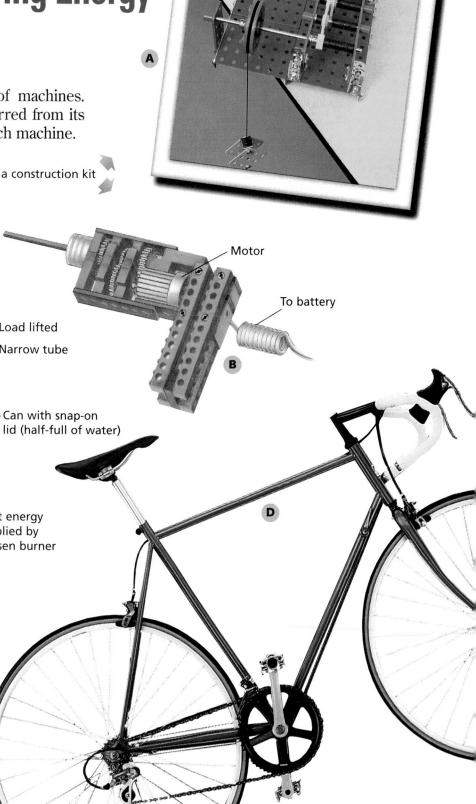

A

Model machines made from a construction kit

Getting Into Gear

Examine the following pictures of machines. Identify how the energy is transferred from its source to the operating parts of each machine.

Nail
Cork
Metal strips
Load lifted
Narrow tube
Can with snap-on lid (half-full of water)

C

Heat energy supplied by Bunsen burner

A homemade steam turbine

Motor

To battery

B

D

A common energy-transfer system

EXPLORATION 4

What Makes It Tick?

What to Do

To do this activity you will need to collect some unused mechanical devices—such as toys, clocks, or watches—that are powered by some form of energy.

1. Working in small groups, first select a device to observe in detail.

2. Disassemble the device to examine the parts necessary for its operation. For example, if you are examining a toy car, does it have a large wheel that continues to spin for some time once it is set in motion? Such a wheel is a *flywheel*. Of what use is the flywheel? You might also find toothed wheels or cylinders that fit into one another. These are *gears*. What do they do?

3. Identify the sequence of movements as energy is transferred through your machine, from the energy source to the final movement of the object.

4. Sketch your machine, and describe how it works.

Joy's Construction Kit

Most hobby and toy shops have technology construction kits; LEGO and Meccano are two popular brands. Joy had such a kit. It contained a motor, wheels, propellers, and all the other items needed to get energy from the battery to these moving parts. Joy put together four assemblies to see how everything worked. She also asked herself questions about each assembly. Joy's assemblies are shown on this and the following page.

ASSEMBLY 1

Assembly 1

1. If I turn the top gear, in what direction does the lower gear turn?
2. How many turns does the lower gear make during one complete turn of the top gear? Is this related to the number of teeth in each gear?
3. What differences would I note if I turned the lower gear instead?

Assembly 2

1. As I turn the small gear, what happens to the large gear?
2. For one turn of the small gear, how far does the large gear move?
3. What differences would I observe if I turned the large gear instead?

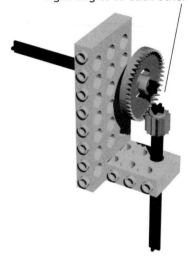

ASSEMBLY 2

Gear wheels of two sizes at right angles to each other

Assembly 3

Joy included a worm gear, another gear, and a wheel in this assembly.

1. As I turn wheel A through one complete turn, how far does the thread of the worm gear move to the left or right?
2. How far does gear B move at the same time?
3. If this system were attached to a motor, how many turns of the motor would be needed to make one turn of gear B?

Assembly 4

1. How does the energy get from the battery to the moving parts?
2. For each turn of the wheel on the motor, how much does each successive part listed below turn?
   * the wheel attached to the motor by a belt
   * the small wheel on the other side of the short axle
   * the wheel on the long axle
   * the tractor wheel

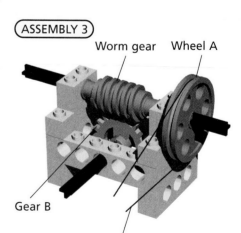

ASSEMBLY 3

Worm gear   Wheel A

Gear B

Elastic band to wheel on motor

Joy first turned the connecting wheels by hand. She marked the top of each wheel to keep track of its motion.

3. How many turns of the motor's axle are needed to produce one turn of the tractor wheel?

4. How does the speed of the motor compare with the speed of the tractor wheel?

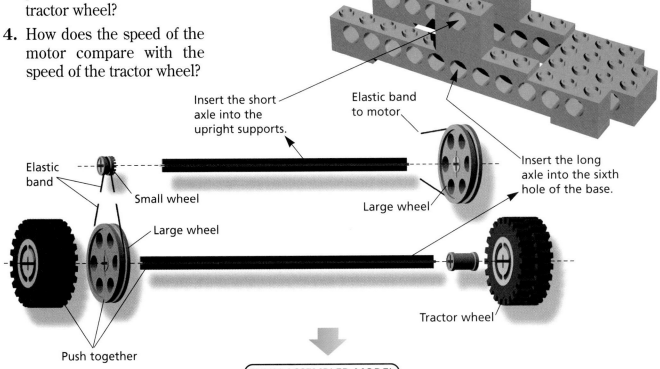

Wires to battery pack

Motor

ASSEMBLY 4

Insert the short axle into the upright supports.

Elastic band to motor

Insert the long axle into the sixth hole of the base.

Elastic band

Small wheel

Large wheel

Large wheel

Tractor wheel

Push together

FULLY ASSEMBLED MODEL

A Project— Model-Building

Now it's your turn to build a complete, working model from a construction set. It could be a device such as a car or tractor. It should have movable parts and be powered by a motor.

After you have completed your model and have done all the troubleshooting to get it working, observe how the energy is transferred from the battery to the working parts. Make a list of all the links in the energy train. Describe how they function and where changes in speed and direction occur.

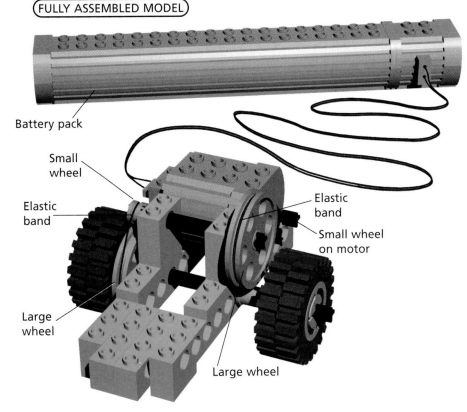

Battery pack

Small wheel

Elastic band

Large wheel

Elastic band

Small wheel on motor

Large wheel

A Closer Look at Wheels, Belts, and Gears

Here is a chance to see what you have learned about wheels and gears. Analyze the four situations below.

Situation 1
Driving Wheels and Driven Wheels

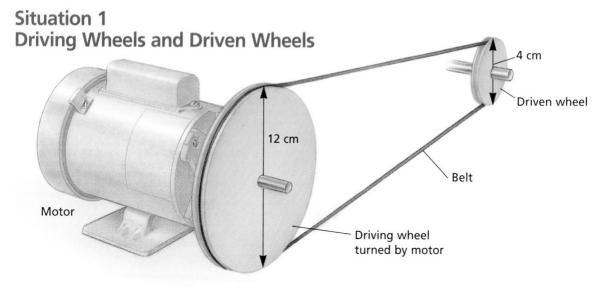

12 cm

4 cm

Driven wheel

Belt

Driving wheel
turned by motor

Motor

Here, the larger wheel is called the *driving* wheel because it drives the smaller wheel. It is turned by a motor. The belt transfers the energy to the smaller wheel, which is called the *driven* wheel.

1. As the large wheel makes one revolution (turn), how many turns does the smaller wheel make?

2. How does the speed of the axle of the driven wheel compare with the speed of the axle of the driving wheel?

3. Which has been increased in the driven wheel—speed or distance? Of what value could this be?

4. If the driven wheel were to become the driving wheel, what difference would this make?

Situation 2
Different-Sized Gear Pairs

1. The driven gear in the photo at right has 36 teeth. How many teeth are in the driving gear?

2. As the driving gear turns counterclockwise, in what direction does the driven gear turn?

3. For each revolution of the driving gear, how many turns does the driven gear make? Which gear axle moves faster?

4. If the speed of the driving gear is 30 turns per second, what is the speed of the driven gear? Which gear exerts less force at its axle?

5. What changes would you make to the gears to increase the speed of the driven gear? to decrease its speed?

Driven gear

Driving gear

Situation 3
Different Kinds of Gears

1. The diagrams at right show three different kinds of gear pairs that perform in a similar way. How do they work?

2. In diagram (a), when the worm gear moves one turn, how far do you estimate the driven gear will turn? Explain how you know. Is the speed of the driven gear greater than, the same as, or less than the speed of the driving gear? Which gear will exert the greater force? The worm gear is almost always the driving gear. Why do you suppose this is so?

3. Estimate the number of teeth in each of the bevel gears shown in diagram (b). If the driving bevel gear makes 7 turns in 1 second, approximately how many turns will the driven gear make?

4. In diagram (c), when the driving gear makes one turn, how many turns does the smaller driven gear make? Has speed or force been multiplied?

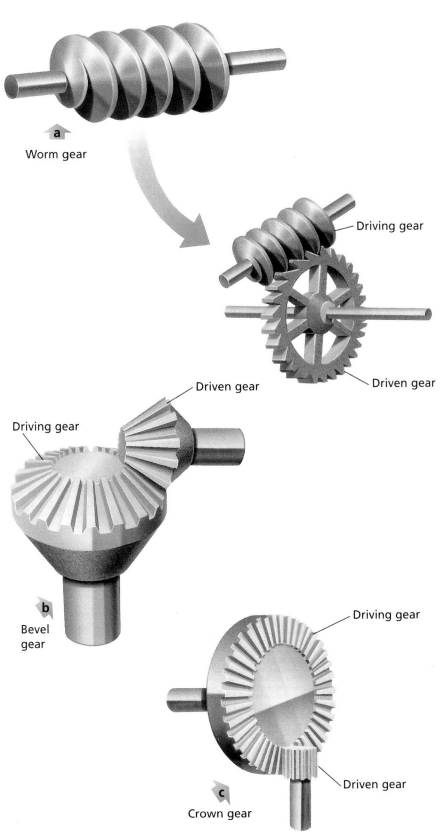

a
Worm gear

Driving gear

Driven gear

Driven gear

Driving gear

b
Bevel gear

Driving gear

Driven gear

c
Crown gear

Situation 4
A Simple Gearbox

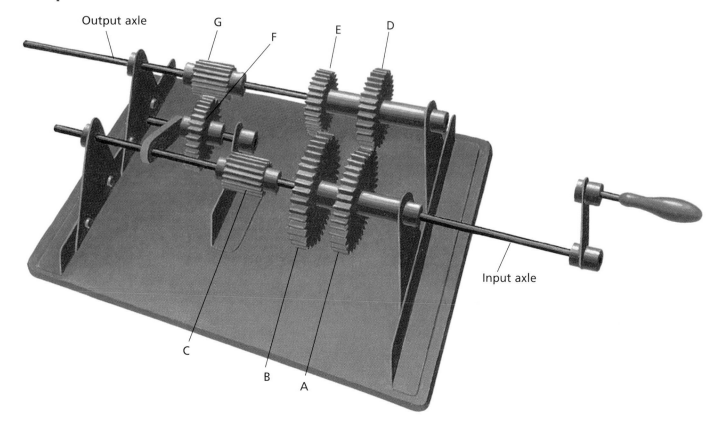

Output axle — G — F — E — D — Input axle — C — B — A

A student made this model of a car transmission (gearbox) from a construction set. The handle can move in and out, causing different gears to engage. There are two "forward speeds" and one "reverse speed." The handle is attached to the input axle. Gears on this axle engage the gears on the output axle.

1. In the illustration, gear A on the input axle is engaged with gear D on the output axle. Suppose the handle is turned counterclockwise. In what direction will gear D on the output axle turn? Which gear, A or D, will move with greater speed?

2. Suppose the handle is pushed to the left so that gear B now engages gear E. Again suppose that the handle is turned counterclockwise at the same speed. How does the speed of the output axle now compare with its speed in question 1? Why?

3. Now suppose that the handle is pushed farther to the left so that gear C engages gear F. As the handle is turned counterclockwise, in what direction will gear G and the output axle move?

4. What pair or group of gears in the model corresponds to low gear? high gear? reverse gear? (In a real car there would be an additional set of gears between the gearbox and the wheels.) Which gear pair gives more force to the forward motion of the car? Which gear pair gives more speed to the car's forward motion?

Mechanical Systems

Most machines that you see and use every day could be properly called *mechanical systems*. This is because they are made up of two or more simpler machines working together. Look back at the mechanical systems depicted throughout this unit. What simple machines do you recognize?

A can opener is one example of a mechanical system that contains several subsystems. Try to identify the following subsystems:

a. a lever that operates like a nut-cracker

b. a kind of gear

c. a wheel and axle

d. a wedge like the blade of an ax

You remember that a mechanical system is designed to do a single task and is made up of groups of parts called *subsystems*. What function does each part of the can-opener system perform? Can you group these four parts into two subsystems? What has been added to one part so that it can serve an additional purpose?

How do you operate the kind of can opener illustrated here? Match the numbered parts in this photo with the lettered subsystems listed above. What simple steps of operation would you describe for someone else to follow? Write them in language that could be understood by an 8-year-old.

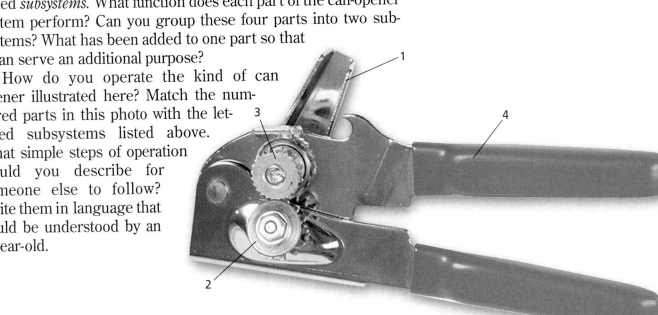

Analyzing Mechanical Systems

Spend a few minutes at each of the following stations. Record all your answers in your ScienceLog.

STATION A

The Write Stuff

1. Check the operation of the pen. How does it write? Is it retractable? Is it refillable? How?

2. Take apart the pen to see its component parts.

3. How do the parts operate? Sketching the parts in operation may help you explain this.

4. Examine another type of pen. How is it similar? How is it different? Is it a better design? Why or why not?

STATION B

A Fasten-ating Device

1. Staple two pieces of paper together. As you do this, carefully observe each aspect of the stapler's operation.

2. Open the stapler and identify each main part. What is the function of each part?

3. Compare a staple before and after stapling. What causes the staple to bend? Could it be bent in another direction? How could you use the stapler without bending the staples?

4. How do you think a stapler like this compares with one used by carpenters or carpet installers?

To learn more about the simple machines in these mechanical systems, read pages S47–S51 in your SourceBook.

STATION C

What a Grind!

1. How does this pencil sharpener operate?

2. Identify all of the pencil sharpener's subsystems. Which of these subsystems are simple machines that you studied earlier?

3. How is this system a more effective design than a hand-held sharpener?

4. Suggest a design for a pencil sharpener that is not powered by hand.

STATION D

Wheel You Look at That!

1. Examine a bicycle closely. Observe how each part and subsystem operates.

2. Where do you apply energy to the bicycle? How is the energy transferred?

3. Compare the mechanical systems of different makes and models of bicycles.

STATION E

A Hole-some Machine

1. What is the overall function of a hand drill? How does it work?

2. Identify at least four simple subsystems in this mechanical system.

3. Describe the speed of the drill compared with that of the handle. What accounts for this?

4. How is the drill powered? In what other ways could it be powered?

One Purpose— Many Designs

The can opener on page 193 is just one device for opening cans. Other devices serve the same purpose.

1. What designs of can openers have you used or seen?

2. Various designs may have different features. What are they?

3. Which designs work best? Which are most practical?

Collect as many can-opening devices as you can. Examine each closely. Identify subsystems and any features that make them useful and effective.

Constructing Your Own Mechanical System

The following Exploration gives you an opportunity to develop a traveling mechanical system. You will design a system and test it.

Constructing a Model Car

Auto manufacturers are researching different energy sources to power the cars of the future. In this Exploration, you will construct your own car and power it with an unusual alternative energy source. Here is your challenge. Design a model car powered by a rubber band, and construct a prototype of the new car. The car must be able to move in a straight line for a distance of at least 5 m. (Slingshot propulsion is not allowed.)

You have been presented with a problem. How are you going to solve it? First review the steps in Technological Problem Solving on page 183. The following suggestions may also help you.

What You Might Want to Use

- rubber bands
- masking tape
- plastic-foam or paper cups
- cup lids
- drinking straws
- index cards
- cardboard tubes
- wooden dowels
- a drawing compass
- straight pins
- a pencil
- corks
- pliers
- scissors
- paper
- paper clips
- white glue
- thumbtacks
- anything else you can think of

What to Do

1. Record your work in your ScienceLog. Note what you do, why you do it, what problems you encounter, and how you solve the problems.

2. Work with one or two others. Begin by discussing the project. Try to think of as many ideas for solving the problem as you can. Sketch them to help you refine the concepts.

3. Review all of your possible designs. Choose the best one and start to work. Decide what materials to use and how large the car should be. Draw a blueprint of your car that shows exact details, identifies subsystems and individual parts, and indicates the scale. (A blueprint should be detailed enough so that someone could use it to build what the blueprint represents.)

4. Make a list of the materials you will need. Collect the parts needed.

5. Build, test, and evaluate your model. Use the results of your tests to make improvements.

6. Give your product a name. Prepare a promotional brochure designed to convince people to buy the product. Be sure to include technical data in your brochure.

Examining One Model Car Design

Jim Louviere, a science teacher, constructed an "air car" that worked surprisingly well. In order to make it very light, he used nothing but paper, index cards, rubber bands, and paper clips. Examine the diagram of the assembled model below. Discuss the following questions with two other students.

1. How does Jim's model car work? How do you get it to move?
2. What is the source of energy that powers the car?
3. What energy changes take place during its operation?
4. What force moves the car?
5. What causes the car to slow down and eventually stop?
6. How might you measure the energy input into the car?
7. What variables might be important in getting the car to work? You might consider the length of the body, the length of the axles, and other characteristics.

Extending Your Design

You may want to build a car of more durable materials and power it in other ways. What materials might you use? How would you power the car? Draw a blueprint of your design. You might construct this car as a science project, test it, modify it, and evaluate it.

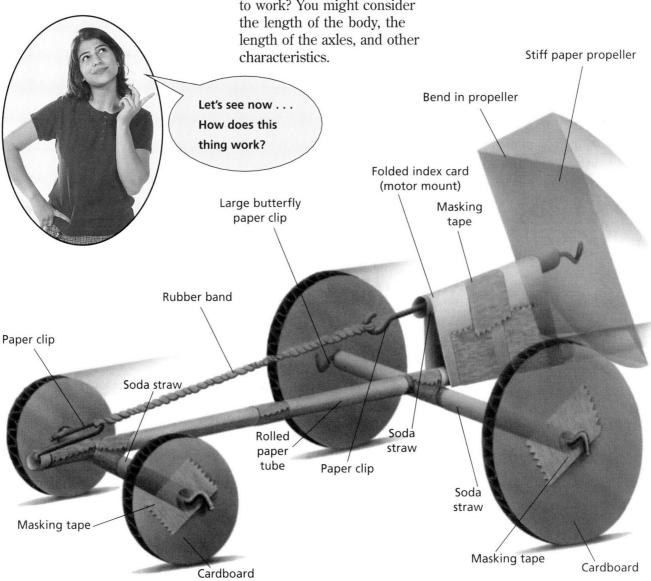

Let's see now . . . How does this thing work?

Stiff paper propeller

Bend in propeller

Folded index card (motor mount)

Masking tape

Large butterfly paper clip

Rubber band

Paper clip

Soda straw

Rolled paper tube

Soda straw

Paper clip

Soda straw

Masking tape

Cardboard

Masking tape

Cardboard

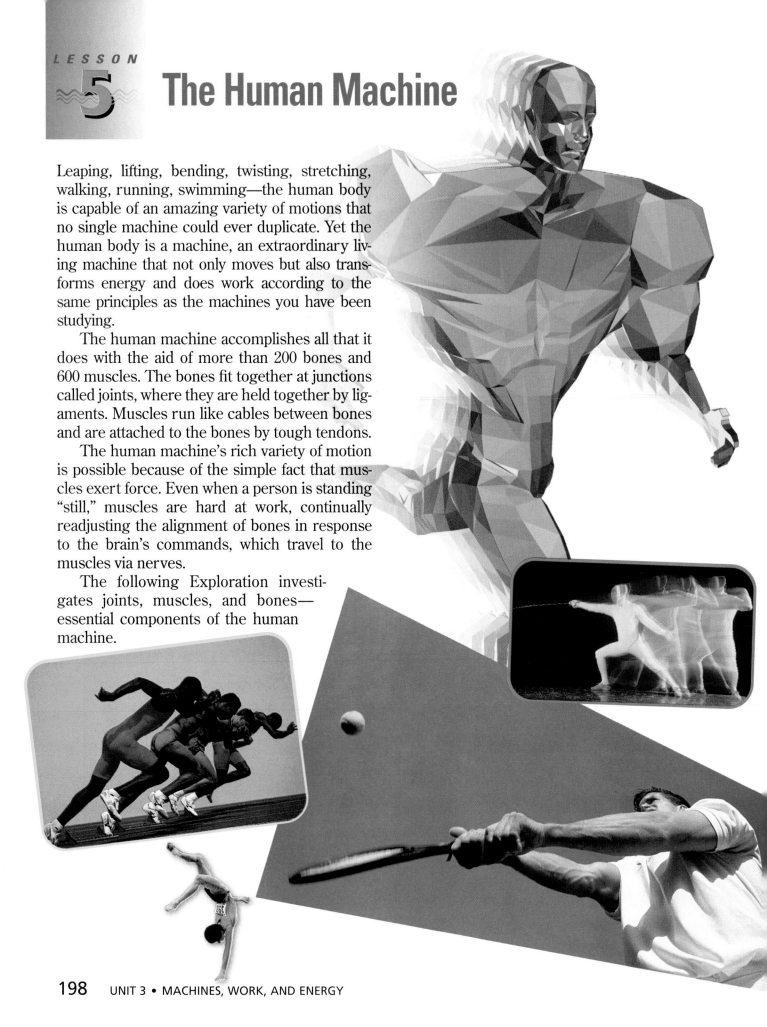

The Human Machine

Leaping, lifting, bending, twisting, stretching, walking, running, swimming—the human body is capable of an amazing variety of motions that no single machine could ever duplicate. Yet the human body is a machine, an extraordinary living machine that not only moves but also transforms energy and does work according to the same principles as the machines you have been studying.

The human machine accomplishes all that it does with the aid of more than 200 bones and 600 muscles. The bones fit together at junctions called joints, where they are held together by ligaments. Muscles run like cables between bones and are attached to the bones by tough tendons.

The human machine's rich variety of motion is possible because of the simple fact that muscles exert force. Even when a person is standing "still," muscles are hard at work, continually readjusting the alignment of bones in response to the brain's commands, which travel to the muscles via nerves.

The following Exploration investigates joints, muscles, and bones—essential components of the human machine.

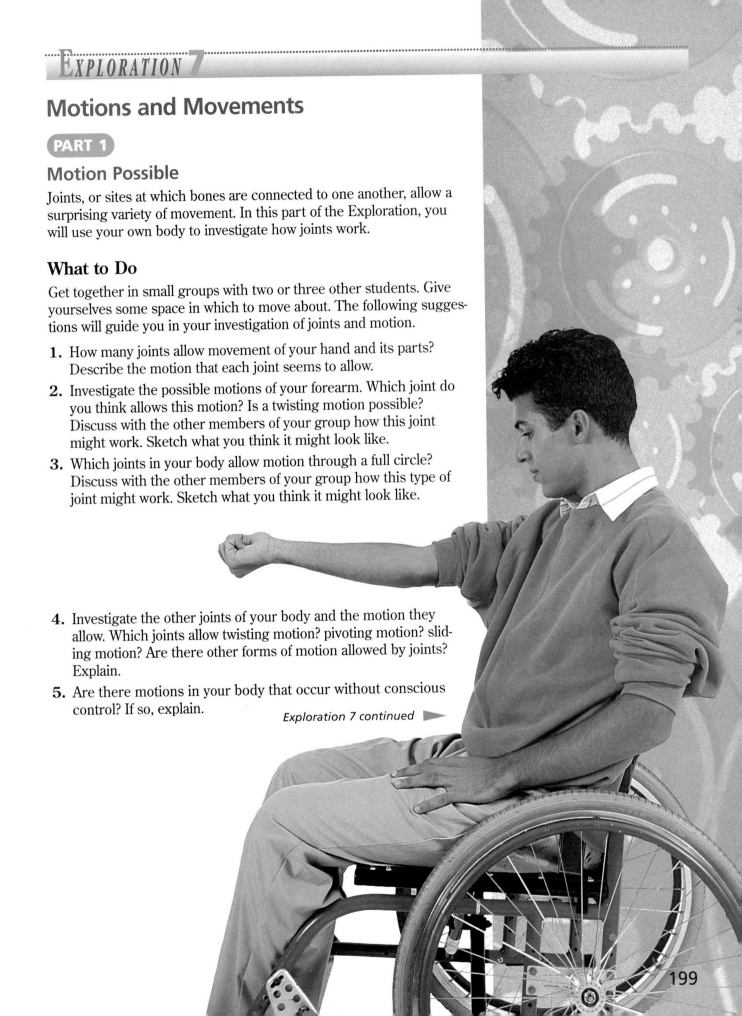

EXPLORATION 7

Motions and Movements

PART 1

Motion Possible

Joints, or sites at which bones are connected to one another, allow a surprising variety of movement. In this part of the Exploration, you will use your own body to investigate how joints work.

What to Do

Get together in small groups with two or three other students. Give yourselves some space in which to move about. The following suggestions will guide you in your investigation of joints and motion.

1. How many joints allow movement of your hand and its parts? Describe the motion that each joint seems to allow.

2. Investigate the possible motions of your forearm. Which joint do you think allows this motion? Is a twisting motion possible? Discuss with the other members of your group how this joint might work. Sketch what you think it might look like.

3. Which joints in your body allow motion through a full circle? Discuss with the other members of your group how this type of joint might work. Sketch what you think it might look like.

4. Investigate the other joints of your body and the motion they allow. Which joints allow twisting motion? pivoting motion? sliding motion? Are there other forms of motion allowed by joints? Explain.

5. Are there motions in your body that occur without conscious control? If so, explain.

Exploration 7 continued ▶

PART 2

Just Wing It

In this activity, you will look at actual bones and muscles from a chicken to help you understand how they work together.

You Will Need

- a fresh chicken wing
- paper towels
- small scissors
- a plastic bag
- a scalpel

What to Do

1. With your scissors, peel back the skin from a chicken wing, as shown in illustration (a). Identify the transparent skin-like *connective tissue* surrounding the muscle bundles and bone. (You may have to cut some of it away to see the muscles better.) Notice how the muscles are arranged in pairs on opposite sides of the bones.
2. Try squeezing each muscle of the pair, in turn, to see how it moves the end of the chicken wing. Muscles always work in *opposing pairs* to cause motion.
3. Carefully separate the tissue. Identify the tiny white *nerves* that activate the muscles, and identify the blood vessels that bring oxygen and nutrients to the muscles. Identify the durable, white tendons that connect the muscle to the bone.
4. Cut away the remaining tissue to expose a joint. Look at the joint. What holds the bones together? Work the joint back and forth, as shown in illustration (b). Is the motion smooth or do you feel a lot of friction? What accounts for this?

Summarize your observations and write any questions you may have in your ScienceLog. Enclose the remains of the chicken wing in a plastic bag for disposal, and wash your hands thoroughly with soap and water.

Questions

1. What purpose do you think the transparent tissue surrounding the muscle bundles might serve?
2. What do you think is the significance of the arrangement of muscles in opposing pairs?
3. Which of the tissues you examined might carry signals from the chicken's brain?
4. In what ways do tendon tissue and muscle tissue differ? How do these differences reflect the functions of each type of tissue?

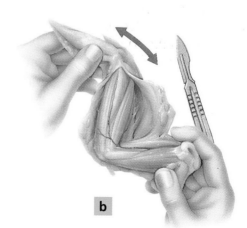

a

b

Feel the Force

You may think that a chicken and a human aren't very similar. But in fact, they contain similar structures. And their muscles, bones, and joints work in almost exactly the same ways. In this activity you will find out more about opposing muscle pairs and how they work, using the human body as an example.

What to Do

1. While seated, grasp a fixed object such as a desk. Have a partner lightly grasp your upper arm so that he or she can feel the action of your muscles. Pull upward as though to lift the desk. (Make sure the desk is secured to the floor.) Report your observations and those of your partner.

2. Without changing your grip on the desk, pull downward on it. Your partner should continue to observe your muscle action. Report your observations and those of your partner.

3. Switch roles with your partner and repeat the experiment.

4. Read the passage that follows and answer the questions it poses.

5. Summarize your findings in your ScienceLog.

A Lever With Dual Controls

You may not think of your limbs as levers, but in fact, they are. Look at the diagram at right. Like all levers, the arm has a fulcrum (pivot point), a load, and an applied force. Identify each of these. Locate the *biceps* muscle on the upper side of the arm. What happens when this muscle contracts? Now locate the *triceps* muscle on the underside of the arm. What happens when this muscle contracts? What has to happen to the biceps muscle before the triceps muscle can do its job? What do you think the triceps does when the biceps contracts?

You might find it interesting to note that muscles never completely relax. They always exert a little tension. This is called *muscle tone*.

Think again about the lever action of the forearm. Is force or distance multiplied? Explain.

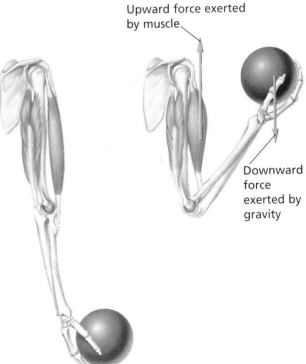

Upward force exerted by muscle

Downward force exerted by gravity

CHALLENGE YOUR THINKING

1. Zach's Jack

Zach suggested another solution to Tony's Problem on page 170. He tried one of the stabilizer jacks used for his family's travel trailer. The measurements he made are shown in his diagram. Zach discovered that when he turned the lever handle through one complete turn, the load was raised a distance equal to the distance between two ridges on the screw bolt.

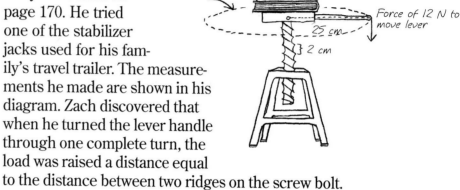

a. What was Zach's work input for one turn of the lever?

b. How much of Zach's work was lost to friction?

2. You Be the Judge

Which of the three machines is the best—Zach's jack above or one of the two machines below? The diagrams and questions below will help you decide.

a. Which machine would be the easiest to use?

b. Which causes you to work through the greatest distance?

c. Which is the most efficient?

d. Which loses the most energy to friction?

e. Which would be the fastest to use?

Based on your answers to the questions above, which do you think is the best machine?

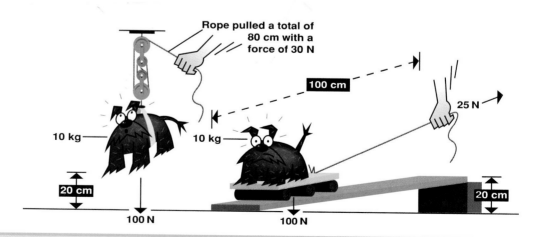

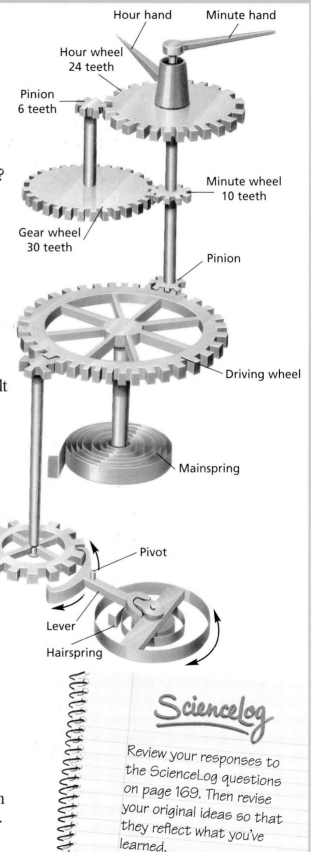

3. Time for a Real Challenge

On the right is a simplified diagram of a common mechanical clock. Use the labels and diagram to answer the following questions:

a. How does this device work? (Hint: Start at the mainspring.)

b. Where does the energy to power the device come from? How is it transferred?

c. Why do the hour and minute hands move at different speeds?

d. The hairspring and balance rock back and forth to mark off the seconds. How does this work?

4. Precision Decision

The distributor in a car engine sends a pulse of electric current to each of the spark plugs. It must turn at *exactly* half the speed of the engine. Would gear pairs or a wheel-and-belt system work better? Why?

5. Ship-Shape Machine

The diagram below shows a device for powering a model boat.

a. How does it work?

b. What is the source of energy?

c. How is the energy transferred?

d. How is friction reduced?

e. Is there a trade-off taking place here? If so, what is it?

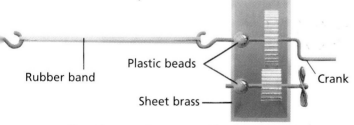

6. Your Invention

Design a mechanical system to perform some task for you. The system must contain three subsystems that are simple machines. Be creative!

ScienceLog

Review your responses to the ScienceLog questions on page 169. Then revise your original ideas so that they reflect what you've learned.

Unit 3

SOURCEBOOK

To find out more about machines and how they make our lives easier, look in the SourceBook. There you will find more information about force, work, and energy and about how even the most complex machines actually consist of many simple machines.

Here's what you'll find in the SourceBook:

The Big Ideas

In your ScienceLog, write a summary of this unit, using the following questions as a guide:

1. What is *work* in a scientific sense? How is it related to energy?
2. How is work measured?
3. How do you distinguish between potential and kinetic energy?
4. How are simple machines and mechanical systems related?
5. How is energy transferred in mechanical systems?
6. Do machines actually save work? Explain.
7. What is efficiency? How is efficiency related to friction?
8. How is energy *conserved* in machines?

Checking Your Understanding

1. Do you agree or disagree with the following statements? Explain your answers.
   a. A machine is a device for converting or transferring energy.
   b. If work is being done, then energy is being converted.
   c. Work is a form of energy.

2. Study the cartoon below. Identify the error or errors in each panel of the cartoon.

3. Here is an interesting collection of gears.
   a. Trace the motion of the components.
   b. Where does the device change the direction of motion?
   c. Where does a trade-off take place?
   d. Which way does energy probably flow through this system? Explain.
   e. What might you use this system for?

4. **concept map** Make a concept map using the following terms or phrases: mechanical systems, work, energy, machines, and simple machines.

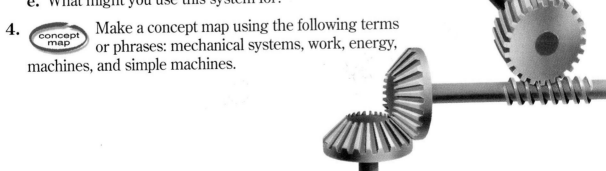

Micromachines

The technology of making things smaller and smaller keeps growing and growing. Powerful computers can now be held in the palm of your hand. But what about motors smaller than grains of pepper? Or gnat-sized robots than can swim through the bloodstream? These are just a couple of the possibilities for micromachines.

Minuscule Motors

Researchers have already built gears, motors, and other devices so small that you could accidentally inhale one! For example, one engineer devised a motor so small that five of the motors would fit on the period at the end of this sentence. This micromotor is powered by static electricity instead of electric current, and it spins at 15,000 revolutions per minute. This is about twice as fast as most automobile engines running at top speed.

▲ The earliest working micromachine had a turning central rotor.

Small Sensors

So far micromachines have been most useful as sensing devices. Micromechanical sensors can go in places too small for ordinary instruments. For example, blood-pressure sensors can fit inside blood vessels and can detect minute changes in blood pressure within a person's body. Each sensor has a patch so thin that it bends when the pressure changes.

Cell-Sized Robots

Some scientists are investigating the possibility of creating cell-sized machines called nanobots. These tiny robots may have many uses in medicine. For instance, if nanobots could be injected into a person's bloodstream, they might be used to destroy disease-causing organisms such as viruses and bacteria. They might also be used to count blood cells or to deliver medicine.

The ultimate in micromachines would be machines created from individual atoms and molecules. Although these machines do not currently exist, scientists are already able to manipulate single atoms and molecules. For example, the "molecular man" shown here is made of individual molecules. These molecules are moved by using a *scanning tunneling microscope*.

▼ "Molecular man," drawn with 28 carbon monoxide molecules

A Nanobot's "Life"

Imagine that you are a nanobot traveling through a person's body. What types of things do you think you would see? What type of work could you do? Write a story that describes what your experiences as a nanobot might be like.

Wheelchair Innovators

Two recent inventions have dramatically improved the technology of wheelchairs. With these new inventions, some wheelchair riders can control their chairs with voice commands, and others can take a cruise over a sandy beach.

Voice-Command Wheelchair

At age 27, Martine Kemph invented a voice-recognition system that allows people without arms or legs to use spoken commands to operate motorized wheelchairs. Here's how it works: The voice-recognition computer translates spoken words into digital commands, which are directed to electric motors. These commands completely control the operating speed and direction of the motors, giving the operator total control over the chair's movement.

Kemph's system can execute spoken commands almost instantly, much faster than previous voice-recognition systems. In addition, the system is easily programmed, so each user can tailor the computer's list of commands to his or her individual needs.

Kemph named the computer Katalvox, using the root words *katal*, which is Greek for "to understand," and *vox*, which is Latin for "voice." Katalvox also has great potential for use in surgery, on assembly lines, and in automobiles.

The Surf Chair

Mike Hensler was a lifeguard at Daytona Beach, Florida, when he realized that it was next to impossible for someone in a wheelchair to come onto the beach. Although he had never invented a machine before, Hensler decided to build a wheelchair that could be driven across sand without getting stuck. He began spending many evenings in his driveway with a pile of lawn-chair parts, designing the chair by trial and error.

The result of Hensler's efforts looks very different from a conventional wheelchair. With huge rubber wheels and a thick frame

▲ **Mike Hensler tries out his Surf Chair. People have found the Surf Chair to be practical, fun, and comfortable.**

of white PVC pipe, the Surf Chair not only moves easily over sandy terrain, but also is weather resistant and easy to clean. The newest models of the Surf Chair come with optional attachments, such as a variety of umbrellas, detachable armrests and footrests, and even places to attach fishing rods.

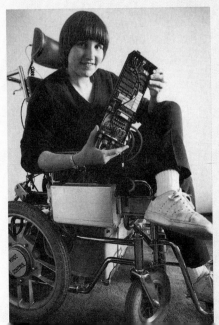

◀ **The voice-controlled wheelchair, pictured with its inventor, Martine Kemph, provides more freedom to people who are unable to use their arms or legs.**

Design One Yourself

Can you think of any other ways to improve wheelchairs? Think about it and put your ideas down on paper. To inspire creative thinking, consider how a wheelchair could be made lighter, faster, safer, or easier to maneuver.

Unbelievable Bicycles

▲ This bicycle incorporates Bill Becoat's two-wheel-drive mechanism. A drive cable transmits power from the rear wheel to the front wheel.

Imagine riding a bicycle that can go more than 104 km/h! Or perhaps you'd like a bike frame that weighs only 1.4 kg. How about a bike that you could ride on snow? These aren't bikes from the future—they're being built today. Read on to find out about ways in which bicycles are being made faster, lighter, and easier to handle than ever before.

Light as a Feather

When you ride a bicycle, you are moving not only your own mass, but also the mass of the bike. The less a bike weighs, the less work you have to do.

▲ This bicycle frame was made with MMCs. It weighs only 1.4 kg.

New, lightweight materials, called metal matrix composites, or MMCs, are now being used to make lighter bicycle frames. MMCs consist of metals that are strengthened and stiffened with nonmetal fibers. MMCs weigh the same as the metals they are made of, but because MMCs are stronger, much less material is needed. Some bicycle frames built with MMCs weigh less than 1 kg.

Through Rain and Sleet and Snow

Riding on slippery surfaces can be both difficult and dangerous. One of the problems is that power is sent only to the back wheel. If the back wheel slips or slides, you lose forward motion.

Inventor Bill Becoat has solved this problem by creating a bike in which both wheels are connected to the drive train. If one wheel slips or slides, the other wheel still receives force from the rider to keep the bike moving.

As Fast as the Wind

Want a faster bike? Maybe you should try lying down! As strange as it may seem, some of the world's fastest bikes are designed to be ridden by someone leaning back or lying down. These positions help reduce air resistance. They also allow the rider to generate more power when pedaling.

◄ An innovative tandem bicycle

The Ideal Bike

What's your dream bike? Create a list of at least 10 features that your ideal bike would have. Try to draw a diagram of a bike that has these features.

Machines as Art

*W*hat do you consider a piece of art—a painting, a sculpture, or a photograph? How about a blender, a fork, or a pencil sharpener? In fact, blenders, forks, and pencil sharpeners are on display in art museums around the world, as are countless other everyday items and machines that you might not normally think of as art. The people who create these pieces are known as industrial designers. Industrial designers work to make machine-made products, such as small appliances, tools, furniture, and vehicles, both useful and beautiful.

◄ The Waring Blender®, introduced over 50 years ago, combined form and function with an elegant design.

The Introduction of Industrial Design

Until the late 1800s, many machine-made items were simply functional. What a particular item looked like was much less important than what it did. Then, in 1919, German architect Walter Gropius founded a school of architecture and design called the Bauhaus. The Bauhaus became a center for artists and architects who combined sharp, clean, geometric styles with new industrial materials and techniques. Although the school closed in the early 1930s, it changed the way that people viewed machine-made items.

Form and Function

After the introduction of industrial design, manufacturers soon realized that more people might buy an item if it looked appealing. Companies began hiring industrial designers to create products that were functional and that had a graceful shape, color, proportion, and texture. Sometimes the results were less than appealing. For example, following World War I, many products had a number of unnecessary decorations and were odd mixes of styles. At other times, however, the design served to underscore and enhance the function of an item. Many of these latter items were pleasing enough to be

▼ The Tizio table lamp was created by Richard Sapper in 1972. It is now part of the permanent collection of The Museum of Modern Art.

considered works of art. In 1934, the Museum of Modern Art decided to exhibit its first "machine art" collection. This exhibit included objects such as glassware, gears, propellers, springs, pans, valves, irons, axes, racks, and ball bearings.

Today, industrial designers not only design useful and beautiful machines and products, but also work with specialists to ensure that their designs are appropriate for human use.

Create Your Own Machine Art

Find a simple, common machine in your house or at school. Think about ways in which you could change the machine to make it more visually pleasing without changing how it works. Draw a poster that shows the original machine as well as the new machine you designed.

Unit 4

OCEANS AND CLIMATES

A cyclonic storm in the North Pacific as photographed from the *Apollo 9* spacecraft

giant cyclonic storm rumbles westward across the North Pacific. Driven by the tremendous heat stored in the ocean, cyclonic storms transfer thermal energy from the ocean to the atmosphere on an enormous scale. As the storm sweeps across the North Pacific, it will continue to pick up energy from the warm ocean water. At its peak, the storm may develop winds in excess of 160 km/h and generate huge waves that can travel thousands of kilometers. In the wintertime, when there is a sharp temperature contrast between ocean and air, storms may form one right after another over the North Pacific.

Like hurricanes, which originate in the tropics, cyclonic storms have a distinctive spiral shape. In the Northern Hemisphere, storms spiral counterclockwise. In the Southern Hemisphere, they spiral clockwise. What could cause such a phenomenon?

Beyond the curve of the horizon looms the black emptiness of space. A more startling contrast would be hard to imagine: the living, ever-changing planet below, and the hostile, empty void above. Out of view 150 million kilometers away is the sun, the source of energy that drives the Earth's weather "engine" and makes life on Earth possible. In this unit you will learn about the ways in which the oceans and atmosphere interact to make Earth a unique oasis of life in the void of space.

Mars

Earth

Mercury

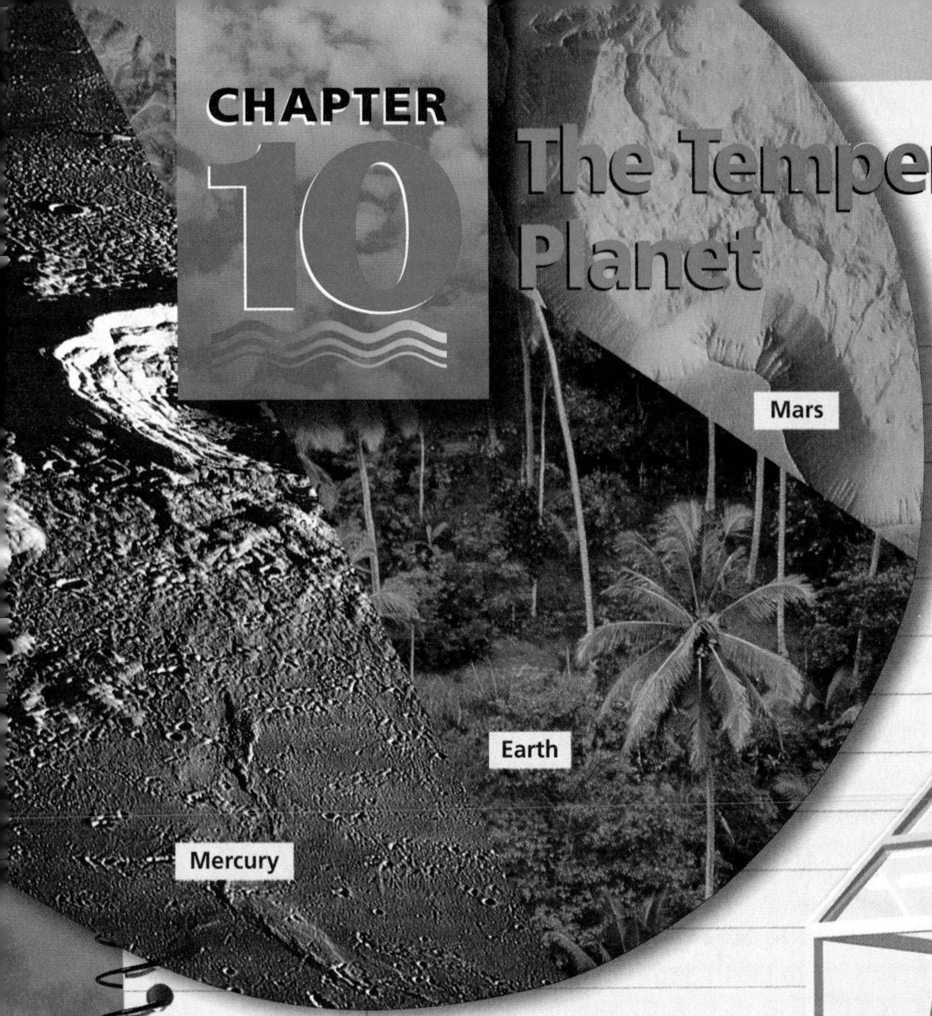

1 **Why is our planet probably the only place in the solar system where life exists?**

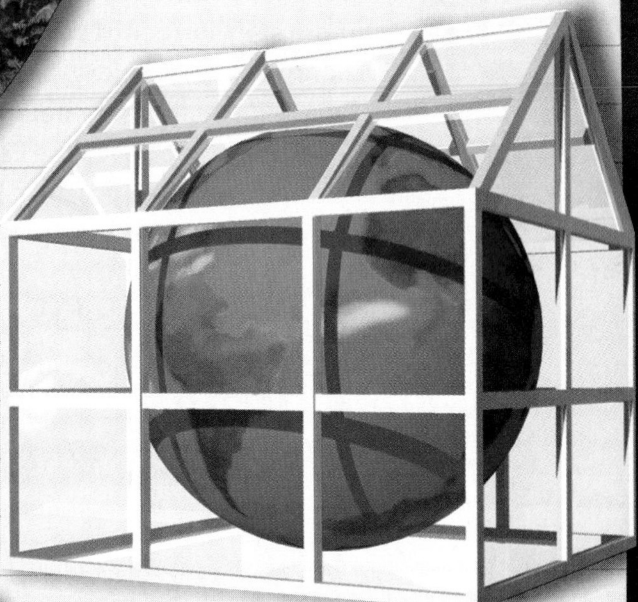

How is the Earth's 2 atmosphere like a greenhouse?

3 **Why does dew form on grass and other objects during the night?**

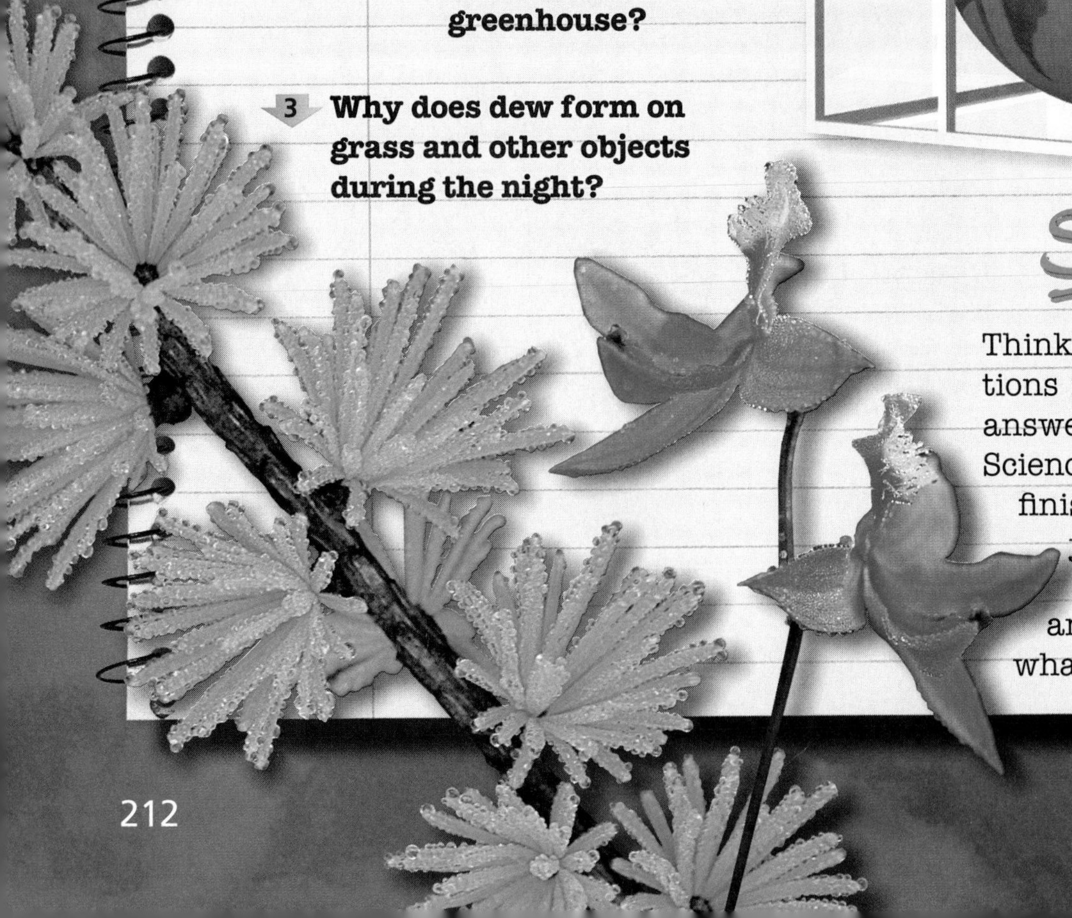

ScienceLog

Think about these questions for a moment, and answer them in your ScienceLog. When you've finished this chapter, you'll have the opportunity to revise your answers based on what you've learned.

An Oasis in Space

You know that the Earth has many different climates—some cold, some warm, some in between. But even when you consider the extremes of its hottest and coldest places, Earth is still a very pleasant place compared with the other planets and satellites in our solar system.

Take, for example, the moon—our nearest neighbor. The moon's surface temperature soars to above 130°C during the day and plummets to below –170°C at night. Does the temperature in Eureka, California, change that much from day to night? over the whole year? How about Lincoln, Nebraska? How about where you live? Do you think any place on Earth has such a wide range of temperatures? What would explain the very large range of temperatures that the moon experiences? Why does Earth experience a more favorable climate?

Earth is unique. It has an average temperature that allows living things to flourish. Why is this so? Is Earth fortunate enough to be just the right distance from the sun, or are there other factors that contribute to its temperate climate?

Examine the data regarding Earth and its closest neighbors on the pages that follow. Compare these planets. How do their average temperatures vary? What factors seem to determine the temperature ranges of the planets?

One group of students made the suggestions that follow. Can you find data to support each claim?

"Distance from the sun is definitely the most important factor in determining how hot or cold a planet is."

"Another factor is whether or not the planet has an atmosphere."

"I think the kinds of gases that make up the atmosphere affect a planet's temperature."

"Another important thing is how dense or thick a planet's atmosphere is."

What other factors might affect the climate of a planet? Check out the astrocorder readings on pages 214 and 215 for some clues.

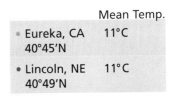

Mean Temp.	
• Eureka, CA 40°45'N	11°C
• Lincoln, NE 40°49'N	11°C

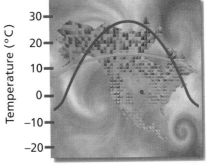

Despite the extreme conditions, an afternoon in the Sahara followed by an evening in Antarctica would be pleasant compared with a day and night on the moon.

Although such devices don't actually exist, these astrocorders provide a quick guide to the four inner planets of the solar system.

MERCURY

Average distance from sun: 58 million km

Average surface temperature: 450°C, day; −173°C, night

Atmosphere: none

Atmospheric pressure: N/A

Water: none

Life: none

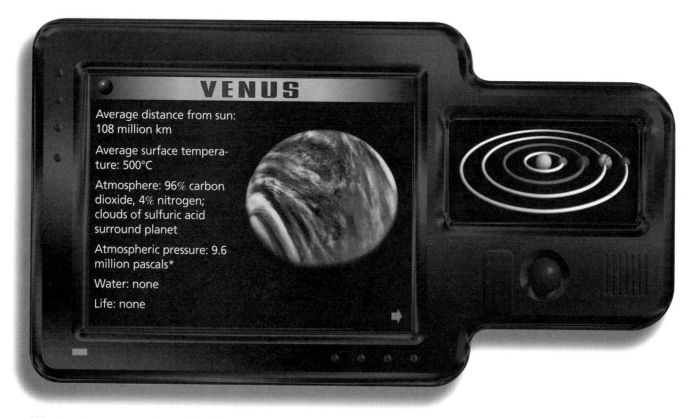

VENUS

Average distance from sun: 108 million km

Average surface temperature: 500°C

Atmosphere: 96% carbon dioxide, 4% nitrogen; clouds of sulfuric acid surround planet

Atmospheric pressure: 9.6 million pascals*

Water: none

Life: none

*Newtons per square meter, a unit of pressure

EARTH

Average distance from sun:
150 million km

Average surface temperature: 15°C

Atmosphere: 77% nitrogen,
21% oxygen, 1% argon,
0%–4% water vapor (variable),
0.03% carbon dioxide; traces
of other gases

Atmospheric pressure:
101,000 pascals

Water: abundant; large oceans,
polar icecaps

Life: extremely abundant and varied

MARS

Average distance from sun:
228 million km

Average surface temperature:
−55°C

Atmosphere: 95% carbon
dioxide, 3% nitrogen

Atmospheric pressure: 1500
pascals

Water: some contained in
polar icecaps and subsurface
permafrost

Life: none detected

Understanding Our Planet

Did you conclude that a planet's temperature is determined by more than just its distance from the sun? What other factors play a role? What is the evidence?

In the following Exploration, you will learn about the heating of the Earth by setting up a simulation. As you learned in Unit 2, a simulation is a type of experiment intended to model actual conditions. Scientists often use simulations to study complex problems. In Exploration 1, a glass jar will represent the Earth and a lamp will represent the sun. If the weather is clear, you can do the simulation using the sun instead of a lamp.

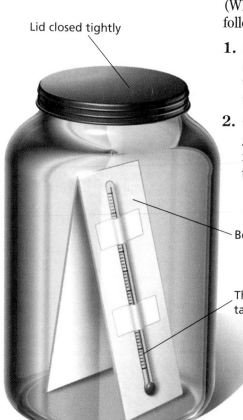

Lid closed tightly

Bent cardboard

Thermometer taped to card

A Simulation

You Will Need

- 2 thermometers
- 2 large glass jars with lids
- a lamp with a 100 W bulb (if not sunny)
- a thin strip of cardboard
- tape

What to Do

Set up two jars, each as shown in the diagram below. One jar is your *control*. The second jar represents Earth and its atmosphere. Your challenge is to modify the second jar in such a way as to cause a change in the temperature. Change anything you want, but change only one variable (such as the distance of the jar from the lamp) at a time. (Why is this necessary?) Now follow the steps below.

1. Get together with two or three other students. Discuss the variables that could influence the temperature in the jar.

2. Choose a variable that your group would like to test, and modify the jar to determine the effect of this variable.

3. Place both jars near the light bulb (or in the sun). Record the temperature changes in both jars.

   What did you do to make this a *controlled experiment*?

Conclusions

1. Did your modified jar reach a different temperature than the control did? Was the temperature higher or lower than that of the control? What do you think caused the difference?

2. Make a list of the different modifications that the class made to the simulation, as well as the results obtained with each. Which modifications had the greatest effect?

3. Did the temperatures in your jars stabilize after a time? Why did this happen? How is this similar to what happens on Earth?

4. What are the shortcomings of this model? Compile a list in your ScienceLog.

5. Suppose you add a layer of dark soil to the bottom of the jar. After 20 minutes the temperature in this jar is 1°C higher than the temperature in the control. What might you conclude?

6. Evidence suggests that the average temperature of the Earth has gone up by about 0.5°C during this century. What might be causing this change? Do the results of any of the simulations suggest possible answers? Explain.

LESSON 2

The Greenhouse Effect

From Today's Headlines

Have you seen headlines like those above? What do they imply? It seems that Earth is becoming a warmer place. What could possibly be causing this? Could the sun's energy output be increasing? Could Earth be moving closer to the sun? Could the warming be part of a natural cycle? (After all, throughout most of its history Earth had a much warmer climate than it has now.)

You have probably heard a great deal of talk about the *greenhouse effect*. Many people are concerned about the greenhouse effect and blame it for the apparent *global warming* trend. What is causing this greenhouse effect? Are people somehow responsible for it?

With all this talk about global warming, we need to find out more about it so that can we understand it and deal effectively with it.

Look at the graph below. This graph shows the estimated average global temperature over the last century. It also shows the average global temperature through the year 2040 as predicted by some scientists. After examining the graph, think about the questions below.

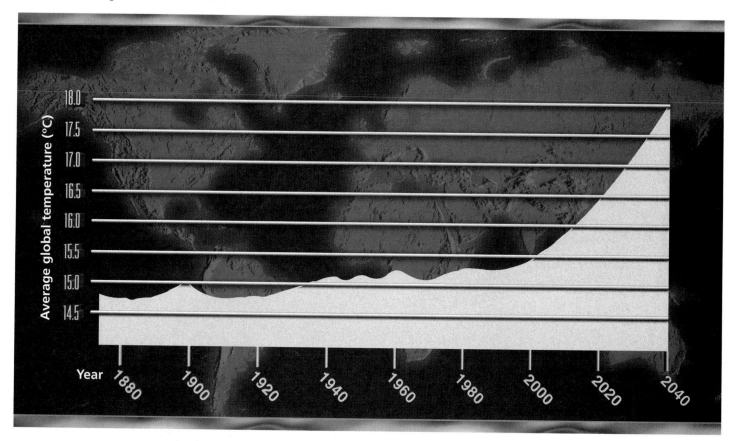

- Is the prediction realistic?
- If the prediction turns out to be right, what might the consequences be?
- How could the greenhouse effect be connected to global warming (if at all)?

Dateline: Earth 2050

Imagine that you could be transported to the year 2050. How might the headlines read then? Make up your own headline and article about the consequences of global warming. Will Earth be a very different and warmer place? Will the icecaps melt, causing the sea levels to rise? Will our temperate Earth become tropical? Perhaps you are skeptical and think that people are overreacting. If that's the case, your headline and article could reflect how all of the predictions about global warming turned out to be wrong.

The Story Behind the Greenhouse Effect

Without the greenhouse effect, Earth would be a cold, lifeless place.

Here's one headline you probably haven't seen. It suggests an aspect of the greenhouse effect that is rarely noted in news stories about the dangers facing our climate. As you will see, the greenhouse effect is something we all depend on more than we realize.

The term **greenhouse effect** was coined in 1822 by Jean Fourier, a French mathematician. Fourier noted that in many ways the Earth's atmosphere behaves like a greenhouse. The glass of a greenhouse lets solar radiation pass through, but it traps the heat given off by the ground and plants inside as they absorb the sun's energy. The glass walls and roof keep the warm air from blowing away. As a result, the greenhouse warms up. If you've ever been inside a greenhouse on a sunny but cold day, you are probably familiar with this effect. In Exploration 1, did the jar in your simulation behave like a greenhouse?

Earth's Special Case

Certain gases in our atmosphere—principally water vapor, carbon dioxide, and methane—are especially good at blocking the flow of heat from the Earth back into space. These gases together make up a small percentage of the atmosphere, but they have an enormous impact on Earth's climate.

The greenhouse effect is essential for maintaining Earth's moderate temperature. If not for the greenhouse effect, Earth's average temperature would be much colder, at least 33°C colder than it is now—too cold for life as we know it to exist.

Compared with Earth, Venus experiences a very strong greenhouse effect, but Mars experiences a weak greenhouse effect. Look back at the astrocorder readings on pages 214–215. How do you think differences in atmospheric composition and atmospheric pressure might account for the differences in the greenhouse effects on Earth, Venus, and Mars?

The Earth's atmosphere is responsible for the greenhouse effect, which I discovered. To find out more about the atmosphere's origin and composition, turn to pages S60–S65 in the SourceBook.

Jean Fourier
1768–1830

Are We Rewriting the Tale?

As you have seen, there is evidence to suggest that Earth is warming up. If this is the case, could human activities be responsible? Are we enhancing the greenhouse effect? (Consider that for at least the last 200 years, humans have had a significant impact on the environment.) Or could the warming trend be attributed to natural causes? Scientists are presently debating these questions.

Form some conclusions of your own. Examine the evidence in the following Exploration. If necessary, do additional research. Work with two or three other students to prepare a position paper presenting your views on the following questions:

- How does the greenhouse effect work?

- Is Earth, in fact, heating up? What evidence supports this interpretation?

- If there is a warming trend, could it be caused by increased amounts of carbon dioxide and other gases in the atmosphere? What evidence is there to support this?

- What else could account for the apparent warming? What evidence suggests this?

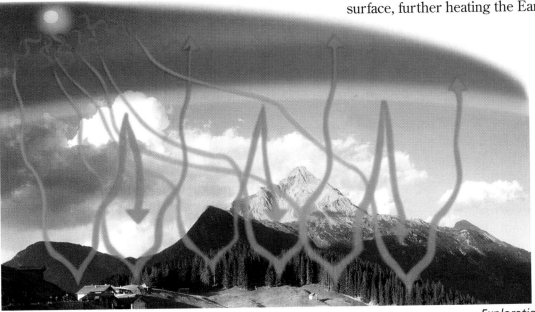

The greenhouse effect illustrated

EXPLORATION 2

Drawing Conclusions

PART 1

How Does the Greenhouse Effect Work?

Refer to the diagram below, which illustrates how greenhouse gases such as water vapor, carbon dioxide, and methane help heat the Earth. Make a rough sketch of this diagram in your ScienceLog and place the letter of each statement below in the appropriate location on the diagram.

a. Sunlight is made up of radiation of different wavelengths.

b. Most of the sun's ultraviolet light and other short-wavelength radiation is absorbed by ozone high in the atmosphere.

c. Radiation with longer wavelengths, such as visible light and infrared light, passes through the atmosphere without being absorbed.

d. This radiation is mostly absorbed by the Earth.

e. The absorbed energy is reradiated at even longer wavelengths (infrared).

f. Some of this radiation escapes back into space.

g. Some of the longer-wavelength radiation is absorbed by greenhouse gases, such as carbon dioxide, water vapor, and methane, thus trapping the energy.

h. Some of this energy is reradiated toward the surface, further heating the Earth.

Exploration 2 continued ▶

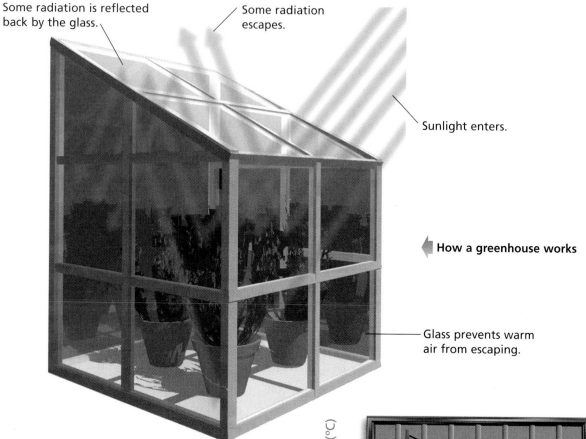

Some radiation is reflected back by the glass.

Some radiation escapes.

Sunlight enters.

How a greenhouse works

Glass prevents warm air from escaping.

An actual greenhouse works a little differently from Earth's "greenhouse." Examine the diagram above and think of some ways in which the two differ. Write down your ideas in your ScienceLog.

PART 2

Temperature Versus Carbon Dioxide Concentration

The concentration of carbon dioxide in the atmosphere is about 350 ppm (parts per million). What does this mean?

Imagine a container filled with a million grains of salt. If you were to replace one of these grains with a grain of pepper, then the concentration of the pepper in the container would be 1 ppm. Imagine replacing 350 salt grains with pepper grains. What would be the concentration of pepper in parts per million?

Using various methods, scientists have been able to estimate the Earth's average temperature, as well as the concentration of carbon dioxide in the atmosphere, over about the last 200,000 years. These are shown in the graph at right.

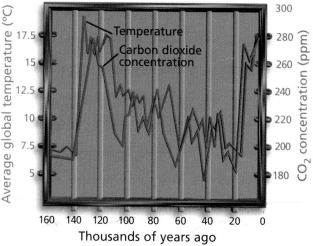

1. When was the carbon dioxide level greatest? When was the temperature highest?

2. Does there seem to be a connection between Earth's temperature and the carbon dioxide concentration in the atmosphere? Explain.

3. At what point during the past 160,000 years do you think there might have been an ice age?

4. Do you think all of the variations in the level of carbon dioxide were caused by human activities? Why or why not?

Carbon Dioxide and Temperature

The graph below contains two sets of data.
Use the data to answer the following questions:

1. What was Earth's average temperature in 1990? in 1890?

2. Does the graph appear to show a relationship between the concentration of carbon dioxide in the atmosphere and the Earth's average temperature?

3. Does the graph *prove* that the change in the atmospheric concentration of carbon dioxide has caused the average temperature to increase? Why or why not?

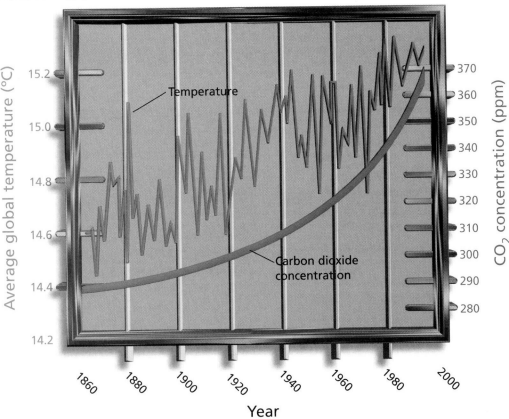

PART 4

Methane and Human Activity

Methane is another greenhouse gas. Methane is present in the atmosphere in smaller quantities than is carbon dioxide. However, a given mass of methane is much more effective as a greenhouse gas than is carbon dioxide. How is the level of methane changing in the atmosphere? Why is it changing? Is there a correlation between methane levels and human population? Using the data at right, prepare a graph in your ScienceLog to help illustrate any possible connection. Then answer the questions on the following page.

Year	Methane concentration (ppm)	Human population (billions)
1990	1.70	5.3
1980	1.50	4.2
1970	1.40	3.5
1960	1.30	3.0
1950	1.25	2.5
1940	1.15	2.0
1900	1.00	1.5
1850	0.85	1.2
1800	0.75	1.0
1750	0.74	0.8
1700	0.72	0.7
1650	0.70	0.6
1600	0.70	0.5

Exploration 2 continued

1. From your graph, predict the world's human population in the year 2000.

2. What will be the expected concentration of methane in the year 2000?

3. Listed below are some sources of methane. How is each related to human activities?

   - Cattle—They produce methane as they digest food.

   - Rice paddies—Bacteria in the flooded soil of rice paddies produce methane, which escapes into the air through the plants' hollow stems.

   - Decomposition—The decomposition of organic material in the absence of oxygen produces methane.

   - Natural gas (its main component is methane)—It leaks constantly from the Earth, from pipelines, and from production and processing facilities.

   - Termites—Methane is produced as termites digest wood. Termites are extremely abundant in the Earth's warm, moist regions.

   Return to the questions (on page 221) that preceded this Exploration. Use them to help you prepare your position paper on the greenhouse effect and global warming. What kind of additional data would be helpful? What other questions arose in the course of your study?

Sources of Carbon Dioxide

Carbon dioxide is perhaps the most important greenhouse gas that we add to the atmosphere. It is produced by a wide variety of natural processes and human activities, including the following:

- Burning—Any time organic material (material that contains carbon, such as gasoline) is burned, carbon dioxide is produced.

- Respiration—Carbon dioxide is a byproduct of respiration in both plants and animals.

- Decomposition of organic matter—Decomposition (in the presence of oxygen) is like slow-motion burning and, like all burning of organic matter, produces carbon dioxide.

- Volcanic eruptions—These spectacular events can release huge amounts, sometimes millions of tons, of carbon dioxide into the atmosphere.

Look at the photographs on this page. Which of these sources of carbon dioxide can people control? Which sources of carbon dioxide can people *not* control?

Global Warming—How Much?

Perhaps you are not yet convinced that global warming is actually occurring. If so, you are not alone. Many scientists disagree about global warming. Using sophisticated computer models, they try to predict the future behavior of the Earth's climate. Different models have produced very different results. Some scientific models predict a significant rise in global temperatures in the near future. Other models predict little or no warming. None of the models is exactly right. The question is, which model is the most accurate?

The accuracy of a scientific model depends on the reliability of the information used to formulate it. An accurate model of the causes and effects of global warming must answer the following questions. Can you think of other questions?

- Would a higher average temperature change the amount of cloud cover across the Earth? How would a change in cloud cover affect the average temperature of the Earth?

- What is the effect of light-reflecting dust and gases that have been pumped into the atmosphere by human activities and natural processes?

- What is the role of the oceans in absorbing and releasing heat energy and carbon dioxide?

- How would plants respond to a higher level of carbon dioxide in the air? How would other organisms, including people, respond?

If the Earth does warm up by a significant amount, what would be some of the consequences? Consider that a range of possible temperature increases (anywhere from 1.5°C to 5°C) has been predicted. With two other students, make a web of changes and effects that might occur. The web below might give you some ideas.

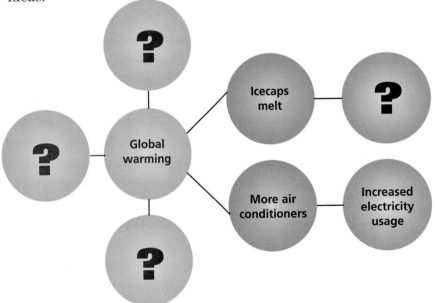

Increased carbon dioxide is absorbed by vegetation.

Warmer temperatures cause faster decay, resulting in faster release of carbon dioxide.

Land and Sea

The Earth's Thermostat

Look at the map on this page. This image, taken from space, shows surface temperatures across the Earth at the time the image was made. What does the map tell us about the heating processes of Earth? Find North America on the map. What is the temperature at your location? Where do you think the sun's rays are most direct? What does this data suggest about the role that the oceans play in distributing heat energy across the globe?

Examine the map to help you answer the following questions:

- Where are the hottest temperatures? the coldest temperatures?

- At what time of year was the image taken? How can you tell?

- Do all locations on land at the same latitude (degrees north or south of the equator) have more or less the same temperature?

- Do all locations on the oceans at the same latitude have more or less the same temperature?

- Does the map suggest that continents and oceans warm up and cool down differently? Explain.

How much of the sun's energy does the Earth absorb? To find out, turn to page S66 in the SourceBook.

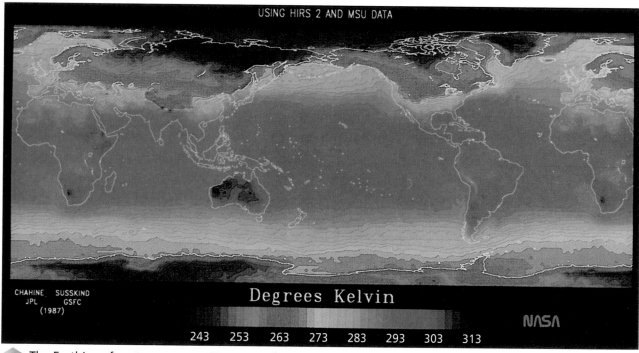

USING HIRS 2 AND MSU DATA

CHAHINE SUSSKIND
JPL GSFC
(1987)

Degrees Kelvin

NASA

243   253   263   273   283   293   303   313

The Earth's surface temperature. To convert from kelvins to degrees Celsius, subtract 273.

Heating of Land and Water

This Exploration is designed to help you answer the questions raised on the previous page. It will be a team effort. Each team will choose an activity, collect the materials, and do the experiment. Later you will give a report on your findings and interpretations. In doing so, make some reference to the photograph of global surface temperatures shown on the facing page. Graph your data to help with your presentation.

You Will Need

- 2 test tubes
- 2 thermometers
- a beaker
- a hot plate or other heat source
- modeling clay
- water
- sand
- aluminum pie pans
- a lamp with a 100 W (or greater) bulb
- charcoal powder
- a graduated cylinder
- test-tube tongs
- a test-tube rack

ACTIVITY 1

Sand Versus Water: Part 1

What to Do

1. Fill one test tube halfway with water. Fill another test tube halfway with sand.

2. Place a thermometer in the sand, and suspend another thermometer at the same depth in the water, using modeling clay to hold each thermometer in place. Place both test tubes in a beaker of hot water. Heat them to a temperature of about 70°C.

3. Remove the test tubes, and record the drop in temperature every 2 minutes for 20 minutes. Did they cool at the same rate?

ACTIVITY 2

Sand Versus Water: Part 2

What to Do

1. Fill one pie pan halfway with water, and fill another pie pan halfway with sand.

2. Place both pie pans an equal distance from the lamp.

3. Every 2 minutes for 20 minutes, record the surface temperature in each pie pan. The thermometer should be laid flat on the sand with the bulb covered by 0.5 cm of sand. The water temperature can be measured by holding a thermometer 0.5 cm below the surface. How did the temperatures change?

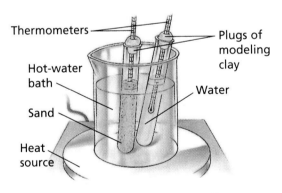

Thermometers — Plugs of modeling clay

Hot-water bath

Sand

Water

Heat source

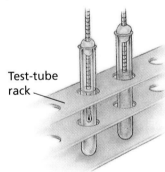

Test-tube rack

▲ Setup for Activity 1

Water

Sand

▲ Setup for Activity 2

Exploration 3 continued ▶

ACTIVITY 3

Heating of Different-Colored Sands

What to Do

1. Measure equal quantities of sand into two separate pie pans. Mix one quantity of sand with enough charcoal powder to make it black.

2. Lay a thermometer on top of the sand in each pan. Cover each thermometer bulb with 0.5 cm of sand.

3. Place the pans so that they are equal distances from the lamp.

4. Take readings every 2 minutes for 20 minutes. What differences did you observe?

ACTIVITY 4

Heating of Wet and Dry Sand

What to Do

1. Measure equal quantities of sand into two identical aluminum pie pans.

2. Add water to one so that the sand is quite damp. The water and sand should be the same temperature before they are combined.

3. Bury the bulb of each thermometer about 0.5 cm below the surface, one in each pan.

4. Place both pans an equal distance from the lamp.

5. Record the temperature of each pan every 2 minutes for 20 minutes. Which one warmed up faster?

Analyzing Activity 4

There is more to Activity 4 than first meets the eye. Certainly, water warms up more slowly than sand. However, there is another factor involved. One group carried Activity 4 a step further. At the end of the 20-minute observation period, they switched off the light and observed what happened. After 15 minutes they were surprised to find that the temperature of the wet sand had actually dropped below that of the room. What happened? Where did the heat energy go? The next Exploration will help you solve this mystery.

Analyzing and Reporting Your Findings

What were your findings? What do these findings suggest about the relative effects of water and land in terms of the heating processes of Earth? How does the image on page 226 relate to your findings? What further questions for investigation does this Activity suggest? Summarize your findings in a brief report.

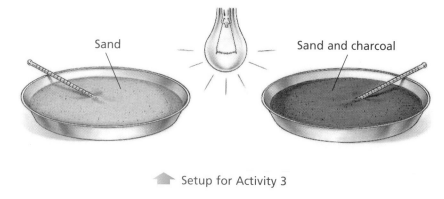

▲ Setup for Activity 3

▲ Setup for Activity 4

EXPLORATION 4

The Water Cycle and Heat Exchange

In Chapter 6 you learned how to make a wet-bulb thermometer. This type of thermometer is used to measure the amount of moisture in the air. In this Exploration you will review how a wet-bulb thermometer works and will discover how this relates to the water present in our atmosphere.

You Will Need

- a thermometer
- a piece of shoelace 5 cm long
- water (room temperature)
- water (warm)
- masking tape
- rubbing alcohol

PART 1

Making a Wet-Bulb Thermometer

Make a wet-bulb thermometer by putting the bulb of the thermometer into the shoelace as you would put your foot into a sock. Secure the shoelace to the thermometer with masking tape, and then read the temperature registered by the thermometer.

Now wet the shoelace and bulb with water that is at or above room temperature. Wave the thermometer in the air (or hold it in front of a fan) for a minute or so and then read the temperature again. What happened? What might have caused this to happen? (Recall what you learned in Exploration 4 of Chapter 6.)

PART 2

Feeling Cool

Dip your finger into warm water and then wave it in the air. What do you feel? What do you think happened to the water on your finger? Could this help to explain what happens with the wet-bulb thermometer? If you wet your finger with alcohol, will it feel cooler than when you wet it with water? Why or why not? Try it to find out!

What explanation do these activities suggest for why wet sand warmed more slowly than dry sand in the previous Exploration?

Insert the thermometer into the shoelace.

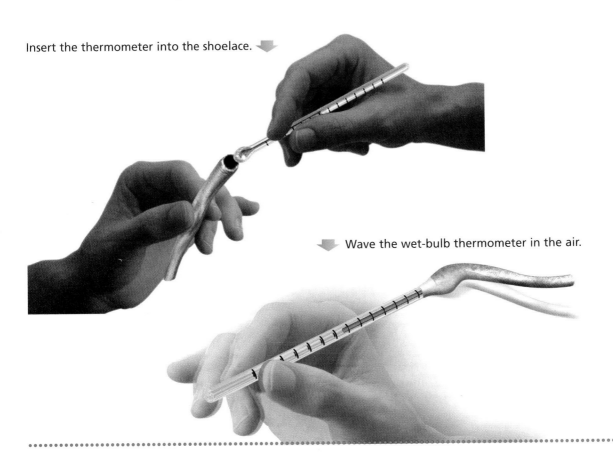

Wave the wet-bulb thermometer in the air.

Explaining Your Findings

Your wet finger feels cool. Heat in your finger must have been absorbed by the evaporating water. What happened to this energy? Did it disappear? No, that would be impossible. As you know, energy can never be created or destroyed, only converted from one form to another. What happened was that some of the heat in your finger and its surroundings was absorbed by the individual particles of water as they evaporated. If this evaporated water were to condense, that heat would be released to the surroundings once again.

Follow the exchange of heat energy in the diagram of the water cycle at right. Where is heat being absorbed from the surroundings? Will this make the surroundings warmer or cooler? Where is heat being released to the surroundings? Does this make the surroundings warmer or cooler?

Examine again the image showing global temperatures on page 226. Where would you expect to see the greatest amount of evaporation taking place? (Hint: Does more evaporation happen on hot days or cold days?) When would the energy stored in the water vapor be released again?

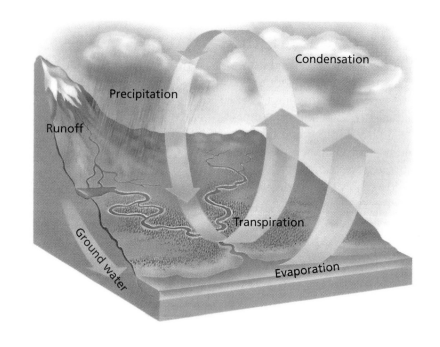

A Mini-Activity in a Bottle

Try this activity. Add some water to a large, plastic soft-drink bottle. Now drop a burning match into the bottle, and screw the cap on tightly. Squeeze the sides of the bottle as hard as you can, and then quickly release it. Do this several times. What do you observe? Explain your observations.

The Dew Point

You may recall from earlier studies that the **dew point** is the temperature at which moisture starts to condense out of the air. The dew point indicates how much moisture there is in the air. The lower the dew point is, the less moisture there is in the air. What happens to the dew point if the amount of moisture in the air increases? The dew point is calculated according to how much a wet-bulb thermometer cools. Can you explain why? Think of a situation in which no amount of shaking of the wet-bulb thermometer would result in a lower reading.

You can calculate the dew point using the table provided below. Find the dry-bulb temperature. Then find the *wet-bulb depression*—the difference between the dry- and wet-bulb temperatures. Use these measurements to find the dew point in the table.

For meteorologists (weather scientists), dew point is actually a very important measurement. In the next Exploration you will use two methods to determine the dew point.

Something to Think About

Trees cool their surroundings by absorbing solar energy and providing shade. But they also provide a cooling effect in another way. (Recall your study in Unit 1 of processes that occur in plant leaves.) How do you think this works? Why would cutting vast areas of forest have a warming effect on the environment? How might the cutting of trees affect the world climate?

Table for Computing Dew Point

This table allows you to compute the dew point. The dew point is plotted for a dry bulb reading of 22°C and a wet-bulb reading of 19°C.

Dry-bulb temperature (°C) vs. Wet-bulb depression (°C)

Dry-bulb	1	2	3	4	5	6	7	8	9	10	11	12	13	14	15	16	17	18	19	20	21	22
-20	-33																					
-18	-28																					
-16	-24																					
-14	-21	-36																				
-12	-18	-28																				
-10	-14	-22																				
-8	-12	-18	-29																			
-6	-10	-14	-22																			
-4	-7	-12	-17	-29																		
-2	-5	-8	-13	-20																		
0	-3	-6	-9	-15	-24																	
2	-1	-3	-6	-11	-17																	
4	1	-1	-4	-7	-11	-19																
6	4	1	-1	-4	-7	-13	-21															
8	6	3	1	-2	-5	-9	-14															
10	8	6	4	1	-2	-5	-9	-14	-28													
12	10	8	6	4	1	-2	-5	-9	-16													
14	12	11	9	6	4	1	-2	-5	-10	-17												
16	14	13	11	9	7	4	1	-1	-6	-10	-17											
18	16	15	13	11	9	7	4	2	-2	-5	-10	-19										
20	19	17	15	14	12	10	7	4	2	-2	-5	-10	-19									
22	21	19	17	16	14	12	10	8	5	3	-1	-5	-10	-19								
24	23	21	20	18	16	14	12	10	8	6	2	-1	-5	-10	-18							
26	25	23	22	20	18	17	15	13	11	9	6	3	0	-4	-9	-18						
28	27	25	24	22	21	19	17	16	14	11	9	7	4	1	-3	-9	-16					
30	29	27	26	24	23	21	19	18	16	14	12	10	8	5	1	-2	-8	-15				
32	31	29	28	27	25	24	22	21	19	17	15	13	11	8	5	2	-2	-7	-14			
34	33	31	30	29	27	26	24	23	21	20	18	16	14	12	9	6	3	-1	-5	-12	-29	
36	35	33	32	31	29	28	27	25	24	22	20	19	17	15	13	10	7	4	0	-4	-10	
38	37	35	34	33	32	30	29	28	26	25	23	21	19	17	15	13	11	8	5	1	-3	-9
40	39	37	36	35	34	32	31	30	28	27	25	24	22	20	18	16	14	12	9	6	2	-2

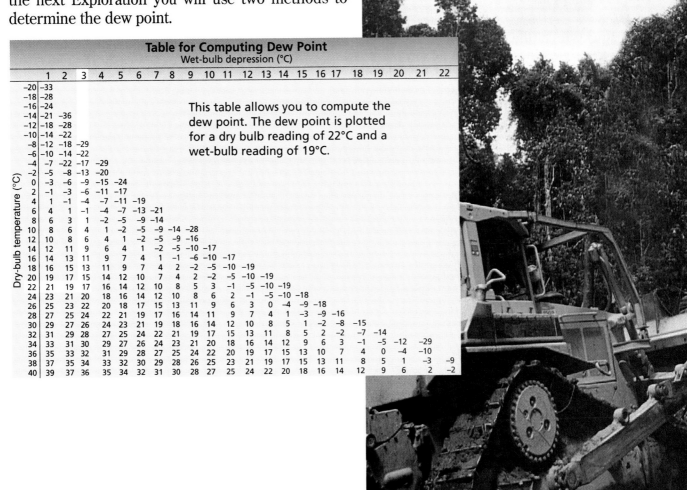

Measuring the Dew Point

PART 1

The Cold-Can Method

You Will Need

- a metal can (250 mL)
- ice cubes
- a thermometer
- a stirring rod
- water (room temperature)

What to Do

1. Add room-temperature water and a few ice cubes to your can until it is half-full. Do this at your workstation so that you can begin to make observations immediately.

2. Place the thermometer in the water. Using a stirring rod, stir the water continuously while observing the sides of the can for any condensation. Record the temperature at which you first notice moisture (dew) forming on the side of the can. This temperature is the dew point. Compare your results with those of your classmates.

PART 2

The Wet-Bulb Thermometer Method

You Will Need

- a thermometer
- a wet-bulb thermometer

What to Do

1. Measure the temperature of the room. Don't use your wet-bulb thermometer to do this because it will give you a false reading. (Do you know why?)

2. Determine the wet-bulb temperature as you did in the previous Exploration.

3. Find the wet-bulb depression by subtracting the wet-bulb temperature from the dry-bulb temperature.

4. Use the table on page 231 to determine the dew point.

How did the results for the two methods compare? Try to explain any differences. What weaknesses, if any, does each method have?

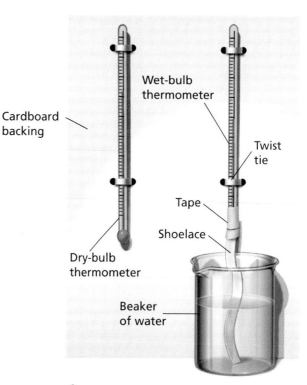

How would you use this homemade setup to calculate the dew point?

Upward Bound

Imagine taking a trip on the aerial tram in Palm Springs, California. In 15 minutes the tram will travel 4 km and climb from 800 m to 2442 m above sea level. A breeze blows against the mountain and forces air to rise along its slope. As you ascend the mountain, you notice that the air grows cooler and that the vegetation changes dramatically. At the base of the mountain, desert plants prevail. As you climb higher, the desert vegetation is gradually replaced by shrubs and small trees. By the time you reach the summit, you are in an evergreen forest. The pine-scented air is crisp and cool.

Why did conditions change so much from the base of the mountain to the summit? Examine the data below for two trips on the tram, one on a clear day and one on a cloudy day. In both cases the temperature at the base was 30°C.

DAY 1 — CLEAR		DAY 2 — CLOUDY	
Height	Temperature	Height	Temperature
1500 m	15° C	1500 m	20° C
1000 m	20° C	1000 m	22° C
500 m	25° C	500 m	25° C
0 m	30° C	0 m	30° C

Base elevation: 800 m

1. By how much did the temperature change on the tram trip on the clear day? on the cloudy day?

2. What was the dew point on the cloudy day?

3. What accounts for the change in vegetation?

As the tram ride clearly shows, the higher you go, the cooler the air becomes. Why does this happen?

Raise the Elevation—Lower the Temperature

Let's summarize what you have just learned.

- Air cools as it rises.

- Rising air cools at a slower rate when condensation occurs.

As you may already have realized, these phenomena are very important in shaping the weather. Pause for a moment to answer the following question: When a mass of air sinks, what happens to its temperature?

Mountain scenes, from base (bottom) to summit (top)

As you may have already reasoned, when air sinks, it heats up. This phenomenon is an important factor in shaping weather.

Read the following situations and think about how the phenomena just described explain each situation:

- On humid days when the wind blows, clouds often form around the peaks of mountains.

- The rainiest (or snowiest) places are usually in the mountains.

- Areas downwind of mountain ranges often receive very little rainfall.

Here is one additional mystery for you to ponder:

It is a bitter cold January morning in Billings, Montana. An icy stillness blankets the land. Suddenly, a breath of wind stirs—a warm wind! Within moments a strong westerly breeze is blowing. The temperature soars 5 . . . 10 . . . 20 degrees. Snow and ice begin to melt rapidly. "Chinook!" you hear someone shout. The wind has a name! What on Earth has happened? What is this chinook wind?

You already have part of the answer to this puzzle. Use the following information, along with the diagram provided, to help you solve the puzzle:

- Rising air cools at a rate of about 10°C for every 1000 m of elevation. Sinking air warms at about the same rate.

- Rising air from which moisture is condensing cools at a rate of about 5°C per 1000 m of elevation.

Can you solve the puzzle? Share your explanation of the chinook wind with a classmate.

Climbing a mountain is like taking a journey northward. Climbing 1000 m is roughly equivalent to traveling 1000 km north, in terms of the change in climate. Higher elevations not only are cooler, but also tend to be wetter. How does this help explain the change in vegetation that you saw in the journey on the tram?

Keeping Track

In this section you have been answering many questions and making many discoveries. In your ScienceLog, make a list of your major findings.

CHALLENGE YOUR THINKING

1. Country Cool

Towns and cities are generally warmer than the surrounding countryside.

a. What are some reasons for this?

b. What are some things that can be done to help lower the temperature within a town or city?

c. The chart at right shows the average yearly temperature recorded at the Green Acres Weather Station over a 50-year span. Describe the pattern shown by the graph. Explain what might have happened to cause such a pattern. (Hint: When Green Acres was first built, it was in a rural area.)

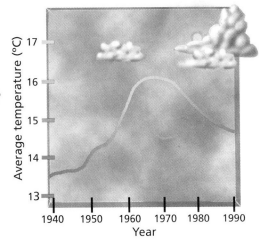

2. Burning Questions

Use your newly acquired knowledge to explain the following phenomena:

a. A car parked in the sun with its windows closed becomes hot very quickly.

b. On a stifling hot, nearly windless day, the slightest breeze feels cool.

c. You are walking barefoot along the waterline at the beach on a summer day. When you walk away from the water, the sand becomes unbearably hot.

3. Care for a Refreshing Swamp Cooler?

Before air conditioners came into widespread use, cooling devices popularly known as "swamp coolers" were widely used. The illustration shows how a swamp cooler works.

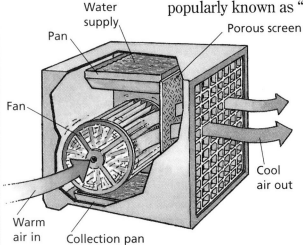
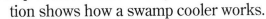

a. Explain the principle of the swamp cooler's operation.

b. Would it work equally well in all climates? Explain.

c. Swamp coolers are still common in certain areas because they are about as effective as air conditioners but much cheaper to operate. Where might you expect to find them in common use?

4. New England News

The graph at right shows temperatures for Montpelier, Vermont, and for Mount Washington, New Hampshire, located a short distance away. Explain what is happening and why.

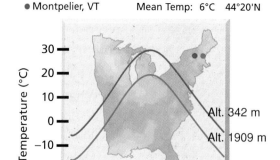

● Mt. Washington, NH   Mean Temp: –3°C   44°20'N
● Montpelier, VT   Mean Temp: 6°C   44°20'N

Alt. 342 m
Alt. 1909 m

Temperature (°C)
30
20
10
0
–10
–20

J F M A M J J A S O N D
Month

5. Something to Dew

Over a few days, keep track of the dew-point temperatures using either of the methods described earlier. Does the dew-point temperature vary from day to day? Why?

6. One Colossal Continent

At the time of the dinosaurs, 200 million years ago, most of the Earth's landmasses were joined together, forming a supercontinent called Pangaea. How would the climate in the interior of this supercontinent compare with the climate along its coastlines? If Pangaea had deserts, where would you expect to find them?

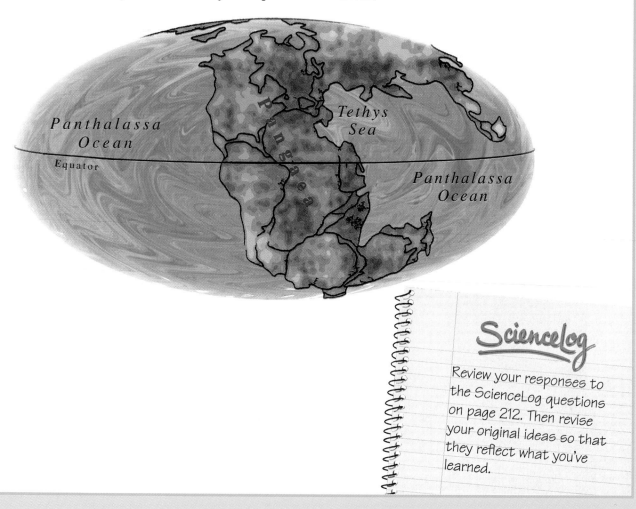

Panthalassa
Ocean

Equator

Tethys
Sea

Panthalassa
Ocean

Pangaea

ScienceLog

Review your responses to the ScienceLog questions on page 212. Then revise your original ideas so that they reflect what you've learned.

CHAPTER 11

Oceans of Water and Air

1 If you could tour the ocean floor, what features might you see?

2 People who live near the ocean experience breezes that blow alternately onshore and offshore. What causes these breezes?

3 A bottle dropped into the ocean may end up on a distant shore because of ocean currents. What causes these currents?

ScienceLog

Think about these questions for a moment, and answer them in your ScienceLog. When you've finished this chapter, you'll have the opportunity to revise your answers based on what you've learned.

Hidden Wonders

A Whirlwind Tour

If you were to plan a tour of Earth's great natural wonders, would you include the Grand Canyon? Mount Everest? the Amazon River? All of these sights are truly spectacular. But did you know that some of the world's great natural wonders are hidden from view? These natural wonders are beneath the oceans, which cover 70 percent of the Earth's surface. The oceans conceal huge canyons, enormous peaks, and mountain ranges that are grander than any on land.

The quick six-stop tour on the next few pages will introduce you to some of the world's great hidden wonders. With each stop there is a question to ponder. After reading about each stop on the tour, suggest at least one new question. These questions may lead to research you can do on your own.

First, look at the map shown here. It shows the ocean floor as it would look if all the water were drained away. What "hidden wonders" can you find on your own?

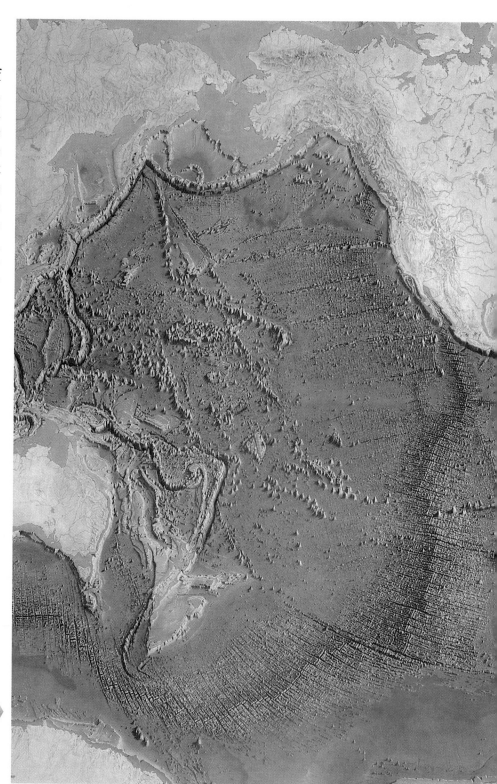

This map shows what the ocean floor would look like if all the water were drained away.

World Ocean Floor by Bruce C. Heezen and Marie Tharp, 1977

Stop 1: Undersea Canyons

Everyone has heard of Arizona's awesome Grand Canyon, but not many people realize that only 350 km from New York City is Hudson Canyon, which is every bit as impressive. Some undersea canyons, with walls 5000 m high, would even dwarf the Grand Canyon.

How do you think undersea canyons were formed?

Stop 2: The Abyssal Plains

Much of the deep ocean bottom is made up of *abyssal plains,* vast stretches of featureless wasteland that are flatter and more barren than any place on land. The abyssal plains are forbidding places of eternal darkness, crushing pressures, freezing temperatures, and absolute stillness. Even so, surprising numbers of living things are found there. The remains of tiny dead plants and animals cover the bottom, accumulating with unimaginable slowness—1 cm or less every thousand years. The slightest disturbance raises blinding clouds of mud.

Why do you think the abyssal plains are so flat?

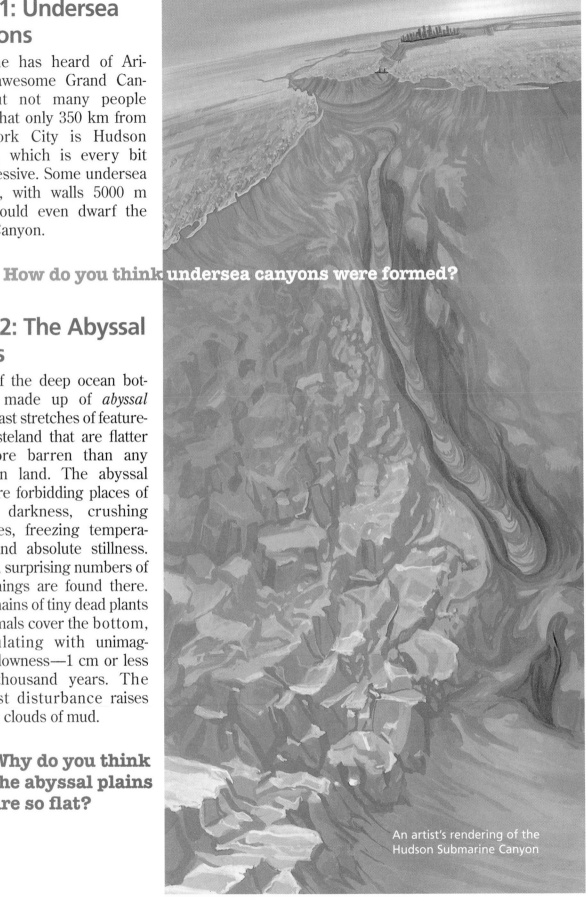

An artist's rendering of the Hudson Submarine Canyon

Stop 3: The Mid-Ocean Ridges

Lacing the Earth's surface like the seams on a baseball, the mid-ocean ridges form the world's largest and longest mountain ranges. Hidden beneath the ocean's surface, these ridges consist of harshly rugged landscapes and deep canyons. The mid-ocean ridges form the boundaries between crustal plates. Here, in the middle of the ridges, huge plates are separating, allowing molten rock to flow from the Earth's interior and create new oceanic crust.

Where is the Mid-Atlantic Ridge on this map of the ocean floor?

Stop 4: Hydrothermal Vents

Along the length of the Mid-Atlantic Ridge, volcanoes and hydrothermal vents (springs that gush hot, mineral-rich waters) are common. Clustered about the hydrothermal vents are strange, previously unknown ecosystems. The existence of these ecosystems disproves the long-held assumption that all life-forms depend ultimately on the energy of the sun. Here, where not even the faintest glimmer of light from the surface can penetrate, dwell giant tube worms and clams, snow-white crabs, weird fish and crustaceans, and other bizarre creatures. This ecosystem is based on a species of bacteria that derives energy and nutrients from the mineral-rich waters of the hydrothermal vents.

What would happen to the organisms around hydrothermal vents if the vents were to stop flowing?

Stop 5: The Mariana Trench

Trenches are the sites at which the ocean crust is slowly being pushed, centimeter by centimeter, back into the interior of the Earth. Here the ocean floor dips sharply downward, forming steep-walled valleys that are deeper, in some cases, than Mount Everest is high. An anvil dropped from a boat floating over the deepest part of the Mariana Trench would take 90 minutes to hit bottom. Incredible pressures and an almost total lack of oxygen make life in any ocean trench a difficult proposition.

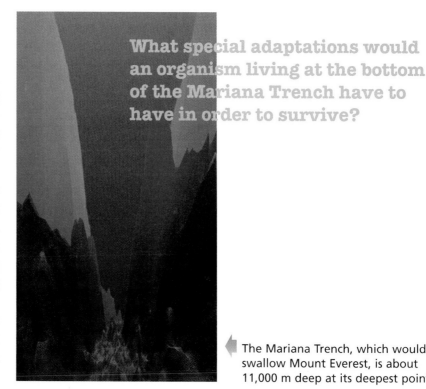

What special adaptations would an organism living at the bottom of the Mariana Trench have to have in order to survive?

The Mariana Trench, which would swallow Mount Everest, is about 11,000 m deep at its deepest point.

Stop 6: Hawaiian Volcanoes

The Hawaiian Islands are a familiar sight, but did you realize that these islands are only the topmost parts of much larger land-masses? The rest lies beneath the ocean. Measured from its base at the ocean bottom, the island of Hawaii (at 9850 m) would be taller than Mount Everest! Volcanoes such as those that formed Hawaii play a major role in the global cycle of greenhouse gases. Many other mountains of the ocean's floor, known as seamounts, do not reach the ocean's surface.

Why do the Hawaiian Islands "trail off" to the northwest?

Lanai

Hawaii   Maui   Molokai   Oahu

The tallest Hawaiian island is almost 10,000 m tall. Only 4000 m extend above sea level.

It would take 25 Empire State Buildings stacked end to end to reach the top of Hawaii.

Pictures of the Deep

Have you thought of any questions that you could explore further? Think of yourself as a travel agent, and develop a poster, brochure, or article to inform others about your tour. Try to include an answer to one of your questions in your travel guide. An encyclopedia in your school library is a good place to start your research.

LESSON 2

The Moving Oceans

The Mediterranean Puzzle

For thousands of years sailors have used winds and ocean currents to propel their ships. Ancient peoples spent much time speculating about causes of winds and ocean currents, but it was left to modern science to solve the puzzle of their origin. What causes these currents in the ocean and atmosphere? Are they somehow related? What role do these currents play in the global climate? These are questions we will investigate in this lesson.

Let's start with an ancient puzzle that was finally solved by Count Luigi Marsili in 1679.

Examine the map of the Mediterranean Sea. The puzzle is this: Sailors had long known that swift currents flowed into the Mediterranean from both the Black Sea and the Atlantic Ocean. Many rivers and streams also empty into it. The Mediterranean has no apparent outlet; therefore, the water level should rise, but it does not. Many explanations were offered—for example, the existence of hidden underground channels to drain the excess water. Can you solve this puzzle?

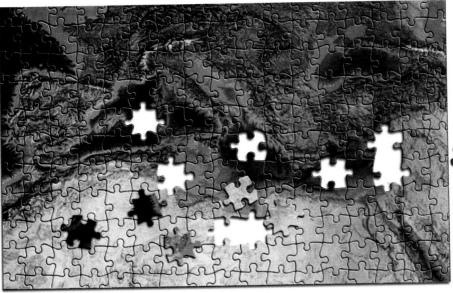

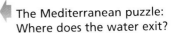
The Mediterranean puzzle: Where does the water exit?

Count Marsili thought he could, and so he set up a model of the Mediterranean to test his idea. In the following Explorations you will trace the steps of Count Marsili as you discover the answer to the Mediterranean puzzle. At the end of Exploration 3, be prepared to describe the Mediterranean puzzle and explain its solution.

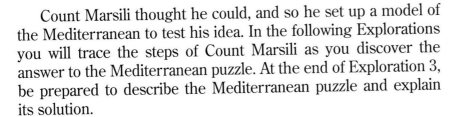

Unraveling the Puzzle

You Will Need

- salt
- an aluminum roasting pan
- aluminum foil
- masking tape
- food coloring
- pepper
- 2 containers that hold at least 1 L of water each
- water
- a sharpened pencil

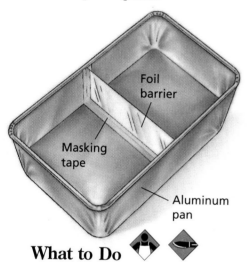

What to Do

1. Cut a piece of aluminum foil slightly wider and taller than the pan. Using masking tape, make a waterproof barrier that divides the pan into two parts.

2. To one container, add 1000 mL of water, 100 mL of salt, and a few drops of food coloring. Stir until the salt is completely dissolved. In the second container, dissolve 50 mL of salt in 1000 mL of water. Do not add food coloring. Save 100 mL of each solution for Exploration 2.

3. With the help of a partner, pour each solution into opposite sides of the pan at the same time (so that the barrier doesn't collapse). Be careful to make the water level on both sides equal. Now sprinkle some pepper into the side containing food coloring.

4. With a pencil point, punch a hole in the aluminum foil just below the surface of the water. Make another hole in the foil near the bottom of the container. Observe the setup for about 10 minutes.

Interpreting Your Findings

1. Make a diagram in your ScienceLog to record what you observed. Use arrows to show current flow. Could this demonstration serve as a model of the Mediterranean Sea? Explain. Using what you have learned, draw a diagram of the Mediterranean showing the inflow and the outflow at both entrances.

2. From your observations, which is saltier—the Atlantic Ocean or the Mediterranean Sea? the Black Sea or the Mediterranean? Which salt solution represents the Atlantic? Which represents the Mediterranean?

3. Why should the saltiness of the Mediterranean differ from that of the Atlantic? (Hint: Most of the Mediterranean basin has a warm and dry climate.)

4. Count Marsili thought that he could explain how the saltiness of ocean water could cause currents. To test his idea he drew water samples from different depths in the Strait of Gibraltar, which connects the Mediterranean and the Atlantic. Using an equal volume of each sample, he found their masses and compared them with the mass of an equal volume of fresh water. What do you think he found?

Test your prediction by doing the next Exploration.

EXPLORATION 2

Marsili's Explanation

You Will Need

- a pill bottle or similar small container
- the colored and uncolored salt solutions from Exploration 1
- tap water
- 3 test tubes
- glass tubing or clear drinking straws
- a balance
- a graduated cylinder (50 mL)
- food coloring of 2 different colors
- a beaker or other container
- a pencil or permanent marker

PART 1

Comparing Masses

1. Determine the volume (in milliliters) and mass (in grams) of the pill container.
2. Fill the container with one of your salt solutions or with tap water. Determine the mass of the container and liquid combined. What is the mass of the liquid?
3. Repeat step 2 with the remaining two liquids. Be sure that you are measuring equal volumes of each liquid. How does the mass of the "Mediterranean" solution compare with that of the "Atlantic" solution? How do both of these masses compare with the mass of tap water?
4. Divide the mass of each solution by its volume. Which solution has the greatest mass per volume? the least mass per volume?
5. Save your solutions for use in Part 3 of this Exploration.

PART 2

Thinking About Density

In step 4 of Part 1 you actually found the density of each solution. As you may recall from Chapter 6, density is mass per unit of volume. Read the following questions about density, and then answer them in your ScienceLog.

Which liquid had the highest density? the lowest density?

The density of any material can be found using this formula:

Density = mass/volume

What are the densities of the following materials?

Material	Mass of 30 mL	Density
water	30.0 g	?
alcohol	21.0 g	?
ice	27.6 g	?
salt water	33.0 g	?
egg	31.5 g	?

Which materials would float on fresh water? on alcohol? on salt water? Devise an experiment to test your predictions.

PART 3

Layering Liquids

Place three empty test tubes marked A, B, and C in a test-tube rack. Fill test tube A with the colored liquid used in Part 1. Color one of the other liquids from Part 1 with a color of food coloring different from that used in liquid A, and fill test tube B with it. Color the remaining liquid used in Part 1 with the third color of food coloring, and fill test tube C. Now use the technique shown below to layer the three liquids in a thin glass tube or drinking straw. Can the liquids be layered in any order?

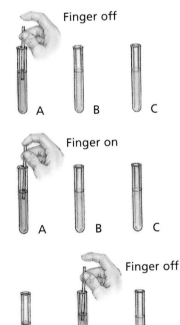

How does this relate to Count Marsili's explanation of the Mediterranean puzzle?

Pill bottle

Balance scale

Predict, Observe, Explain

Could density differences in the oceans cause currents similar to those in the Mediterranean? Perhaps the following Exploration will answer this question.

You Will Need

- a 10% salt solution
- tap water (cold and hot)
- food coloring
- 2 small jars
- a large pan or bucket
- index cards or stiff paper

What to Do

In Activities 1–4 you will invert one water-filled jar over another to study how solutions of different densities interact. Be sure each jar is filled to the rim with the correct type of water.

Set up each Activity as shown in the illustration at top right; also follow the additional instructions for each Activity. Note that after setting up the apparatus, you will turn the jars on their side only in Activities 1 and 3.

To catch any spills, carry out this Exploration over a pan or bucket. For each Activity in this Exploration, do the following:

- Predict what you expect to happen, observe what does happen, and finally, give an explanation for your observations.
- Re-examine the map of global temperatures on page 226. Where in the oceans might the phenomenon you observed in this Exploration be occurring?

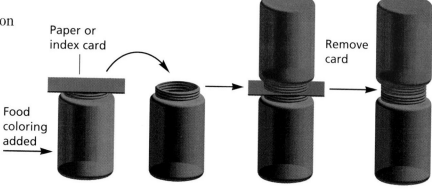

ACTIVITY 1 **ACTIVITY 2**

a. Allow jars to stand upright for 40 seconds after removing the card or paper.

b. Turn jars on their side while holding them together.

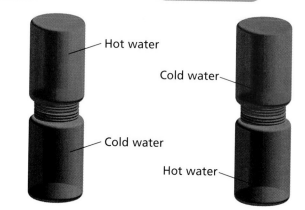

What conclusions would you draw from Activities 1 and 2?

ACTIVITY 3 **ACTIVITY 4**

Follow the same procedure as in Activity 1.

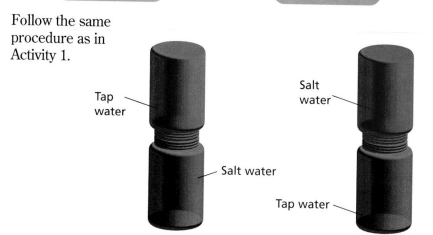

Count Marsili's Letter

After reading the letter at right, write Count Marsili's follow-up letter. Include experiences and diagrams that make your explanation clear. Keep in mind that Marco is less knowledgeable than you about density and currents.

Follow the Flow

How many factors have you discovered that affect the density of ocean water? The previous Exploration suggests two factors. The water of the Mediterranean is denser than that of the Atlantic Ocean because of evaporation. When water evaporates, the salt in it is left behind. Look at the map below. Where else in the world might this be happening? Sea water in the tropics generally has a different density from that of the sea water in cold regions. Would tropical water be more or less dense than polar water? Why?

The Norwegian Sea bordering Greenland and the Weddell Sea bordering Antarctica are both chilled by the cold winds that blow off glacier-covered landmasses. Does this increase or decrease the density of the surface water?

You may be surprised to find that ice formed from sea water is not very salty. Why? The dissolved salt does not fit well into the crystal structure of ice, so it is "squeezed" out as the water freezes. The water left behind becomes a little more salty and therefore more dense.

> *Dear Marco,*
> *I have recently returned from the Mediterranean, where I observed the strong surface currents at both entrances.*
> *I believe I have found an explanation for the Mediterranean mystery. I shall elaborate in my next letter.*
>
> *Sincerely,*
> *Luigi*

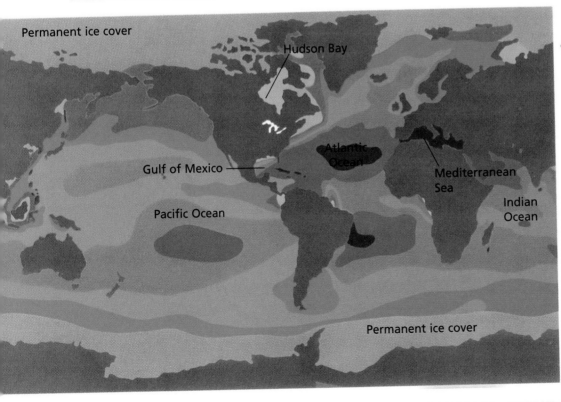

This map shows the average surface salinity of the oceans. How do you account for the differences in salinity?

Permanent ice cover

Hudson Bay

Gulf of Mexico

Atlantic Ocean

Mediterranean Sea

Indian Ocean

Pacific Ocean

Permanent ice cover

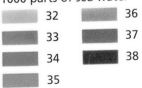

Proportion of salt per 1000 parts of sea water

32	36
33	37
34	38
35	

Examine the diagram of deep-ocean currents shown below. These currents are caused by density differences in ocean water. Where do they originate? Where do they go? Is there evidence in the diagram that density currents occur at different depths and go in different directions?

These deep-ocean currents play an important role in controlling Earth's temperature. Carbon dioxide dissolves well in water (think of carbonated soft drinks). The surface water of the ocean absorbs huge amounts of carbon dioxide from the atmosphere. When currents push the surface water to the bottom, the carbon dioxide becomes concentrated in the deep-ocean waters. The deep-ocean currents circulate very slowly. The water and carbon dioxide caught in a deep-ocean current may not resurface for centuries. When deep-ocean currents finally resurface, they bring with them nutrients collected over the years. Sites of upwelling deep currents (such as along the west coast of South America) are rich in life and make excellent fisheries.

These penguins are able to live very close to the equator thanks to a cold, upwelling current off the western coast of South America.

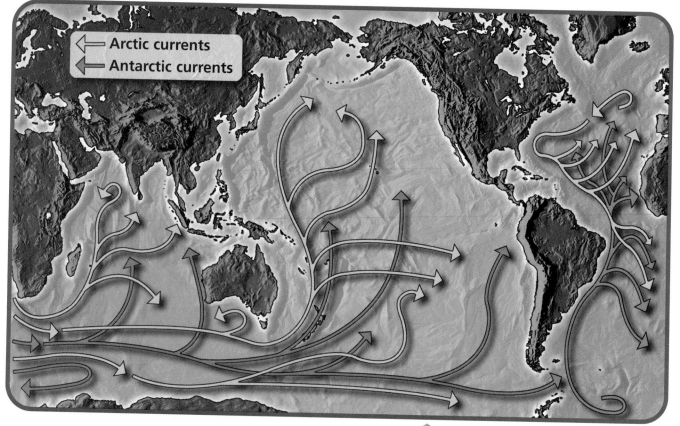

Arctic currents
Antarctic currents

Deep-ocean currents

The Composition and Density of Sea Water

The substances that make up sea water remain quite constant no matter where the ocean is sampled. Most sea water contains the substances shown at right.

Here is a simpler recipe for sea water:

Add 9.0 g of salt to 250 mL of water.

This solution will have the same density as the solution made with the more complicated recipe. What is the density of sea water with this salt content? How can a device such as the one in the photos at right be used to determine the density of a salt solution? Such a device is called a *hydrometer.* The hydrometer is calibrated in such a way that it reads the density of the liquid in which it is placed.

Believe it or not, it is easier for you to float in salt water than in fresh water. In fact, some bodies of water such as the Dead Sea and the Great Salt Lake are so salty that a person would find it almost impossible to sink. In the following Exploration you will use this idea to develop your own device for measuring the density of salt water.

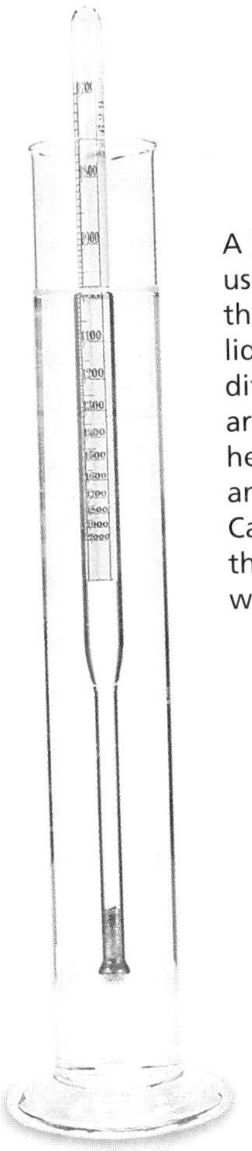

A Recipe for Sea Water

Mix:

Sodium chloride	23.48 g
Magnesium chloride	4.98 g
Sodium sulfate	3.92 g
Calcium chloride	1.10 g
Other compounds	1.00 g

Add enough water to form 1000 g of solution.

A hydrometer is used to determine the density of a liquid. Two different liquids are being tested here: fresh water and salt water. Can you tell from the readings which is which?

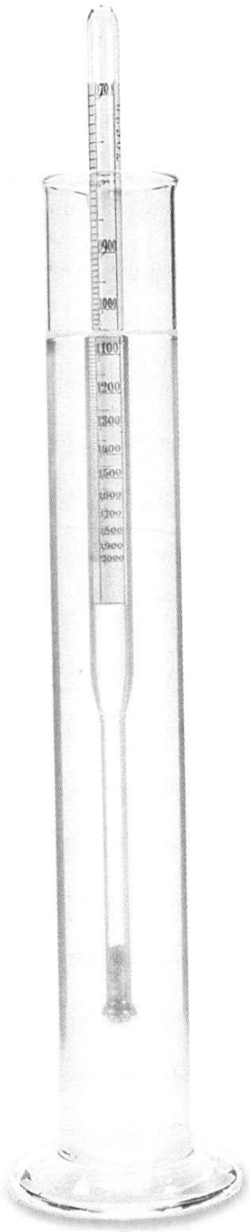

The Great Hydrometer Challenge

You Will Need

- an assortment of common materials such as a pencil, modeling clay, a drinking straw, cork, wire, and thumbtacks
- a graduated cylinder
- water
- a "standard" salt solution with a density of 1.05 g/mL (made by adding 12.5 g of salt to 250 mL of water)
- artificial sea water (using either recipe on page 249)
- a permanent marker

What to Do

1. Using the materials listed here or others of your choice, make three hydrometers. Test your hydrometers for their ability to float in a graduated cylinder filled with water.

2. You can *calibrate* your hydrometers by placing them in solutions of different known densities and marking the level to which they sink. First place the hydrometer in tap water and mark the level to which it sinks. Follow this same procedure using the standard salt solution. Using the results of your calibration, can you extend the scale above and below your marks? Which of your hydrometers appears to be the most sensitive?

3. Use your best hydrometer to measure the density of the artificial sea water. How could you confirm this reading?

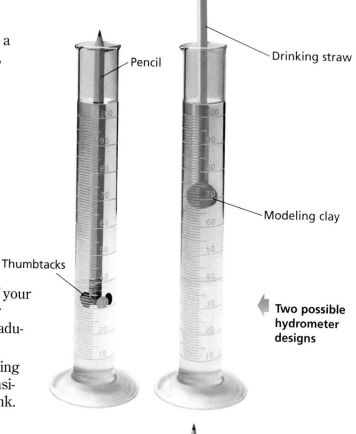

Pencil

Drinking straw

Modeling clay

Thumbtacks

Two possible hydrometer designs

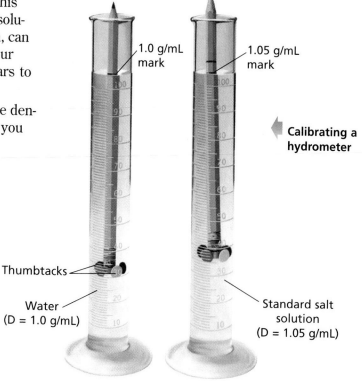

1.0 g/mL mark

1.05 g/mL mark

Calibrating a hydrometer

Thumbtacks

Water (D = 1.0 g/mL)

Standard salt solution (D = 1.05 g/mL)

The Atmosphere

Density Currents in the Atmosphere

Differences in density cause water to flow. Can differences in density also cause air to flow? What would cause density differences to develop in air? Examine the diagram below, which shows a breeze blowing from the ocean to the land during the day. Before reading further, give your explanation of what causes such breezes to blow. Then review the activities you have done in this unit for evidence that each of the following statements is true.

1. In the daytime, land heats up more quickly than does water.

2. The air over land warms and becomes less dense *(A)*. This air rises above (floats on) the denser, surrounding air.

3. Cooler and denser air moves in from the ocean *(B)*, taking the place of the rising air.

4. As air rises, it spreads out and cools *(C)*, thus becoming more dense. The air then sinks *(D)*, completing the cycle.

Draw a similar labeled diagram to show what happens at night, when there is no sun to warm either the land or sea and both begin to cool off.

Storm clouds build over the tropics. Why do these clouds form?

Where else in the world would you find similar kinds of density currents in the atmosphere, only on a much larger scale? Think about the tropics. What happens to the air over the tropics as it warms? It rises, of course, and as you would expect, cooler air from farther north and south moves in to take its place. This pattern is seen throughout the tropics. The winds that result are called *trade winds*. Trade winds blow steadily from season to season and year to year.

Air Masses

You have seen that bodies of water with different densities do not mix readily. Would you expect air to behave similarly?

There is constant interaction between a body of air and the land or water it lies over. Air over warm water becomes warm and moisture-laden, while air over a cold landmass becomes cold and dry. What characteristics would air over the Sahara have? How about air over Siberia in the winter? Would these two *air masses* have about the same density?

An **air mass** is a body of air that covers a large area and has nearly uniform temperature and humidity throughout. The diagram at right shows the air masses that affect weather in North America in summer and winter. Use it to answer the following questions:

1. Air masses are either polar or tropical and either maritime or continental. The name of an air mass has two parts, for example, *continental polar*. Here is what each part indicates:

   continental—dry   polar—cold

   maritime—moist   tropical—warm

   Suggest how each type of air mass compares in terms of its temperature, moisture content, and density, based on its location and name. Look back over the last Exploration for observations to support your answer to this question.

2. Would the different types of air masses readily mix together? Which activity that you performed suggests that they do not?

3. What type of air mass is causing your weather today?

4. What type of air mass(es) will probably be causing your weather 6 months from now?

5. What type of air mass do you think you would find over the Atlantic Ocean near the equator?

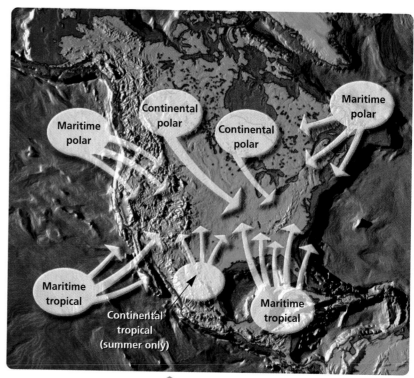

Maritime polar

Continental polar

Continental polar

Maritime polar

Maritime tropical

Continental tropical (summer only)

Maritime tropical

Air masses that affect our weather

To find out more about the behavior of the air and water in our atmosphere, see pages S66–S74 in the SourceBook.

Air Masses in Motion

If a weather forecaster said, "Look for a Canadian air mass to sweep into our area tomorrow," what kind of weather would you expect? Since much of our weather is the result of collisions between air masses, what would you see if you could see the boundary between air masses of different densities? These boundaries are called *fronts*. The following diagrams show two types of fronts. Answer the questions that go with each diagram.

A Cold Front

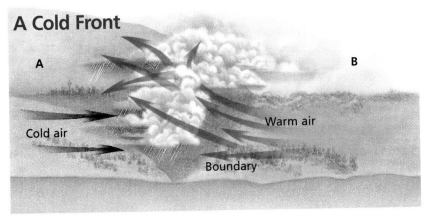

A cold front The blue arrows indicate the overall movement of the cold air mass. The red arrows indicate the movement of the air within the warm air mass.

1. Which air mass is more dense?

2. What happens to the warm air as the cold air approaches?

3. What causes the clouds to form?

4. Compare the temperature at *A* with that at *B*. How will the temperature at *B* change shortly?

5. The boundary between the cold air and the warm air is called a cold front, not a warm front. Why?

A Warm Front

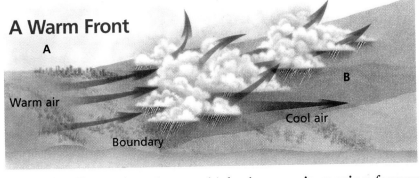

A warm front The red arrows indicate the movement of the air within the warm air mass. The blue arrow indicates the overall movement of the cold air mass.

1. In the illustration above, which air mass is moving forward faster—the cold air mass or the warm air mass?

2. What is happening to the warm air as it overtakes the cold air mass?

3. Why are clouds forming?

4. How does the temperature at *A* compare with that at *B*? How will it change?

5. Why is the boundary between the warm air and the cold air called a warm front, not a cold front?

Putting It All Together: Cold Fronts and Warm Fronts

Weather forecasters often make predictions based on their understanding of the movement of cold fronts and warm fronts. Study the diagram at lower left, which shows the interaction of two large air masses, and then answer the following questions:

1. Where is the warm front? the cold front? Approximately where will each front be tomorrow?

2. How will the weather at *A* compare with the weather at *D* for the next day or so? How will the weather at *B* compare with the weather at *C* during the same period of time?

3. How does the interaction between air masses at a warm front differ from the interaction of air masses at a cold front?

4. In many respects, cold fronts are "active" phenomena, while warm fronts are "passive." Explain why these terms apply.

Weather Maps

Weather maps are simplified representations of the interactions between air masses. Through the use of standard symbols, much information about the weather can be depicted on a single map. Study the weather map below, and then answer the questions that follow.

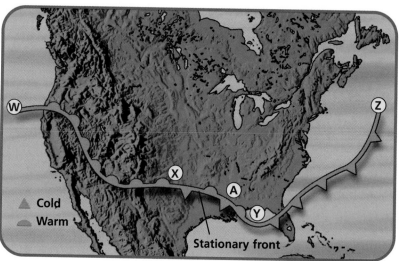

Cold ▲
Warm ◠
Stationary front

This weather map shows two air masses interacting in three different ways. How does this occur?

1. What do you think the triangles on the line indicate? What do the half-moons indicate?

2. What does the line *WX* on the weather map represent? the line *XY*? the line *YZ*?

3. On tomorrow's weather map, approximately where will the warm front be?

4. As the map shows, one section of the front is not moving. This is called a *stationary* front. How is it represented?

5. How does the air-mass diagram at lower left help explain why the frontal boundary at *A* is stationary?

An At-Home Task

Follow your local weather forecasts for a week. What elements of weather are included in the forecasts? How are forecasters (meteorologists) able to make predictions about future weather?

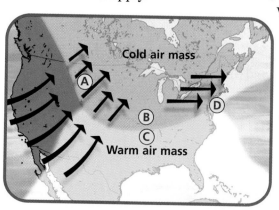

Cold air mass

Warm air mass

This diagram shows how air masses interact. The black arrows indicate the motion of the air masses.

A Map Sequence

The weather maps below represent a span of 6 days. Examine the maps. Then answer the following questions:

1. Describe, in general terms, what happens during this sequence. How do the various weather systems move?
2. Identify any fronts. What types of fronts are present?
3. What time of year do you think the maps probably represent?
4. What types of air masses are present in each map?
5. What happens at the boundaries between air masses?
6. What happens to the weather conditions of an area after one air mass overtakes another?

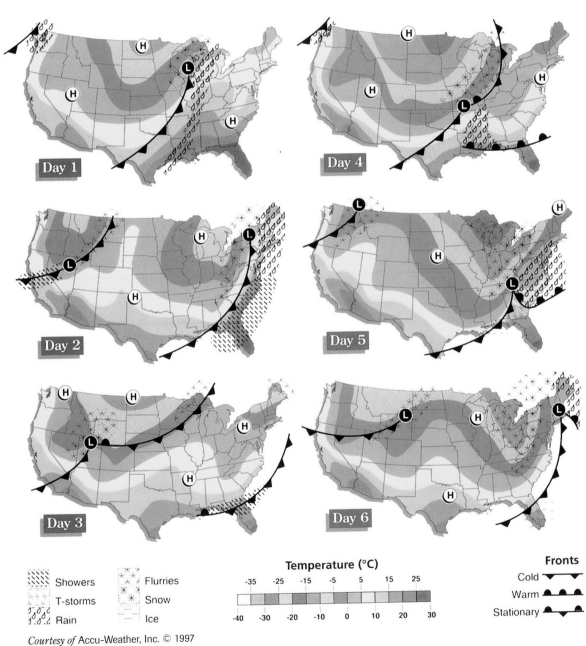

Courtesy of Accu-Weather, Inc. © 1997

A Climate Sampler

Reykjavik, Iceland, and Yellowknife, Canada, are both at about the same latitude, but, as the graph shows, during most months these cities have very different weather. In fact, both cities have very different climates. Why? What clues does the map provide?

Climate is a word we use quite often, but what exactly is a climate? Could you define it? How does it differ from weather?

Write each of the following sentences in your ScienceLog. In each blank, write whichever term best applies, *weather* or *climate*.

1. "What? Rain again! This kind of ___?___ depresses me."
2. "In January we normally have clear, cold ___?___."
3. During the ice age, the Earth had a much colder ___?___.
4. The ___?___ of the Amazon rain forest hasn't changed in millions of years. The ___?___ is almost the same every day.

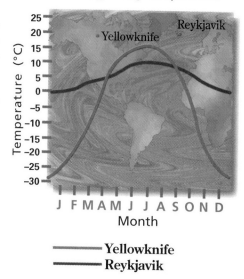

As you probably realize, weather is the condition of the atmosphere at a particular time and place. On the other hand, climate is the average weather conditions over a long period of time—years, decades, or even centuries. The diagram at right shows some of the world's major climatic regions. What seem to be the most important factors in determining climate? Why do you think this is so?

Let's go back to our earlier example. Why do Yellowknife and Reykjavik have such different climates?

Reykjavik has a *maritime climate*, that is, a climate typical of the ocean. Maritime climates are moist and exhibit relatively little change in temperature from day to night and from season to season. Reykjavik's climate is also influenced by the warm waters of the Gulf Stream, which you will learn about later.

Yellowknife has a *continental climate*. Continental climates are drier than maritime climates and typically are marked by large daily and seasonal temperature ranges. Why are continental and maritime climates so different? Where else in this unit have you seen evidence of this?

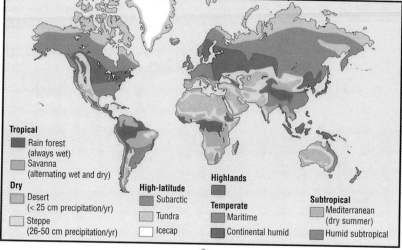

Major climatic regions of the world

Keeping Track

What have been your major findings in this chapter? Continue the record that you started in the last chapter.

CHALLENGE YOUR THINKING

1. Earth's Tragic Thaw?

The Arctic Ocean is covered year-round by a layer of ice. If this ice melted, would you expect sea levels to rise worldwide? Why or why not?

2. The Case of the Baffling Bath Water

Robert filled a tub with water. The water ran cold for about a minute before it finally got hot. Robert expected the water in the tub to be uniformly warm throughout, but instead the hot water floated on top and the cold water stayed underneath. He had to slosh the water around before it felt comfortable.

a. Robert thought, "This is like the Mediterranean effect." What did he mean by this?

b. How do you explain this occurrence?

3. Saline Sailing

Ships loaded to the limit in cold, northern waters would be sailing into danger if they headed for the tropics without first unloading some cargo. To prevent this problem, international marine regulations require most ships to bear *Plimsoll marks.* Plimsoll marks are a kind of scale that shows the maximum safe load for a given water condition. The diagram below shows a typical set of Plimsoll marks. Freshwater marks are on the left, and saltwater marks are on the right.

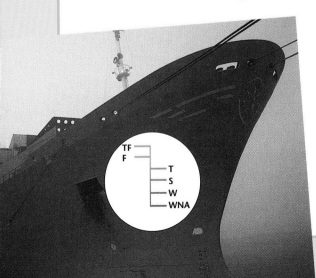

a. Why are the saltwater marks lower than the freshwater marks?

b. Why is tropical fresh (TF) the highest mark?

c. Why is winter north Atlantic (WNA) the lowest mark?

d. In January, a captain plans to sail his fully loaded ship from Oslo, Norway, across the Atlantic to Manaus, Brazil, 1500 km up the Amazon River. How should the captain determine how much cargo the ship should carry?

4. Massive Project

Find out as much as you can about the air mass presently over your area. What is the temperature of the air mass? Is it a tropical air mass or a polar air mass? Is it a dry air mass, or is the humidity high? Will the same air mass be over your area tomorrow?

5. A Mysterious Matter

An object of unknown material has a mass of 46 g and a volume of 42 mL. Will this object float or sink in water? How do you know?

6. Surprising Similarities

Air and water seem to be completely different substances, but in some ways they are very much alike. Suggest ways in which the two substances are similar.

ScienceLog

Review your responses to the ScienceLog questions on page 237. Then revise your original ideas so that they reflect what you've learned.

Winds and Currents

1

What is the cause of winds that can do this much damage?

What are the trade winds?

3

2

The word *pressure* has a number of meanings. How would you define *pressure*?

ScienceLog

Think about these questions for a moment, and answer them in your ScienceLog. When you've finished this chapter, you'll have the opportunity to revise your answers based on what you've learned.

LESSON 1

Pressure Differences

Structure of the Atmosphere

From the moon, space looks black. From the Earth, the sky looks blue. The difference is due to the thick blanket of air that surrounds the Earth. It extends upward with decreasing density until it merges with space at about 500 km. From a space shuttle, the atmosphere appears as a hazy blue band against the blackness of space.

Imagine taking a trip through the atmosphere aboard a space shuttle. The trip would take about 10 minutes. You would pass through the various layers shown in the diagram at right.

The boundaries between layers are not as distinct as those shown here, nor are the elevations of the layers the same in all parts of the world. As you travel in the shuttle, at what altitude do you think one-half of the atmosphere would be beneath you? At what altitude would three-quarters of the atmosphere be beneath you? Use the data in the table below to make a graph that will give you the answer. Record this graph in your ScienceLog. The data give the weight of the atmosphere pressing on an area of 1 m² at different elevations. The weight is expressed in *newtons* (one newton is equal to the gravitational force on a 100 g mass).

Altitude (km)	Weight per square meter (N/m²)
0 (sea level)	100,000
5.6	50,000
16.2	10,000
31.2	1000
48.1	100
65.1	10
79.2	1
100	0.1

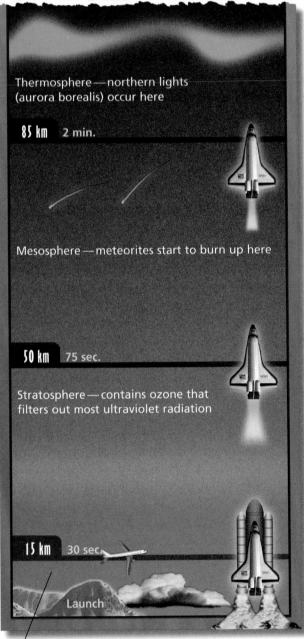

Exosphere — extends into outer space

500 km   10 min.

Thermosphere — northern lights (aurora borealis) occur here

85 km   2 min.

Mesosphere — meteorites start to burn up here

50 km   75 sec.

Stratosphere — contains ozone that filters out most ultraviolet radiation

15 km   30 sec.

Launch

Troposphere — weather occurs here

Understanding Pressure

You are beginning to work with the concept of **pressure**. The pressure of the atmosphere at sea level is about 101,000 N/m². This means that a force of 101,000 N presses down on every square meter of the Earth at this elevation. Is this pressure higher or lower at 10 km above sea level? Review the graph you made using the data in the table on the previous page. What would be the atmospheric pressure at 10 km? at 50 km? How does pressure vary with altitude?

How would you define *pressure* now? How is pressure related to force? After completing the next Exploration, return to these questions. Then provide a scientific definition of pressure.

EXPLORATION 1

Pressure Situations

PART 1

Talk About It

Pressure is an important factor in the following situation. With a classmate, discuss the explanation for this situation.

You can press on a balloon with your finger, and the balloon bends but does not break. But when you apply the same amount of force to the balloon with a needle, the balloon pops. Why?

Now share a pressure situation of your own with other students.

Exploration 1 continued ▶

PART 2

A Brick Trick

You Will Need

- a brick with a string (1 m long) tied securely around it
- a spring scale or force meter
- a metric ruler

What to Do

Predict which face the brick must rest on in order to exert (a) the least pressure against a table and (b) the most pressure. Test your predictions. Determine the pressure of the brick against a table when it is lying on each of its faces. Here is the information you will need.

> Area of a face of a brick = length × width
>
> Weight = reading on the spring scale in newtons
>
> Pressure = force (weight of brick) ÷ area

Questions

1. In what units is pressure measured?
2. Does the weight of the brick change when the face it rests on changes?
3. Does the pressure that the brick exerts change depending on which face is resting on the table? Why or why not?
4. If the brick were placed on a piece of foam rubber, on which of its faces, large or small, would it sink deepest?

What is the force in newtons?

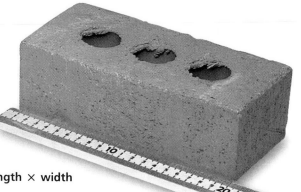

Area = length × width

Pressure in the Ocean

How does water pressure change as you travel deep into the ocean? Is it similar to the changes in atmospheric pressure at different altitudes? The following Exploration focuses on the way in which pressure is exerted by liquids.

ACTIVITY 1

Water Pressure and Depth

You Will Need

- a large can with a hole punched in it close to the bottom
- a meter stick
- a tripod stand
- water
- a stream table
- a ruler

What to Do

Place the can on a stand in a setup like the one shown below. Fill the can with water and observe what happens. Record the height of the water in the can and the distance that the water squirts.

Thinking About It

1. The distance that a volume of water squirts from the hole at the bottom of a can is a measure of the pressure exerted by water. How can you use this information to compare the pressures of different depths of water?

2. Place your data in a table similar to the one shown. Prepare a graph of your data.

Height of water in can (cm)	Distance water squirts (cm)

3. Compare the graph you just made with the graph of atmospheric pressure that you made at the beginning of this lesson. Do you think there is an altitude where the atmospheric pressure is equal to zero? Is there a depth of water for which the pressure in the can is equal to zero? Can you think of an important way in which the air in the atmosphere is different from the water in the can? What is it?

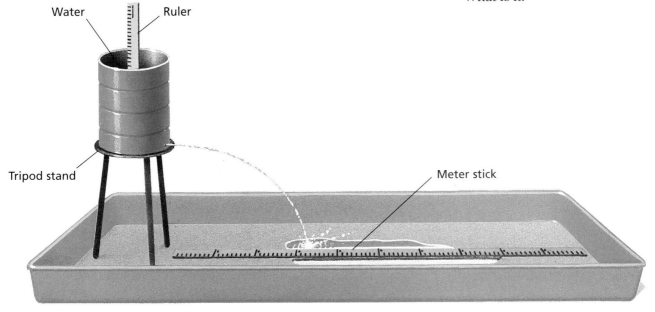

Water   Ruler

Tripod stand

Meter stick

Exploration 2 continued ▶

ACTIVITY 2

The Effect of Shape and Volume

You Will Need

- 2 cans of different diameters, each with a hole punched close to the bottom
- a meter stick
- a tripod stand
- a stream table
- water
- a graduated cylinder

What to Do

Arrange the materials as shown below. Measure and record a volume of water. Place one of the cans on a tripod stand and pour in the water. Measure the distance the water squirts. Replace the first can with the second can and repeat the experiment, using the same volume of water. Record and then compare your results.

Thinking About It

1. What conclusions can you make from this Activity?

2. What would you observe if you filled both cans with water so that the water was at the same depth in each can?

3. Would there be any difference in water pressure if you swam 2 m below the surface of the water in a swimming pool as opposed to 2 m below the surface in a large lake? Does pressure depend on the volume or the depth of water? Explain your reasoning.

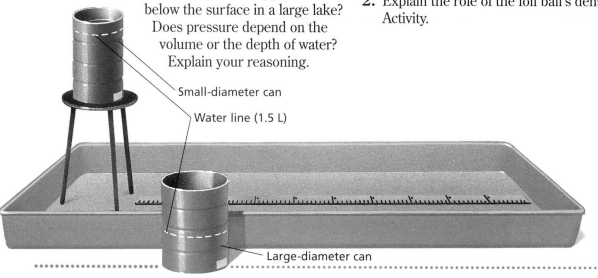

Small-diameter can

Water line (1.5 L)

Large-diameter can

ACTIVITY 3

Is Water Pressure Exerted Equally in All Directions?

You Will Need

- a plastic bottle
- a small piece of foil (4 cm × 4 cm)
- water

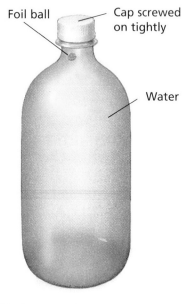

Foil ball

Cap screwed on tightly

Water

What to Do

Fill the plastic bottle all the way to the top with water. Wad the piece of foil into a small ball, and drop it into the container. Screw the cap on tightly. Wait 20 seconds and then squeeze the bottle. Can you make the foil ball rise and fall in the water? Does it matter where you squeeze the bottle?

Now lay the bottle on its side. Can you control the movement of the foil ball? Does it matter where you squeeze the bottle?

Thinking About It

1. How would you answer the question asked in the title of this Activity?

2. Explain the role of the foil ball's density in this Activity.

Historical Flashbacks

Here are three episodes from history. Each illustrates something about the pressure exerted by the atmosphere. Test your understanding by responding to the questions and completing the tasks that follow each episode.

Holy Hemispheres! The Magdeburg Experiment

During the 1650s, Otto von Guericke was both an inventive scientist and the mayor of the German town of Magdeburg. In 1652 Emperor Ferdinand III heard of his experiments and asked

to see them. So Otto von Guericke gave him a dramatic example. He placed two copper hemispheres together so that they formed a hollow sphere with a diameter of about 45 cm. He removed as much air from inside the sphere as he could by using a crude vacuum pump. After creating a partial vacuum inside the sphere, how did the air pressure inside the sphere compare with the air pressure outside?

 A depiction of Otto von Guericke's most famous experiment

Now comes the part that astonished the emperor. Two teams of horses—one team attached to each hemisphere—could not pull the two hemispheres apart! What does this tell you about the magnitude of atmospheric pressure?

Consider this: Air moves from regions of high pressure to regions of low pressure. Where have you heard of high- and low-pressure areas before? What do you think would happen in von Guericke's experiment if a small hole were punched through the copper sphere?

Try It Yourself

The following simulation will help you understand the effect of air-pressure differences in the atmosphere. Pour a small amount of water into an aluminum can. Heat the can until you see steam escaping from it. Using oven mitts, turn the can upside down in a large container of cold water. What happens? How do you explain this in terms of pressure?

Blazing a Scientific Trail: Pascal's Experiment

Grazie, Evangelista! (*thanks)

Prego, Blaise! (*no problem)

Just before von Guericke performed his demonstration, Blaise Pascal, a French scientist and mathematician, expanded the study of atmospheric pressure. He used a new invention created by the Italian scientist Evangelista Torricelli—the mercury **barometer**. Torricelli had filled a tube with mercury and turned it upside down in a pool of mercury. Not all of the mercury ran out. Instead, a column 76 cm high remained, supported by atmospheric pressure. Whenever the pressure of the atmosphere changed, so did the height of the mercury column.

In 1648, Pascal had his brother-in-law carry a barometer to the top of a 1500 m mountain. At the bottom of the mountain, the column of mercury in the barometer was 71.1 cm high. At the top of the mountain, the mercury column was 62.6 cm high. Why was the measurement at the bottom of the mountain different from the one at the top? What had Pascal proved?

Because of Pascal's contribution to science, he had a unit of measurement named after him. A pressure of 1 N/m² is equal to 1 **pascal** (Pa). Normal atmospheric pressure at sea-level is about 101,000 N/m², or 101,000 Pa.

Try It Yourself

Below is one design for a homemade barometer that will detect daily changes in atmospheric pressure. Build this barometer or one of your own design, and keep a record of the increase or decrease in atmospheric pressure each day. Also, keep a record of the kind of weather you experience each day. Is it sunny or cloudy? hot or cold? humid or dry? Can you find a relationship between pressure changes and changes in the weather?

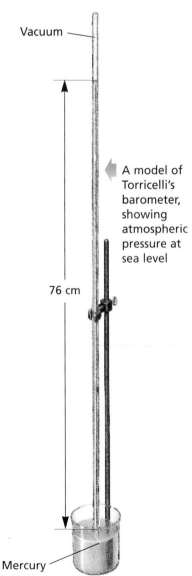

Vacuum

A model of Torricelli's barometer, showing atmospheric pressure at sea level

76 cm

Mercury

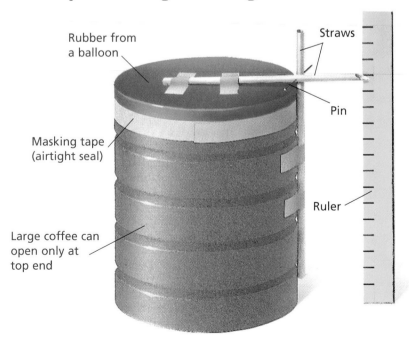

Rubber from a balloon

Straws

Pin

Masking tape (airtight seal)

Ruler

Large coffee can open only at top end

Pumped Up: Boyle's Experiment

In 1667 Robert Boyle made an improved pump that created a nearly perfect vacuum in a container. When he pumped air out of a pipe that was placed in a container of water, he was able to raise the water in the pipe a little over 10 m—but not any higher.

- What caused the water to move up the pipe?
- Why was he able to lift about 10 m of water this way? (Hint: Water is 13.6 times less dense than the mercury used in the barometer on the previous page.)

Try It Yourself

Place an index card over a full glass of water. Turn the glass upside down over a bucket or sink, and remove the hand holding the card in place. Why does the water stay in the glass? According to Boyle's experiment, what is the maximum height that a water column can reach in an experiment of this kind?

Maximum water level—10 m

Robert Boyle was able to pump water to a height of only 10 m.

Under Pressure

In addition to pascals (Pa), another unit commonly used to measure pressure is the *atmosphere* (atm). One atmosphere is the amount of pressure exerted by the Earth's atmosphere at sea level. You can use the following formula to convert from pascals to atmospheres:

$$101 \text{ kPa} = 1 \text{ atm}$$

As you discovered in Exploration 2, pressure in the ocean increases with depth. In fact, it increases by 1 atmosphere for every 10.4 m of depth. How does this relate to Boyle's experiment on page 267?

Imagine that you dove 200 m beneath the surface of the ocean. How much pressure would you experience? How much pressure do deep-dwelling organisms experience? Study the diagram at right, and then answer the following questions:

1. How much pressure do giant squids normally experience? viper fish?

2. The sperm whale spends most of its time near the surface of the ocean, but on occasion it dives to depths of over 1500 m. What change in pressure does the whale experience?

3. While crossing the Atlantic Ocean on its maiden voyage, the passenger ship *Titanic* struck an iceberg and sank. What is the water pressure in pascals and atmospheres at the *Titanic's* final resting place, nearly 3700 m beneath the surface of the ocean?

4. The deepest part of the Mariana Trench, about 11,000 m deep, was visited by the bathyscaph (deep-water vessel) *Trieste* in 1960. At this depth, how great was the pressure experienced by the *Trieste*? The crew compartment of the *Trieste* was a sphere about 2 m in diameter. What was the *total* force on the crew compartment? (The formula for the area of a sphere is $4\pi r^2$.)

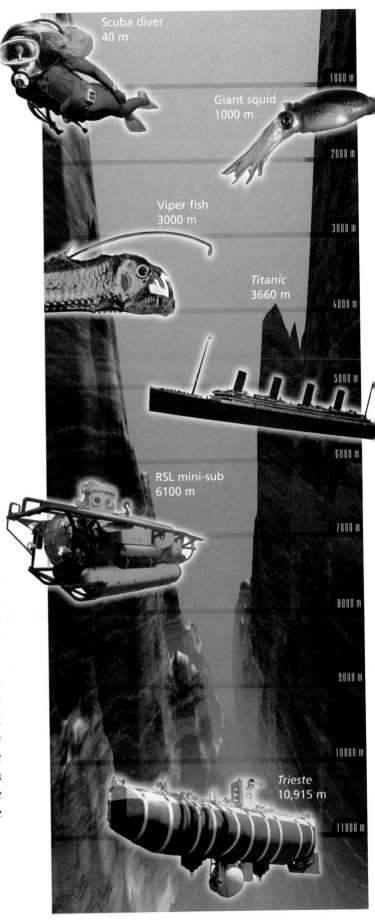

Scuba diver
40 m

Giant squid
1000 m

Viper fish
3000 m

Titanic
3660 m

RSL mini-sub
6100 m

Trieste
10,915 m

1000 m
2000 m
3000 m
4000 m
5000 m
6000 m
7000 m
8000 m
9000 m
10000 m
11000 m

The Direction of Flow

Mystery 1

On August 3, 1492, Christopher Columbus lifted anchor and set sail across the Atlantic. He stopped first at the Canary Islands, located just off the coast of Africa. There he knew he could catch steady, northeasterly winds—these later earned the name *trade winds.* It was his idea to use these winds to blow him all the way to China. About 1 month later, he reached the Caribbean and the New World instead.

After 3 months, Columbus sailed north hoping to find suitable winds to carry him home. Luckily, he found the *westerlies,* which carried him back to Spain. No previous explorer had documented the great winds that blow in different directions across the oceans.

What was the mystery? No one knew what caused the winds or why they blew in such a consistent and predictable manner. What do you think causes the trade winds?

Portrait of Christopher Columbus

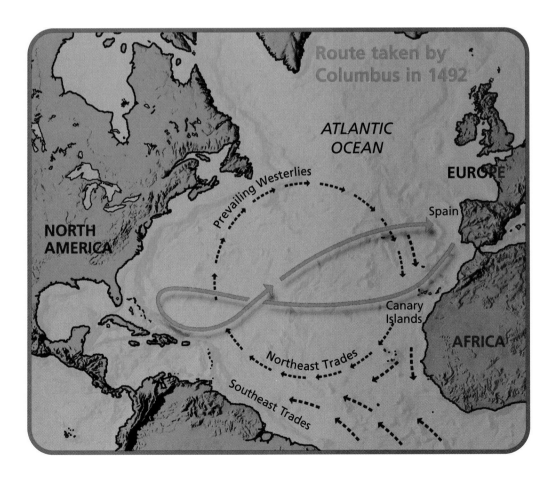

Route taken by Columbus in 1492

Mystery 2

In the mid-1700s another maritime mystery unfolded. Ships sailing from England to New York took much longer to make the journey than ships sailing from England to Rhode Island. And the two destinations were only a single day's sailing apart! Benjamin Franklin solved the mystery by talking to Atlantic whalers. They told him about a strong surface current that flowed across the Atlantic Ocean. This current was named the Gulf Stream because it was believed to originate in the Gulf of Mexico.

While traveling between Europe and America, Franklin discovered that this flow of water was warmer than the water surrounding it. He used this information to draw the first map that included the Gulf Stream. With Franklin's map, sailors could make use of the stream—or avoid it.

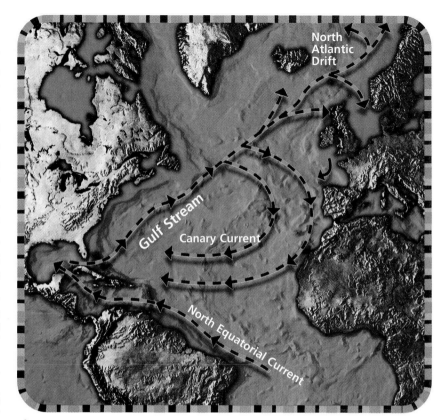

The path of the Gulf Stream

This satellite image shows the temperature of the Gulf Stream and surrounding waters.

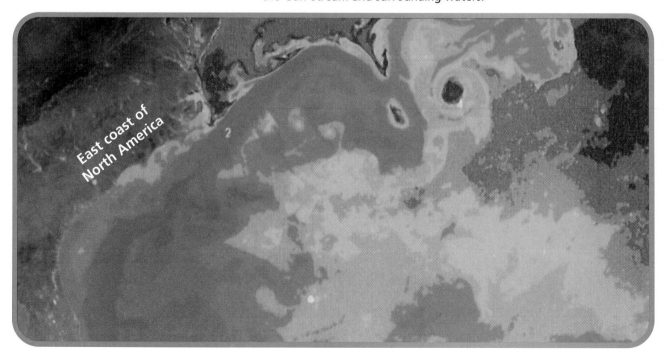

A Windy Solution

How are these two mysteries connected? It turns out that the same winds that blew Columbus back to Europe are also responsible for pushing the ocean water toward Europe. Thus, the ocean's surface currents are wind-driven.

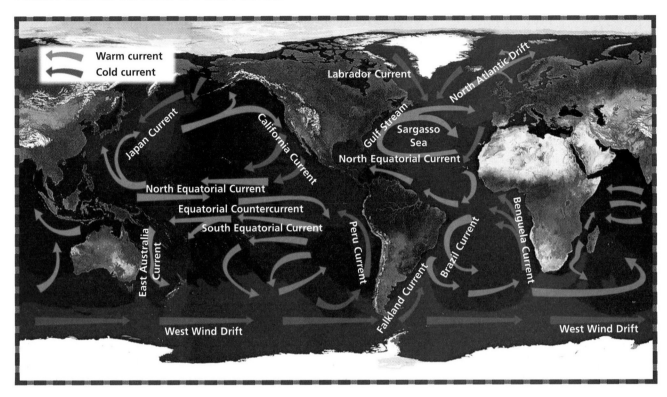

Examine the map above, which shows the surface currents of the world's oceans. What do you notice about the direction of flow of the major currents north of the equator? What about the currents south of the equator? If surface currents such as the Gulf Stream are created by winds, what can you infer about the direction of these winds? Can you think of any reason for the circular motions of the ocean currents?

In 1835, a French scientist, Gustave-Gaspard de Coriolis, published a paper that solved this mystery. He determined that the circular motions of the currents are related to the rotation of the Earth. In the next Exploration you will see exactly how this works.

I, Gustave-Gaspard de Coriolis, solved the mystery that had sailors and scientists alike scratching their heads! Read on to find out more.

The Coriolis Effect

You Will Need

- $\frac{1}{4}$ sheet of poster board
- baking soda (in a saltshaker)
- a large ball bearing or marble
- a pushpin or thumbtack
- a piece of wall paneling or a slab of plastic foam (larger than the piece of poster board)

PART 1

Direction of Rotation

What to Do

1. In the center of the poster board, write the letter *N* as a symbol for the North Pole of Earth. On the backside of the poster board, write an *S* to symbolize the South Pole.

2. The Earth rotates counterclockwise when viewed from above the North Pole. Slowly rotate your poster-board model of Earth to simulate this.

3. Continue rotating the model in this direction, but raise it so that you can see the backside of the poster board. You are now viewing the Earth's rotation from above the South Pole. In what direction is it rotating now?

PART 2

Coriolis's Explanation

What to Do

1. Near the edge of your poster board, cut a hole large enough for your finger to fit.

2. For backing, place the poster board on the piece of paneling or plastic foam. Locate the center of the poster board. Push a pin or thumbtack through this spot into the backing. Lightly dust the poster board with baking soda.

3. Without rotating the model, roll a wet marble across it. Describe its path.

4. Now place your finger in the hole you cut earlier and spin the poster board (not too fast!) to simulate the Earth's rotation as observed from above the North Pole. As the poster board is rotated, have someone roll the wet marble across its surface, as shown below. (The throw must be timed so that your hand doesn't get in the way as you turn the poster board.) Describe the path the marble takes.

5. Try rolling the marble in different directions across the rotating model. As you look in the direction of the marble's motion, does its path always swing in one particular direction?

6. Rotate the model Earth clockwise. Now what pole does the center of the poster board represent?

7. Roll the marble across the model Earth while it is rotating clockwise. As you look in the direction of the marble's motion, does its path swing one way or another?

Your Analysis

From your observations in this Exploration, find proof for the following conclusions. Then write a statement in your ScienceLog that explains how you came to each conclusion.

1. When viewed from above the North Pole, the Earth rotates in a counterclockwise direction.

2. When viewed from above the South Pole, the Earth rotates in a clockwise direction.

3. A moving object in the Northern Hemisphere swings to the right as an observer looks in the direction of the object's motion.

4. A moving object in the Southern Hemisphere swings to the left as an observer looks in the direction of the object's motion.

Dust the poster board with baking soda, and rotate the poster board to simulate the Earth's rotation. Roll a wet marble across the poster board's surface and observe its path.

Explaining Your Discoveries

If you found evidence to support the statements on the previous page, congratulations! You have rediscovered what Coriolis first explained in 1835—the **Coriolis effect**.

Return to the map of the ocean's surface currents on page 271. Apply your knowledge of the Coriolis effect to an analysis of the Gulf Stream. Can you now explain why the current in the Atlantic Ocean moves in a clockwise pattern? Why do currents in the Southern Hemisphere follow a counterclockwise pattern? Remember that the major factor that drives surface currents is the wind. What can we infer about the direction of the winds from the direction of the surface currents?

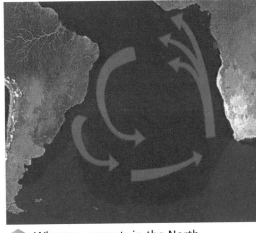

Whereas currents in the North Atlantic flows clockwise, currents in the South Atlantic flow counterclockwise.

The Sargasso Sea

As the currents of the North Atlantic Ocean slowly flow in their clockwise pattern, a calm area forms at the center of the rotation. This area is called the Sargasso Sea. Under constantly clear skies, the sea is warmed by the sun, which increases the evaporation of water in the area. Thus, the Sargasso Sea is saltier than the surrounding ocean waters.

The area has its own ecosystem. A seaweed called sargassum weed has adapted to this unique environment. It floats in the warm water, often forming patches that cover large areas of open sea. The Sargasso Sea is also home to a number of unique animals. One of these, the silver eel, lives in fresh water but

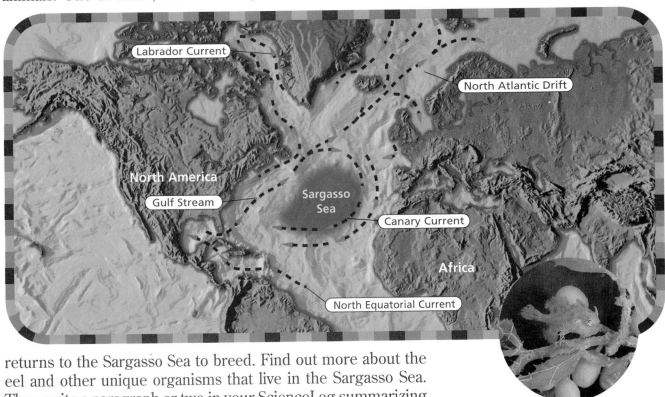

Sargasso Sea life

returns to the Sargasso Sea to breed. Find out more about the eel and other unique organisms that live in the Sargasso Sea. Then write a paragraph or two in your ScienceLog summarizing what you've learned.

World Winds Explained

What causes the winds that helped Columbus reach America and return home to Spain? These same winds also drive the currents in the Atlantic Ocean. In what direction do they blow? Why? The answers lie in ideas that you have already investigated. First, you know that air moves from areas of high pressure to areas of low pressure. Second, the movement of air is influenced by the Coriolis effect. Keep these ideas in mind as you complete the Exploration.

What to Do

In small groups, discuss the following information. Copy each diagram into your ScienceLog and respond to the questions.

1. Permanent high- and low-pressure areas exist at certain latitudes because of temperature differences. Why is a low-pressure area located at the equator? Why are high-pressure areas located at the poles?

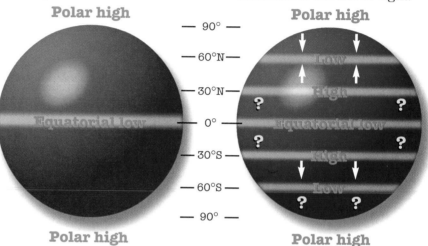

2. Other factors, especially the rising and sinking of air at certain latitudes, create other permanent high- and low-pressure areas.

   What is the approximate line of latitude for the United States?

3. Winds blow from areas of high pressure to areas of low pressure. If Earth did not rotate and there were no Coriolis effect, winds would blow from high- to low-pressure areas with no apparent deflection to the left or right.

A few wind patterns are shown on the diagram above. Sketch the diagram and draw in the missing arrows.

4. Because the Earth rotates, the winds are deflected to the right or left, depending on the hemisphere. A few wind directions are shown below. Sketch the diagram in your ScienceLog and draw in the missing information.

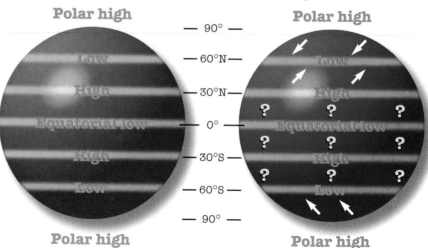

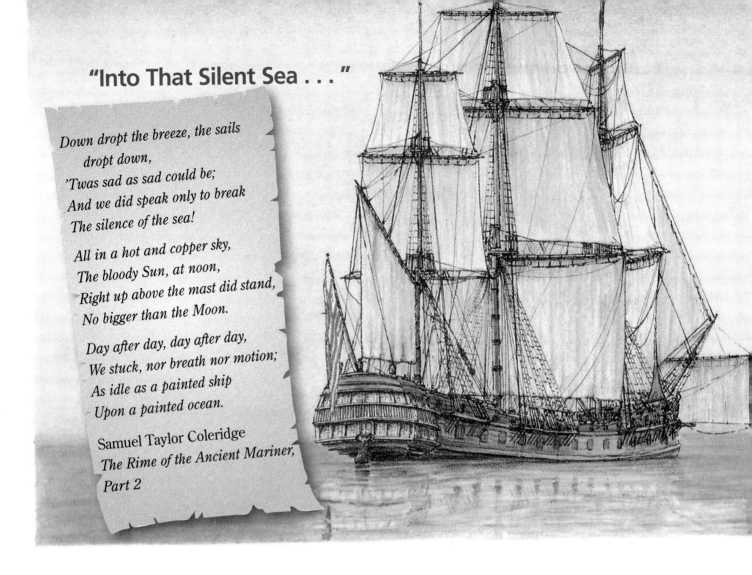

"Into That Silent Sea . . ."

Down dropt the breeze, the sails
      dropt down,
'Twas sad as sad could be;
And we did speak only to break
The silence of the sea!

All in a hot and copper sky,
The bloody Sun, at noon,
Right up above the mast did stand,
No bigger than the Moon.

Day after day, day after day,
We stuck, nor breath nor motion;
As idle as a painted ship
Upon a painted ocean.

Samuel Taylor Coleridge
The Rime of the Ancient Mariner,
Part 2

Coleridge published this poem in 1798. At that time sailing ships were the only means of travel across the vast oceans. How did he refer to the characteristics of winds and currents in his poetry? Where do you think the sailors are stranded? (Hint: See the second stanza, which describes the position and appearance of the sun.) Did sailors really have cause to worry that the wind would stop blowing?

During the time when ships were powered by the wind, certain regions of the ocean presented serious dangers. Near the equator is an area where the air rises due to intense heating, so there is very little wind. Ships could sit for days or even weeks with no wind to fill their sails. This region became known as the **doldrums.**

And how would you like to be stuck in the **horse latitudes**? This is a region of calm located at the northern edge of the northeast trade winds. In the 1700s, ships sailing through this area often carried horses. If a ship was slowed or stranded and supplies ran low, the horses were thrown overboard in order to conserve drinking water—hence the name horse latitudes.

El Niño

What is El Niño? Consider the following news items, and then answer the questions below.

Experts Ponder Impact of El Nino Phenomenon

North America could be facing dramatic weather changes due to El Nino, a variation of temperature and pressure patterns over the eastern Pacific Ocean.

This phenomenon occurs every three to five years — its severity changing from year to year. Weather patterns worldwide are affected. The strong El Nino of 1982-83 was blamed for a devastating drought in Africa and Australia. Also, severe winter storms lashed California, and parts of South America were deluged with torrential rains. The El Nino in 1986-87, however, was barely noticed.

What Causes El Nino?

El Nino (Spanish for "the Christ Child") occurs in the equatorial waters off the west coast of South America around Christmas time. Normally, trade winds push surface waters away from the coastline of South America, causing cold water to well up from below. The cold surface waters suppress the formation of clouds and rain. Every few years these winds die down and then reverse, causing warm water to be pushed toward the South American coast. The warmer waters produce abundant rainfall. At the same time, normal weather patterns are disrupted over a wide area.

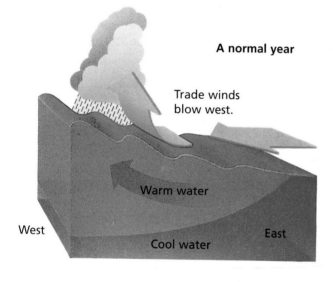

A normal year

Trade winds blow west.

Warm water

West

East

Cool water

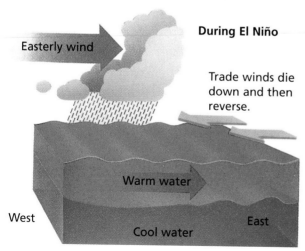

During El Niño

Easterly wind

Trade winds die down and then reverse.

Warm water

West

East

Cool water

In a normal year (top), westward-blowing winds push warm waters toward the western Pacific, where they "pile up" (although only to a height of a meter or two). When El Niño occurs (bottom), these winds reverse, causing warm waters to pool along the western coast of South America.

Questions

1. The El Niño effect illustrates the link between oceans and the atmosphere. Use information from the articles above to describe this link.

2. Examine the map on page 271 that shows the ocean's surface currents. What is the name of the surface current that flows up the coast of South America?

3. Examine the picture of global temperatures on page 226. What evidence suggests that this colder current flows up the west coast of South America?

4. What kind of damage can El Niño cause? What regions suffer the most? How are plant and animal life affected? Research the El Niño phenomenon to learn more about its effects.

LESSON 3

Hurricanes

Have you been following daily pressure changes on your homemade barometer? What kind of pressure change is associated with calm, sunny weather? with unsettled, stormy, or cloudy weather? Did you find that calm weather is usually associated with increasing atmospheric pressure and that stormy weather is associated with decreasing pressure? What is the reason for this? Through the diagrams and pictures that follow, you'll examine the relationship between low atmospheric pressure and stormy weather.

A hurricane, seen from space

Picture Study 1

The satellite picture above shows an unusual weather system—a *hurricane.* Hurricanes vary in size, but most are 150 to 450 km across. Some have winds up to 300 km/h.

1. Assuming that there is a low-pressure area at the center of the hurricane, which way are the winds blowing—inward or outward?

2. Are the winds moving in a straight line, or are they spiraling around the hurricane's center? How does the photograph suggest an answer to this question?

3. Is the wind moving clockwise or counterclockwise? Why does it blow in this direction? (In the Southern Hemisphere, the wind direction around low-pressure areas is the opposite of that in the Northern Hemisphere.)

Picture Study 2

Compare the satellite picture on page 277 with the cross-sectional view of a hurricane shown below. As you read the following, match the italicized terms with the numbers in the diagram and answer the accompanying questions.

1. Most hurricanes are carried toward the eastern coast of the United States by *westward-flowing winds*.

   What are these winds called?

2. The *surface winds* pick up heat energy from the warm waters they pass over. In addition, a large amount of water evaporates—another mechanism of gathering energy that will later be released.

   How is heat energy stored by the evaporation process?

3. As the *warm, moisture-laden air* spirals toward the center of the hurricane, it also rises. Condensation creates clouds. This process releases heat energy.

   Why does moisture condense out of the air as the air rises?

4. Some of the rising air is carried away by *high-altitude winds*.

   Which way do these high-altitude winds spiral?

5. Some of the rising air is drawn back down into the *eye of the hurricane*. Air in the eye sinks, warming as it does so and causing clear skies and calm conditions. Why doesn't the descending air form clouds like moist, rising air does?

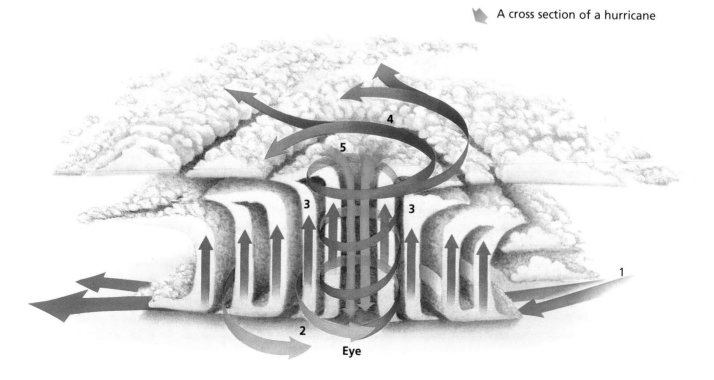

A cross section of a hurricane

Eye

Picture Study 3

A hurricane is an extreme example of a low-pressure system. Normal atmospheric pressures are close to 100 kPa. The eye of a hurricane deviates sharply from this norm.

The diagram at right shows a hurricane as it might be symbolized on a weather map. A series of solid lines called *isobars* connect points of equal pressure. The same pressure difference separates each isobar. Wind directions have been included on this map as well.

1. What is the wind-direction pattern around a low-pressure system such as a hurricane—clockwise or counterclockwise?

2. How does the pressure change as you progress farther from the eye of a low-pressure system?

3. At each station, what number was dropped from the complete pressure reading? (Why?)

4. Why are the dew point and the air temperature readings the same?

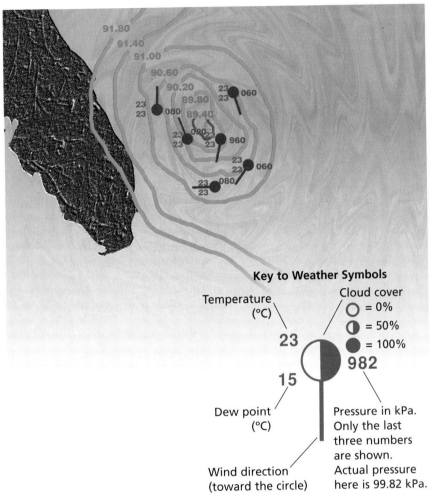

Key to Weather Symbols

Temperature (°C)

Cloud cover
○ = 0%
◐ = 50%
● = 100%

23

982

15

Dew point (°C)

Wind direction (toward the circle)

Pressure in kPa. Only the last three numbers are shown. Actual pressure here is 99.82 kPa.

Keeping Track

Make a list of the major ideas and findings you have encountered in this chapter. What have you discovered about pressure in the atmosphere and ocean? What causes global ocean currents and wind? How does a hurricane form? Your list should answer these questions and more. Compare your list with one from a classmate.

CHALLENGE YOUR THINKING

1. Treading Lightly

Jack's mass is 60 kg (600 N). His snowshoes cover an area of 1.5 m². How much pressure does he exert against the snow?

2. Pressure Points

Another name for a low-pressure system is a cyclone. High-pressure systems are called anticyclones.

a. Is a hurricane a cyclone or an anticyclone?

b. Describe the airflow around a cyclone and around an anticyclone.

3. Weather Forecast

Looking at the weather map shown below, identify the following:

a. wind direction at point *A*

b. temperature differences between points *B* and *C*

c. wind direction at point *D*

d. regions of cloudy skies and rain

e. areas of high pressure

f. areas of low pressure

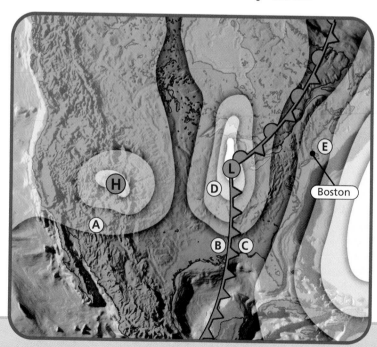

Over the next day the low-pressure system will move over point *E* (Boston). How do you think the pressure readings in Boston will change? What about the temperature? How will wind direction change?

4. The Motion's Over the Ocean

Once a hurricane reaches land, it rapidly loses energy. Can you think of a reason for this?

5. Pressing Issues

Here are some everyday experiences and observations. They all have something to do with atmospheric pressure. What is the connection?

a. On Nora's last plane trip, she noticed that the tops of the small sealed creamer containers bulged outward. But in restaurants on the ground they do not.

b. While drinking a milkshake through a straw at the ice cream shop, Janis wondered exactly how the straw worked.

c. Rick always had trouble getting frozen orange-juice concentrate out of the can, until someone suggested punching a small hole in the bottom of the can.

6. Global Confusion

What do you think would happen to the pattern of global winds and currents if the Earth were suddenly to start spinning in the opposite direction—from east to west instead of from west to east? How would low- and high-pressure systems behave differently?

7. Touchy Question

When you place your finger over the end of a plastic syringe, it becomes very difficult to pull out the plunger. Why is this?

ScienceLog

Review your responses to the ScienceLog questions on page 259. Then revise your original ideas so that they reflect what you've learned.

Making Connections

Unit 4

SourceBook

Look in the SourceBook to read more about the atmosphere and oceans that make Earth a temperate planet. The SourceBook also contains information about a subject that's always topical—the weather.

Here's what you'll find in the SourceBook:

The Big Ideas

In your ScienceLog, write a summary of this unit, using the following questions as a guide:

1. What is the greenhouse effect, and what gases are responsible for it? How might it relate to global warming?
2. What influence do the oceans have on the climate of the Earth?
3. What is the dew point?
4. What causes ocean currents? What causes winds?
5. What role does density play in driving winds and ocean currents?
6. What are air masses, and how do they form?
7. Why do winds and the ocean's surface currents move in predictable patterns?
8. How does pressure affect the weather?

Checking Your Understanding

1. Every summer, the average concentration of atmospheric carbon dioxide drops significantly in the Northern Hemisphere. What reasons can you suggest for this?

2. The diagram at right shows the path of a sailplane (a type of glider) on a sunny day.

   a. Use your understanding of air currents to explain the glider's movements.

   b. How would the glider's path differ on a cloudy day? at night? over a city?

   c. How do the effects shown here relate to weather and climate?

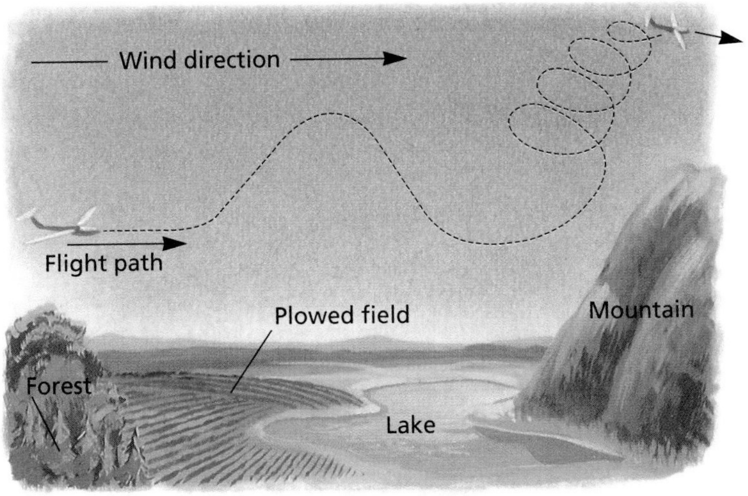

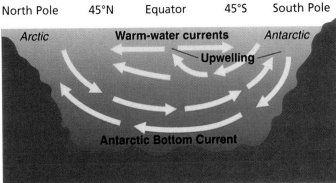

Wind direction

Flight path

Plowed field

Mountain

Forest

Lake

3. Examine the simplified diagram of deep-ocean currents shown below.

   a. What causes these currents?

   b. How do these currents help to absorb the world's "excess" carbon dioxide?

   c. If glaciers melted, what effect might this have on deep-ocean currents?

North Pole   45°N   Equator   45°S   South Pole

Arctic   Warm-water currents   Antarctic

Upwelling

Antarctic Bottom Current

4.  **concept map** Copy the concept map at right into your ScienceLog. Then complete the map using the following words and phrases: currents, oceans, atmosphere, trade winds, Antarctic currents, and density differences.

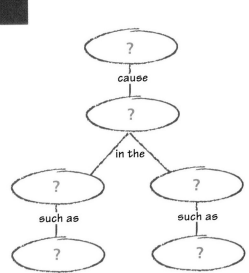

?

cause

?

in the

?   ?

such as   such as

?   ?

WEiRD Science

Weather From Fire

In 1993 a devastating wildfire took place near Santa Barbara, California. As the fire burned, huge storm clouds formed in an otherwise cloudless sky. Fast-moving fiery whirlwinds danced over the ground. The fire wasn't only destroying everything in its path—it was creating its own weather!

Fire-Made Clouds

Hot air rising from a forest fire can create tremendous updrafts. Surrounding air rushes in to take the place of the rising air, stirring up huge plumes of ash, smoke, hot air, and noxious gases. Cool, dry air normally sinks down from above the fire and stops these plumes from developing further. But if the conditions are just right, a surprising thing happens.

If the upper atmosphere contains warm, moist air, the moisture begins to condense on the ash and smoke particles. This creates droplets that can develop into clouds. As the clouds grow, the droplets begin to collide and combine until they are heavy enough to fall as rain. The result is an isolated rainstorm complete with thunder and lightning.

Whirlwinds of Fire

Forest fires can also create whirlwinds. These small, tornado-like funnels can be extremely dan-

▲ **A towering whirlwind sucks burning debris from a forest fire in Idaho**

gerous. Whirlwinds are similar to dust devils that dance across desert sands. They are created by upward-moving currents of air. Their circular motion is started by an updraft that is forced to curve after striking some obstacle such as a cliff or slope of land. Whirlwinds move across the ground at 8 to 11 km/h, sometimes growing up to 120 m high and 15 m wide.

Most whirlwinds last no longer than a minute, but within this time they can cause some big problems. Firefighters caught in the path of whirlwinds have been severely injured and even killed. Also, if a whirlwind is hot enough, it can suck up tremendous amounts of air into its vortex. The resulting updraft can pull burning debris, called firebrands, up through the

whirlwind. In some cases, the updraft has been so strong that burning trees have been uprooted and shot into the air above. When such firebrands land, they often start new fires hundreds of meters away.

Think About It

Many scientists agree that fires are a natural part of the growth of a forest. For example, some tree seeds are released only under the extreme temperatures that occur in a fire. Some people argue that forest fires should be allowed to run their natural course. Others argue that the effects of a forest fire are just too damaging and that forest fires should be put out as soon as possible. Which side do you agree with? Why?

Space-Age Subs

*T*he Mariana Trench in the Pacific Ocean near Guam plunges to a bone-crushing depth of 11 km. This is the deepest point of the ocean, and the water pressure at this depth is over 110,000 kilopascals (kPa). This is enough pressure to collapse a conventional submarine. In fact, due to these great pressures, this dark ocean abyss remains shrouded in mystery. But a new generation of space-age submersibles may change all that!

Deep-Diving Robots

One approach to exploring the deepest parts of the ocean is to use unmanned submersibles. Japan's Marine Science and Technology Center has perfected such a vessel, called *Kaiko*.

Kaiko is a 5.4 ton, 3.5 m aquatic robot that operates at more than 10,000 m below the ocean's surface. It is equipped with high-tech TV cameras, sonar, and a pair of 1.8 m robotic arms. However, like most robotic submersibles, *Kaiko* must be tethered to a research ship by 11 km of power and communications cables.

As *Kaiko* sinks to within a few meters of the ocean floor, a small robotic vehicle separates from the bottom of the main craft. Although attached to the main craft by a control cable, the robotic vehicle descends still deeper to explore the ocean floor, transmitting pictures to

▼ The *Kaiko* submarine. *Kaiko* means "trench" in Japanese.

▲ The *Deep Flight* submarine. The pilot lies in this sub in a prone position.

the surface and gathering samples. When its work is done, the robotic vehicle returns to the main craft and is taken back to the surface.

Nothing Beats Being There

The Ocean Exploration Group in Richmond, California, has designed a submersible that looks more like a fighter plane than a submarine. Unlike conventional submersibles, which slowly sink to the bottom, this one-person sub, called *Deep Flight*, is designed to "fly" to the sea floor under it own power. The pilot determines the sub's flight path by adjusting its wing flaps and tail fins. The craft could dive vertically, but this would be uncomfortable for the pilot. Instead, it descends at an angle of about 30 degrees.

Deep Flight's 3.5 m hull is made of a hard ceramic material that is extremely strong under pressure. In addition to the pilot, the vessel houses the controls, an instrument panel, a life-support system, and a 24-volt, battery-operated power supply. The batteries feed a pair of electric motors at the back of the craft that propel it up to 25 km/h (14 knots).

Currently we know more about the surfaces of other planets than we know about our own ocean floor. But with submersibles like *Kaiko* and *Deep Flight*, we may some day unlock the mysteries of the deep.

Find Out for Yourself

Scientists divide the ocean into several layers, or zones—the sublittoral zone, bathyal zone, abyssal zone, and hadal zone. Find out how each of these zones is defined and what its characteristics are.

The Wildest of Waves

*I*magine a foamy wave sloshing onto a peaceful beach. Now imagine the same wave three stories high and moving as fast as a truck on a highway. Sound impossible? It's not. It's called a *tsunami*, and it's very real and very danger-ous. *Tsunami* (soo NAH me) means "harbor wave" in Japanese.

The Birth and Death of Giants

Where do these immense waves come from? Tsunamis are created far out in the ocean. A tremor on the ocean floor caused by an underwater land-slide or earthquake can make a huge volume of water rise and fall quickly, creating a wave. Out in the middle of the ocean, the wave may not look very large or powerful. However, as the wave nears the shore and the water gets shallower, the wave changes. It slows down and gets taller and taller, sometimes reaching as high as 10 m by the time it crashes onto the shore-line. A large tsunami can demol-ish trees, houses, cars, and anything else in its path. One tsunami hit Sanriku, Japan, in 1896 and killed between 22,000 and 27,000 people.

▲ **Monster waves are well known in many communities along the Pacific coast. Hiroo Kanamori (inset) thinks that mud can make a difference.**

Predicting Tsunamis

Since most tsunamis are caused by underwater earthquakes, sci-entists can monitor earthquakes to predict when and where a tsunami will hit land. But the predictions are not always accu-rate. Some earthquakes are very weak and should not create powerful tsunamis, yet they do. The reason for this has scientists puzzled.

A scientist named Hiroo Kanamori has an idea. He thinks that when the ocean floor shifts, sediment can sometimes act as a lubricant, allowing a smooth movement of the floor instead of an abrupt jolt. Thus, even if the earthquake does not create a big jolt, it may still create a powerful tsunami. Kanamori calls these special events *tsunami earthquakes*. By modi-fying traditional meas-uring devices to ac-count for this effect, he has had more success in pre-dicting the birth of a tsunami. Accurate early warnings can give people enough time to get safely out of the tsunami's way.

Wavy Water

The speed of water waves, including tsunamis, depends on the depth of the water. Check it out for yourself by filling a tub that is about 0.5 m long and at least 5 cm deep with about 1 cm of water. Then nudge the tub. How long does it take for the wave to go back and forth? Add more water, and nudge the tub again. How fast does the wave move the second time? What do you think would happen to the speed of the wave if you made the water deeper still? Try it!

From Thermometer to Satellite

▲ Photograph of Hurricane Elena taken from the space shuttle *Discovery* in September 1985. Cristy Mitchell (inset) uses computer technology to analyze the weather.

Predicting floods, observing the path of a tornado, watching the growth of a hurricane, and issuing flood warnings are all in a day's work for Cristy Mitchell. As a meteorologist for the National Weather Service, Cristy spends each working day observing the powerful forces of nature.

Forces in the Atmosphere

When asked what made her job interesting, Cristy Mitchell replied, "There's nothing like the adrenaline rush you get when you see a tornado coming! I would say that witnessing the powerful forces of nature is what really makes my job interesting."

Meteorology is the study of natural forces in the Earth's atmosphere. Perhaps the most familiar field of meteorology is weather forecasting. However, meteorology is also an important aspect of air-pollution control, agricultural planning, and air and sea transportation. Meteorologists also study trends in the Earth's climate, such as global warming and ozone depletion. In fact, anything that relies on or relates to weather is a meteorologist's concern.

Collecting the Data

Meteorologists collect data on air pressure, temperature, humidity, and wind velocity.

Then, by applying what they know about the physical properties of the atmosphere and analyzing the mathematical relationships in the data, they are able to forecast the weather.

The data that a meteorologist needs in order to make accurate weather forecasts comes from a variety of sources. When asked what kinds of instruments she uses, Cristy said, "The computer is an invaluable tool for me. Through it, I receive maps and detailed information, including temperature, wind speed, air pressure, and general sky conditions for a specific region."

In addition to computers, Cristy also relies on radar and satellite imagery to show the "whole picture" of a region and the nation's weather. Meteorologists also use sophisticated computer models of the world's atmosphere to help forecast the weather and weather trends.

Find Out for Yourself

Using the barometer you made in Chapter 12, Lesson 1, graph the barometric pressure for 7 days. At the same time, graph the barometric pressure reported by a local meteorologist. How do the graphs compare? What kind of forecast would you make based on the results of your barometer? How does your forecast compare with that of the meteorologist?

Unit 5

Electromagnetic Systems

Like some fantastic mechanical ballet, sophisticated industrial robots put together cars on an assembly line. Working at high speed and with extreme accuracy, the robots perform repetitive tasks without ever growing tired.

These complex mechanical devices are dependent on electricity in many ways. The energy that permits the robots' motion is supplied by electric motors. At the same time, powerful pulses of electric current are used to weld together the car parts as they pass down the assembly line. Finally, electrical signals from a computer direct the motions of the robots' mechanical parts.

Magnets also play a role in this industrial drama. In fact, all electric motors contain magnets and could not operate without them. Furthermore, the generators that supply the electricity to make this scenario possible contain giant magnets. As you might have guessed, electricity and magnetism are related. What is some evidence of this relationship?

Electricity and magnetism are key elements of our modern industrial society. In the unit that follows, you will learn more about the nature of electricity and magnetism, the relationship between them, and the ways in which they are utilized and controlled.

Cars being assembled on a fully automated assembly line

289

CHAPTER 13

Energy in a Wire

What is electricity? 1

How would you arrange the materials pictured below to make a *circuit*? 2

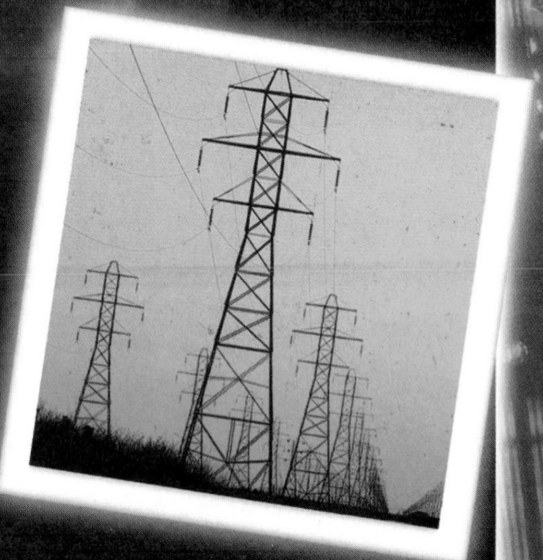

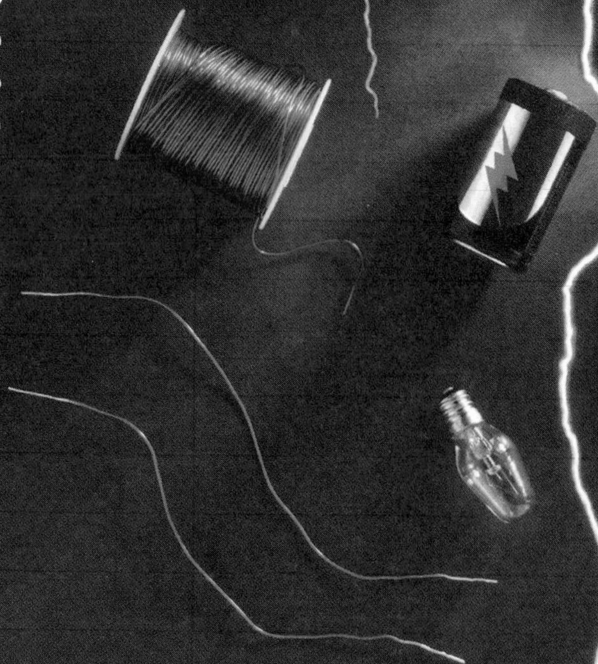

3 What causes electricity to flow?

ScienceLog

Think about these questions for a moment, and answer them in your ScienceLog. When you've finished this chapter, you'll have the opportunity to revise your answers based on what you've learned.

Indispensable Energy

Your school is planning an exhibition called the Museum of Today. The idea behind the exhibition is a simple one: to introduce the technology of today to imaginary visitors from the past. It is your job to plan the exhibition. Include as many examples as you can of the technology and gadgetry that make life what it is today.

Where do you start?

You could start with simple things like electric lights. We take them completely for granted, yet we could hardly imagine life without them. Electric lights were a revolutionary development. With their invention, the 24-hour-a-day city suddenly became a possibility. Think of the tremendous variety of different types of lights and their uses. What are some of them?

Then, of course, there's the telephone. Who could imagine life without the telephone? A hundred and fifty years ago, to be able to punch a few buttons and talk to someone far away was inconceivable.

And don't forget radio. How wonderful it is to be able to flip a switch and hear music, talk shows, or news—all of it originating many kilometers away.

Television! Now there's an invention! Life would be far different without television. Who would have imagined that you could send pictures through the air and have them appear on a screen? That would seem like magic to someone from the past.

Here are a few examples of technology to get you started. What other examples would you include in your exhibit?

The Museum Is Open!

As you might imagine, most of the exhibits in the Museum of Today would perplex visitors from the past. In fact, the lights, sounds, and moving parts of the items on display might even scare them—a situation that could lead to chaos, since even the idea of turning something "off" would be unfamiliar to most of your guests.

Of course today everybody knows how to flip a switch on and off. But how many people stop to consider what actually happens when something is turned on or off? What happens when that button is pushed or that switch is flipped or that knob is turned? Fortunately, we do not need to know the answer to these questions in order to use our microwave ovens, home computers, lamps, dishwashers, power tools, and so on.

An Electrical Connection

What do all the devices mentioned so far in this chapter have in common? They all use electricity. Most of the devices use magnetism as well. Obviously, electricity and magnetism serve us in many ways. Can you identify some of these ways?

Clearly, life without electricity would be very different. Imagine going an entire day without it. How would you get by? Write a scenario about a day in your life without electricity.

What exactly is electricity? How would you explain it to your visitors from the past? Although you can see what electricity does, you can't see electricity itself. In this unit you will be introduced to this mysterious, indispensable form of energy and its close relative, magnetism.

Electricity and You: Case Studies

It takes only a little knowledge and skill to make use of many common electrical devices. It takes more knowledge to understand how they work—and still more to design them. With one or two classmates, discuss how electricity works in one of the following situations. Then make up a case study of your own.

Case Study A

"Let's see," thought Bill. "Colors go in cold water and whites go in hot—I think." Bill has many options for setting the automatic washer. What are some of them? How does he control them? How many different kinds of functions does the machine perform or control? For example, if Bill had only a small load of laundry to do, could the electrical system control the amount of water needed? Does he need to stay beside the washer throughout the cycle? Why or why not? What takes place during the wash cycle? How do you think these events occur? Are there any safety features? Sometimes a system provides you with information as it operates. This is called feedback. Can you identify any controls that serve as feedback mechanisms? What is the source of electricity for the washer?

Case Study B

Jenny arrives at the hospital to visit her friend Kim, who is on the fourth floor. Jenny uses the elevator. What information can she get from the elevator lights? What control does Jenny have over the elevator's operation? Are there other controls in addition to those that she uses? For example, are there safety controls to prevent people from walking into an empty shaft? What features of the electrical design are specifically for the passengers' benefit? for their safety? What feedback systems are there? The elevator is a complex electrical system. Although you cannot see the subsystems, you know that they exist because of the functions they perform.

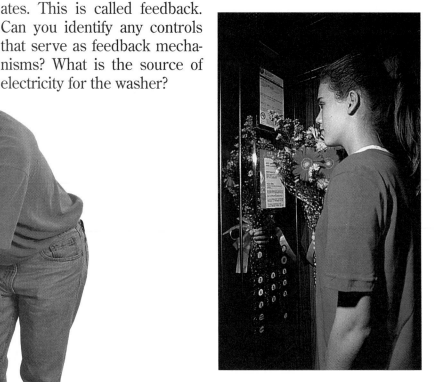

Case Study C

With much anticipation, Dylan turned on his computer. After it started up, he inserted the CD-ROM he'd just bought. The computer beeped and whirred. Then the words *American History* appeared on the screen while music played. After a few clicks of the mouse, Dylan was listening to a speech given by John F. Kennedy almost four decades ago. After a few more clicks of the mouse, he was watching Neil Armstrong walk on the moon. Then Dylan used the computer to send a message to his schoolmate across town: "Come on over—I got some great reference material for our history project."

Electricity is certainly at work in a computer. What are some things that a computer can do for you? What are some ways in which a computer provides feedback? The computer is quite complex. However, you do not need to know how it works in order to operate it. How do you control the computer's various operations?

Case Study D

"Drivers start your engines!" Alex turned the key, expecting to feel her race car's powerful engine roar to life. Instead she heard only the "clunk" that signals a dead battery. Later, Alex lifted the hood and stared at the engine. "What could possibly be wrong?" she wondered. "It started fine in practice an hour ago . . ."

What could be wrong with Alex's car? What components make up a car's electrical system? Where does the power needed to start a car come from? What kind of electrical feedback or control systems do cars have? How are different forms of energy converted to electricity, and vice versa, in a car?

Your Own Case Study

In your ScienceLog, describe a situation in which electricity performs some sort of function for you. Describe any devices or processes involved, what takes place, how the device or process is controlled, any energy conversions that take place, and so on.

EXPLORATION 1

Electricity Working for You

You have probably seen warnings like the one below on appliances or power tools. As useful as it is, electricity in large amounts is deadly and must be carefully controlled.

The familiar situations that you have been analyzing in the Case Studies involve complex electrical parts and arrangements. Large quantities of electricity are used in the devices in Case Studies A and B. This is true for the operation of most appliances in homes, stores, or industry. A moderate amount is used in the devices in Case Studies C and D. Care must be exercised in using these amounts of electricity.

Most of the Explorations in this unit require only a small amount of electricity. They are quite safe to do. Here are four experiments in which a small amount of electricity works for you by producing other forms of energy. Watch for these energy forms.

In these experiments the electricity is supplied by dry cells such as those used in flashlights.

ACTIVITY 1

Shedding a Little Light

You Will Need

- a length of magnet wire
- sandpaper
- a flashlight bulb
- a D-cell
- a rubber band

What to Do

1. Begin by sanding the enamel off the last few centimeters of each end of the wire to expose the copper underneath. Then, using only the other items listed, find all of the different arrangements that will light the bulb.

2. Sketch each arrangement.

3. What form(s) of energy does the electricity produce?

4. What are other examples of electricity being used in this way?

5. You have constructed an electric *circuit*. What parts make up this circuit? Check the dictionary to find out the origin of the word *circuit*. How is it significant? What would you say an electric circuit is?

The Heat Is On

You Will Need

- two 30 cm lengths of magnet wire
- sandpaper
- modeling clay
- a clothespin
- 2 D-cells
- a wide rubber band
- a strand of steel wool
- aluminum foil
- a thin nichrome wire (10 cm long)

What to Do

1. Sand the enamel off of the last 5 cm of the ends of each length of magnet wire. Then make a small loop at one end of each wire.

2. Bend the wires and support them with modeling clay, as shown.

3. Attach the ends of the wires to the D-cell, securing them with a wide rubber band.

4. Place a strand of steel wool through the loops and let it rest in contact with each loop. What do you observe? Repeat with a rolled-up length of aluminum foil and then with a length of nichrome wire. Bring your hand close to each piece being tested, but do not touch any of them. What do you feel?

5. What form of energy does the electricity produce?

6. What are some examples of devices in which this type of electrical energy is used?

7. Do Activities 1 and 2 demonstrate the same principle? Explain.

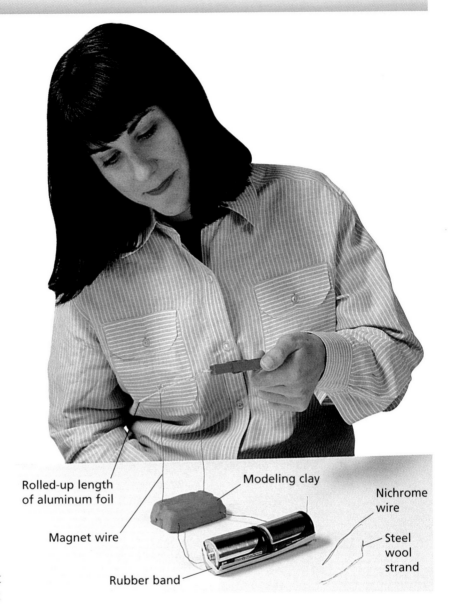

Rolled-up length of aluminum foil

Modeling clay

Nichrome wire

Magnet wire

Steel wool strand

Rubber band

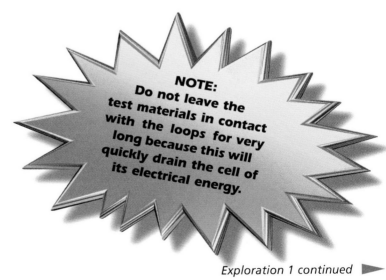

NOTE:
Do not leave the test materials in contact with the loops for very long because this will quickly drain the cell of its electrical energy.

Exploration 1 continued ▶

As you may have guessed from the title of this unit, there is a connection between electricity and magnetism. The following Activities will help illustrate that connection.

ACTIVITY 3

The Electricity-Magnetism Connection

You Will Need

- a D-cell
- a compass
- 2 thumbtacks or screws
- a wood block
- a paper clip
- 2 lengths of magnet wire (15 cm and 25 cm long)
- a rubber band
- sandpaper

What to Do

1. Sand the enamel off of the last 2 or 3 cm of the ends of each wire. Then set up the apparatus as shown. Align the compass needle, and place the wire over the compass in a north-south direction so that the wire lines up with the compass needle.

2. Close the electrical circuit by pressing the contact switch. What happens?

Caution: Don't keep the switch closed for very long.

3. What kind of energy does the electricity produce in this experiment?

4. Can you think of any everyday applications that make use of electricity in this way?

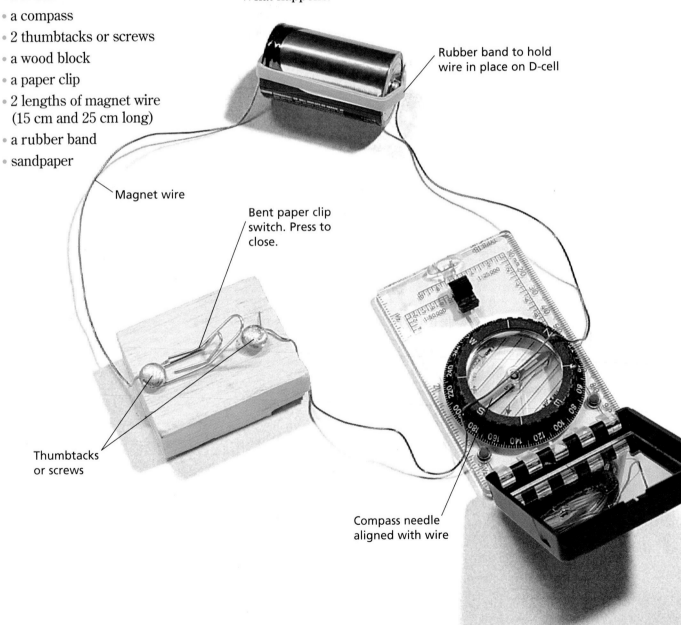

Rubber band to hold wire in place on D-cell

Magnet wire

Bent paper clip switch. Press to close.

Thumbtacks or screws

Compass needle aligned with wire

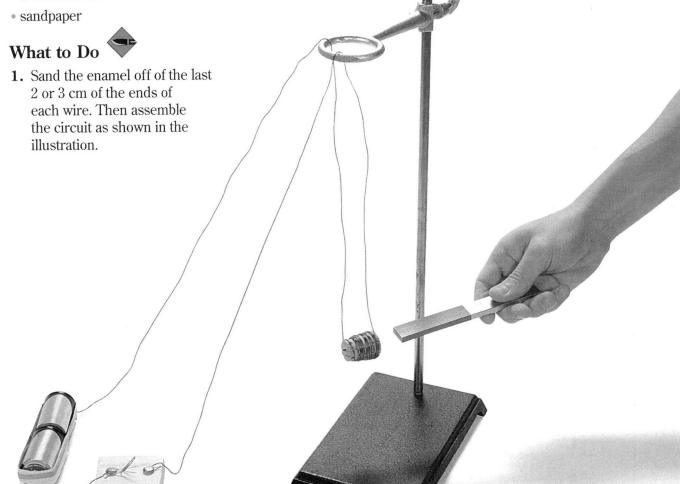

ACTIVITY 4

Let's Get Moving!

You Will Need

- 2 D-cells
- a paper clip
- 2 lengths of magnet wire (10 cm and 250 cm long)
- bar magnet
- a support stand with ring clamp
- a cork
- a wide rubber band
- a narrow strip of cardboard (2 cm × 30 cm)
- 2 thumbtacks or screws
- a wood block
- sandpaper

What to Do

1. Sand the enamel off of the last 2 or 3 cm of the ends of each wire. Then assemble the circuit as shown in the illustration.

2. Have one person hold a strong magnet near the cork while the other person presses the contact switch to complete the circuit. Observe what happens. Open the switch. What happens?

Caution: Don't keep the switch closed for very long.

3. Using the other end of the magnet, repeat step 2. What happens?

4. What kind of energy is produced by the electricity?

5. What are some practical examples of how energy works for you in this way?

A Home Project

Make a battery tester. Use one of the arrangements you discovered in Activity 1 of Exploration 1. Devise a tester that consists of a light bulb with two wires connected to it. Touch the ends of the wire to the battery. The brightness of the light bulb will indicate the strength of the battery.

2 What Is Electricity?

In Exploration 1 you used electricity to produce other forms of energy. What were they? Later in the unit you will discover that each of these energy forms can, in turn, produce electricity. Remembering that energy can be transformed from one form into another, does it seem that electricity itself must be a form of energy, just as heat and light are?

Electricity accomplishes these changes in energy when an **electric current** is flowing in a circuit. But what is a current? What flows or moves in the circuit?

Have you ever gotten a shock when you touched someone or something after walking across a carpet? If so, you have become *electrically charged.* Have you ever rubbed an inflated balloon on your hair and noticed how your hair sticks to it? In this case, the balloon has become electrically charged. This charge can be so strong that you can even stick the balloon to a wall. What might these charges be? In the demonstration that follows you will see evidence of these charges in action.

A Classical Current Demonstration

Here is a demonstration similar to one first done hundreds of years ago. Make a setup like the one shown below, and try it yourself. Follow the steps closely.

1. Vigorously rub a plastic strip, such as a plastic ruler, with plastic wrap.

2. Quickly touch the strip or ruler to the end of the nail. What happens to the wheat puff?

3. Rub the plastic strip again and bring it close to the end of the nail without quite touching it. Observe the wheat puff.

4. Repeat the experiment using a vinyl strip rubbed vigorously with flannel cloth. What happens to the wheat puff now?

How do you explain the events in this demonstration?

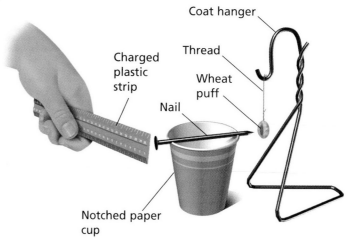

Coat hanger

Thread

Charged plastic strip

Wheat puff

Nail

Notched paper cup

A Theory of Charged Particles

Scientists explain events such as those seen in the previous demonstration in this way:

1. When one material is rubbed against another, friction causes charged particles to move from one material to the other. The accumulated charge is indicated by the charged material's ability to attract lightweight or finely powdered substances. When did this happen in the previous demonstration?

2. There are two kinds of charged particles: positively charged particles and negatively charged particles. Why might you conclude that the charges on the plastic strip are different from those on the vinyl strip?

3. Objects that have the same kind of charge tend to repel one another. Objects that have different charges tend to attract one another. When did you observe these effects in the demonstration?

4. Charged particles pass easily through certain materials, called **conductors**, but pass with difficulty or not at all through other materials, called **insulators** (nonconductors). What evidence is there that iron is a conductor of electricity? You might experiment by replacing the nail in the demonstration with a glass rod, a wooden stick, a copper wire, or objects made of other materials. Which are conductors? Which are insulators?

The drawings below apply the theory just stated to the previous demonstration. Express in your own words what is taking place in each drawing.

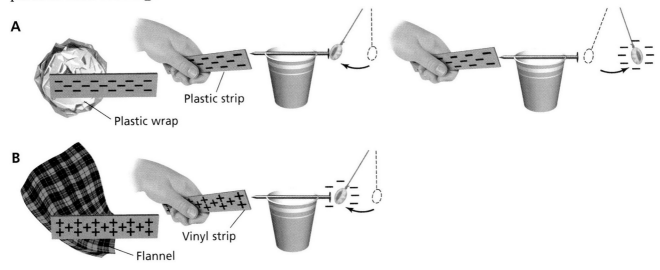

A

Plastic wrap

Plastic strip

B

Flannel

Vinyl strip

The theory we have been considering is an interesting one. It helps to explain many observations about electricity. But how do scientists know about these charged particles? Read the following account from the pages of history.

The Discovery of Charged Particles

It is the year 1900 at Cambridge University in England. J. J. Thomson is talking to a small group of students at the famous Cavendish Laboratory. Listen to what he might be saying:

"We believe that we have discovered the smallest particle of negative electricity. We have obtained a stream of identical negative particles from many metals. In fact, we believe that every bit of matter contains these same particles. We call them **electrons.**

"Obviously, most substances are uncharged. Therefore, most substances must have some kind of particle that neutralizes the charge of the electron. We have discovered just that kind of particle in other experiments, and we call it a **proton.** Each proton has a positive charge that exactly counteracts an electron's negative charge.

"It appears that the fundamental particles of matter are composed of electrons and protons in equal amounts so that no overall charge occurs. Electrons are relatively light and move freely in conductors. Protons, being much heavier, remain fixed in their positions."

J. J. Thomson and his colleagues expanded the theory of charged particles. Did you follow what he was saying to the students? Can you relate it to what an electric current is—such as the current you got when you correctly connected the dry cell, wires, and flashlight bulb in Exploration 1?

Don't forget who coined the terms *positive* and *negative* for nature's basic charges—me, Ben Franklin. To find out more about my work with electricity, see page S82 of the SourceBook.

Charged or Uncharged?

In terms of the theory of charged particles, what are positively charged, negatively charged, and uncharged objects? What makes them that way? Note that in the illustrations on page 301, there are no charges shown on the plastic wrap or the flannel. Should there be? The illustrations at right give a more complete picture of what happens when a plastic ruler is rubbed with plastic wrap. Count the number of positive and negative charges in the "Before" and "After" situations.

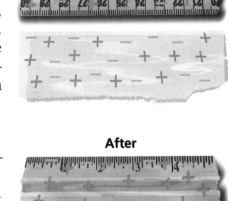

Before

After

a. Explain the movement of the electrically charged particles.

b. Now make before and after drawings that illustrate what happens when you rub a vinyl strip with flannel.

c. Each of the balloons shown below is charged differently. Which balloon is slightly negative? strongly negative? uncharged? slightly positive? strongly positive? What remains the same in each balloon? Why?

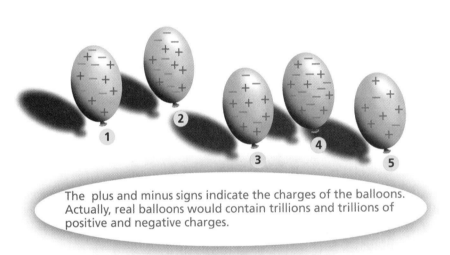

The plus and minus signs indicate the charges of the balloons. Actually, real balloons would contain trillions and trillions of positive and negative charges.

Charges on the Move

The buildup of electric charges on surfaces is called *static electricity*. You have seen several examples of this phenomenon on the previous pages. For example, when a person walks across a carpet, charges accumulate on the surface of his or her body and create an *electrostatic* charge. But what happens to this static electricity when the person nears a conductor like a metal doorknob? Something shocking! The built-up charges move from the person, through the air, and into the metal knob, causing a momentary flow of charges, or **electric current**.

Current Thinking

What causes electrons to move in a conductor to produce a current?

Do you recall the wheat puff in the classical demonstration? You brought a plastic ruler that you had rubbed with plastic wrap near the end of a nail. The wheat puff at the other end of the nail moved away. The puff moved because charged particles passed through the nail to the puff, causing the puff to jump away. The passing of the particles through the nail created a momentary electric current. But how, exactly, did it happen?

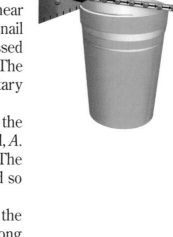

Wheat puff

Consider the labeled diagram below. First, electrons in the plastic ruler repel (push away) electrons in the point of the nail, *A*.

Moving to the right, these electrons repel electrons at *B*. The electrons at *B* move to the right, repelling electrons at *C*, and so on all the way through the nail to the other end, *Z*.

The electrons at *Z* are forced onto the wheat puff. When the puff is sufficiently charged, the repelling force becomes strong enough to push it away from the nail. The current then stops flowing because there is nowhere for the electrons to go.

Think of electrons as behaving somewhat like falling dominoes. Each electron pushes the one next to it. If there is a gap in the chain of electrons, no electric current can flow.

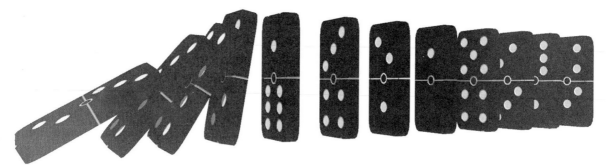

What causes a continuous flow of current?

Let's summarize. We need two things for a continuous current: a continuous supply of charges and an uninterrupted conducting pathway to carry the charges. Compare current flow to the flow of water to your home. For water to get to your home, you need a source (a reservoir) and a conducting path (pipes).

An electrical conducting path usually includes wire made of a metal such as copper and at least one device that makes use of the electricity. Trace the conducting path in the diagram, starting at the cell and coming back to the cell.

A complete circuit consists of the following:

a. a source of electric charges

b. a conducting path

c. a device that uses the electrical energy

Match (a), (b), and (c) with the numbered labels in the diagram. What would you add to the circuit so that you could control the continuous flow of charges, that is, start and stop the flow at will? What does it mean to make, or close, a circuit? What does it mean to break, or open, a circuit?

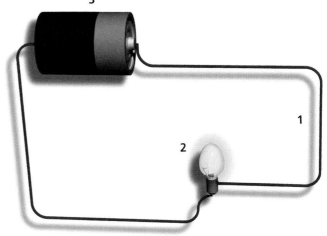

A Continuous Supply

One way to obtain a continuous supply of charges is to use a **chemical cell**. People often call a cell a battery. However, a battery is actually a group of connected cells. A car battery, for example, is made up of six separate but connected cells.

As you have observed, a cell always has positive and negative components. These are the *electrodes* of the cell. They are also conductors. The two electrodes of a cell are made of different materials and are in contact with a solution of chemicals called an **electrolyte**.

The electrodes and the electrolyte interact chemically with each other. The result is that electrons accumulate on one electrode and are removed from the other electrode. The electrode with the accumulation of electrons becomes negatively charged. (Why?) The electrode from which electrons are removed becomes positively charged. (Why?) In a typical dry cell, a zinc casing constitutes the negative electrode. The positive electrode is a rod of carbon (graphite) in the center of the cell.

When the negative electrode is connected to the positive electrode by a conducting path, electrons repel each other along that path toward the positive electrode. The positive electrode helps by adding an attractive force on the moving electrons. Thus, a current flows. The chemicals in the cell keep taking electrons from the positive electrode and giving electrons to the negative electrode. The cell, therefore, causes a continuous current to flow.

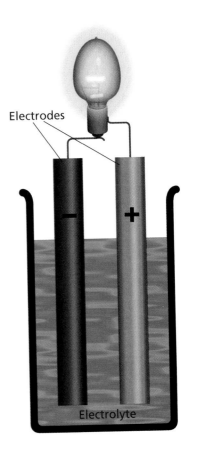

Detecting Electric Current

How can you tell if something's there if you can't see it?

Well, you can try to detect its effects.

Yeah, you can't see heat but you can feel it when you hold your hand over the stove.

And I can turn on a switch and the light tells me that the electricity's on.

Yes, but what if you had only the tiniest bit of electricity, not enough to light a bulb?

Hmmm. How *would* you be able to detect a really small current?

What these students seem to be saying is, "You can't see electricity, but you know that it's there by its effects." You can feel the heat produced in an electric stove and you know that an electric current is flowing. Likewise, when you see the light given off by a flashlight bulb, you know that an electric current is flowing. In this case, a small amount of current is enough to produce a visible effect.

Later, you will generate even smaller amounts of electricity. How will you detect them? Look back at Activity 3 of Exploration 1 on page 298. What effect did a current of electricity produce in that Activity? This effect can be increased many times if you wrap the wire (which must be insulated) around the compass. This involves the same principles of operation as sensitive current detectors called **galvanometers**.

Constructing a Current Detector

Your task is to design and construct a homemade galvanometer. The device consists of a small magnetic compass, insulated wire, and anything else you need to hold the parts in place.

Hints

1. Remember the results of Exploration 1, Activity 3.
2. Try coiling the wire around the compass. Use different numbers of turns and observe the effect.
3. Leave the two ends of the wire free so that you can attach them to the source of the small current.
4. Position the galvanometer so that the compass needle and the coil of wire are parallel to one another.
5. Test your galvanometer using a small current, which can be obtained from a lemon-juice cell. To make a lemon-juice cell, place a straightened paper clip and a piece of sanded magnet wire into a small cup of concentrated lemon juice. The paper clip and the wire are the electrodes, and the lemon juice is the electrolyte. Hook the free ends of the galvanometer wire to the electrodes. What happens?
6. Will your galvanometer be able to give you any information about the size of the current? Explain.

1. All Charged Up

a. One by one, a negatively charged plastic ruler is brought near three light, foil-covered spheres suspended by nylon threads. The ruler repels sphere *A* and attracts spheres *B* and *C*. Sphere *A* attracts sphere *B*, and sphere *C* attracts sphere *B*. Do you have enough information to determine the charges on spheres *B* and *C*? Why or why not?

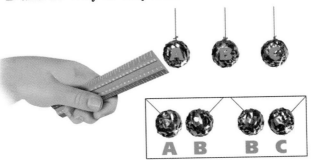

b. Jeff rubbed two pieces of plastic wrap with a sock and then suspended them (below left). Note what he observed.

c. Jeff then brought one of the pieces near a table (below right). Again observe what he saw.

d. Explain what is happening. Then try Jeff's experiment yourself.

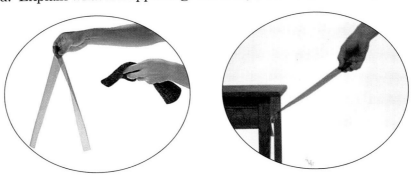

2. Sure Shot

When spray painting a screen, the screen is given an electric charge. Why doesn't the spray paint go through or around the screen?

Look, no paint!

3. Current Puzzle

Copy the puzzle at right into your ScienceLog. Then complete the statements below, and locate the answers in the puzzle by searching horizontally, vertically, or in a combination of both directions. Cross out the letters of each answer in the puzzle. The letters that remain will tell you something about an electric current.

a. A material that allows charges to go through it is a(n) _____?_____.

b. A material that does not allow charges to go through it is a(n) _____?_____.

c. J. J. Thomson discovered a charged particle that moves readily; it is called a(n) _____?_____.

d. This type of particle has a(n) _____?_____ charge.

e. The other charged particle in materials, which is more massive and does not move, is called a(n) _____?_____ and has a(n) _____?_____ charge.

f. If a material has equal numbers of these two kinds of particles, the material is _____?_____.

E	N	N	O	R	T	C	U
G	R	E	L	E	C	R	E
A	T	I	V	E	N	T	R
T	C	U	D	N	O	C	O
O	I	N	S	U	L	A	T
R	=	C	H	I	T	I	V
P	A	R	G	S	U	E	E
R	O	T	O	O	N	D	E
S	I	N	N	P	C	M	G
O	T	I	O	N	H	A	R

4. That's a Wrap

Some kinds of plastic wrap can be stretched tightly over a container and down its sides. The plastic wrap sticks to the sides of the container. Why?

ScienceLog

Review your responses to the ScienceLog questions on page 290. Then revise your original ideas so that they reflect what you've learned.

Sources of Electricity

1 What is the difference between alternating and direct currents?

2 How might you get an electric current using these materials?

3 How does a battery work?

ScienceLog

Think about these questions for a moment, and answer them in your ScienceLog. When you've finished this chapter, you'll have the opportunity to revise your answers based on what you've learned.

LESSON 1

Electricity From Chemicals

A surprising number of things that we use every day are powered by cells or batteries. These devices convert the energy stored in chemicals into a form of energy we can use—electricity. What are the advantages of chemical cells? the disadvantages? Are all chemical cells alike, or are some better than others?

You have already made one kind of chemical cell—from lemon juice (page 306). In the Explorations that follow, you will construct several chemical cells, which you can test using a galvanometer.

EXPLORATION 1

Chemical Cells

EXPERIMENT 1

Dry Cells

You Will Need

- a homemade or commercial galvanometer
- a 40 cm length of magnet wire
- masking tape
- rubber bands
- 2 zinc strips (3 cm × 8 cm)
- 2 copper strips (3 cm × 8 cm)
- blotting paper or filter paper
- calcium chloride solution
- a container to hold the calcium chloride solution
- forceps
- latex gloves

What to Do

Caution: Wear goggles and latex gloves when working with calcium chloride.

1. Make a single-cell sandwich like the one shown below.

2. Measure the amount of deflection this sandwich produces in your galvanometer.

3. Now make a double-cell sandwich.

4. Measure the deflection it produces in the galvanometer. How does it compare with the deflection caused by the single-cell sandwich?

Because the double-cell sandwich consists of more than one cell, it is known as a **battery**.

Questions

1. What accounts for the difference in the galvanometer readings for the single-cell and double-cell sandwiches? What would be the effect of adding more layers to the sandwich?

Single-Cell Sandwich

Tape
Rubber bands
Homemade galvanometer
Magnet wire
Plastic-foam cup

Copper (becomes positively charged)
Blotting paper soaked in calcium chloride solution
Zinc (becomes negatively charged)

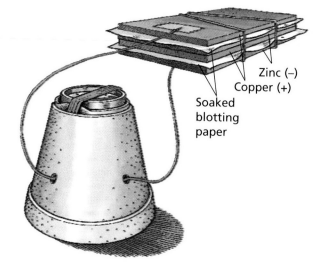

Double-Cell Sandwich

Zinc (−)
Copper (+)
Soaked blotting paper

2. Which electrodes were linked together in converting a single-cell sandwich to a double-cell sandwich?

EXPERIMENT 2

Wet Cells

You Will Need

- a commercial galvanometer
- a zinc strip (2 cm × 15 cm)
- a copper strip (2 cm × 15 cm)
- salt solution
- a 250 mL beaker
- 2 pieces of magnet wire (each 20 cm long)
- 2 alligator clips or clothespins

What to Do

1. Make a setup like that shown below.
2. Add enough salt solution to cover about half of the metal strips. Connect the strips to the galvanometer as shown.
3. Observe the galvanometer. Record the highest reading reached. What happens to the reading? Observe each electrode carefully. What happens to each electrode?
4. Add enough salt solution to fill the beaker.
5. Record the galvanometer reading once again.

Questions

1. How would you connect two wet cells to get more current? Draw a sketch showing your answer.
2. What is one way of increasing the current in a wet cell? How would you explain this?
3. What other factors might be altered in a wet cell to increase its current output?
4. What energy changes take place in the operation of dry and wet chemical cells?

Chemical Cell Technology

You have made some simple chemical cells. The current produced was very small—enough to be detected by a sensitive galvanometer, but not enough to light a bulb. Many different types of chemical cells have been invented—some tiny, and some large and powerful. Chemical cells provide the small amounts of current needed to run calculators, radios, flashlights, pacemakers, hearing aids, and portable telephones. Chemical cells are also used to provide larger amounts of current to operate systems in cars and spaceships. You will find out about how cells are constructed and used in the next Exploration.

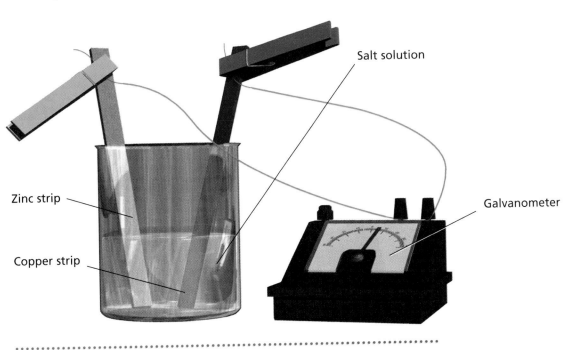

Salt solution

Zinc strip

Copper strip

Galvanometer

Commercial Electric Cells

In this Exploration you will examine some common chemical cells. You have probably seen most of them. You may have even wondered how they work. Here's your chance to find out.

PART 1

More Dry Cells

Dry cells were invented to overcome the disadvantages of wet cells. However, dry cells are not really dry. Rather, the solution in them (the electrolyte) is blended with other substances to make it thick and pasty.

1. Look at the cells pictured at right. Which would you use in a standard-sized flashlight? in a penlight? in a watch?

2. The voltage of each cell is marked. Notice how several cells of different size have the same voltage. How can that be? What does *voltage* mean to you?

Check out my lemon-powered flashlight! The lemons are like dry cells, so I don't have to worry about spilling any juice.

Several types of chemical cells and batteries

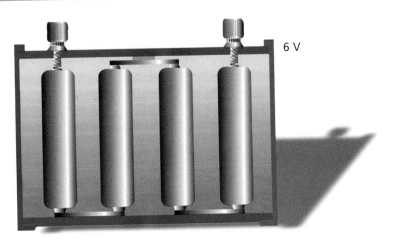

An inside view of cells and batteries. Compare the AAA, AA, and D-cells with the 9 V and 6 V batteries. What differences do you see?

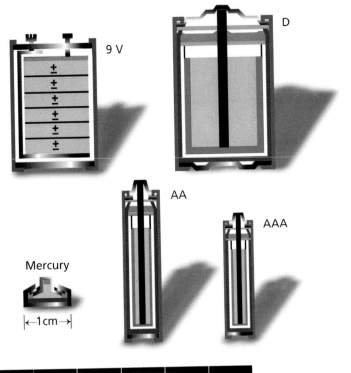

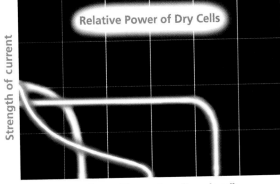

3. Look at the graph below. Which type of cell gradually "winds down"? Which type loses power quickly? Which type of cell would you probably use to power devices that require a steady current?

4. In your ScienceLog, sketch one of the cells and one of the batteries shown on this page. Then use the descriptions below to help you label their parts.

Like all chemical cells, the ordinary dry cell has two electrodes, or conductors, and an electrolyte solution. The *positive electrode* has two parts—a *graphite rod* in the center of the cell and a mixture of *manganese oxide* and *powdered carbon* surrounding the graphite rod. The *negative electrode* is zinc; it makes up the sides and bottom of the cell. The *electrolyte* fills the space between the electrodes. It consists of *ammonium chloride* paste. At the top of the cell is an *insulator*. *Batteries* consist of at least two *individual cells* joined together by *conducting strips*.

In the mercury cell, the *positive electrode* consists of a small block of zinc. The *negative electrode* is a layer of *mercury oxide*. The *electrolyte* is *potassium hydroxide*.

5. The *alkaline cell* differs from an ordinary dry cell in two major ways. First, the negative electrode is made of spongy zinc. Second, the electrolyte is potassium hydroxide, a strong base. What effect do these differences have on the power output of the alkaline cell?

Exploration 2 continued ▶

PART 2

Other Cells

1. A powerful surge of electric current is needed to crank an automobile engine. This surge of current is provided by a group of cells joined together in a battery. Study the drawings at right to discover or infer the answers to these questions:

   a. What substances make up (1) the two electrodes and (2) the electrolyte in a car battery?

   b. Why is such a large battery needed for a car?

2. Automobile batteries have a limited life span, and not all batteries last the same amount of time. Why do batteries wear out? Why do some wear out sooner than others?

3. Research "maintenance-free" batteries. How do they work? How are they different from standard batteries?

4. For many applications, the *nickel-cadmium* cell is replacing both lead-acid batteries and dry cells. Find out how this type of cell works.

5. Unlike dry cells, nickle-cadmium cells and lead-acid batteries can be *recharged*. What does this mean? How is this property useful?

Intercell connectors

Lead grills filled with lead oxide (positive electrode)

Single cell in a lead storage battery (grills separated to show construction)

Lead grills filled with spongy lead (negative electrode)

CONCENTRATED SULFURIC ACID CAUTION

Highly corrosive. Avoid spillage when pouring.

750 mL

LESSON

Electricity From Magnetism

You have discovered that magnetic effects can be caused by a current flowing in a wire or coil. Could the reverse be true? Could a magnet produce electrical effects in a wire? Try the following Exploration to find out. The Exploration has three parts—an activity that you can do and two completed experiments for you to analyze.

EXPLORATION 3

Moving Magnets and Wire Coils

PART 1

Building Your Own

You Will Need

- a commercial galvanometer
- a 150 cm length of magnet wire
- a cardboard tube
- a strong bar magnet
- sandpaper

What to Do

1. Sand the enamel off of the last 2 or 3 cm of the ends of the magnet wire. Wrap the magnet wire around the tube to make a coil as illustrated below. Attach the bare ends of the wire to a commercial galvanometer.

2. While watching the galvanometer, move a bar magnet into the coil, hold it there for a moment, and then remove it. Is the galvanometer needle affected?

3. Repeat step 2 several times, moving the magnet at different speeds. What do you observe? Does moving the magnet into the coil have a different effect on the galvanometer than moving the magnet out of the coil? What might this suggest about the direction of current flow through the coil?

4. Disconnect the galvanometer and move the magnet to see whether the magnet itself is affecting the galvanometer.

5. Connect the galvanometer to the coil again. This time, hold the magnet still, but pass the coil over the magnet. What do you observe?

6. Use your observations to answer the following questions:

   a. How can a magnet help produce electricity?

   b. How is the direction of the current affected by the motion of the magnet?

   c. How does the speed of the magnet's motion affect the amount of electricity generated?

   d. What changes in forms of energy occur in this investigation?

   e. Could a stationary magnet ever produce electricity? Explain.

   f. Would current still be generated in the wire if the wire were broken at some point? Why or why not?

Exploration 3 continued ▶

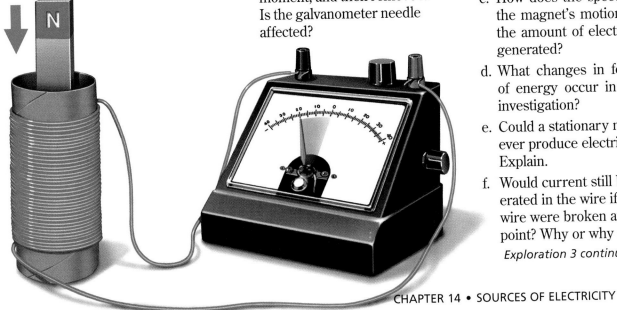

PART 2

Francesca's Experiment

Francesca devised an experiment to answer questions raised by Part 1 of this Exploration. She started with a wire coil of 15 turns. Illustrations (a) through (c) show the galvanometer readings she recorded. Then she used a coil with twice as many turns. Illustrations (d) and (e) show these readings.

Francesca tried each part of this experiment three times and obtained similar results each time. What conclusions do you think she drew for each part? The illustrations provide some hints.

Analysis

1. When a magnet is moved inside a coil of wire, _____?_____ is detected in the wire, which _____?_____ its direction when the magnet is moved in the opposite direction inside the coil.

2. A larger current is produced if _____?_____ or if _____?_____.

3. Suppose Francesca moved the magnet into and out of the coil 15 times in a minute. What would happen to the current? How many times per minute would the current go first in one direction and then in the opposite direction?

4. What do you think would happen if Francesca held the magnet stationary and moved the wire coil instead? Why?

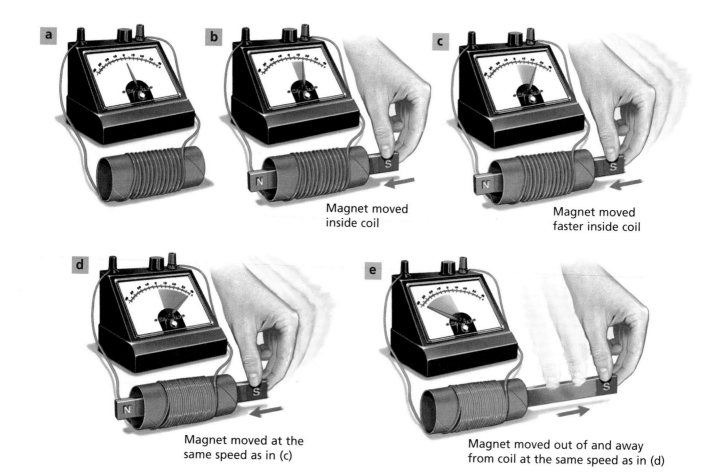

Magnet moved inside coil

Magnet moved faster inside coil

Magnet moved at the same speed as in (c)

Magnet moved out of and away from coil at the same speed as in (d)

A Related Experiment

Francesca made an important discovery: When a magnet is moved through a coil of conducting wire, electricity is generated. Both the number of coils and the speed of movement of the magnet affect the amount of current produced. Francesca also discovered that the direction in which the magnet was moved made a difference. If the magnet was moved in one direction, current flowed one way. If the magnet was moved in the other direction, current flowed the other way. Let's examine the findings of a related experiment.

But before you begin, think a little bit about how a magnet exerts its influence. Does the magnet have to touch something to have an effect, or does its force act through space? Look at the photo at the upper right. It shows a magnet on which iron filings have been sprinkled. Do you see evidence that (a) the iron filings have been attracted and that (b) the *magnetic force* is exerted through space along curved paths? We call these paths *magnetic lines of force.*

Look at the series of illustrations at right, which represent the results of the experiment. The wire is being moved while the magnet is held stationary. The arrows between the north and south poles of the magnet represent the lines of magnetic force.

Analysis

1. How do the results compare with those of Francesca's experiment?
2. What happens when the wire is momentarily motionless as it changes direction, as in (a)?
3. What happens when the wire is moved parallel to the magnetic lines of force, as in (d)?
4. What role do the magnetic lines of force appear to play in the generation of electricity?

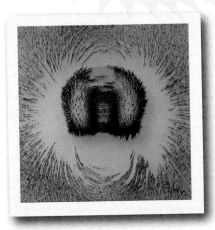

Does this magnet need a shave? No, it's just showing off its magnetic force! To learn more about magnetism, see pages S91–S96 of the SourceBook.

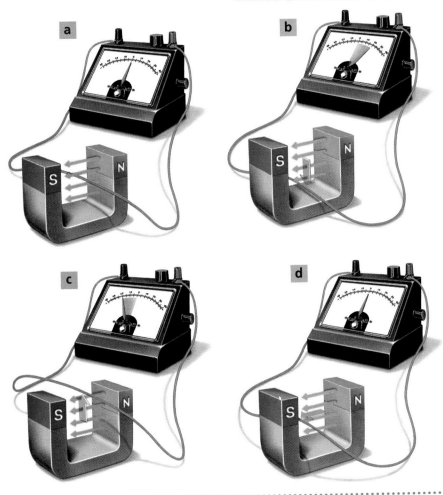

Generators—Small and Large

Our way of life requires large amounts of energy. For example, a medium-sized city requires enormous amounts of electricity to operate normally. Do you think that the electricity-generating systems you have seen so far in this unit could meet such demands? Could they be adapted to do so?

Examine the electricity-generating systems on this and the next page. Although they vary greatly in size, the operating principles of each are essentially the same.

Tiny Dynamo

A bicycle generator is a practical application of the *electromagnetic principle* that you discovered in Exploration 3. Whenever a magnet's influence sweeps across a wire that is part of a closed circuit, an electric current is generated. It does not matter whether the magnet or the wire moves to cause this to happen; the effect is the same. In Exploration 3 you saw electricity generated by back-and-forth motion. However, it is easier to generate electricity by rotating either the magnet or the wire coil. In the generator shown here, look for a magnet that rotates near a coil of wire. The generator is shown with its parts separated to help you see how it works.

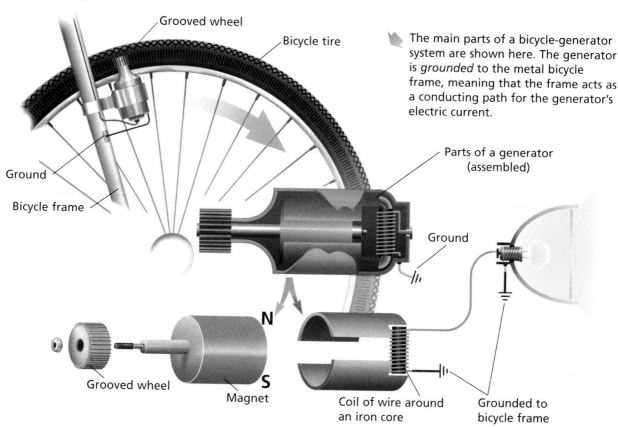

The main parts of a bicycle-generator system are shown here. The generator is *grounded* to the metal bicycle frame, meaning that the frame acts as a conducting path for the generator's electric current.

Grooved wheel

Bicycle tire

Ground

Bicycle frame

Parts of a generator (assembled)

Ground

N

S

Grooved wheel

Magnet

Coil of wire around an iron core

Grounded to bicycle frame

Analyze the generator's construction and operation.

1. Locate the magnet in the generator. How does it move? What causes it to move?

2. Locate the conducting wire that is wound around an iron core. The core is attached to curved metal plates, which help transmit the effect of the magnet coil just as if the magnet were actually moving into and out of the coil. Trace the complete electric circuit.

3. The current Francesca got was very small—it could be detected only by a sensitive galvanometer. The current developed in the generator is hundreds of times greater. What factors in the generator design could account for the larger current? Which of Francesca's experiments support your answers?

4. Write an entry for a student encyclopedia explaining how the bicycle generator works to power the headlight. Write it in a style understandable to a sixth-grader.

Large Generators

Generators usually use moving magnets to generate electricity in coils of wire. Study the illustration, which shows a large hydroelectric generator (*hydro* means "water"). What features of the generator account for the great amount of electric energy it can produce? How do Francesca's results support your answer?

Tracing the Flow of Energy

Use the diagrams on this page to complete the following energy-flow story.

Water in the reservoir has ____1____ energy. As it flows down, this energy changes into ____2____ energy of the moving water. The moving water forces the ____3____ to turn, providing it with ____4____. The attached ____5____ turn inside a stationary ____6____, in which ____7____ is produced. The resulting energy of the generator operation is ____8____ energy.

What is the energy story suggested by this diagram?

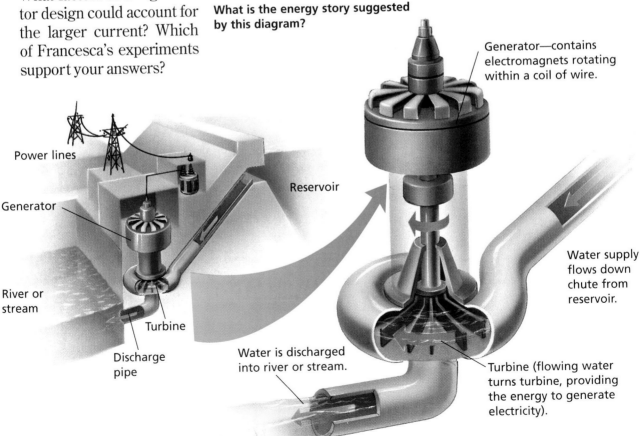

Power lines

Generator

River or stream

Turbine

Discharge pipe

Reservoir

Water is discharged into river or stream.

Generator—contains electromagnets rotating within a coil of wire.

Water supply flows down chute from reservoir.

Turbine (flowing water turns turbine, providing the energy to generate electricity).

Alternating Current

The principles shown by the systems on the previous page can be used to generate a special kind of electric current called **alternating current** (AC for short). The device below is an example of a simple AC generator. The handle sets the device in motion. As the wire loop moves through the magnetic field, electric current is generated.

Carefully study the diagram to figure out how the device works. In your ScienceLog, write a description of what is happening in the diagram. Then answer the following questions:

1. How is current conducted from the moving wire loop to the rest of the circuit, which includes the lamp and galvanometer?

2. As the device rotates, the current changes direction. Why? How do we know this?

3. In what direction is the current flowing in illustration (b)? in illustration (d)? Why does it differ?

4. How often does the current reverse direction with each complete turn?

5. Why is the light bulb unaffected by the change in current direction?

6. What's happening in illustration (c)? Why does this happen? (Remember the third experiment in Exploration 3.)

7. Suppose that you rotated the handle 5 times per second for 1 second. How many times does the current go first in one direction and then in the other? This is a current of 5 cycles per second. What do you think a cycle is?

8. What happens to the current output if you turn the handle faster and faster?

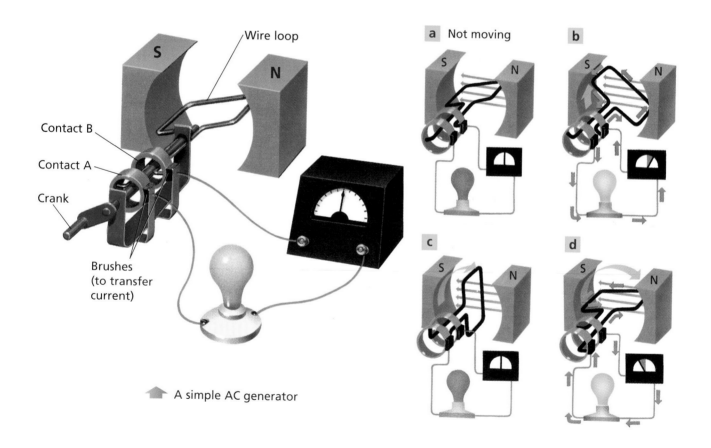

A simple AC generator

9. How is this device different from the two electricity-generating systems shown in Exploration 3? How is it similar to each system?

10. Why is the type of current produced by this device called *alternating current*?

Normal house current makes 60 complete cycles every second. Have you ever noticed "60 Hz" marked on tools or appliances? Look at the label from the electric drill shown below. The abbreviation *Hz* stands for *hertz*, a unit meaning "one cycle per second." This mark on a device means that the device is designed to run on 60-cycles-per-second alternating current.

Every time alternating current switches direction, for an instant no electric current flows. If this is so, why don't we notice it? We don't notice it

because it happens so quickly that our senses can't detect it. There are ways to detect this change indirectly, though. Here is one way. Wave a meter stick back and forth in the light from a single fluorescent (tube-type) or neon light. An ordinary incandescent light will not work. The room should be dark except for the single light. It also helps to face away from the light. As you sweep the meter stick quickly back and forth, you should see several repeated images of the meter stick. Each image represents the time during which the light is on as the current flows in one direction or another. Every time the current drops to zero as it changes direction, the light actually goes off for an instant. When this happens, you see no image.

Many electrical devices work with either alternating current or *direct current* (current that flows in only one direction). Why? What kind of systems do you think produce direct current?

Direct Current

Previously, you were introduced to chemical cells. Why does the current produced by a chemical cell go in just one direction? Study the diagram below to help you answer this question.

Chemical cells produce **direct current** (DC), or current that flows in one direction. Direct current is needed instead of AC for many circuits, such as those in a car. Why can't you charge a battery with alternating current?

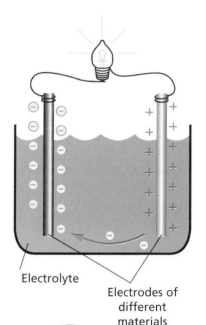

Electrolyte

Electrodes of different materials

NO. 7190
120 VOLTS 50/60 Hz
AMPS 3.0 • RPM 0-1200
DOUBLE INSULATED
CAUTION: FOR SAFE OPERATION SEE OWNER'S MANUAL. WHEN SERVICING USE ONLY IDENTICAL REPLACEMENT PARTS. WEAR EYE PROTECTION.

TYPE 1

LESSON 3

Other Sources of Electricity

So far you have produced electricity on a small scale from (a) kinetic energy alone, (b) chemical energy, and (c) a combination of kinetic and magnetic energies. You have also investigated applications of (b) and (c). In this lesson you will find that electricity can also be produced from light energy, heat energy, and mechanical energy (pressure).

You may not have heard of these last two methods of generating electricity. In fact, these methods can generate only tiny amounts of electric current. Nevertheless, they have specialized applications, as you will see.

EXPLORATION 4

Research Projects

PROJECT 1

Solar Cells

Solar panels generate electricity from light energy. You may be familiar with solar-generated electricity. Many calculators are powered by light energy alone, for example.

Try some experiments with a solar cell using different levels of light at various distances from the solar cell. Check on the amount of electricity being produced in each case.

Generating large amounts of electricity using only solar energy is difficult for many reasons. First of all, solar panels take up a great deal of space. Advances in technology will not shrink them beyond a certain size because there is only so much energy available in a given amount of sunlight.

Investigate sites where solar generation of electricity occurs. What are some advantages and disadvantages of each site? Would power generation by solar panels on the roofs of individual homes be feasible? Find out where this is being done and how solar panels work. Do you think that solar energy is the answer to some of our energy needs? Why is solar power a good source of energy for a satellite or space station?

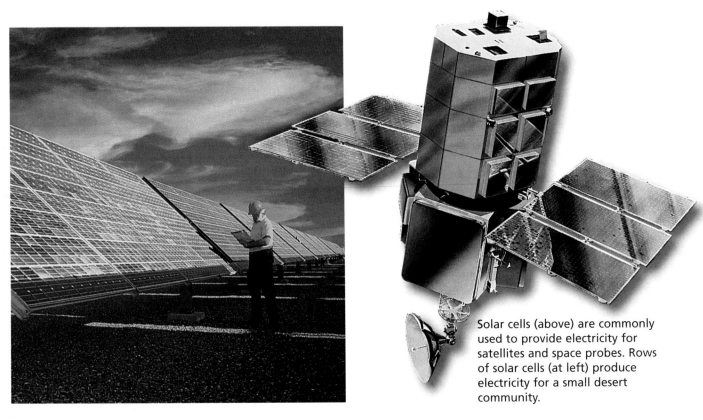

Solar cells (above) are commonly used to provide electricity for satellites and space probes. Rows of solar cells (at left) produce electricity for a small desert community.

Thermocouples

A *thermocouple* converts heat energy into electricity. It consists of wires of two different metals joined together. When heat is applied to the point where the metals are joined, an electric current flows. Make a simple thermocouple using the diagram below as a guide. Apply heat to one end of it (the other end is not heated), and check the current output with a galvanometer.

Thermocouples have only limited use as sources of usable electric energy. They are most commonly used as high-temperature thermometers or thermostats (devices for maintaining a set temperature). How do you think they work? How do you think industries might make use of these devices?

Piezoelectricity

Certain types of crystals produce electric currents when squeezed or stretched. This is called the *piezoelectric effect*. Piezoelectric crystals are useful for turning vibrations into electrical signals.

A classic example of the usefulness of such crystals is the record player, which was the standard home-audio device before compact-disc technology was developed. The needle of a record player uses a tiny quartz crystal. This crystal converts the vibrations created by the grooves in a record into an electrical signal suitable for amplification. How does it do this? As the needle rides along in the groove of a record, the tiny bumps in the groove, which represent the recorded sound, cause the crystal to be squeezed and stretched. Thus, the crystal produces an electric current—a current that mirrors the original recorded sound. Piezoelectric crystals also vibrate when an electric current is passed through them.

Computers, radios, watches, microphones, and many other devices could not operate without piezoelectric crystals. Research how piezoelectric crystals are used in some of these devices.

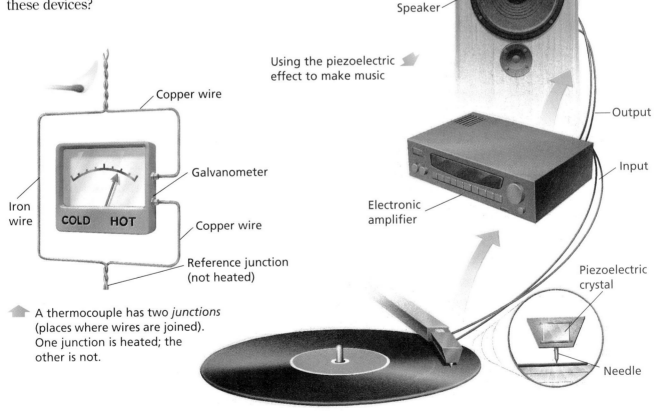

A thermocouple has two *junctions* (places where wires are joined). One junction is heated; the other is not.

CHALLENGE YOUR THINKING

1. Generator X

The sequence of pictures below shows a type of generator in action. Use the pictures to help you answer the questions that follow.

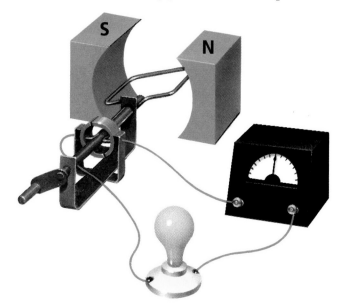

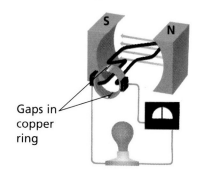

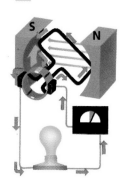

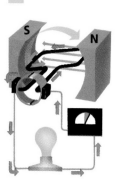

a Not moving

b

c

d

Gaps in copper ring

a. Examine the construction of this generator. How does this generator work?

b. Study the galvanometer readings. What kind of current does this generator produce?

c. How does this generator differ from the generator shown on page 320?

d. Explain what is happening in each illustration in the sequence.

2. Current Events

Electricity is related in some way to each of the following energy forms: light, heat, magnetic, chemical, kinetic, and vibrational. Identify the relationship among the energy forms in each of the following converters: dry cell, solar cell, wet cell, light bulb, generator, piezoelectric crystal, and thermocouple.

3. Play It Either Way

A light bulb can use either AC or DC. Hypothesize why this is so.

4. Irregular Exercise

Stan made a jump rope out of a loop of wire and then performed the activity pictured. The galvanometer showed that a current was being generated.

a. Explain what happened. (Hint: What makes a compass work?)

b. Did Stan generate direct current or alternating current? Explain.

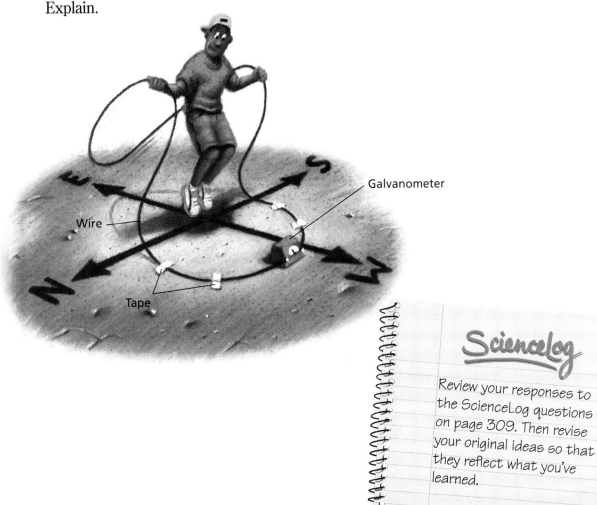

Galvanometer

Wire

Tape

ScienceLog

Review your responses to the ScienceLog questions on page 309. Then revise your original ideas so that they reflect what you've learned.

Currents and Circuits

What factors affect the flow of electricity in a circuit? ▶ 3

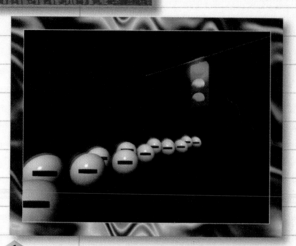

1

Explain how a switch stops and starts the flow of current.

Which circuit would have more *electrical resistance*— one made with thin wire or one made with thick wire? ◀ 2

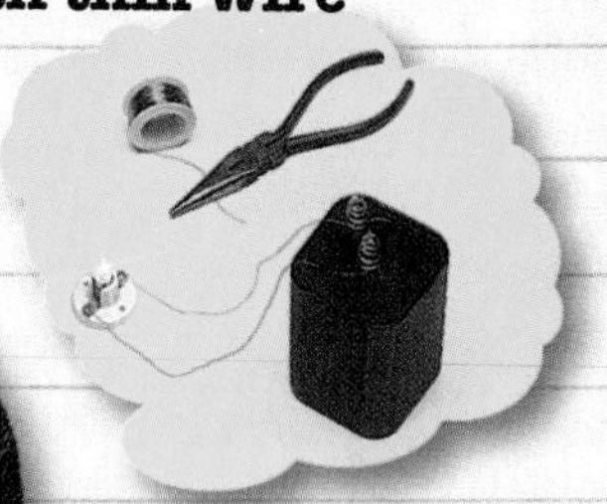

ScienceLog

Think about these questions for a moment, and answer them in your ScienceLog. When you've finished this chapter, you'll have the opportunity to revise your answers based on what you've learned.

Circuits: Channeling the Flow

In examining many of the circuits that follow, you will see bulbs, cells, and switches. If you have not already designed bulb and cell holders, switches, and connecting clamps, it would be good to do so now. All of these are available commercially, but you may wish to build your own.

The illustrations show some circuit components designed and built by students.

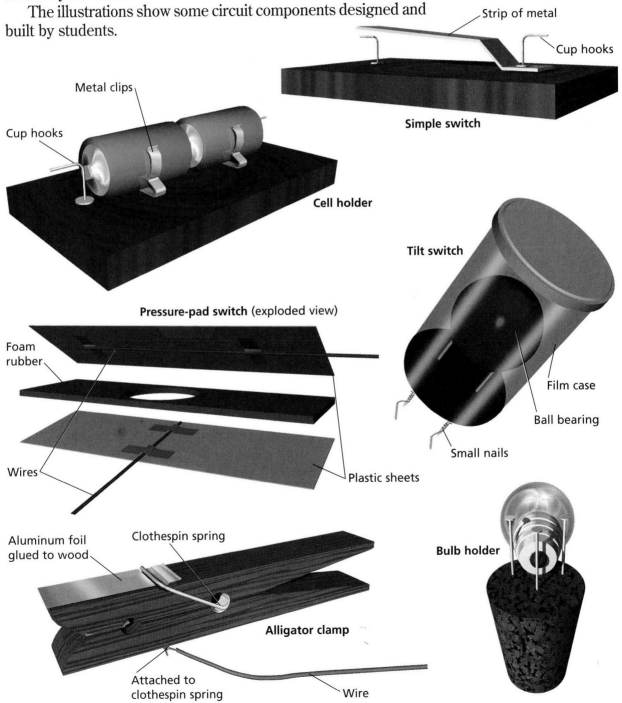

Strip of metal

Cup hooks

Simple switch

Metal clips

Cup hooks

Cell holder

Tilt switch

Pressure-pad switch (exploded view)

Foam rubber

Film case

Ball bearing

Wires

Small nails

Plastic sheets

Aluminum foil glued to wood

Clothespin spring

Bulb holder

Alligator clamp

Attached to clothespin spring

Wire

Circuit Symbols

Imagine using words alone to describe the circuitry of a radio. It would be almost impossible! For this reason, electricians, engineers, and circuit designers use symbols to represent circuits. Familiarize yourself with the following symbols.

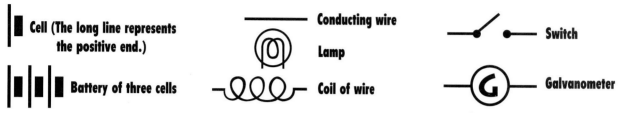

Cell (The long line represents the positive end.)

Battery of three cells

Conducting wire

Lamp

Coil of wire

Switch

Galvanometer

Observe how these symbols are used to represent this circuit.

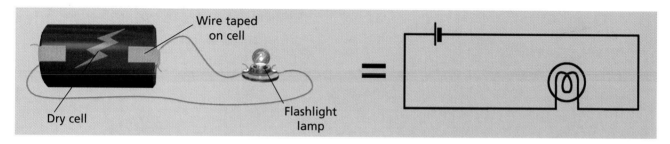

Wire taped on cell

Dry cell

Flashlight lamp

How would you draw a circuit diagram for the circuit below?

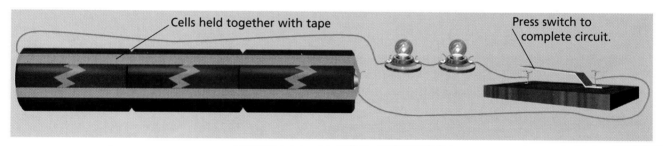

Cells held together with tape

Press switch to complete circuit.

Try drawing circuit diagrams for several circuits from earlier in this unit.

An Incredible Journey

The world at the scale of the electron is far different from our everyday world. We would find little that is familiar there. Imagine that you could shrink to the size of an electron and accompany this electron as it made its way through a circuit. Choose a circuit from this unit or one that you've devised. Write about your experiences. Feel free to include any ideas you've learned about electrons, charges, energy, and parts of a circuit.

A Conduction Problem to Investigate

The Problem: Do all wires conduct an electric current equally well? Here are some questions to answer as you investigate this problem.

a. Does the kind of metal affect the transmission of current?

b. What is the effect of having different thicknesses of wire?

c. Does the length of wire influence the current?

d. What happens if a wire resists the flow of current?

You Will Need

- magnet wire
- sandpaper
- thin and thick nichrome wire
- a wooden dowel
- D-cells
- a flashlight bulb
- newspaper
- a coin
- a rubber band
- wire cutters

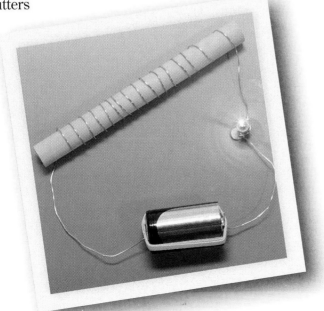

PART 1

Investigating Questions (a) and (b)

1. Prepare equal lengths of the three wires that you will test. Sand the enamel off of the last 2 or 3 cm of each end of the magnet wire.

2. Set up the circuit as shown in the photo below using one of the three wires. Observe the brightness of the light.

3. Do the same for each of the other wires. Observe the bulb in each case. Does the intensity of the light vary? Double-check your results.

Conclusions

1. Is it easier for a current to flow through thin nichrome wire or thin copper wire?

2. Is it easier for a current to flow through thin nichrome wire or thick nichrome wire?

3. How might you explain your observations?

PART 2

Investigating Question (c)

Vary the length of the thin nichrome wire in the circuit by placing the contact wires at different points along the wire, as shown below. Observe the bulb.

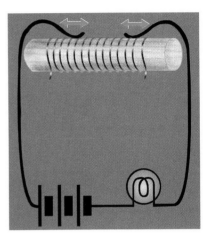

Conclusions

1. Is it easier for a current to pass through a long piece or a short piece of nichrome wire?

2. How might you explain this observation?

Interpreting Parts 1 and 2

Some wires do not allow electric charges to move through them as readily as do other wires. In other words, these wires offer more *resistance* to the flow of the charges. Which offers more resistance: nichrome or copper wire? thin or thick wire? long or short wire?

Resistance can be compared to friction. In what ways do you think they are similar?

Exploration 1 continued

Friction

Flick a coin across a table with your finger. It moves a little, slows down, and stops. What causes the coin to slow down? Where does the kinetic energy of the moving coin go? Here's how to find out. Place your finger firmly on the coin, and rub it back and forth a dozen times or so against a tabletop. Now touch the coin to your chin. What kind of energy was produced? What caused it to be produced?

Resistance

Resistance is like friction. Electrons flow because they receive electric energy from a cell. As the electrons flow through a piece of nichrome wire, the wire resists the flow of the electrons—in much the same way that the nails resist the rolling marbles in the photo below.

If the nails were a little closer together, how would this affect the rolling balls? Would this situation represent a wire with more resistance or less?

What form of energy do you predict will be produced from the electric (kinetic) energy of the electrons as they slow down? You will check your prediction in Part 3.

PART 3

Investigating Question (d)

Connect the circuit as shown. Wrap the wire with newspaper, and then watch it for 1 minute. Now disconnect the circuit, unwrap the newspaper, and carefully touch the wire.

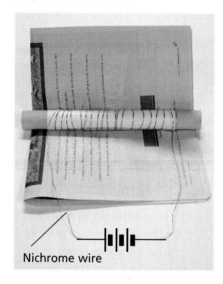

Nichrome wire

Conclusions

1. When a wire resists the flow of current, what happens?

2. What energy change is taking place in the nichrome wire?

Any conductor that offers considerable resistance is called a **resistor** and is represented by the symbol -⋀⋀⋀-. Using circuit symbols, draw a circuit containing two D-cells, a coil of nichrome wire, a switch, and a bulb.

Applications of Resistance

You have found that resistors produce heat, and you have inferred that they reduce the flow of current. Both of these characteristics are useful.

- Resistors are used to produce heat in appliances like toasters or irons. Appliances that produce heat when an electric current flows through them are called *thermoelectric* devices.

  Why is this a good name for them? Make a survey of the thermoelectric devices that are used in your home. How much power do they use? (This will be the number, in watts, marked on the appliance.)

- Resistors also reduce current flow. A *variable resistor*, such as the one used in Part 2 of this Exploration, varies the current. Such resistors are common— the volume control on a radio is one example.

  Here is a project for you to try. Turn an ordinary graphite pencil into a usable rheostat (device for varying the current). What might you use it for?

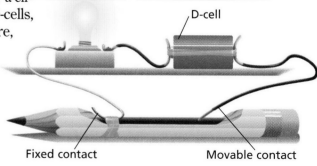

D-cell

Fixed contact   Movable contact

A graphite-pencil rheostat. How does it work?

Making Circuits Work for You

How have you made use of electric circuits in the last 24 hours? How have they worked for you?

In the following Exploration, you will build and test a number of simple yet functional circuits.

Dry cells are quickly drained if left in a closed circuit without something to provide resistance (for example, a bulb). Placing a switch in the circuit and keeping it open until you check the circuit's operation will help to conserve the cells' energy.

EXPLORATION 2

Constructing Circuits

You Will Need

- 6 pieces of copper wire (each 10 cm long)
- masking tape
- 3 D-cells
- 3 flashlight bulbs in holders
- 2 contact switches

PART 1

Exploring

You can devise any circuits you wish. Use some or all of the equipment listed above to construct your circuits. Make any arrangements desired. If you want to use more than one dry cell and do not have a holder, you can use masking tape to hold them together. If the bulbs light up, you have complete circuits. After constructing your circuits, draw them in your ScienceLog using circuit symbols. Suppose that a certain number of electrons flow out of the cell(s) in a given time. Describe the path(s) taken by these electrons.

PART 2

Solving Circuit Problems

The following are a series of circuit problems. What arrangement of circuit components would you make to accomplish the functions described in each problem?

First, make a circuit diagram of your proposed solution. Then construct each circuit.

1. A circuit that lights two bulbs, A and B, when the switch is closed. If either bulb burns out or is unscrewed, the other bulb goes out too. This type of circuit is called a **series circuit**.

2. A circuit that lights two bulbs, A and B, when the switch is closed. If either bulb burns out or is unscrewed, the other bulb stays lit. This type of circuit is called a **parallel circuit**.

3. A circuit that contains three bulbs, A, B, and C. If A is unscrewed, then B and C go out. If B is unscrewed, A and C stay lit. If C is unscrewed, A and B stay lit.

4. A circuit with two bulbs and two switches, P and Q. When P and Q are closed, both bulbs light up. If either switch is opened, neither bulb lights up.

5. A circuit that contains two switches and two bulbs. If both switches are open, neither bulb lights up. If either one of the switches is closed, both bulbs light up.

6. Analyze your findings.

   a. Identify the series and parallel circuits in each of your designs.

   b. Is there any parallel circuitry in the room where you are now? How could you find out without having to expose any wiring?

Exploration 2 continued

Current Questions to Investigate

Remember: The brightness of the light bulb is a measure of the amount of current flowing.

You Will Need

- 6 light bulbs
- 3 contact switches
- 3 D-cells
- copper wire
- wire cutters

What to Do

Construct each of the circuits shown in the table, and record the results in a similar table in your ScienceLog. Make certain that switches are included in the circuits you construct.

Question	Experimental design	Results/conclusions of experiment (Select from the list on the next page.)
1. How is the amount of current affected by the number of cells in a circuit?	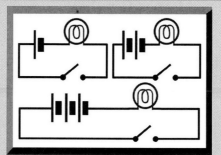	
2. How is the amount of current affected by the number of bulbs connected *in series*, that is, one right after the other?	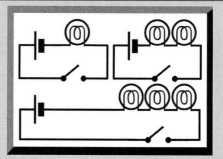	
3. How does the current flowing through each bulb in a *parallel*, or branched, circuit compare with the current flowing through each bulb in a series circuit?	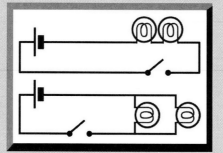	
4. What difference is there between the current flowing through the battery when two bulbs are connected in series and the current flowing through the battery when two bulbs are connected in parallel?	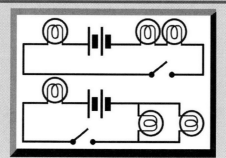	

Putting It Together

Below are a number of statements with options. Each statement, with the correct option, is a valid conclusion for one of the four experiments you just did. Choose appropriate conclusions for each experiment.

a. Connecting bulbs one after another in a circuit (decreases, increases) the amount of current flowing in the circuit.

b. If the number of cells is increased in the circuit, the amount of current is (decreased, increased).

c. If more bulbs are connected in series in a circuit, the resistance of a circuit is (decreased, increased).

d. If two bulbs are placed in a branched circuit rather than in an unbranched circuit, the current through the battery is (decreased, increased).

e. The current flowing through each bulb in a parallel circuit is (greater than, less than) the current flowing through each bulb in a series circuit.

f. The resistance of a circuit is decreased when the bulbs are placed in (series, parallel).

g. The resistance of a circuit is (more, less) with three bulbs in series than with two bulbs in parallel connected to a third in series.

How Bright Are You?

How well can you apply what you discovered in Exploration 2? In the circuits shown on the next page, choose the correct brightness (*S* for standard brightness, *L* for less than standard, *M* for more than standard) for each of the 20 numbered bulbs. Assume that all bulbs are identical.

First, note the three illustrated degrees of brightness: *L, S,* and *M.* The standard brightness, *S,* to which *L* and *M* are compared, is the brightness of a single bulb connected to a single dry cell.

In your ScienceLog, make a table similar to the one below. Record your choices in it. Be prepared to defend your choices. The first situation is done for you.

S—standard brightness

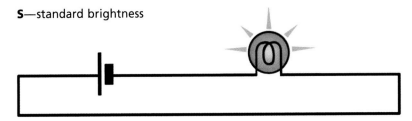

L—less than standard brightness

M—more than standard brightness

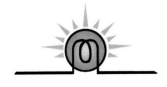

Bulb	L, S, or M
1	S
2	
3	
4	

SCORECARD

Over 15   You dazzle me!

10 – 15   You are bright!

Under 10   You need enlightenment

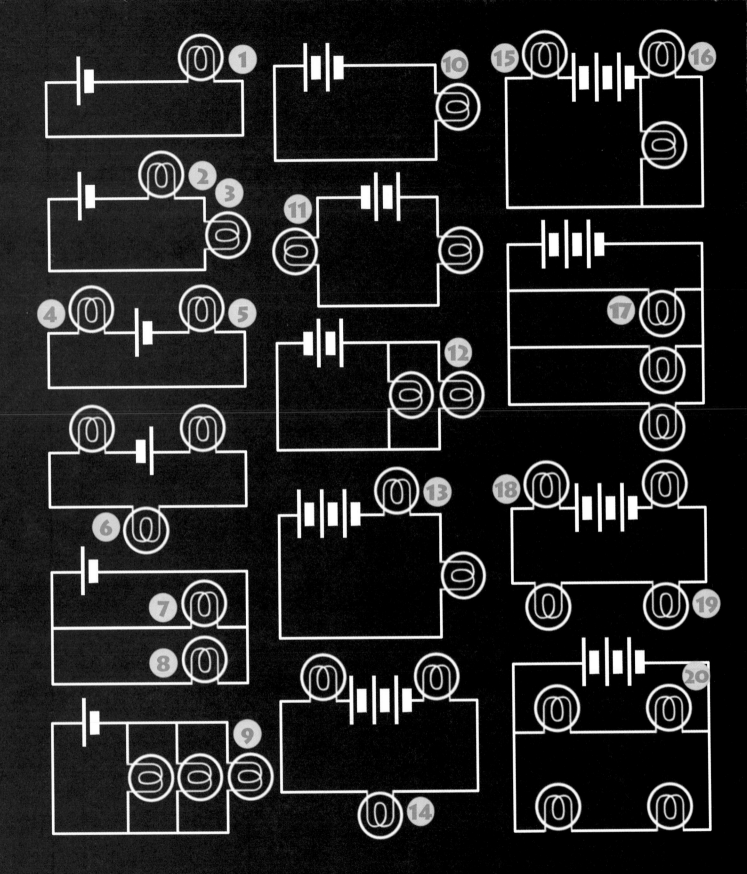

Controlling the Current

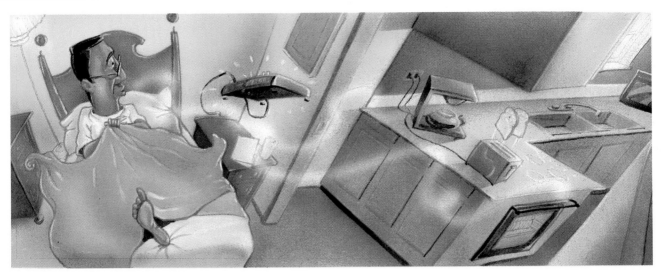

The radio blared suddenly to life, jolting Nick from a sound sleep. He sat up, looked at the time on the clock radio that had just come on, and jumped out of bed. Nick could smell the aroma of coffee coming from the kitchen, indicating that the timer on his new automatic coffee maker had worked properly. "Right on time," he thought to himself, looking at his watch as he entered the kitchen. Nick put some bread in the toaster. After a few moments, a couple of pieces of golden brown toast popped up. He then set the oven to come on at 4 P.M. to cook his casserole for 1 hour, knowing he would come home about 5:30. "I'm hungry already," he said to himself as he pulled out of the driveway. When Nick returned home that night, his apartment was already lit up. A light-sensitive switch had triggered the lights to come on at sunset. Best of all, there was a delectable smell in the air. He dimmed the kitchen lights and sat down to a scrumptious meal.

Automation! People depend on various types of circuits, switches, and controls to make life easier and more convenient. Although you may not know how they work, you are still able to make use of them.

1. How many automatic devices are mentioned in this story?
2. Which of these devices have switches that merely turn electricity on or off?
3. What are some of the kinds of controls that affect what is being done by the appliances and electrical devices?
4. Which of these switches or controls operate by mechanical means? by some other means?

Switched On!

PART 1

Under Control

Team up with two classmates. Together, study and discuss the circuits shown here and on the following page. Identify the switches in each circuit.

Can you determine how each switch works to control the current? In your ScienceLog, write down what you think is happening. Share your responses with other groups, and discuss any questions you still have about the operation of the circuit switches.

What practical use might be made of each circuit? Identify where Nick made use of similar current controls in the preceding story. Which current controls do you use at home?

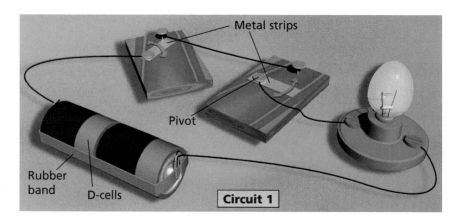

Metal strips

Pivot

Rubber band   D-cells

Circuit 1

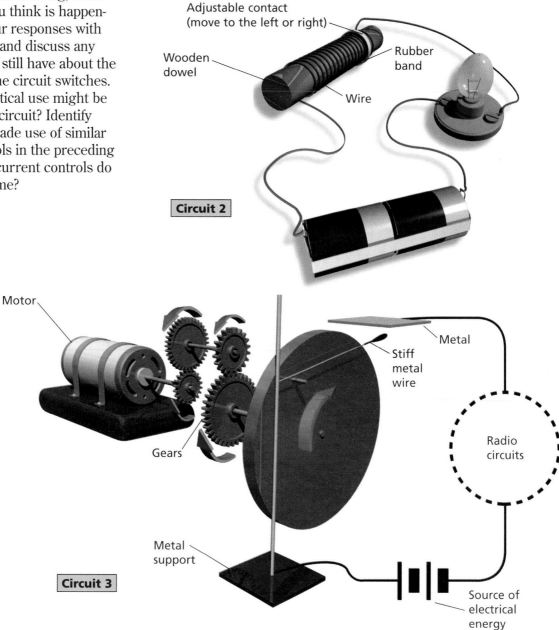

Adjustable contact
(move to the left or right)

Wooden dowel

Rubber band

Wire

Circuit 2

Motor

Metal

Stiff metal wire

Gears

Radio circuits

Metal support

Circuit 3

Source of electrical energy

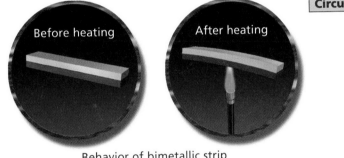

Resistance wire
that heats up

Bimetallic strip

Source of electricity

Circuit 4

Before heating

After heating

Behavior of bimetallic strip

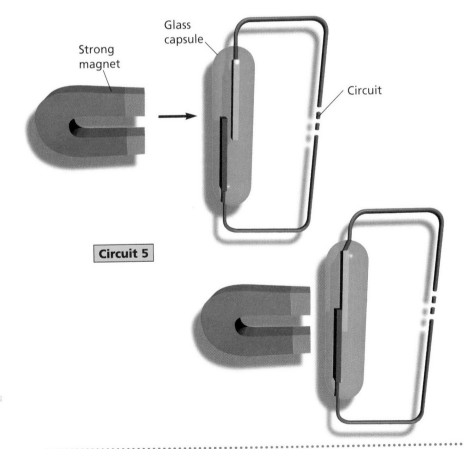

Strong
magnet

Glass
capsule

Circuit

Circuit 5

PART 2

To the Drawing Board!

Applying what you have learned
so far, design and construct two
circuits from the following list.
Construct a circuit that

a. has two lamps and two
   switches. Each switch operates only one lamp.

b. has two lamps and three
   switches. One switch turns on
   both lamps; the other two
   switches can turn off the
   lamps one at a time.

c. has a lamp, a switch, and a
   brightness control.

d. is a model circuit of a doorbell
   operated by two different
   push buttons.

e. contains a simple switch
   that could be operated by a
   magnet.

f. closes when a given temperature is reached.

For each circuit, follow these
steps:

1. Brainstorm possible solutions
   with your two partners.

2. Choose the best solution.

3. Draw your design.

4. Assemble the materials and
   construct a model of your
   design.

5. Suggest and try possible
   improvements.

6. Decide on a good application
   for the circuit.

Electromagnets

Electromagnets are magnets created by flowing electric currents. You made a kind of electromagnet when you constructed your galvanometer. The current-bearing coil of wire became magnetized and deflected the magnetic compass needle. Look at the circuit in the diagram below. This arrangement of a spike and a coil of wire is a simple electromagnet.

A typical electromagnet consists of a core of iron or soft steel surrounded by a coil of insulated wire. (Why must the wire be insulated?) The magnet wire you have been using in the Explorations in this unit is an insulated copper wire that is often used in making electromagnets. When the current flows through the wire, the core quickly becomes a temporary magnet. When the current is interrupted, though, the core loses its magnetism. Note that there is a switch in the circuit that allows the circuit to be easily disconnected.

EXPLORATION 4

Constructing an Electromagnet

You Will Need

- a paper clip
- washers
- some light, insulated wire
- 2 D-cells
- an iron spike
- a switch

What to Do

Get together with one or two classmates. Your task will be to make a functioning electromagnet and then to determine how its strength can be increased.

Making the Electromagnet

Use the materials listed above to make an electromagnet capable of supporting a paper clip from which several washers are hanging. Identify the parts of the circuit, and trace the path of the current. Are the spike, paper clip, and washers part of the circuit? How many washers can your first design hold?

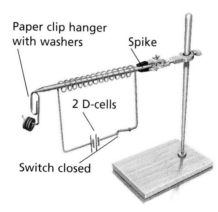

Paper clip hanger with washers

Spike

2 D-cells

Switch closed

Increasing the Electromagnet's Strength

How can you make your electromagnet hold more washers? Take some time to discuss the following:

- factors or variables that might be altered
- ways to measure the magnet's strength
- safety precautions
- the apparatus

Get your design approved by your teacher. Then assemble the necessary apparatus, do the experiment, record the results, and draw conclusions based on your results. Share your results with others.

How does this electromagnet work?

Figuring Out Electromagnetic Circuits

No doubt you have used diagrams when putting something together or figuring out how something works. A diagram can show how something operates or how parts should be assembled. The following are diagrams of circuits containing electromagnets. Working in small groups, determine how several of these circuits work. A few hints are provided in some of the drawings.

A good plan is to begin with the source of the electricity and follow the path of the complete circuit back to its source. For example, in the doorbell circuit, when the button is pushed down, the circuit is completed. Electrons flow along the parts of the circuit: battery, switch, contact screw, springy metal strip, coil, and back to the battery. Now retrace the path and think about what is happening in each part of the circuit, especially the electromagnet.

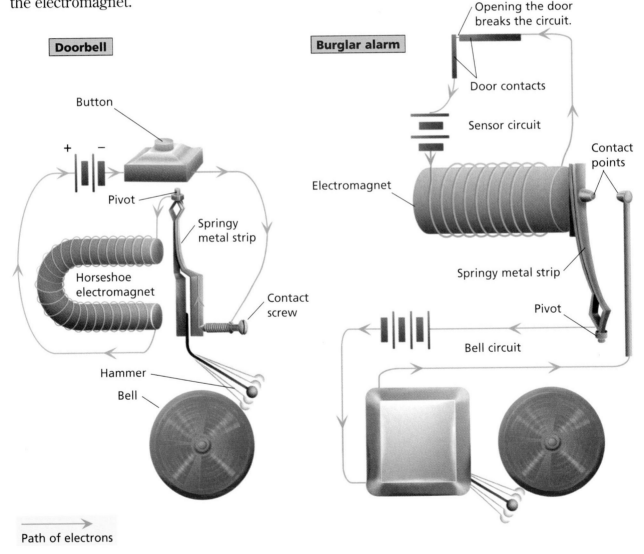

Doorbell

Button
+ −
Pivot
Springy metal strip
Horseshoe electromagnet
Contact screw
Hammer
Bell

Burglar alarm

Opening the door breaks the circuit.
Door contacts
Sensor circuit
Electromagnet
Contact points
Springy metal strip
Pivot
Bell circuit

Path of electrons

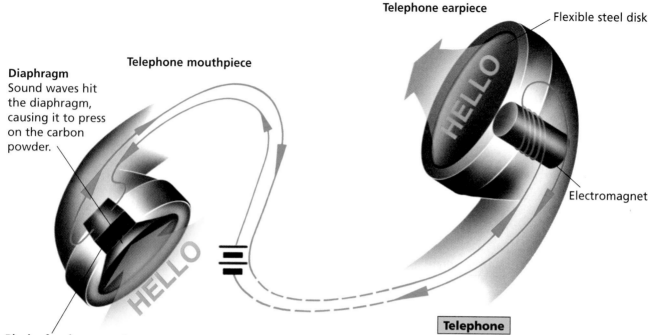

Telephone mouthpiece

Diaphragm
Sound waves hit the diaphragm, causing it to press on the carbon powder.

Telephone earpiece

Flexible steel disk

Electromagnet

Telephone

Block of carbon powder
When carbon powder is squeezed, its resistance decreases and more current can flow in the circuit. The amount of current varies with the strength of the sound waves.

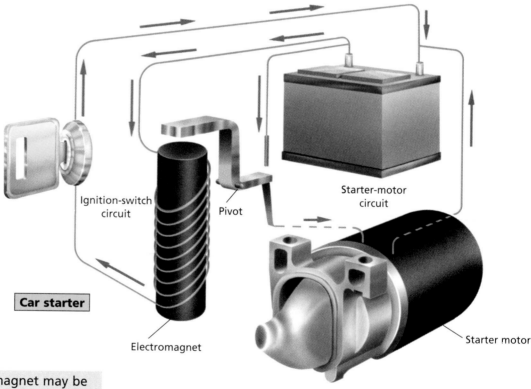

Ignition-switch circuit

Pivot

Starter-motor circuit

Car starter

Electromagnet

Starter motor

Hint: An electromagnet may be used to cause a small current in one complete circuit to initiate a large current in a second circuit. This is an example of a *relay*.

Moving Coils—Revolutionary!

One of the most common applications of electromagnets is the motor. Can you imagine life without motors? The invention of the motor was revolutionary in more than one way.

Take a look at an electric motor, such as the one in a washing machine or furnace blower, and examine the metal tag attached to it. You may see information similar to that shown here. Some of these words and symbols may be unfamiliar to you. What do you think "RPM" means? "RPM 3450" means that in 1 minute the motor makes 3450 complete turns, or *revolutions,* about a central axis.

What part does an electromagnet play in a motor? What part of a motor revolves? To find out, assemble the device below and close the circuit. What happens?

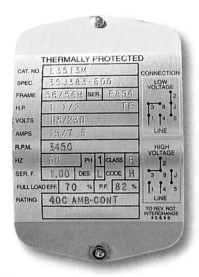

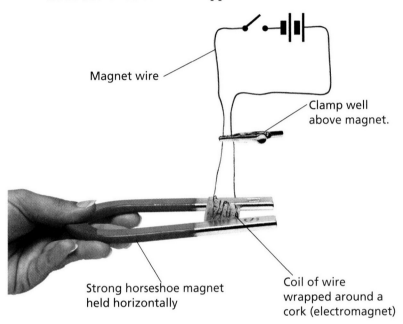

Magnet wire

Clamp well above magnet.

Strong horseshoe magnet held horizontally

Coil of wire wrapped around a cork (electromagnet)

This motor will make 3450 revolutions in 1 minute.

The Polarity Rule

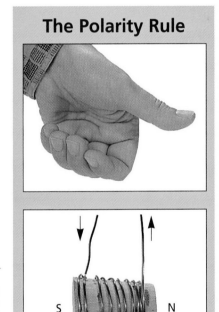

The coil of wire acts like a magnet. When a current flows, the coil has north and south poles. If, as in the photo at right, you curl the fingers of your left hand in the direction of the flow of electrons through the wire coil, the position of your thumb gives you the north pole of the coil.

Locate the N and S poles of the coil above. What will be the interaction between the N pole of the magnet and the N pole of the coil? between the S pole of the magnet and the N pole of the coil? What other interactions are there? You should realize that the coil will rotate in the presence of the horseshoe magnet when a current is flowing. Will it turn clockwise or counterclockwise? Can the coil make a complete turn? If you reverse the coil's connections with the battery after the coil has turned halfway, what happens? What causes a continuous rotary motion in a motor? Constructing a motor will help you see the importance of coil connections with the wires of the circuit.

Try the polarity rule on this one.

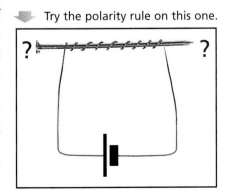

A Model Motor

Before building this motor, read the labels and comments in each step.

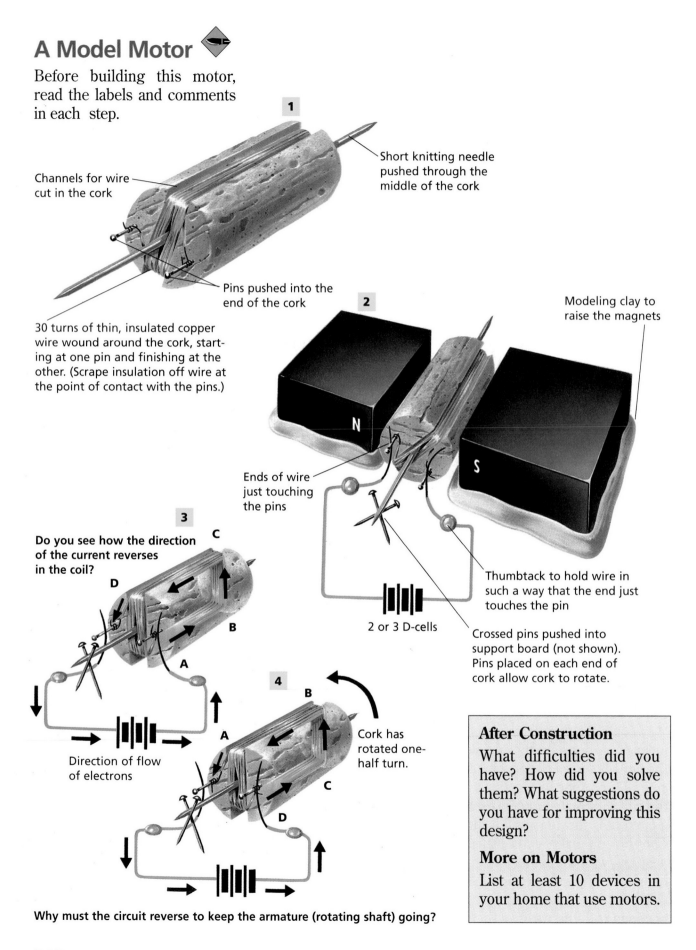

1

Channels for wire cut in the cork

Short knitting needle pushed through the middle of the cork

Pins pushed into the end of the cork

30 turns of thin, insulated copper wire wound around the cork, starting at one pin and finishing at the other. (Scrape insulation off wire at the point of contact with the pins.)

2

Modeling clay to raise the magnets

N

S

Ends of wire just touching the pins

Thumbtack to hold wire in such a way that the end just touches the pin

2 or 3 D-cells

Crossed pins pushed into support board (not shown). Pins placed on each end of cork allow cork to rotate.

3

Do you see how the direction of the current reverses in the coil?

C

D

B

A

Direction of flow of electrons

4

B

A

C

D

Cork has rotated one-half turn.

Why must the circuit reverse to keep the armature (rotating shaft) going?

After Construction

What difficulties did you have? How did you solve them? What suggestions do you have for improving this design?

More on Motors

List at least 10 devices in your home that use motors.

LESSON 5
How Much Electricity?

Look again at the motor tag pictured on page 341. This tag bears information about the motor. We already know what RPM and Hz stand for, but what do the terms *amps* and *volts* mean? These terms relate to certain characteristics of electricity. In this case, these terms and the numbers that accompany them indicate how much electricity the motor needs to run properly.

How would you go about measuring electric current? Electric currents are often compared to flowing water. We can use this as a model.

How Much Charge?

If you were to talk about amounts of water, you might use words like *liters, cubic centimeters,* or *milliliters.* In describing amounts of electricity, you might say, "How many electrons?" However, electrons are far too small to be easily used as a unit of measurement. Charge is measured in *coulombs* (pronounced KOO lahms). One coulomb (C) is the charge carried by 6.24 quintillion (billion billion) electrons.

The more charges you have, the more current that can flow. How would you measure the amount of electricity available from a given source? Let's use the water analogy. Which of the containers shown below do you think contains the greater "charge"? How did you arrive at this decision?

How Much Current?

Look at the diagram at right. Which hose has a greater rate of flow? That is, which hose would carry a greater amount of water in a given amount of time? What is your clue?

The electrical equivalent of flow rate is *current*. Current is measured in *amperes* (A), or *amps* for short. A current of 1 ampere is the rate of flow at which 1 coulomb of charge passes a given point in 1 second. The more coulombs passing a given point in 1 second, the greater the rate of electrical flow, or amperage. Which hose has the greater "amperage"? What is your clue?

How Much Energy?

Have you ever tried to hold your thumb over the mouth of a garden hose with the water turned on full blast? Most likely, the water pushed your thumb out of the way and you got wet. Why was it so hard to close off the hose with your thumb? The answer, of course, is pressure.

50 Coulombs

5 Coulombs

Charge is like a quantity of water.

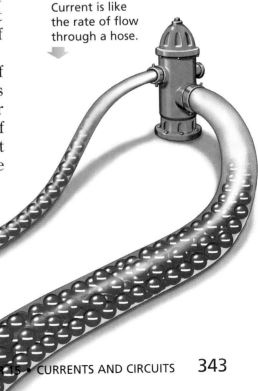

Current is like the rate of flow through a hose.

High-speed pump

Low-speed pump

Pressure is given to the water by a pump or by a reservoir in an elevated position. The greater the force given to the water, the greater the pressure. Electrical **voltage** is somewhat like water pressure. Voltage is measured in *volts* (V). Voltage is a measure of the energy given to the charge flowing in a circuit. The greater the voltage, the greater the force, or "pressure," that drives the charge through a circuit. How do you think voltage and current are related?

Which hose in the illustration at left has the higher "voltage"? What is your clue?

Putting It All Together

Charge, current, voltage—how do they all work together? So far, our water model has worked well. The flow of electricity behaves remarkably like that of water. However, we have to complete the analogy by considering that electricity flows in circuits. Look at the diagram below, which shows a "water circuit." Let's analyze how it works. (A mathematical formula is provided for key steps.)

1. The pump raises a quantity of water to the top. In doing so, it boosts the water's potential energy level (J/L).

2. As the water flows downward, its potential energy is converted to kinetic energy in the form of moving water—a "water current" (L/s).

3. At the water wheel, work is done as most of the energy in the falling water is converted to mechanical energy.

4. The water returns to the pump. The pump once again puts energy into the water, beginning the cycle again.

Let's compare our water circuit to an electric circuit.

1. A cell, battery, or generator provides a voltage, which boosts the potential energy (J/C or V) of the negative charges, or electrons, to a high level.

2. Closing the circuit allows the potential energy of the electrons to be converted to kinetic energy. The moving electrons create a current (C/s or A).

3. As the current flows through circuits, the electrons do work. The kinetic energy of the electrons is changed into other forms of energy.

4. The electrons, their energy expended, return to the electrical source, where their energy is boosted and the cycle begins again.

Voltage is similar to water pressure.

A "water circuit"

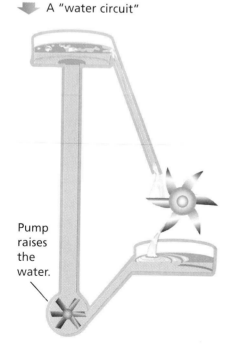

Pump raises the water.

Using Electrical Units

Here is a summary of all you need to know to make practical use of electrical units.

- The quantity of electric charge is measured in coulombs.
- The rate of electric flow, or current, is measured in amperes. One ampere of current is the same as one coulomb of charge passing a given point in one second.
- Voltage is the electric energy given to a unit of charge flowing in a circuit. One volt is the same as one joule of electric energy given to one coulomb of charge.
- Electricity, like other forms of energy, is measured in joules.
- Power, measured in watts, is the rate at which electric energy is used. One watt is one joule of electric energy used in one second.

You will draw on these definitions to complete the activities that follow.

Understanding Amperes, Volts, and Coulombs

1. Imagine using a toaster. Suppose that the toaster draws 5 A of current and that it takes 3 minutes to toast a slice of bread.

   a. How many coulombs flow through the toaster in 1 second? during the time it takes to toast the bread?

   b. To calculate the number of coulombs, multiply the number of ___?___ by the number of ___?___. This can be represented by the following word equation:

$$\text{coulombs} = \text{amperes} \times \underline{\quad ? \quad}.$$

2. The toaster is connected to a house's 110 V electrical source.

   a. The house's electrical source supplies ___?___ J to 1 C of charge, and ___?___ J to the 900 C required to toast the bread.

   b. The calculation of the number of joules of energy used may be represented by the following word equation:

$$\text{joules} = \underline{\quad ? \quad} \times \underline{\quad ? \quad}.$$

Understanding Watts

The rate at which appliances use energy is measured in watts. This is also known as the power of the appliance.

3. If something uses 1 W, it uses ___?___ J of electric energy every second.

4. An 800 W iron uses ___?___ J of electric energy every second. If the iron is used for 15 minutes, it uses ___?___ J of electric energy.

One coulomb of charge is like a bag containing 6.24 billion billion electrons.

10 C of electrons passing a point in 1 s equals 10 A of current.

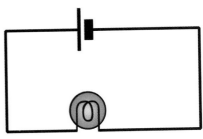

1.5 J of electric potential energy given to 1 C of charge by the cell

Reddy Kilowatt says if you want to read more about amperes, volts, and electric currents, turn to pages S86–S88 of the SourceBook.

Reddy Kilowatt®, The Reddy Corporation, International, 1996

5. To calculate the number of joules of energy used, multiply the number of watts by the number of ____?____. This can be represented by the following word equation:

joules = watts × ____?____.

6. In questions 2b and 5, you found that total electrical energy, measured in joules, is the product of two different measures of electrical energy. By setting these two different measures equal to one another, see whether you can show that watts = volts × amps.

7. A 1320 W electric heater is plugged into a 110 V household circuit. A current of ____?____ A flows through the heater. Explain in your own words what each of the numbered quantities in this question means.

Power by the Hour

As you saw earlier, multiplying power by time gives you the total amount of energy used. For example, watts multiplied by seconds equals joules. Utility companies use another unit to measure the energy used by consumers—the kilowatt-hour (kWh). It is obtained by multiplying the power in kilowatts by the time of energy usage in hours.

Imagine that you ran a 2000 W electrical heater for 5 hours.

8. A kilowatt is equal to 1000 W. The power of the heater is ____?____ kW.

9. In 5 hours an electric heater uses ____?____ kWh.

10. The heater used ____?____ J of energy. (Hint: First figure out how many seconds are in 5 hours.)

11. In questions 9 and 10 you found the energy used by the heater in both joules and kilowatt-hours. Therefore, 1 kWh = ____?____ J.

12. Why do you think kilowatt-hours, instead of joules, are commonly used to measure energy usage?

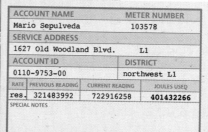

ACCOUNT NAME		METER NUMBER	
Mario Sepulveda		103578	
SERVICE ADDRESS			
1627 Old Woodland Blvd.		L1	
ACCOUNT ID		DISTRICT	
0110-9753-00		northwest L1	
RATE	PREVIOUS READING	CURRENT READING	JOULES USED
res.	321483992	722916258	401432266
SPECIAL NOTES			

City Electric Cooperative, Inc.
SUMMARY OF CHARGES

JOULES USED	401432266
PREVIOUS BALANCE	$0.00
CURRENT BALANCE	$59.11
	PAID
BALANCE DUE	$59.11
DATE DUE	10/17/97

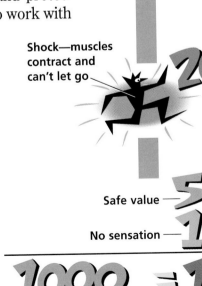

500 times higher than wet-skin resistance

Volts, Amperes, and the Human Body

You probably don't worry about electric shock when you use dry cells. Why not? First of all, dry cells cannot produce enough current to be very harmful. Second, the resistance of the skin is great enough to prevent the dry cell's small current from flowing through it. What happens if the resistance is decreased? Look at the graph to the left. What happens when the skin gets wet? What implications does this have for working with electricity? How high should the bar for the resistance of dry skin be if the bar for wet skin is 5 cm high?

Now look at the graph at right to see what amount of current is safe for the body to receive. Are you surprised at the small amount of current that our bodies are able to receive internally without causing havoc? The resistance of our skin is very important when you consider the amount of current flowing through most household electric devices.

What about voltage? The voltage of a dry cell is 1.5 V, that of a car battery is 12 V, and the voltage of household current is generally 110 V. What is safe? Generally, any voltage above 30 V can overcome the skin's resistance. If the supply of current is large, higher voltages can force enough current through the skin to cause shock or more serious damage. What kinds of procedures and protection would you suggest for people who work with high-voltage circuits?

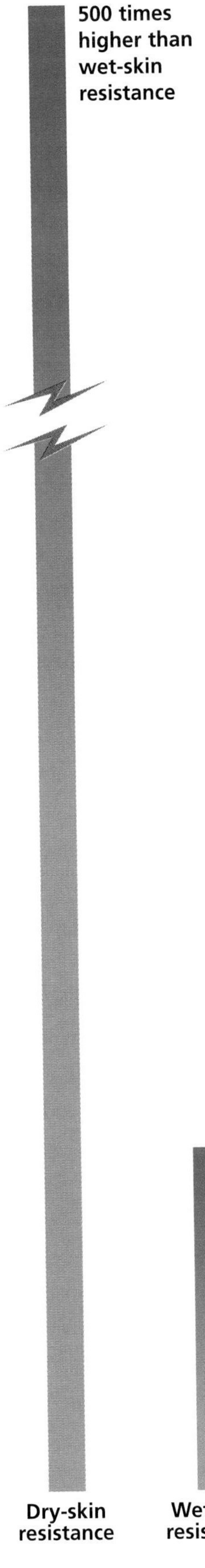

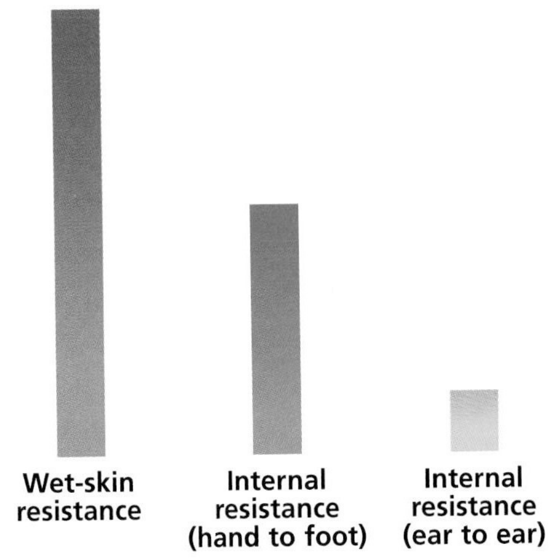

1 A

Heart convulsions

50 mA

Shock—muscles contract and can't let go

20 mA

Safe value — **5** mA

No sensation — **1** mA

1000 mA **= 1** A

Dry-skin resistance | Wet-skin resistance | Internal resistance (hand to foot) | Internal resistance (ear to ear)

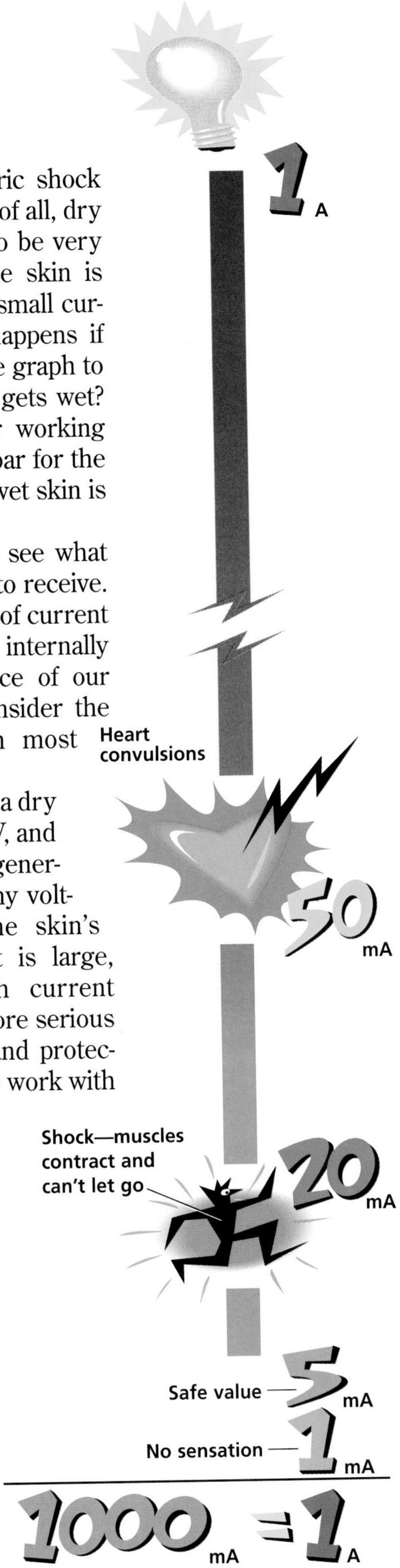

CHALLENGE YOUR THINKING

1. Safe Circuit

Many apartment buildings have security doors that can be opened from each apartment. The diagram below shows the circuitry of one such security door. Explain how it works.

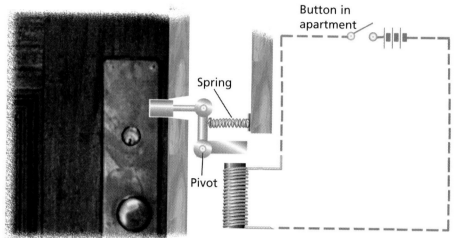

2. Go With the Flow

This drawing from a sixth-grade science book illustrates a lesson about electricity. Write a paragraph to go with this lesson. What words should be used to fill in the blanks?

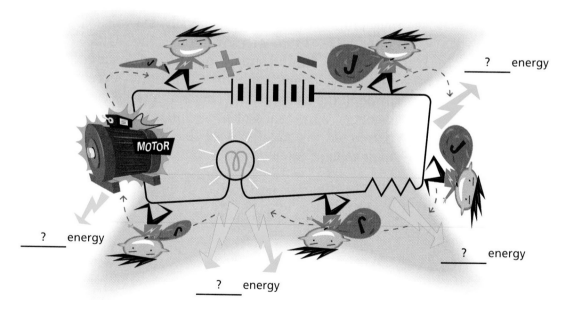

3. The Inside Story

A common type of switch is the *dimmer switch*. Turn the knob clockwise, and the light becomes brighter. Turn the knob counterclockwise, and the light becomes dimmer. Draw a sketch showing the circuit that may be inside.

4. You Be the Teacher

Ms. Alvarado asked her class to design a circuit containing two switches (S_1 and S_2), two light bulbs (L_1 and L_2), and one dry cell. She asked them to arrange the circuit so that it does the following:

a. If only S_1 is closed, L_1 will light.

b. If only S_2 is closed, nothing will happen.

c. If S_1 and S_2 are closed, both lamps will light.

Their work is shown below.

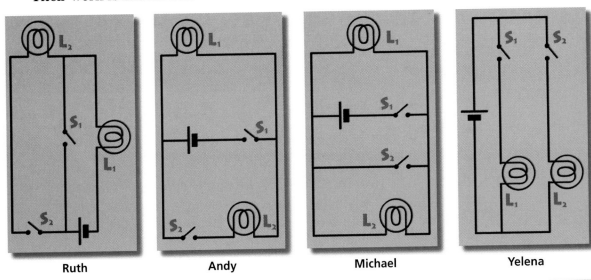

Ruth   Andy   Michael   Yelena

Play the role of Ms. Alvarado, and grade the students' work. Indicate how you know whether each design is right or wrong.

5. The Ol' Double Switcheroo

In the circuit shown at right, the upper switch moves from contact *A* to contact *C* and the lower switch moves from contact *B* to contact *D*. Suggest a function for this circuit.

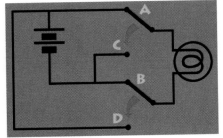

6. Safe at Home

Fuses are designed to protect household circuits against damage due to electrical overload. Fuses are made of metal alloys that have low melting points. How do you think fuses work? (Hint: Think about the effects of resistance.)

ScienceLog

Review your responses to the ScienceLog questions on page 326. Then revise your original ideas so that they reflect what you've learned.

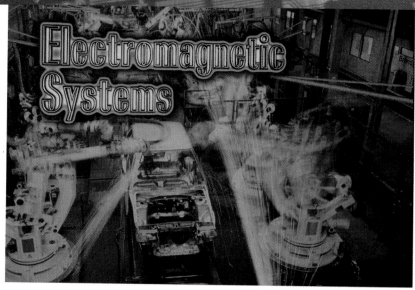

The Big Ideas

In your ScienceLog, write a summary of this unit, using the following questions as a guide:

1. How may electricity be produced?
2. What is the difference between a cell and a battery?
3. What parts are common to all chemical cells, and how do they operate?
4. How do AC and DC currents differ?
5. What factors affect the size of currents in circuits?
6. What are some differences between series and parallel circuits?
7. What are electromagnets? How can their strength be increased?
8. How is the flow of water like that of electricity?
9. By what units is electricity measured? How are these units related?

SOURCEBOOK

Electricity and magnetism are constantly at work, both in nature and in the modern technological world. To find out more about these phenomena, look in the SourceBook. You'll discover more about electric charges, currents, and magnets.

Here's what you'll find in the SourceBook:

Checking Your Understanding

Top view of house fuse

Side view of house fuse

Fuse strip

15

1. Recall your introduction to fuses in Challenge Your Thinking, Chapter 15. Suppose a fuse is rated at 15 amps (it melts when the circuit carries more than 15 amps). How many 100 W light bulbs would it take to blow the fuse? Assume a 110 V current. (Don't try this yourself!) You will find questions 6 and 7 on page 346 to be helpful.

2. At right is a diagram of a circuit breaker, a device that mechanically performs the same task as a fuse, breaking the circuit when too much current flows through it. How does this device work?

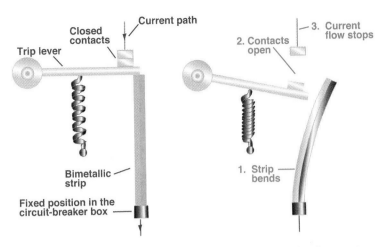

A. Normal current

B. Overload current

3. Electricity is often compared to flowing water. Read the following examples and decide whether each suggests high or low amperage, high or low voltage, or any combination of these.

   a. the Mississippi River
   b. Niagara Falls
   c. the blast from the spray-gun nozzle at a do-it-yourself carwash
   d. a dripping faucet

4. The diagrams below show two different types of microphones. Use the diagrams and your knowledge of the principles of electricity to explain how each of these microphones works.

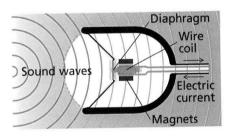 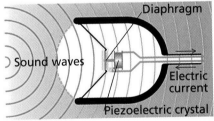

5. (concept map) Make a concept map using the following terms: volts, current, electricity, amperes, voltage, coulombs, and charge.

Science Snapshot

Paul Chu (1941–)

Imagine a puck whizzing around on an air-hockey table. The puck glides effortlessly on little jets of air. Now imagine a train moving just like the hockey puck, zooming along a frictionless track. Paul Chu and other physicists are investigating ways to take friction out of the transportation picture.

The Future in a Deep Freeze

Chu is a physicist who works with *superconductors*, materials that have no electrical resistance. Electric currents traveling through a superconductor can speed along forever. This sounds perfect, but there's a catch. Superconductors have to be cooled to extremely low temperatures in order to work. That makes them too expensive to be manufactured on a large scale. Chu has made a discovery that may help solve this problem.

▶ Japanese researchers have developed working models of super-conducting levitated trains called *maglev* trains. These models have reached speeds over 480 km/h!

▲ A magnet levitating over a superconducting material. One day scientists might be able to levitate trains over superconducting tracks.

Cheap as a Can of Cola

Chu was born in China and grew up in Taiwan. He came to the United States and earned a doctorate from the University of California at San Diego. He then went to work at the University of Houston, where he and his colleagues began to test ceramics for superconductivity. In the mid-1980s, they finally created a ceramic material, called compound 1-2-3, that superconducts at a then record-high temperature of –175°C.

Having to cool compound 1-2-3 to only –175°C means that scientists can use liquid nitrogen, a resource that is as cheap as a can of cola, as a coolant.

Chu, the director of the Texas Center for Superconductivity at the University of Houston, has discovered super-conductors at temperatures as high as –109°C. Yet he continues to work toward the ultimate goal: finding a material that superconducts at room temperature. If Chu were to find such a material, levitated trains and electric cars could soon become common modes of transportation.

The Path of Least Resistance

Model the passage of an electric current through a normal conductor by randomly sticking 10–15 pushpins through the bottom of a shoe box. The pushpins represent the molecules of the conductor. Now lift up one end of the box and roll a few marbles—which represent the electric current—down the slope. Resistance occurs when the marbles hit the pushpins and lose energy. Now model a superconductor by moving the pushpins into two rows running down the slope, and again roll the marbles down the slope. Why is this situation more efficient in terms of energy loss?

A Cleaner Ride

At least half the air pollution in this country comes from automobile exhaust. Many state governments are now creating new regulations to reduce the amount of air pollution generated by cars. For example, by 2003, 10 percent of all new cars sold in California must produce no exhaust gases. Government officials in other states are looking at similar measures. But how can automobile companies produce cars that don't pollute? One answer may be electric cars.

Batteries Included

Electric cars are powered by electric motors that get their energy from a large bank of on-board batteries. This means that electric cars are quiet and exhaust-free. In addition, electric cars are more dependable, more durable, and cheaper to operate than gasoline-powered cars.

The batteries in electric cars are similar to the batteries in gasoline-powered cars, but are much bigger and much more expensive. And, like all batteries, electric-car batteries go dead. Currently, the batteries must be recharged every 190 km (120 mi.) or so. This process can take 3 hours or more. The batteries also have a limited life span and must be replaced every 40,000 km (25,000 mi.).

▼ Some electric cars use electric power in combination with solar power. The solar cells on top of the car help recharge the batteries.

Using Energy Wisely

At present, batteries provide very little energy for their weight. Therefore, electric-car manufacturers must use that limited energy wisely. One electric-car model has a number of special features designed to help make the most of the batteries' energy. For example, the car uses a special braking system called a *regenerative braking system*. When the accelerator is released, the car's motor becomes a generator. The turning of the car's wheels drives the generator, which both produces energy and helps slow down the car. The electrical energy produced by the generator helps recharge the batteries.

Dreams and Drawbacks

Widespread use of battery-operated cars could result in many positive changes. For example, city skies would be much cleaner, and traffic noise would be greatly reduced. It sounds exciting, doesn't it? Electric-car manufacturers think so too! But electric cars are far from perfect. Currently, electric cars are expensive to produce. Also, the electricity used to power these cars usually comes from fossil fuels, which also pollute the environment. In addition, because they are large and must be replaced frequently, electric-car batteries will be costly and will pose disposal problems. Finally, many of the materials used in existing batteries are harmful to the environment. However, automobile companies are hopeful that new batteries will be more powerful, smaller, and less toxic.

Think About It

Believe it or not, at one time in the United States, more electric cars were sold than any other type of car. Do a little research about this time in the late 1800s. Write a short paper on what you find out.

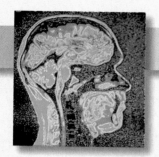
Magnets in Medicine

Think about what it would be like to peer inside the human body to quickly locate a tumor, find tiny blockages in blood vessels, or even identify damage to the brain. Medical technology known as magnetic resonance imaging (MRI) allows doctors to see these things and more. MRI is a quick, painless, and safe way for doctors to look inside the body to diagnose a variety of ailments.

Magnetic Images

Like X rays, MRI creates pictures of a person's internal organs and skeleton. But MRI produces clearer pictures, and it does not expose the body to potentially harmful radiation as X rays do. Instead, MRI uses powerful electromagnets and radio waves to create the images.

The patient is placed in a large, tunnel-shaped machine that contains four electromagnetic coils and a device that transmits radio signals. When electric current flows through the coils, a powerful magnetic field is created around the patient. Because the human body is made mostly of water, there are a lot of hydrogen atoms in the body. The magnetic field in the machine causes the hydrogen atoms to line up. At the same time, a brief radio signal is transmitted, knocking the atoms out of alignment. After

the signal ends, a computer measures the activity of the hydrogen atoms as they return to their original lineup. The computer analyzes this activity to produce an image.

A Diagnostic Device

MRI is particularly useful for locating small tumors, revealing subtle changes in the brain, pinpointing blockages in blood vessels, and showing damage to the spinal cord. This technology also allows doctors to observe the function of specific body parts, such as the ear, the heart, muscles, tendons, and blood vessels.

Researchers are experimenting with more powerful magnets that work on other types of atoms. This technology is known as magnetic resonance spectroscopy (MRS). One current use of MRS is to monitor the effectiveness of chemotherapy in cancer patients. Doctors analyze MRS images to find chemical changes that might indicate whether the therapy is successful.

◄ This color-enhanced MRI image of a brain shows a tumor (tinted yellow). The tumor was removed, and the patient resumed a healthy life.

▲ This patient, entering an MRI tunnel, is wearing a cylinder on his knee to provide greater detail of the bone and soft tissue in that area.

Picture This

The images of an MRI scan appear on the screen as cross sections. The doctor can adjust the point of view and depth of the cross section to focus on a specific organ or tissue. Look again at the color-enhanced MRI image on this page. What do you think the different colors represent? What do you think produces the differences in color?

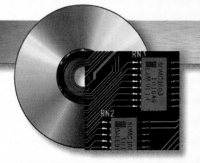

TECHNOLOGY *in* SOCIETY

Riding the Electric Rails

*F*or more than 100 years, the trolley, or streetcar, was a popular way to travel around a city. Then, beginning in the 1950s, most cities ripped up their trolley tracks to make way for automobiles. Today, trolleys are making a big comeback around the world.

From Horse Power to Electric Power

In 1832, the first trolleys, called *horsecars,* were pulled by horses through the streets of New York City. Soon horsecars were used in most large cities in the United States. However, using horses for power presented several problems. Among other things, the horses were slow and required special attention and constant care. So inventors began looking for other sources of power.

In 1888, Frank J. Sprague developed a way to operate trolleys with electricity. These electric trolleys ran on a metal track and were connected by a pole to an overhead power line. Electric current flowed down the pole to motors in the trolley. A wheel at the top of the pole, called a *shoe,* rolled along the power line, allowing the trolley to move along its track without losing contact with its power source. The current passed through the motor and then returned to a power generator by way of the metal track.

Taking It to the Streets

By World War I, more than 40,000 km of electric-trolley tracks were in use in the United States. The trolley's popularity helped shape American cities because businesses were built along the trolley lines. But competition from cars and buses grew over the next decade, and many trolley lines were abandoned.

By the 1980s, nearly all of the trolley lines had been shut down. But by then people were looking for new ways to cut

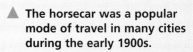

▲ The horsecar was a popular mode of travel in many cities during the early 1900s.

down on the pollution, noise, and traffic problems caused by automobiles and buses. Trolleys provided one possible solution. Because they run on electricity, they create little pollution, and because many people can ride on a single trolley, they cut down on traffic.

Today, a new form of trolley is being used in a number of major cities. Called light-rail transit vehicles, these trolleys are quieter, faster, and more economical than the older versions. They usually run on rails alongside the road and contain new systems such as automated brakes and speed controls.

◄ Many cities across the country now use light-rail systems for public transportation.

Think About It!

Just because trolleys operate on electricity, does this mean that they don't create any pollution? Explain your answer.

Unit 6

Sound

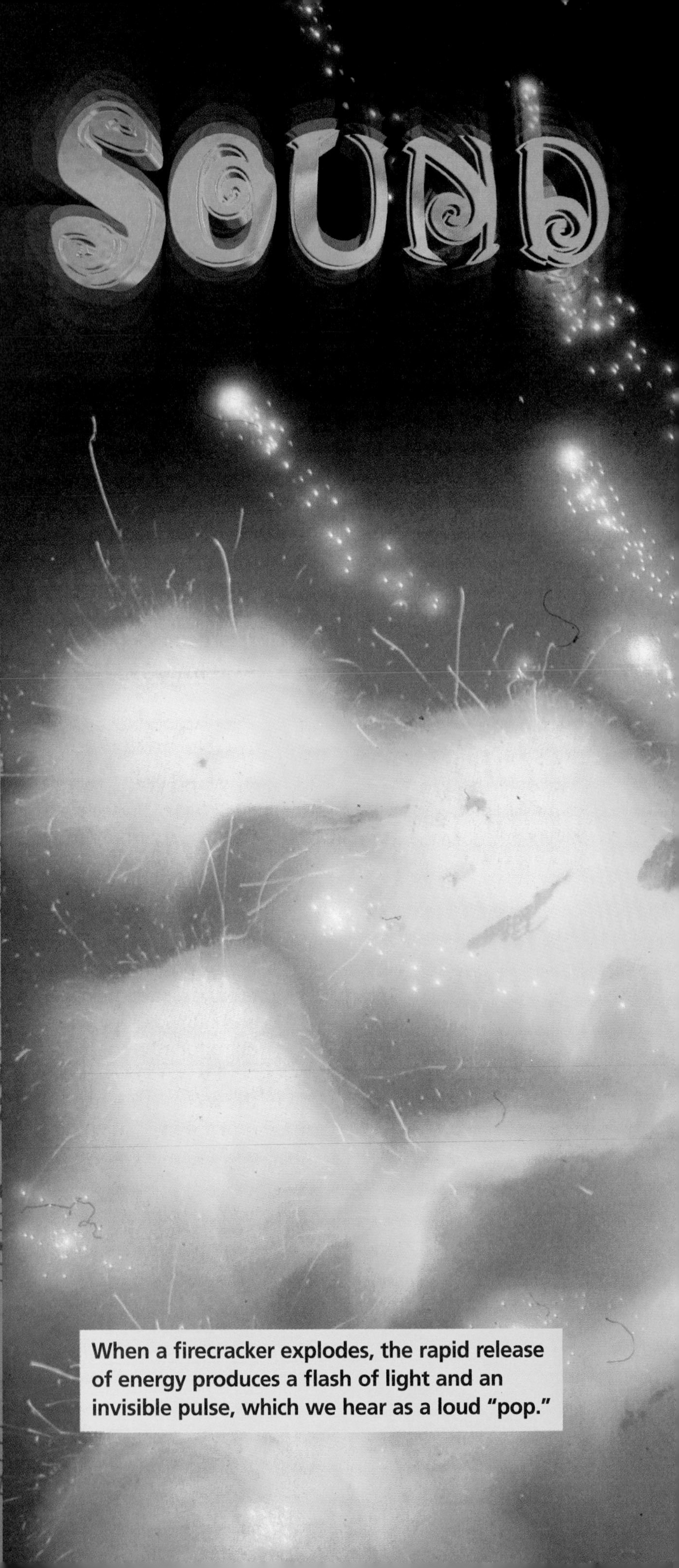

When a firecracker explodes, the rapid release of energy produces a flash of light and an invisible pulse, which we hear as a loud "pop."

K ABOOM!

The ignition of each of these colorful
energy converters sparks a spectacular transfor-
mation. In an instant, the chemical energy stored
within the firecracker's powder is transformed into an ear-
battering bang accompanied by a tiny tempest
of light, heat, and blackened bits of paper. Some
of the released energy even causes motion in the
remaining firecrackers, which recoil with each blast.
It would be dangerous to be close to this explosion.
Among other dangers, the intensity of such a loud sound could cause
pain and even damage to your ears. Although you cannot always prevent
loud noises, you can play a part in determining their effect on you.
What are some factors that you think affect the loudness of a sound?
Even from a safe distance, the noise of an exploding firecracker
may cause a feeling of pressure in your ears. This has to do with the
way in which sounds are created, transmitted, and detected. In this unit, you'll
discover how sounds originate and how they travel,
which will help you understand why you'd
actually see the flash of one of these
explosions before hearing the noise it produced.

What Is Sound?

How will the **1** sound from the banjo change if the player moves her finger halfway up the neck of the banjo?

2 How would you describe the difference between music and noise?

3 What causes sound?

ScienceLog

Think about these questions for a moment, and answer them in your ScienceLog. When you've finished this chapter, you'll have the opportunity to revise your answers based on what you've learned.

A World of Sounds

Sounds All Around

You are surrounded by a sea of sound. There is a tremendous variety of sounds in this sea. At times, some of them become part of the background. For example, did you *really* hear the following sounds this morning?

- the hiss of water running from the tap
- the orange juice splashing into the glass
- the rattle of dishes
- the slam of the door
- the din of distant traffic

How many other sounds can you think of that are part of your everyday life but that you don't really notice? Why do you think we filter out certain sounds?

Investigating Sound

Studying sound raises many interesting questions. Think about the following questions:

- What was the loudest sound ever heard?
- Are there sounds you can't hear?
- How can a faint whisper sometimes be heard many meters away?
- What is an echo? What causes it?
- Why is it so silent after a snowfall?
- How do blind people use sounds to "see"?
- How do musical instruments make their sounds?
- How do music and noise differ?

Can you add to this list of questions? With your friends, generate a list of other questions about sound that you would like to investigate. At the end of your study of sound, see how many you have answered.

Wynton Marsalis playing jazz on his trumpet

The Sound of Silence

Is there such a thing as total silence? For instance, if you were put in a completely sound-proof room, would you hear anything? Actually, you would. People placed in soundproof rooms hear two distinctly different sounds. Sound engineers have discovered that one sound is made by the person's nervous system and the other sound is made by the person's blood circulating. Can you hear *your* body's sounds right now?

This specially designed room eliminates almost all sound. It is being used to check the amount of static transmitted over a telephone system.

Noise

In many cities today, the continuous background of sound is increasing to alarming levels. Studies show that many people suffer physically and mentally when exposed to excessive noise. What are some sources of offending sounds? What are some possible harmful effects?

At the other extreme, people can become nervous and irritable when there is too little sound. Modern buildings are often so quiet that even ordinary noises can startle people working in them. To overcome this problem, sound is piped back into the rooms! It sounds something like the hiss of escaping steam and is called "white noise." Why do you suppose it is given this name?

Describing Sounds

Make or obtain an audiotape of various sounds. Have others guess what is making each sound. Then brainstorm to think of words or word combinations that describe each sound, such as *harsh, loud, sweet, shrill, grating, high,* or *low.* Afterward, think of all the ways you might classify these word descriptions.

Yikes! Sometimes sounds can be too loud. For information about sound and safety, see page S124 of the SourceBook.

Classifying Sounds

As you read the following article, write a descriptive heading for each paragraph in your ScienceLog. Use no more than four words for each heading.

Which sound has a higher pitch?

Which sound is louder?

SQUAWK!

Which sound is "sweeter"?

Why is this music, rather than noise?

(a)

Sounds have certain characteristics that can be thought of as dimensions. We also speak of dimensions when measuring objects. One such dimension is height. In a way, sounds also have height. A sound may be high like the chirp of a cricket, lower like the croak of a frog, or still lower like the grunt of a pig. The "highness" or "lowness" of a sound is called its pitch.

(b)

Sounds like rain falling may be quite soft. Other sounds, like a door slamming, may be louder. A speeding train is louder still. Loudness *is another dimension of sound.*

(c)

The rat-a-tat-tat of a jackhammer is a sequence of very short sounds; each lasts a fraction of a second. Other sounds linger, such as that made by a large bell. Are musical sounds generally long or short? The length of time, or duration, *for which a sound can be heard is another dimension of sound.*

(d)

When we try to further classify sounds, we may use words like sweet, dull, bright, blaring, or harsh. We may also use words such as crackling, buzzing, clanging, and tinkling. We use these descriptive words to describe another dimension of sound—its quality. *A bell makes a sound of a different quality from that of a car horn, even if the loudness, pitch, and duration of both sounds are the same.*

(e)

In music, tonal quality is called timbre, *or* tone. *Why is there such variation in tonal quality? When you mix paints, the shade you get depends on the colors you start with and how much of each color you use. This is also true of sounds. The quality you hear depends on the mix of component sounds that form the overall sound.*

(f)

What is music, and what is noise? Think of the musical tones of the flute, the less musical sound of a ringing telephone, and the totally unmusical noise of the lawn mower. Musical sounds have fewer component sounds; they are "purer."

The dimensions of sound are pitch, loudness, duration, and quality. Changes in these dimensions create the incredible variety of sounds that surround us.

Making Sounds

EXPLORATION 1

A Symphony of Sound

It's not difficult to make sounds, but it is sometimes difficult to see what is happening when sounds are made. Do several of the Activities that follow. For each Activity, record answers to the following questions:

1. How do I make the object produce the sound?
2. What is the object doing as it produces the sound?
3. How long does the sound last?
4. How can I stop the sound?
5. Can I change any characteristics (dimensions) of the sound, such as loudness and pitch? If so, how?

You Will Need

- tuning forks of different sizes
- a rubber stopper
- a glass container of water
- 2 rulers—one wooden, one metal
- a drinking glass
- a cardboard box
- 2 rubber bands
- a pencil
- a plastic drinking straw
- scissors
- a balloon
- a cardboard tube (12 cm long)
- rubbing alcohol
- cotton balls
- paper
- miscellaneous objects

ACTIVITY 1

Tuning-Fork Sounds

1. Strike a tuning fork on the edge of a rubber stopper.
2. Hold the tuning fork close to your ear. What do you observe?
3. Lightly touch the prongs to various parts of your body.

Caution: Do not touch the prongs to your eyes or eyeglasses. Clean the tuning fork with rubbing alcohol and a cotton ball before another student uses it.

After striking the fork again, touch the prongs to the surface of a glass container of water and then to a loosely held piece of paper. What do you observe?

4. Next, while the tuning fork is sounding, touch its base to your teeth, to the table, to a cup held over your ear, and to other objects.
5. Try some of these same experiments with a tuning fork of a different size.

ACTIVITY 2

Ruler Sounds

1. Hold one end of a wooden ruler firmly on the edge of a table. Push down on the other end, and then let it go. Try this several times, with various lengths of the ruler extending over the end of the table. What do you observe?
2. Substitute a metal ruler for the wooden one, and repeat the activity. What do you observe now?

ACTIVITY 3

Rubber-Band Sounds

1. Pluck a rubber band that has been stretched across a cardboard box. Listen carefully and observe what happens.

2. Tighten the part of the rubber band that is on the top of the box. Pluck it again. Is there any difference in sound? Why or why not?

3. Put a pencil across the top of the box (the short way), under the rubber band, and pluck again. Do you hear or see any differences? Why or why not?

ACTIVITY 4

Straw Sounds

1. Make a "straw saxophone," as shown below.

2. Adjust the position of the "sax" in your mouth until a steady sound is produced. Try producing different sounds by blowing in different ways.

Be careful: Use only your own straw, and dispose of it after completing the Exploration.

3. While blowing a steady sound, use scissors to cut the straw shorter and shorter. What happens?

ACTIVITY 5

Simulated Voice Sounds

1. Place your fingers on your neck near your vocal cords. Say "Ah" loudly. What do you feel? Try some high sounds and some low sounds.

2. Blow up a balloon. Make the balloon squeal as you slowly release air from it. Try for variations of sound. What must you do to get a higher sound? a lower sound? This is similar to what happens in your throat when you speak.

3. Make a working model of human vocal cords. Stretch a piece of balloon over the end of a cardboard tube, but not too tightly. Cut a narrow slit in the piece of balloon, and blow into the tube from the opposite end. Now tighten the balloon so that the slit becomes longer, and blow again. How does increasing the tension affect the pitch?

Caution: Blow only into your own tube. Dispose of tube after use.

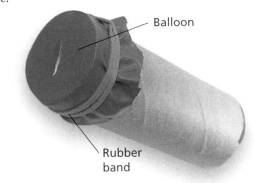

Balloon

Rubber band

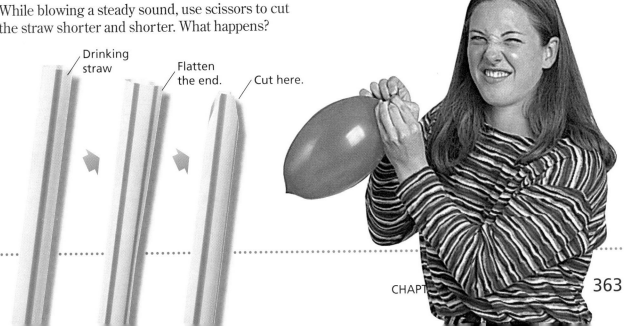

Drinking straw

Flatten the end.

Cut here.

Making Sense of Your Observations

After doing Exploration 1, you now know the following facts about sound:

1. *Sound is caused by vibrating objects.* What vibrates in each of the experiments in Exploration 1?

2. *Energy must always be added to an object in order for sound to be produced.* A force is exerted on a part of an object to move it a certain distance and to make it start vibrating. This means that work is done on the object. Whenever work is done on an object, energy is given to it. In the case of sound sources, this energy causes a vibration, which results in sound—one of the forms of energy produced by the vibrating object.

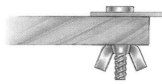

How did you provide energy in each of the Activities in Exploration 1?

3. *If more energy is added to an object, the object passes through a greater distance as it vibrates, and the sound is louder.*

Ruler motionless—no sound produced

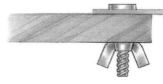

Ruler pushed down and released—sound produced

Ruler pushed farther down and released—louder sound produced

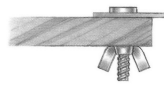

How did you produce a louder sound in each experiment in Exploration 1?

4. *The more often an object vibrates within a given period of time, the higher the pitch of the sound produced.*

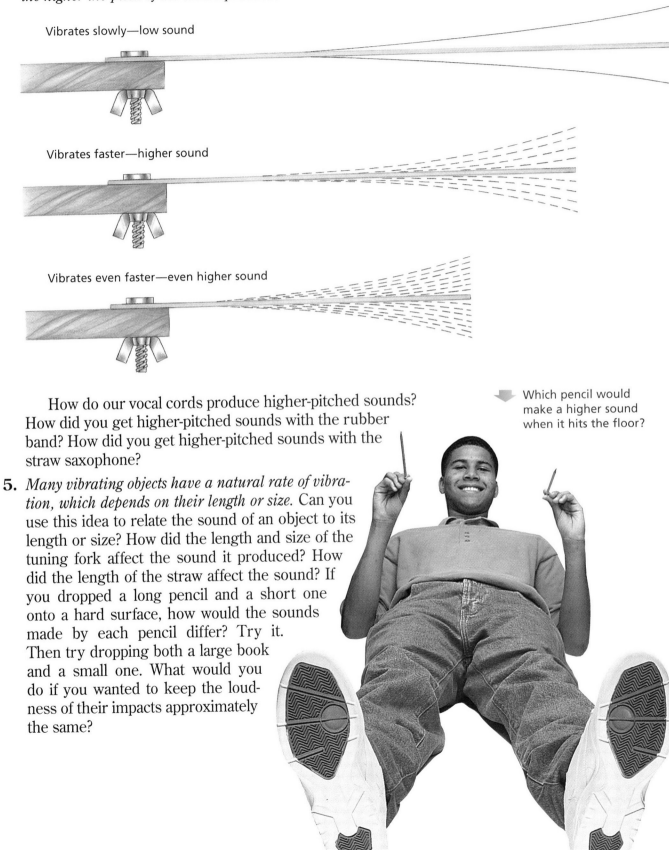

Vibrates slowly—low sound

Vibrates faster—higher sound

Vibrates even faster—even higher sound

How do our vocal cords produce higher-pitched sounds? How did you get higher-pitched sounds with the rubber band? How did you get higher-pitched sounds with the straw saxophone?

Which pencil would make a higher sound when it hits the floor?

5. *Many vibrating objects have a natural rate of vibration, which depends on their length or size.* Can you use this idea to relate the sound of an object to its length or size? How did the length and size of the tuning fork affect the sound it produced? How did the length of the straw affect the sound? If you dropped a long pencil and a short one onto a hard surface, how would the sounds made by each pencil differ? Try it. Then try dropping both a large book and a small one. What would you do if you wanted to keep the loudness of their impacts approximately the same?

Some Puzzlers

1. What is vibrating when you perform each of the following actions?

   a. blow over the mouth of a soft-drink bottle

   b. tear a piece of paper

   c. whistle

   d. shut a door

   e. listen on a phone

   f. tap-dance on an uncarpeted floor

2. What is vibrating when you play each of the following musical instruments?

   a. a piano

   b. a harmonica

   c. a trumpet or bugle

   d. a drum

   e. an upright bass

   f. a flute

Sound Questions to Investigate

PART 1

Design an experiment that uses a meter stick to answer the following questions:

- Must vibrations occur at a certain rate to be audible (capable of being heard)?
- Must vibrations move a certain distance to be audible?
- Is a minimum amount of energy necessary in order for sound to be audible?

Write your procedure in your ScienceLog, and then perform the experiment. The diagrams in Making Sense of Your Observations (pages 364 and 365) may give you some ideas.

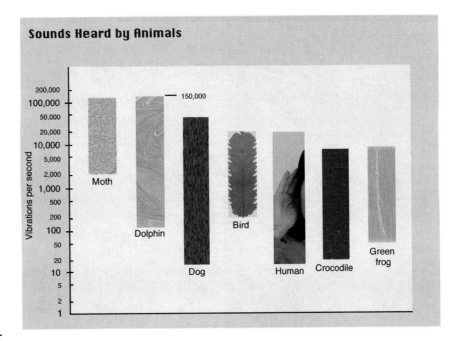

Sounds Heard by Animals

PART 2

The graph above shows the approximate vibration rates required to produce sounds that various kinds of animals can hear. From the graph, find the answers to the questions below. (Hint: First write down the approximate range of vibration rates heard by each animal.)

1. Note that a vibrating object must move back and forth at least 20 times per second to produce a sound that humans can hear. Above what rate of vibration are sounds inaudible to humans?

2. Which animal(s) can hear higher sounds that aren't audible to people? Which animals can hear lower sounds?

3. Which animal can hear the widest range of vibration rates?

4. What sounds can be heard by a green frog but not by a bird?

5. Which animal can't hear most of the sounds that a human can hear? What range of vibration rates can both humans and that animal hear?

6. Wanda blows a whistle. She hears nothing, but her dog begins to howl. Why?

Sounds, Naturally

Animals produce sounds of many kinds and make them in many ways. A dog, for instance, whines, whimpers, barks, growls, and howls. To make these sounds, it uses vocal cords, just as humans do. In fact, this is true of most mammals as well as other animals, including frogs. Many kinds of frogs also have balloon-like air sacs that amplify the sound made by their vocal cords.

Some other mammals (as well as humans) actually make sounds that we might describe as songs. Perhaps the most interesting songster is the humpback whale. Roger Payne, an American naturalist, investigated whale sounds with underwater microphones. Imagine his surprise when he discovered that humpbacks sing, producing what he described as "hauntingly beautiful sounds." Lasting from 6 to 30 minutes, the songs consist of low moaning, wailing, and lowing (cow-like) sounds, as well as high-pitched whistles and screeches.

Humpbacks' songs are the loudest sounds made by any animal. Scientists infer that before the time of steam- and oil-powered ships with noisy propellers, the song of a whale could be heard by other whales as far away as 1600 km. Even today, despite the continuous background noise from ships' propellers, a whale can hear another whale's song up to 160 km away.

Humpback whales have no vocal cords. Their sounds, or songs, are probably produced by forcing air back and forth through the **larynx,** or "voice box."

The most familiar animal songs are those produced by birds. Bird songs are made up of chirps, trills, whistles, and, sometimes, quick mechanical sounds made by clicking the beak. The non-mechanical sounds are produced by an organ called the **syrinx,** or "song box." The syrinx is located at the bottom of the windpipe (the tube connecting the mouth to the lungs). It contains a pair of membranes that vibrate when air passes over them, producing sound. Muscles attached to the membranes change the pitch of the sound.

 Humpback whale

Eastern meadowlark

Grasshoppers produce their mechanical sounds by rubbing, or "bowing," their long, hard, rough legs against their hardened wing covers. Mosquitoes produce their high-pitched whine by flapping their wings in the air many hundreds of times per second. All male crickets and females of some species chirp by moving a body part that looks like a toothed file, and is fastened under one wing, against a scratcher located under the other wing. There are approximately 4000 kinds of crickets, each producing its own distinct pattern of chirps. This is how members of the same species recognize each other. Even if 50 kinds of crickets are singing in a meadow, female crickets can recognize the call of the male crickets of their species.

Remember that sound is energy—and that it takes energy to make sound. For each chirp of a cricket, 19 pairs of muscles are doing work! Where does the energy come from when *you* make sounds?

Vibrations: How Fast and How Far?

When a bird sings, the membranes of its syrinx may vibrate as many as 20,000 times per second. When you pluck middle G on a guitar, it vibrates about 400 times in 1 second. The larynx of the humpback whale vibrates about 20 times per second when making its lowest sound.

Objects that produce sound generally vibrate too quickly for the vibrations to be seen. How might you observe the individual vibrations? One way is to take pictures of them with a high-speed motion picture camera and then project the pictures at a slow speed. You would then see the vibrations in slow motion.

In the following Exploration, you will analyze a slow, "soundless" vibrating situation. This will help you understand the much faster vibrations that produce sound.

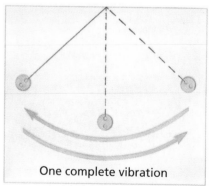

EXPLORATION 3

Investigating an Object Vibrating Very Slowly

You Will Need

- a heavy button, washer, or large paper clip
- a metric ruler
- a watch or clock with a second hand
- fine thread
- scissors

What to Do

1. Make a pendulum by hanging a button from a thread that is 25 cm long. Pull the button 10 cm to one side and let it go.

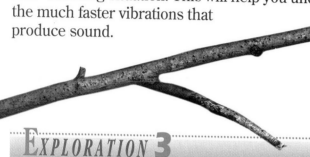

2. A **vibration** is defined as one complete back-and-forth movement of an object, from one side to the other and back again. Approximately how long does one vibration take? Does the time seem to change from one swing to the next?

Devise a way to find the time one vibration takes. About how many vibrations are there in 1 second?

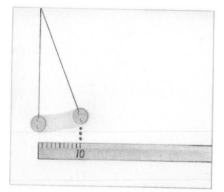

One complete vibration

3. Repeat step 2, but pull the button 20 cm to one side. The size of the swing—the horizontal distance from the side position to the central position, as shown below—is called the **amplitude**. Does the time for one vibration appear to change when the button swings twice as far?

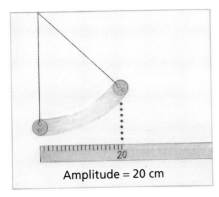
Amplitude = 20 cm

4. Shorten the thread to 6 cm. Pull the button 3 cm to one side and let it go. How much time does one vibration take? How many vibrations are there in 1 second? The number of vibrations in 1 second is called the **frequency** of vibration.

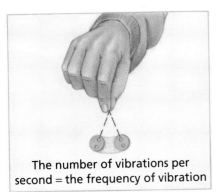
The number of vibrations per second = the frequency of vibration

Vibration
Amplitude
Frequency

Mason's Problem

Suppose you had to explain the meaning of *vibration, amplitude,* and *frequency* to Mason, a fourth-grader. How would you do this, using a metal ruler as your teaching tool? In your ScienceLog, write down what you would say.

Check It Out

Review what you discovered in doing the previous Exploration. From the statements below, identify those that you agree with, those that you disagree with, and those that you are unsure about.

1. Changing the amplitude of the swing did not change the time needed for one swing very much. If I could do the experiment accurately enough, it probably wouldn't change the time at all.

2. If I double the distance the button swings, I can make a big change in the time it takes for one vibration to occur.

3. If I were to start the swing only 5 cm out from the center, there would be fewer vibrations in 1 second.

4. If I shorten the thread, the time needed for one vibration to occur will be reduced.

5. If I shorten the thread, the frequency will be greater (the button will complete more swings in 1 second).

6. If I use about one-quarter of the length of thread, the time it takes for one vibration to occur will be about half that required when using the whole length of thread.

7. Changing the length of the thread does not change the number of vibrations that occur in 1 second.

8. If I were to make the thread 100 cm long, the time of one vibration might be 2 seconds.

A Sound Experiment

Jennifer wanted to count the vibrations of some metal rulers. She took pictures of them with a high-speed movie camera and then projected the film at a much slower speed—about 1/200 of the original speed. Suppose one back-and-forth motion (one vibration) took 1 second when her film was shown at this slower speed. Satisfy yourself that the ruler's actual frequency of vibration would be 200 vibrations per second.

Jennifer counted the number of vibrations in a given time from her screen projections, as observed in slow motion.

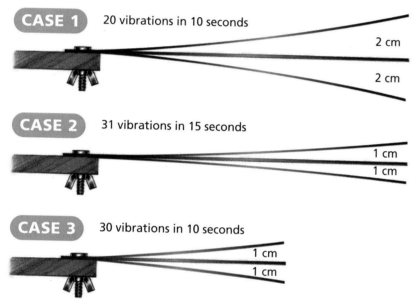

CASE 1   20 vibrations in 10 seconds   2 cm   2 cm

CASE 2   31 vibrations in 15 seconds   1 cm   1 cm

CASE 3   30 vibrations in 10 seconds   1 cm   1 cm

For Case 1

What is the frequency in slow motion? Try calculating the actual frequency.

For Case 2

a. What is the frequency in slow motion? in real motion?

b. How do the amplitudes in Cases 1 and 2 compare?

c. Does the amplitude affect the frequency of vibration?

For Case 3

a. What is the frequency in slow motion? in real motion?

b. How do the lengths of the two rulers in Cases 1 and 3 compare?

c. Do their lengths affect the frequency of vibration?

Conclusions

a. In which Case is the loudest sound produced?

b. In which Case is the highest-pitched sound produced?

CHALLENGE YOUR THINKING

1. Silent Communication

Sara is totally deaf and has never heard a sound. You are able to communicate with her only by writing. What would you write to help Sara understand the ideas of high- and low-pitched sounds, soft and loud sounds, and how we distinguish one sound from another?

2. Liner Notes

Ryan attached a piece of nylon fishing line to a nail at one end of a board. He stretched it along the board and hung a pail of sand from the other end. When he plucked the string, he got noise. When he slipped a wooden dowel between the string and the board, though, he got a musical sound. Why? What different ways could he devise to make the musical sound (a) higher pitched? (b) louder? (c) of a different quality?

3. Sound Reasoning

How would you demonstrate or explain convincingly each of the following to someone who has never studied science before?

a. Energy is needed to produce sounds.

b. Sound is produced from vibrating objects.

c. Not all vibrating objects produce sounds that you can hear.

d. Some sounds are too high or low for humans to hear.

e. The vibrations that produce noise and music are different.

4. I Call This One "Kerploosh"

Words often sound like the sounds they describe. For example, the words *gurgling* and *splash* convey certain sounds made by water. How many more watery-sounding words can you think of? What are they? Coin some new words that describe the action of water in various situations.

5. Acrostics

Copy the acrostic puzzle shown at right into your ScienceLog. Can you fill in the blanks with the five appropriate words and then determine the mystery word? The questions should be answered in order.

a. An object whose frequency is small vibrates ____?____.

b. If a violin string is slightly loosened, the musical tone will be ____?____.

c. Two sounds of the same pitch and loudness may still differ in ____?____.

d. The loudness of a sound depends on the ____?____.

e. If the amplitude of a vibrating object is increased, its sound will be ____?____.

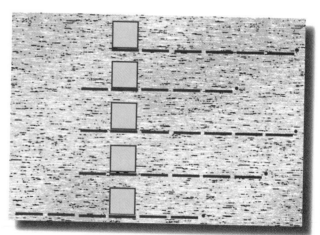

Now make an acrostic puzzle of your own using the ideas you have learned about sound so far. Give it to someone else to solve.

ScienceLog

Review your responses to the ScienceLog questions on page 358. Then revise your original ideas so that they reflect what you've learned.

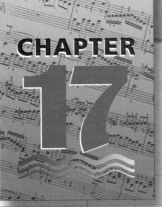

How Sound Moves

10:06:17

1 You see a flash of lightning, and several seconds later you hear the roll of thunder. How do you explain this?

10:06:19

BOOM!

2 Is there sound in outer space?

3 How does sound travel from source to listener?

ScienceLog

Think about these questions for a moment, and answer them in your ScienceLog. When you've finished this chapter, you'll have the opportunity to revise your answers based on what you've learned.

Sound Travel

How does sound energy travel from the sound source to your ear? Do you have a theory that might explain this process? What evidence do you have for your theory? The six experiments in the next Exploration will provide evidence to help you build a theory of sound travel.

EXPLORATION 1

Transmitting Sounds

EXPERIMENT 1

How are other forms of energy transmitted?

1. Place five pennies flat on the table.

2. Flick a sixth penny so that it hits penny *E*.

3. What happens to penny *A*? What kind of energy was transmitted from *E* to *A*? What was the energy transmitted through?

Scientists tell us that air is made of tiny particles. Is sound energy transmitted through these particles in the same way that kinetic energy (energy of motion) is transmitted through the pennies?

E   D   C   B   A

Have you ever experimented with a device like this one? It operates on a principle similar to that demonstrated by the pennies. Can you explain what happens? ➡

EXPERIMENT 2

If there were no air (or anything else) between a sound source and your ear, would sound still be transmitted to you?

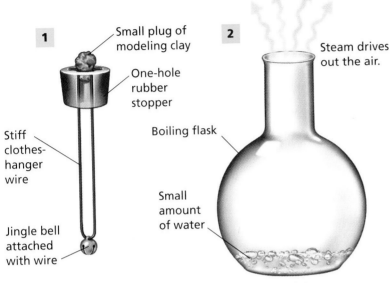

1
Small plug of modeling clay

One-hole rubber stopper

Stiff clothes-hanger wire

Jingle bell attached with wire

2
Steam drives out the air.

Boiling flask

Small amount of water

Boil water and then remove from heat source. Do NOT boil away all the water.

3
Quickly insert the bell assembly. Let flask cool and then shake.

What do you observe?

4
Remove the clay plug to allow air to enter. Then replace plug and shake.

What do you observe?

EXPERIMENT 3

Does the air between a sound source and your ear move?

1. Touch the piece of stretched balloon lightly with your finger as you hold the mailing tube steady. What happens? Is air moving in the tube?

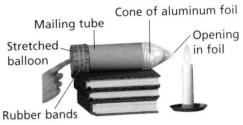

Mailing tube

Cone of aluminum foil

Stretched balloon

Opening in foil

Rubber bands

2. Make a loud sound near the piece of stretched balloon by hitting the bottom of a wastebasket with your hand. Observe the flame.

3. Try counting out loud next to the piece of stretched balloon. What happens?

4. As a control, remove the apparatus and repeat steps 2 and 3.

When sound energy moves from a vibrating object through the air to the ear, what do you think happens to the air?

EXPERIMENT 4

PART 1

A Model of Air

How is energy transmitted through a coiled spring, for example, a Slinky?

1. Hold one end of a spring and have a classmate hold the other end. Send pushes and pulls along the spring. Do you see energy being transmitted?

Exploration 1 continued

2. What do you feel when the "push" or "pull" reaches your hand?

3. How fast is the energy transferred along the spring?

4. Tie a ribbon to the middle of the spring. What happens to the ribbon as energy passes along the coiled spring?

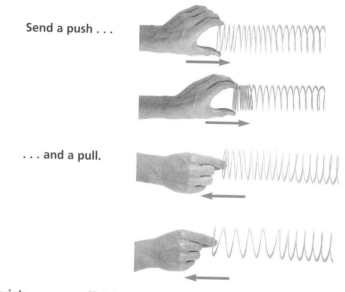

Send a push . . .

. . . and a pull.

PART 2

Spring Fling

The steps below match the diagram at lower right.

a. Mario and Kate hold the stretched spring on the floor.

b. Mario gives a quick push on the coiled spring, toward Kate. He then pulls it back quickly to its original position. Do you see the energy being transferred?

c. When the coiled spring is still again, Mario quickly pulls the spring toward himself. He then returns it quickly to its original position. Do you observe the energy being transferred?

d. Mario pushes his end of the spring quickly. He then pulls the spring back quickly as far past the starting position as he had pushed it forward. He then quickly pushes the spring back to the starting position. Mario has produced one complete vibration of the end of the spring. Kate says, "I think this is like what happens when a sound is made. One compression and one expansion together create a sound wave."

e. Mario produces two complete vibrations of the end of the spring, one right after the other.

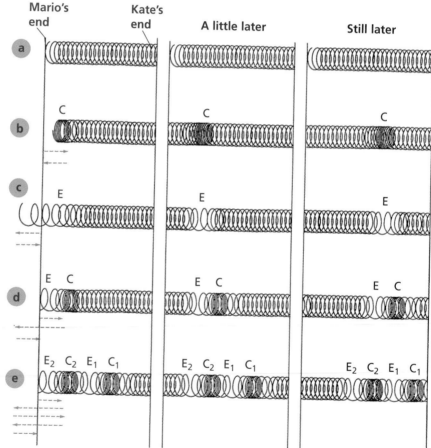

Key

C = Compression of the spring. The coils or turns of the spring are closer together than normal.

E = Expansion of the spring. The coils or turns are spread farther apart than normal.

Use what Kate and Mario observed on page 378 to help you answer the following questions:

Questions

1. When Mario pushes, as in (b), what happens to the coils of wire?

2. What happens to the spring compression after Mario stops pushing?

3. When Mario pushes on the end of the spring, as in (b), he gives energy to the spring. What happens to Kate's hand when she receives the compression?

4. When Mario pulls, as in (c), what happens to the coils of wire? What happens to the spring expansion after he stops pulling?

5. Every time Mario makes his end of the wire go through one vibration, as in (d), what is given to the coiled spring?

6. How is the energy that Mario gives to his end of the spring passed along to Kate?

7. What is happening to each coil of wire as Mario sends a series of "waves" along the spring? (Review your observations of the ribbon.)

8. Suppose that air particles act like the turns of wire in a coiled spring. How would sound energy from a vibrating tuning fork pass through the air to the ear?

EXPERIMENT 5

Elastic Air

Are air particles elastic? That is, do they act like the turns of wire in a coiled spring?

1. Fill a syringe halfway with air, and plug its tip with modeling clay.

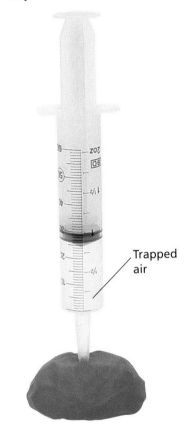

Trapped air

2. Push down on the piston, compressing the trapped air. Let the piston go. What happens?

3. Now pull up on the piston, letting the trapped air expand. Let the piston go. What happens?

4. Try compressing and extending a ball of clay. Which behaves more like a spring, clay or air?

EXPERIMENT 6

Viewing Moving Waves

Here's another way to view what may be happening in air. Cut a narrow slit 10 cm long and 1–2 mm wide in a piece of cardboard. Place the slit over the diagram on the next page and move it down at a constant speed.

Questions

1. In which direction does the compression move? (It's difficult to see expansions move.)

2. How can you make compressions move in the opposite direction?

3. How can you make compressions move faster?

4. How many compressions can you see at one time?

5. What do you think the **wavelength** is? Measure it.

6. As compressions move along, what is happening to each line as the card goes down the page?

Exploration 1 continued

Drawing Conclusions

From which experiment might you have drawn the following conclusions?

1. Sound energy is not transmitted through a vacuum. In other words, sound cannot be transmitted where there is no air.

2. Air is elastic, like a spring.

3. Energy often uses a **medium** (consisting of particles) for its transmission.

4. When sound moves through air, the air moves.

5. One way in which energy may be transmitted is in the form of waves of compression and expansion.

Using this evidence, construct a theory of how sound energy is transmitted through air from a sound source to your ear.

On the next page is a series of diagrams that will help you visualize what is happening in the air around a vibrating object. Later, you will be able to compare the theory that these diagrams suggest with the theory you constructed here.

Sound Thinking

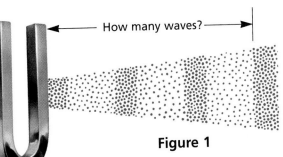

How many waves?

Figure 1

Figure 2

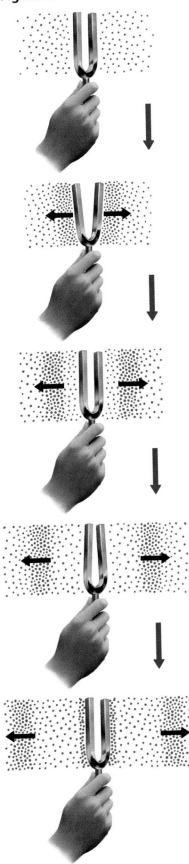

1. Here are illustrations of what you might see if the air around a vibrating tuning fork were visible. Locate the places where the particles are pushed together (compression) and where they are much farther apart (expansion). In Figure 1, how many waves are shown in the indicated region? How many times has the fork vibrated within that region?

2. Measure the wavelength in Figure 1 with a ruler. For waves of the length shown, how many centimeters would a compression move to the right when the tuning fork vibrates once? when it vibrates five times?

3. Figure 1 is a scale drawing of what would happen in air with a tuning fork of a certain frequency. In the figure, 1 cm along the wave represents approximately 40 cm. How far would the sound travel during four complete vibrations of the tuning fork? during one vibration?

4. The tuning fork in the illustration vibrates 440 times in 1 second. How far (in centimeters) would a wave travel through the air in 1 second? How many meters would this be?

5. In question 4 you calculated the distance that the wave traveled through the air in 1 second. In so doing, you estimated the speed of the sound wave. Complete the following equation:

wavelength $\times$ _____?_____ = speed of sound waves

6. Suppose a tuning fork vibrates at the rate of 220 times per second. What would be its wavelength, given the value for the speed of sound that you obtained in question 5? How would an illustration of the sound waves produced by this tuning fork differ from Figure 1 above?

7. Figure 2 shows what happens during one complete vibration of the fork. Add four more drawings to show what happens during the second vibration of the fork. Include the motion of the first vibration in your new drawings.

Investigations of Sound Transmission

Is sound transmitted through wood, metal, cardboard, plastic, and other materials? Find out by doing the following activities.

1. Hold one end of a wooden rod (at least 2 cm in diameter) close to your ear. While doing so, scratch the other end with a tack. Can you hear the sound through the wood? Could you hear the same sound in air?

2. Hold a vibrating tuning fork against the end of the rod away from your ear. Then repeat the experiment, substituting plastic, metal, and paper objects for wood.

Be Careful: Use objects whose diameters are larger than that of your ear canal. Carefully place each object so that it is very close to the outside of your ear canal but not touching.

Which material transmits sounds most loudly? In other words, which material is most efficient in conducting sound energy? Is there a material that will not transmit sound?

3. Which transmits sounds more efficiently—air or water? Try the following activity: Hit two spoons together in air and then in water, listening carefully to the sounds each time. (Use a homemade stethoscope like the one shown below to listen to the sound in water.)

4. Try the following combinations at home in the bathtub.

Caution: Do not submerge your head under water, and keep all electrical devices away from the water.

   a. sound source in water—ears in air, ears in water

   b. sound source in air—ears in air, ears in water

5. You've probably seen a movie in which someone put his or her ear to the ground to hear the sound of horse hoofs before the sound could be heard through the air. Design an experiment to test this idea, using the floor. You will need a very quiet room.

6. Do some humming. While you hum, plug your ears. What difference(s) do you notice in the sound? Why does this occur?

The Speed of Sound

What Is Speed?

Whew. That was fun. How'd I do, Travis?

Is that a good speed?

Wow, Lena, you lapped the track in only **1 minute**. That's 3 kilometers in **60 seconds**.

Well, you traveled 3000 meters in 60 seconds, so your speed was 50 m/s. That means you broke the track record of 48 m/s. Congratulations, Lena!

How did Travis calculate the speed? What facts did he need to know? Travis divided the distance in meters by the time in seconds and found the speed in meters per second. Lena's motorcycle speed tells us that she traveled 50 m in 1 second. (To change meters per second to kilometers per hour, which is how speed is customarily measured, multiply by 3.6.) How would you define *speed*?

An Unusual Race!

Was Lena's speed really that good? Suppose Lena raced for 5 seconds with each of the following:

a. a cheetah (the world's fastest animal sprinter)

b. a peregrine falcon (the world's fastest bird)

c. the fastest human runner

d. the *Bluebird* (a record-holding race car)

e. the sound from a rifle shot

f. the Concorde (a supersonic jet airplane)

g. a space shuttle

h. the flash of light from a rifle shot

i. the bullet from a rifle

Scale: 1 cm = 50 m
Drawn to scale, the distances shown represent how far each competitor went in 5 seconds.

Lena

a 250 m

b 125 m

c 260 m

55 m

d

e

f

g

h

i

Starting Line

1. Which of these would Lena outrace?

2. Which would outrace Lena?

3. Which of all these has the fastest speed?

4. A rifle is fired 100 m away from a target. Which reaches the target first—the bullet, the flash of light, or the sound of the rifle shot? Which arrives last?

5. What is the speed record (in meters per second) for a race car?

6. What is the speed of the fastest human runner in meters per second?

7. What is the speed of sound?

8. Objects that approach the speed of sound compress the air ahead of them until the air is almost like a solid wall. At the speed of sound, a shock wave is created that makes a bang that sounds like a clap of thunder. Such a bang is known as a **sonic boom**. Which of the race contestants might cause a sonic boom?

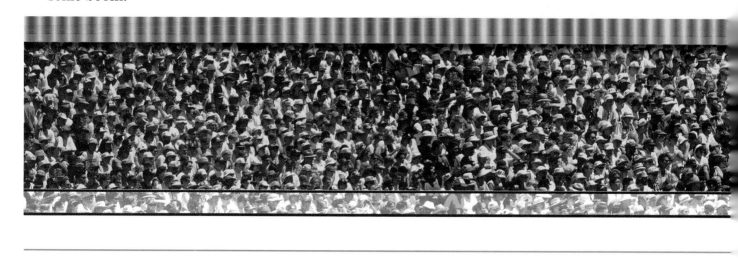

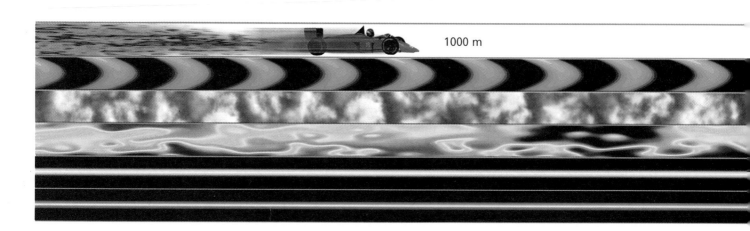

1000 m

An Echo Experiment to Measure the Speed of Sound

You Will Need

- a meter stick or tape measure
- a watch or clock with a second hand

What to Do

1. Stand about 50 m from the wall. Clap your hands. Listen for the echo. Clap your hands twice. Listen for two echoes.

   Try clapping your hands at such a speed that you begin a new clap just when you hear the echo from the previous clap. You won't hear the echoes any more.

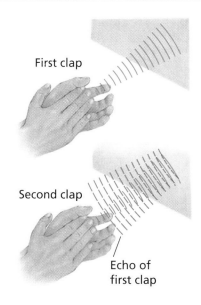

First clap

Second clap

Echo of first clap

What does the time between claps tell you?

2. Have someone time how long it takes to clap 20 times in such a way that you cannot hear the echo. (Don't count the first clap.) Try this several times.

   What is the average time between claps? Where has sound traveled during this time?

3. Measure the distance to the wall, in meters, by pacing the distance (once you've found the length of your pace). Have two or three people check the distance. What is the total distance traveled by the sound of the clapping—the distance from your hand to the wall and back to your ear?

4. From the distance and time of travel, calculate the sound's speed.

   Blake used the above procedure to calculate the speed of sound. Here are his results.

   20 claps in 8 seconds
   115 paces to wall
   10 paces are 6 m long
   Distance sound traveled
     = 138 m
   Time per clap = 0.4 seconds
   Speed of sound = 345 m/s

   Did you notice that the speed of sound calculated by Blake was considerably higher than the speed you calculated from the race on page 383? Blake performed his experiment at 25°C. The speed from the race is what it would have been at 0°C. What effect does temperature appear to have on the speed of sound? What is the average increase in the speed of sound per degree Celsius?

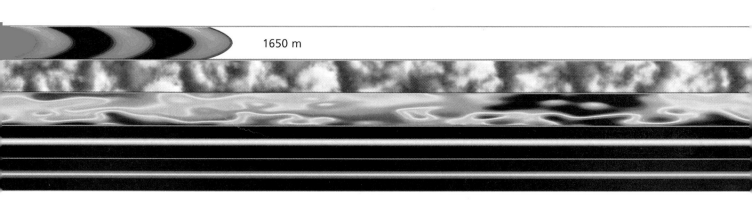

1650 m

Some Sound Puzzlers

1. What causes thunder to rumble, sometimes for quite a long while?

2. The speed of sound at 25°C = 345 m/s, and the speed of light = 300,000,000 m/s. Can you explain the different observations below? How far from the lightning bolt are the two people in the middle of the drawing? Why doesn't the last person hear the thunder?

That sure was close—a lightning flash and a thunderclap at almost the same time!

I wonder how far away the lightning is. I heard the thunder about a second after the flash.

The time between the flash and the thunder was about 4 seconds. The distance must be . . .

No thunder. The storm must be pretty far away.

3. Suppose you are riding in a car down a city street. Close your eyes. Can you tell when you are passing a building, passing a space where there are no buildings, or passing an intersection? How does the sound of the car help you?

4. The Colorado River winds its way through the deep gorge of the Grand Canyon. In this canyon, an outboard motor on a raft can be heard 15 minutes before the raft actually arrives. Why?

5. Why is it so silent after a snowfall?

6. Why does your singing in the shower sound so great?

7. A challenge! How far away from a wall should you be if you want to hear a complete, distinct echo of your voice saying "hello"? (Hint: To solve this, you'll need to measure how long it takes to say "hello.")

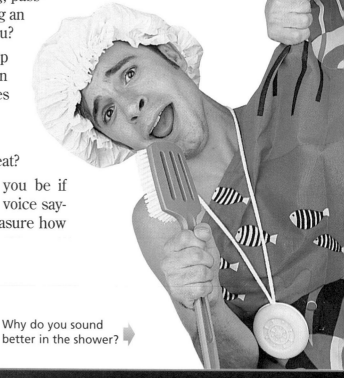

Why do you sound better in the shower?

2900 m

LESSON 3

More About Echoes

How Do the Blind "See"?

Ask a blind person how he or she is able to sense and avoid obstacles in his or her path. You may get answers like these:

"I feel the presence of the object."
"There are pressures on my skin that tell me an object is there."
"I sense danger."
"I have facial vision."

There is a story about a six-year-old blind boy who learned how to ride his tricycle on the sidewalks near his home. He never had an injury or an accident. He could veer around people, and he knew when to turn corners without going into the street. How did he "see"?

For years, there was a blind cyclist who rode his bike in downtown Toronto, Canada. With less than 10 percent vision, he could not see traffic lights, jaywalkers, or road repair crews. How did he "see"?

How might blind individuals use the tapping sound of their cane to identify objects around them?

A Research Project

A Cornell University professor and two of his students, one of whom was blind, wondered how blind people are able to avoid obstacles even though they can't see them. The research team designed a project to find out.

Think like a researcher: What steps might you follow? The questions below will assist you. Each question is followed by an explanation of what the researchers actually did. Cover these explanations with a card, revealing them only after you have suggested a possible answer.

The researchers were particularly interested in testing the following hypotheses:

- *Skin-Pressure Hypothesis:* Obstacles send out signals that produce pressure on the skin, enabling the blind person to be aware of the obstacles.

- *Sound Hypothesis:* Sounds hit obstacles and bounce back, alerting the blind person to their presence.

a. What kind of people would you use for the experiment—blind, blindfolded, or both?

The Cornell team decided to use both blind and blindfolded subjects. They gave special training to the blindfolded people. After a little practice, they too could avoid obstacles placed in their path.

4000 m

b. What kind of experimental situation would you set up to test the avoidance of obstacles?

The experimenters walked down a long hallway toward a fiberboard screen. The experimenters changed the position of the screen from trial to trial. Before long, each subject could walk down the hallway, detect the presence of the screen, and stop. On average, subjects detected the screen when they were about 2 m away from it.

c. Which hypothesis would you test first?

The researchers tested the hypothesis that blind people mention most often, the skin-pressure hypothesis.

d. What would you do to the subjects to prevent their skin from being affected by the screen?

The subjects had to wear an "armor" of thick felt over their heads and shoulders and heavy leather gloves on their hands.

While clothed this way, they couldn't even feel the air from an electric fan, but they could still hear sound.

e. If they bumped into the obstacle when wearing the armor but didn't bump into the obstacle when they were without it, what would you conclude? If they avoided the obstacle when wearing the armor, what would you conclude?

The subjects wearing the armor avoided the obstacle. However, they tended to approach it more closely, on average, than they did before. They stopped 1.6 m away, as compared with 2 m without the armor. The researchers concluded that the skin-pressure hypothesis was not correct. Blind people do not "feel" the pressure of obstacles with their hands and face.

f. What would you do to test the sound hypothesis?

Ear coverings of wax, cotton, earmuffs, and padding were worn by the subjects. They could not even hear their own footsteps. Their faces and hands were left uncovered.

g. If they bumped into the screen while their ears were plugged, what would you conclude? If they stopped short of the obstacle, what would you conclude?

Spectacular results! Both blind and blindfolded people bumped into the obstacle. The blind people said that all sensation of "feeling" had gone. The sound hypothesis best explained the researchers' observations.

h. Some scientists objected to this conclusion. Perhaps the ear coverings changed the pressure on the ear and ear canal—pressure that would have resulted from the obstacle's presence. What could you do now to determine that avoidance of the obstacle was due to sound rather than to some sort of pressure on the ears?

The subjects were placed in a soundproof room some distance away from a hallway where an experimenter walked with a microphone. The sounds of the experimenter's footsteps were picked up by the microphone and carried by a telephone line to the subjects. The subjects were asked to locate the screen by listening to the sounds picked up by the microphone.

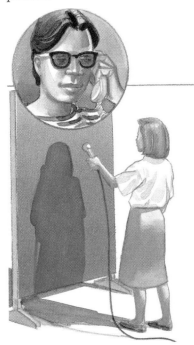

i. **What results would you expect if the sound hypothesis were true?**

The subjects were able to detect the screen from the sounds picked up by the microphone. The closest approach was 1.9 m. That's more proof for the sound hypothesis! The ability of blind people to avoid obstacles is apparently due to sound—not to pressure on the ear.

j. **Some scientists saw weaknesses in this approach as well. Can you think of any?**

The experimenter's pace and breathing might have given the subjects hints about the obstacle's presence.

k. **What other experiment might you design to check this possible weakness?**

The experimenters decided to use a motor-driven cart to carry the microphone toward the screen, and a loudspeaker to make sounds. The movements of the cart were controlled remotely by the subjects in the soundproof room.

l. **What conclusions would you draw if the subjects detected the screen?**

Although the average closest approach was a little closer than in previous experiments, the subjects always detected the screen. Therefore, signals from the breathing or pace of the walker could not explain why the subjects were able to detect the screen. The experimenters made three general conclusions.

1. Blind people locate obstacles by sound and by the reflection of this sound—that is, by echoes from the obstacles.

2. With practice, sighted people can also learn to detect obstacles by this means.

3. We are often not aware of how our senses and brain are working. Blind people sense that they "feel" the presence of objects around them. But scientific experiments show that they are, instead, hearing changes in sound as they approach obstacles.

m. **What further experiments might you now do?**

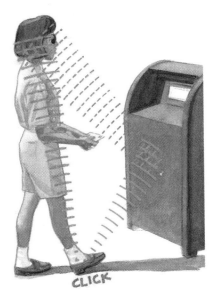

CLICK

The researchers are still interested in sending different kinds of sounds through the loudspeaker. Through experimentation, they may discover what sounds are best suited for the **echolocation** of objects.

Reporting Your Findings

Often, the final step of a research project is a written report in a scientific journal. In addition, scientists, like many people, enjoy sharing their interests with friends by writing letters. (Galileo, for instance, wrote many fascinating letters about his work to his friends.) Write a letter to a friend, telling him or her of your exciting discoveries in your role as a researcher into how blind people "see."

Echolocation at Its Best!

Whales and dolphins make clicks and other noises that reflect off objects as echoes. Using echolocation, river-dwelling dolphins can thread their way among logs and fallen trees in muddy rivers. In the open sea, dolphins accurately echolocate the fish they eat.

Bats, too, use echolocation to find food. They hunt flying insects in the dark by sending out pulses of high-frequency sound waves (about 45,000–100,000 vibrations per second—too high to be audible to humans). The waves bounce off of the insects, and the bats find their prey by listening for the echoes. In one experiment, a North American brown bat was able to echolocate and catch 175 mosquitoes in 15 minutes. If a bat were unable to echolocate, it might have to fly all night with its mouth open before catching one mosquito.

A bat usually chirps about 20 times per second. This number increases to 200 shorter chirps per second as it approaches its prey. How far away from its prey could a bat be and still hear a separate echo?

Many ships have a navigation system that uses echoes to find the depth of the water, the presence of schools of fish, and even the structure of the rock on the ocean bottom. This system is called *sonar*, which stands for **so**und **n**avigation **a**nd **r**anging. Sonar works in much the same way that echolocation works for a bat. The sonar device sends short pulses of sound waves through the water. When the sound waves hit the ocean floor, some of the waves are reflected back as an echo. The echo is then detected by a receiver. The gathered information can then be displayed on video monitors.

Shown here is a screen image from a fisherman's depth finder. This device uses the principles of echolocation to map the ocean bottom and to spot schools of fish.

Dolphins are champions at echolocation.

Completing the Sound Story

You know that sounds start with vibrating objects, which start sound waves in a medium (such as water, wood, or air). But the sound story isn't complete until sound waves activate your ear or some other *receiver*. You will now complete the sound story by looking at what happens between the time that sound waves reach a receiver and the time that the receiver registers, or "hears," the sound.

Designing an Instrument for Hearing

Can you design a device that can

- detect air particles moving back and forth in a sound wave,
- detect particles hitting it with only 0.000001 J of energy (by comparison, a square centimeter of paper dropped from a height of 50 cm hits with 0.001 J of energy),
- function when subjected to very great air pressure (very loud sounds)—up to 5 J of energy,
- detect air particles vibrating as slowly as 20 times per second and as quickly as 20,000 times per second,
- distinguish between air particles vibrating 500 times per second and those vibrating 505 times per second (compare this with the difference in vibrations per second between notes B and C on the piano keyboard at right),
- change the information it has detected into electrical impulses and carry this information to a central recording place,
- and detect its own orientation in space—whether level, on a slant, or upside down?

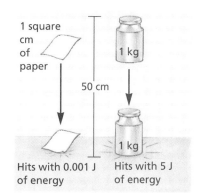

1 square cm of paper

50 cm

1 kg

1 kg

Hits with 0.001 J of energy

Hits with 5 J of energy

If your device can do all this, congratulations! You have designed the human ear!

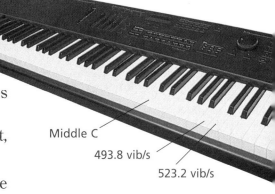

Middle C

493.8 vib/s

523.2 vib/s

The Human Ear

Here are three diagrams of an amazing instrument—the human ear. Figure 1 shows the whole ear, while the two smaller figures show sections of the ear, enlarged for easier viewing. You'll understand how the ear works after reading The Story of Hearing on the next two pages.

Notice that The Story of Hearing is divided into two parts. The left-hand column describes the structures of the ear, while the right-hand column describes the journey of a sound wave through the ear. In your ScienceLog, match the italicized terms in the story with the numbered items in Figures 1, 2, and 3, using the information provided. For each term, write a brief description summarizing its function.

When you have finished, you'll use your understanding of how we hear to answer some questions.

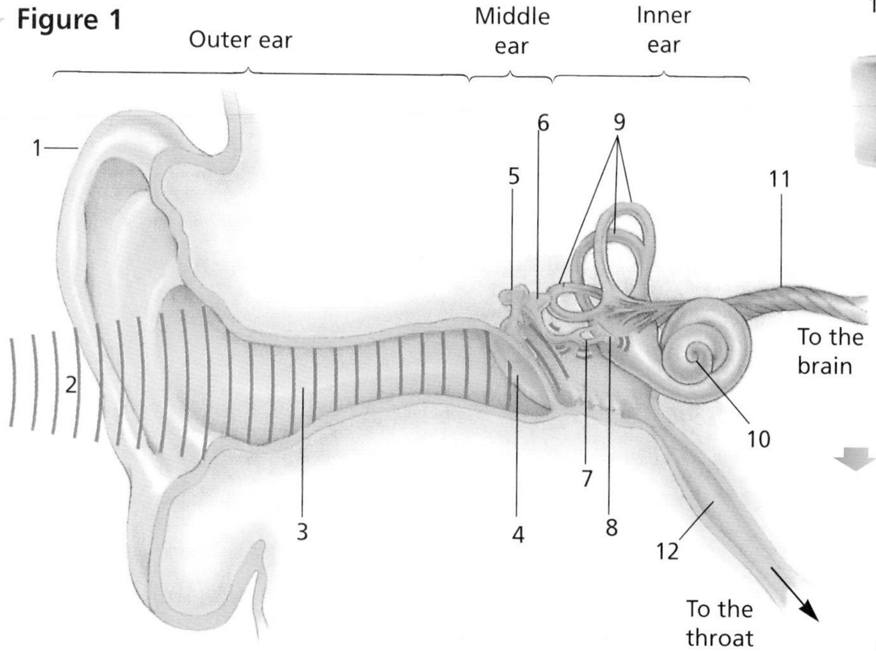

Figure 2
Close-up of middle ear and part of inner ear

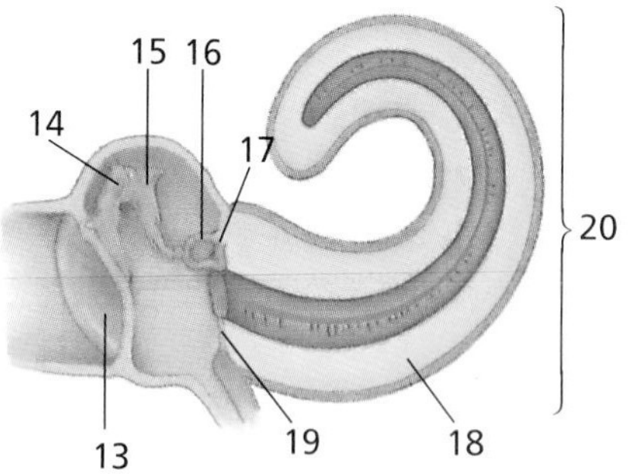

Figure 1

Outer ear   Middle ear   Inner ear

To the brain

To the throat

Figure 3
Cross section of the cochlea

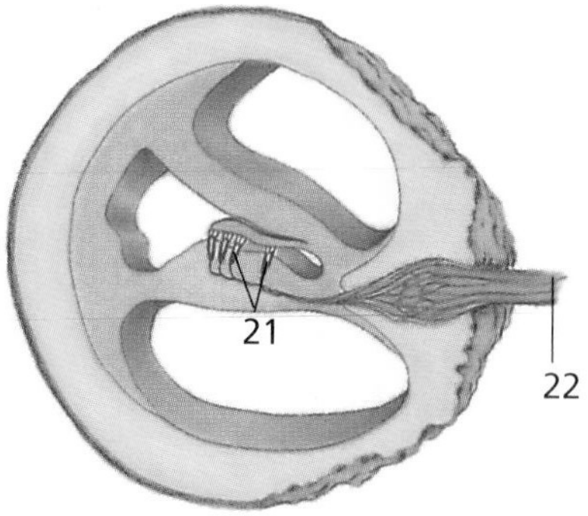

The Story of Hearing

In Figure 1, the brackets mark off the three regions of the ear.

Vibrating source of sound

The story of hearing tells how sound waves are passed from one region of the ear to the next. The story begins when a vibrating object, for example, a tuning fork, produces vibrations in the air particles surrounding it.

The outer ear (which is what people are usually referring to when they say "ear") includes the large external flap of cartilage and skin called the *pinna*. It leads into a narrow tube called the *canal.*

Sound waves enter the ear.

The pinna catches the sound waves and directs them into the canal.

At the end of the canal is the *eardrum.* This is the beginning of the middle ear. The middle ear is connected to the throat by the *Eustachian tube.*

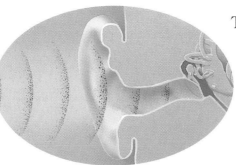

Eardrum

Eustachian tube

Sound waves cause the eardrum to vibrate.

The vibrations of the air particles cause the eardrum to vibrate. The energy of the sound waves determines how far and how fast the eardrum vibrates. Sometimes the eardrum is pushed far into the middle ear by the excessive pressure of a loud sound. When this happens, some of the pressure is released into the throat through the Eustachian tube.

Next to the eardrum are the three bones of the middle ear: the *hammer,* the *anvil,* and the *stirrup.* They fit snugly into each other. Pictured above the stirrup are the three *semicircular canals* of the inner ear.

Vibrations are passed on to the bones of the middle ear.

The eardrum passes the vibrations on to the three bones of the middle ear, which in turn vibrate. The bones act like levers, multiplying the force of the vibrations (at the expense of distance). The stirrup then passes the vibrations on to the inner ear. The semicircular canals have nothing to do with hearing. They are the organs of balance.

The stirrup fits into the small *oval window* in the inner ear (see Figures 1 and 2). The inner ear is a cavity in the bone of the skull that is filled with *watery liquid.* The cavity in the bone is made up of the semicircular canals as well as a snail-shaped part called the *cochlea.* (A close-up view of the cochlea is shown in Figure 2.) Located below the oval window is the *round window,* another membrane-covered opening of the inner ear.

Stirrup transfers the vibrations to the liquid of the inner ear.

When the stirrup vibrates, the energy passes through the oval window into the cochlea, where it causes waves in the fluid there. The round window vibrates at the same time, maintaining a constant pressure in the fluid.

The cochlea contains a membrane that runs the length of the tube, down its middle. Located on this membrane are a large number of little *hair cells.* Figure 3 shows a cross section of the cochlea.

Vibrating hair cells stimulate the nerve endings of the inner ear.

Movement of the liquid in the cochlea causes the little hairs there to vibrate, stimulating nerve endings within the inner ear. The short hairs at the beginning of the cochlea vibrate in response to sound waves of high frequency (which are caused by high-pitched sounds). Farther down the cochlea, longer hairs respond to sound waves of lower frequency.

Located at the base of the hairs in the cochlea are nerve cells that join together to form the *auditory nerve,* which is shown in Figure 1. The auditory nerve connects the ear to the brain.

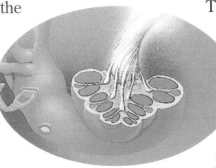

Nerve impulses are sent to the brain.

The nerve cells change the movements of the hair cells into electrical impulses (which are much like vibrations). The nerve impulses travel through nerve fibers to the auditory nerve, which carries the impulses to the brain. The brain then interprets the impulses as various kinds of sound. It is the brain that ultimately allows us to hear.

Questions to Think About

1. When a compression in a sound wave in the air hits the eardrum, in which direction does the eardrum move? In which direction does the eardrum move when an expansion of a sound wave arrives?

2. If you hear a bird sing a note with a frequency of 2000 vibrations per second, how many times per second does each air particle vibrate? How many compressions (and expansions) reach the eardrum per second? How many times does the eardrum vibrate per second?

3. How does the ear's response to a loud sound differ from its response to a soft sound? to a high sound and a low sound?

4. How does the ear strengthen the sound waves so that they will be strong enough to affect the liquid of the inner ear?

5. You turn around fast a few times and find that you can hardly stand up. Suggest what might be happening in your ear to "upset" you. (Hint: Swirl water in a glass and then set it down. What happens?)

6. Francine's model of the inner ear is shown in the photograph below. Which part of the inner ear is represented by each of these items?

    a. the tubes of the stethoscope

    b. the water in the container

    c. the stretched rubber sheet

    d. the tuning fork

Although they can make you feel like dancing, musical instruments actually produce *standing sound waves*. To find out more about standing waves, see page S117 of the SourceBook.

Does Francine's model work? Try it!

A Centuries-Long Debate

An ancient philosophical question asks, If a tree falls in the forest when no one is around to hear it, is there a "sound"? Write your opinion, and back it up with some principles and facts that you have learned in this unit, such as the following:

- Sounds are produced by vibrating objects.
- Sounds are transferred from place to place by vibrating particles.
- Sounds are heard as a result of vibrations transmitted through parts of the ear and the nerves to the brain.
- Dogs hear sounds that people cannot hear.
- Bats make sounds that they can hear but that people cannot hear.

More Items for Debate

Is sound produced in each of these situations? Give reasons for your answers.

1. Huge explosions occur on the sun's surface.
2. In the middle of a quiet night, you are in a house all alone.
3. A tiny square of tissue paper falls to the rug.
4. Your heart beats.
5. There is a running brook in the middle of the woods.
6. An electric bell rings in an airless bell jar.
7. Moths make clicks to confuse the echolocation sounds of bats. These clicks are inaudible to humans.
8. An electronic audio oscillator vibrates at 25,000 vibrations per second (25,000 Hz).
9. Particles of the gases in air constantly hit all of the objects around us.
10. A plane moving faster than the speed of sound has just passed you.

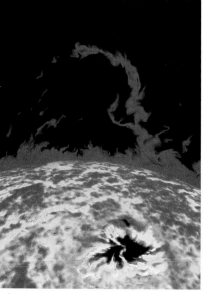

An explosion on the sun's surface

You already know that the *frequency* of vibration is the number of vibrations produced by an object in 1 second. Frequency is measured in units called **hertz**. One hertz (Hz) is the frequency of one complete vibration, or wave, per second.

The space shuttle would travel 50,000 m in 5 seconds. That's another 53 book pages. But what about the flash of light? This book would need another 20,000,000 pages!

50,000 m

1,500,000 m

CHALLENGE YOUR THINKING

1. Listen Up

How might you explain each of the following?

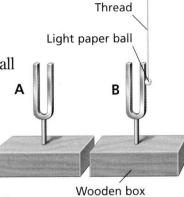

Thread

Light paper ball

A   B

Wooden box

a. When you strike tuning fork *A*, the paper ball jumps off its resting spot on an identical tuning fork, *B*. Why?

b. A grasshopper produces sound by rubbing its rough legs against its hard wing covers. This sound can be heard almost 100 m away. There are over 1000 tons of air in 1 cubic hectometer of air (1 hectometer = 100 m). How does a grasshopper have the strength to move that much air as it chirps?

c. How is sound communicated in outer space between two astronauts outside their spaceship?

d. The sound of a dentist's drill is louder to the patient than to the dentist. Why?

e. In the mountains, why does the echo of a shout last so much longer than the original shout?

2. Beat It!

Yale was watching Janelle play her bass drum on the football field. Janelle kept a steady rhythm going—about 20 beats every 10 seconds. When Yale backed away from Janelle to a distance of 85 m, he noticed that each *Boom!* came precisely when the drumstick was farthest away from the drum.

a. How do you explain this phenomenon?

b. From the mathematical information given, find (1) the time between drumbeats, (2) the time it takes for the drumstick to get to the farthest point from the drum, and (3) the speed of sound in air.

3. A Simple Tune

Jim blows the note "A" on his pitch pipe. The barbershop singers all hum the note. "Right on key," says the conductor.

Can you describe the whole story of what is going on, from the note Jim blew to the hum that pleased the conductor?

4. One Hand Clapping in a Forest?

After considering the debate questions on page 396, you may have realized that your answers depended on how you defined *sound*. Complete the sentence "Sound is . . . " in as many words as you wish, in order to create a definition that satisfies you. State why this is *your* definition.

5. A Model With Impact

Suppose you had a row of dominoes and you hit the first domino so that it fell backward. What would happen? Is this a good model of wave motion? Why? How would you depict a louder sound? a sound of longer duration?

Sciencelog

Review your responses to the ScienceLog questions on page 375. Then revise your original ideas so that they reflect what you've learned.

Listening Closely

1 All other factors being equal, these instruments will play notes of different pitch. What accounts for this?

2 Will everyone in the audience hear the sound at the same volume? Why or why not?

3 Why do a violin and a trumpet playing the same note at the same loudness sound different?

ScienceLog

Think about these questions for a moment, and answer them in your ScienceLog. When you've finished this chapter, you'll have the opportunity to revise your answers based on what you've learned.

A Closer Look at Pitch

EXPLORATION 1

An Investigation: Strings and Pitch

Before you do this Exploration, think about these questions:

- Many musical instruments use strings to produce their sound. What are some of these instruments?
- How can you make high or low musical sounds (tones) with these instruments?
- What characteristics of the strings can be changed to make musical tones higher or lower?

You Will Need

The diagram below shows one setup that will enable you to do this Exploration. You may, however, find other materials that work just as well. Try to design your own apparatus.

Be Careful: Do not let the plywood extend over the table's edge.

What to Do

Consider the following questions:

> Does the thickness of an instrument's string affect its pitch?

1. Stretch the three different strings or fishing lines using the same amount of tension. (Why?) You can do this by adding the same amount of water to each container.

2. Pluck the strings. What do you observe? By sounding the musical scale (*doh, re, mi,* etc.) you may be able to locate the tones produced by these strings on the scale. What conclusions can you draw?

> Does a string's tension (the amount it is stretched) affect pitch?

3. Test this hypothesis by using the strongest of the three nylon strings (a 9-kg-test fishing line works well). While one person plucks the string, another person can gradually add water to the jug.

4. What do you observe? What conclusions can you draw from your observations?

Try this: Add enough water to a jug so that the jug and water have a total mass of 1 kg. Hang the jug from the strongest string, and pluck the string. Sing this tone as *doh*. Then sing the scale *doh-re-mi-fa-sol-la-ti-doh'*. (The scale from *doh* to *doh'* forms an **octave**.) Try to keep in mind the sound of *doh'* (or record it).

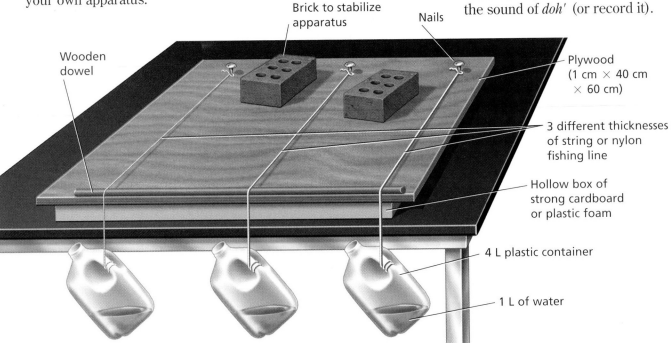

Brick to stabilize apparatus

Nails

Wooden dowel

Plywood (1 cm × 40 cm × 60 cm)

3 different thicknesses of string or nylon fishing line

Hollow box of strong cardboard or plastic foam

4 L plastic container

1 L of water

Now add 3 L of water to the jug (that is, 3 kg of water). What is the total mass of the jug and water now? Again, pluck the string. Does the tone sound like the note you sang earlier? If not, can you tell how far away it is on the scale?

You may have discovered an important idea. If the tension of a string is quadrupled, the frequency is doubled; in other words, *doh* becomes high *doh*, or *doh'*.

Does the length of a string affect pitch?

5. You can make a vibrating string shorter by using a pencil as shown below. How does the musical tone change as you shorten the vibrating part of the string?

6. Suppose the tone produced by the whole string vibrating is *doh*. Find the length of string that gives *doh'* (twice the frequency). How does this length compare with the whole length?

7. By moving the pencil along one string, locate the tones of the entire octave. Label them *d, r, m, f, s, l, t, d'* on a piece of paper taped under the string, as shown above.

Analysis

1. What effect does the thickness of a string have on the pitch of a sound?

2. What effect does the tension of a string have on the pitch of a sound?

3. What change in tension causes the frequency to double?

4. What effect does the length of a string have on pitch?

5. What change in the length of a string causes the frequency to double?

6. What change in length might cause the frequency to be halved?

An Orchestra

Find out more about the main groups of instruments in an orchestra—strings, woodwinds, and brass.

- How is sound made in each musical instrument?
- How are sounds made higher? lower? louder?
- How is the size of each instrument related to the pitch of its musical tones?

What are some factors that explain the different pitches of these instruments?

Loudness

Making Sounds Louder: A Problem

You are given a tuning fork. When you strike it, you can scarcely hear it. How can you make the sound of the fork louder? Before you read any further, write down some suggestions in your ScienceLog.

Some Solutions to the Problem

Darrel came up with a simple solution. He asked himself a question: How is the loudness of a sound related to the energy put into the object that produces the sound? His answer can be seen in the illustrations at right.

Nicole also found a simple solution. She wondered about the relationship between the loudness of the sound and the distance from the ear to the source of the sound. Her answer can be seen in the illustrations below, which she found in a reference book.

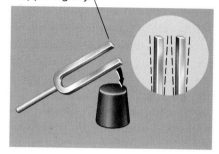

Hit the rubber stopper lightly.

Small amplitude

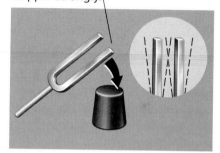

Hit the rubber stopper strongly.

Large amplitude

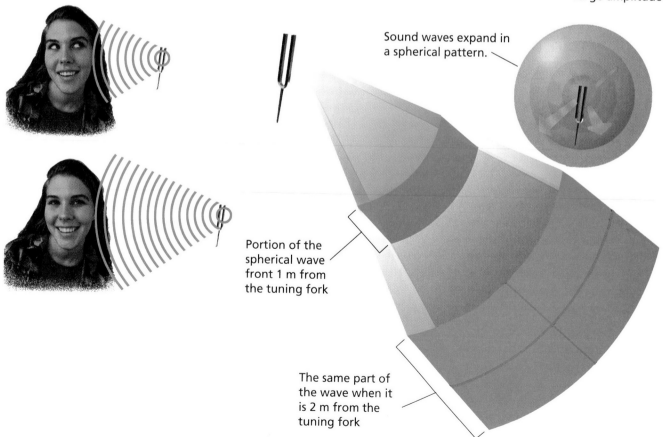

Sound waves expand in a spherical pattern.

Portion of the spherical wave front 1 m from the tuning fork

The same part of the wave when it is 2 m from the tuning fork

Study the illustration that Nicole found. Notice that when the distance of the sound wave from the tuning fork doubles, the wave spreads out over four times as much area. How loud will the sound be to a listener who is 1 m from the tuning fork, as compared with the loudness at 2 m? at 3 m? Try to derive a formula showing the relationship between the area of the sound energy and its distance from the sound source. Write this formula in your ScienceLog.

Diana discovered that placing the base of the tuning fork on a tabletop or a hollow wooden box makes the sound louder. Doug discovered that he could get a similar effect by using a cymbal rather than a box.

Diana and Doug have discovered **forced vibration**—getting an object to vibrate by touching it with an already vibrating object. What is the loudest sound you can get from a tuning fork by using forced vibration? To find out, try touching the fork to objects made of various materials, such as glass, metal, and plastic. How important is the size of the object? the material it is made of?

Yolanda found that the cone-shaped apparatus shown below helped make the sound of the tuning fork louder. How did it work? She concluded that the cone shape prevented some of the sound waves from spreading out in all directions. The cone concentrates the sound energy in a specific direction. Therefore, the sound appears louder to a listener in the path of the sound energy.

Have you ever used a **megaphone**? What are some situations in which a megaphone is used? Design an experiment to show how much farther a sound can travel when a megaphone is used.

A simple megaphone

Examples of forced vibration

Leonard discovered that he could make the sound louder by holding the tuning fork next to a cardboard tube. If the air column within the tube is just the right length—so that it has the same natural rate of vibration as the tuning fork—it will start to vibrate along with the fork. As a result, the sound is louder. The air column is vibrating in **resonance** with the tuning fork. What do you think causes this to happen?

Resonance also occurs when you strike a certain key on the piano and something in the room begins to vibrate. What would you say about the frequency of the piano string and the natural frequency of the object in the room? What other experiences have you had that involved resonance? Try writing the meaning of *resonance* in your own words. How is *resonance* different from *forced vibration*?

Leonard's design for making the tuning fork louder is quite simple. To try it, cut a cardboard tube to the right length using this approximate formula:

$$\text{length of tube in centimeters} = \frac{17{,}200}{\text{frequency of fork (Hz)}}$$

For a 512 Hz tuning fork, to what length would you cut the tube? You might want to cut the tube a little long and then shorten it as needed until it is just right.

After you get the tube to work, place it vertically on the table so that one end is closed. Next, hold the vibrating fork above the tube. Does the air column still vibrate in resonance with the fork?

Now cut the tube in half and try again, first with the tube open at both ends and then with it closed at one end. What do you observe? Why do you think this happens?

Who Said It?

Which of the students might have made each of the following comments based on his or her findings?

a. As more and more air particles are affected, their back-and-forth movement decreases.

b. If two objects are vibrating at the same frequency (their natural frequency), the sound will be louder.

c. If the disturbance of the particles is directed, their energy lasts longer.

d. If the air particles vibrate a large amount, the sound is loud.

e. The more air particles that are set into vibration, the more likely they are to hit the eardrum.

Measuring Loudness

The loudness of a sound can be measured with a *sound meter*. The scale on such a device is marked in units called **decibels (dB)**. On the meter shown below, what would happen to the reading if the meter were moved closer to the source of sound? farther from the source?

Estimating Loudness

How sensitive are your ears? How well can you tell whether one sound is louder than another? On the next page, List A contains a number of sounds. List B gives the approximate loudness of each of these sounds but not in an order that corresponds to List A. Match each sound with its loudness. Then examine the answers at the bottom of the next page to check your work.

List A

1. circular saw (nearby)
2. leaves rustling in a breeze
3. telephone ringing
4. water at the foot of Niagara Falls
5. jet taking off (nearby)
6. jackhammer breaking concrete (nearby)
7. rock-and-roll band at a concert
8. high-powered rifle shot (nearby)
9. quiet restaurant
10. ordinary conversation
11. quiet neighborhood at night
12. silence in a soundproof room
13. vacuum cleaner
14. whisper
15. ordinary breathing

List B

0 dB, 10 dB, 20 dB, 30 dB, 40 dB, 50 dB, 60 dB, 70 dB, 80 dB, 90 dB, 100 dB, 110 dB, 120 dB, 130 dB, 140 dB

The Decibel—An Unusual Unit

The decibel scale of loudness is an unusual scale. For example, the sound of a person breathing registers about 10 dB, while the rustle of a newspaper registers about 30 dB. Yet the sound energy provided by the newspaper is about 100 times that of breathing. The table below shows the relationship between sound energy and decibels.

Sound Energy and Decibels—Some Simple Rules	
Change in sound energy	Increase in loudness
2 × sound energy	3 dB more
4 × sound energy	6 dB more
8 × sound energy	9 dB more
10 × sound energy	10 dB more
100 × sound energy	20 dB more
1000 × sound energy	30 dB more
? × sound energy	40 dB more
1,000,000 × sound energy	? dB more

A Puzzler
One sound has a loudness of 30 dB. Another sound has 400 times as much energy. What is the loudness of the second sound, in decibels? (Use the table to find the answer.)

How loud is a ringing telephone in decibels?

Noise Pollution—The Effect of Sound on Health

As you read Now Hear This! find clues that will help you answer these questions:

1. What is crammed into the small space of your ear? (You know this from material you studied earlier in this unit.)

2. What does the author think are the most damaging sounds to a person's ear?

3. As you grow older, what kind of change will occur in your hearing range?

4. What kinds of sounds will be harder for you to hear as you get older?

5. What causes ringing in the ears?

6. What causes middle-ear infections?

7. What can you do to protect your hearing now?

8. What are four types of noise pollution that can really hurt your ears?

The human ear extends deep into the skull. Its main parts are the outer ear, the middle ear, and the inner ear.

Now Hear This!

Got your headphones on? Going to the concert and gonna stand right in front of the speakers? Got the TV, the radio, the stereo, or the CD player turned all the way up? Well you'd better hear this—that's right, I'm talking to YOU! And just who am I? I am your high-powered, anatomically amazing right ear.

I think you and I need to have a little talk. I mean, we've hung out together all our lives, but how much do you really know about me? Did you know that there's enough circuitry packed inside me to light a city? Or at least to give it phone service. And that's no joke!

Without me, you'd be missing out on the world of sound—the sounds of leaves rustling, waves crashing, music playing, and friends talking. And I do all of this in a space so small that I rival computer chips. But our lines of communication seem to be breaking down. Those tiny parts and delicate pieces that make me so incredible are beginning to deteriorate. So you gotta take care of me if you want to keep me working for you.

I was at my best the day you were born. From that time on, my abilities have been on the decline. What causes this? Loud sounds, mainly. Sounds that are too loud, occur too often, and last too long. As I get older, my tissues begin to lose their flexibility. We may be pretty young, but just 10 or 15 years down the road, we're going to feel the effects of any abuse now. Besides the loss of elasticity, hair cells begin to fall apart, and deposits of calcium build up. I'm sure you didn't know it, but that's all going on right now, right here, inside me.

When you were born, I had a hearing range of 16 to 30,000 cycles (vibrations) per second. But by now, my upper limit is just about 20,000 cycles per second. By the time we get to be as old as our grandparents, I could be down to as low as 4000 cycles per second. Think of all the sounds—good and bad—you'd be missing out on.

I tell you I'm cool, but I'm also fragile. Drum punctures happen very easily. Fortunately, these usually heal themselves. Ringing in here is another problem—the doctors call it tinnitus, and it can come from almost anything: drugs, fever, circulation changes, or tumors on my acoustic nerve. Sometimes the ringing can be stopped—but not always. I can also get infected, usually inside the middle ear. The Eustachian tube exposes me to those nasty infections. The Eustachian tube goes from the middle ear to the throat. In the throat there are a lot of microbes. Those microbes can make their way up to me, causing a painful middle-ear infection.

Another way I get damaged is by an overgrowth of bone from my middle ear. This freezes the motion of the bones, causing conduction deafness. If you have conduction deafness, a hearing aid might help. A surgeon can also go in and replace the stirrup bone with a metal duplicate. This is effective about 80 percent of the time. But bone overgrowth is not the most common cause of hearing

loss. I think you know what that is. You got it—LOUDNESS!

When you crank up the volume, my tiny muscles tighten up. I'm better at standing up to sounds like thunder or a bass guitar. But high-pitched, piercing sounds from things like jet airplanes, factory machines, and lead guitarists in rock bands destroy my delicate hairs and wreck my tiny muscles. I mean, when I am at my best, I am a finely tuned organ. But when you make me listen to high-pitched, glass-shattering screeches over and over again—well, then I'm done for.

So that's why I'm asking you to listen to me. We've got to do something to protect ourselves. Hearing is too precious for you to lose—take care of your ears. Keep them away from damaging sounds—plug 'em if you have to. Don't use cotton, though; it doesn't work. Use special earplugs instead. Hey, I'm the only right ear you got. And I'll keep hearing for you if you look out for me. Deal?

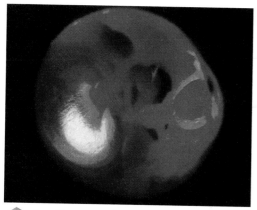

See if you can locate which part of the middle ear shown here could be surgically replaced by stainless steel.

Effects of Noise

Listed below are the decibel levels at which you'd experience certain effects.

70 dB: difficulty in hearing conversation
80 dB: annoyance at the noise level
90 dB: hearing damage if noise is continuous over time
120 dB: permanent hearing loss if noise is sustained
130 dB: beginning of pain in the ear
140 dB: sharp pain in the ear
170 dB: total deafness

LESSON 3

The Quality of Sounds

"Seeing" Sounds

Take a look at the paint smears on the palette. In what way are they similar? In what way do they differ? Although all of them are blue, each represents a different shade of the color blue. Which might be described as navy blue? ocean blue? Just how many shades of blue might there be?

Just as there are shades of color, there are "shades" of sound as well. Your ear (and brain) can distinguish between the different sounds produced by a trumpet, a piano, a violin, and a guitar all playing middle C at about the same loudness. But what makes this difference? If you could see sounds the way you can see different shades of color, you would better understand what gives each of these instruments a unique sound.

There is, in fact, a way to change sound waves into shapes that can be seen using a device called an **oscilloscope**. Study this method by examining the images below.

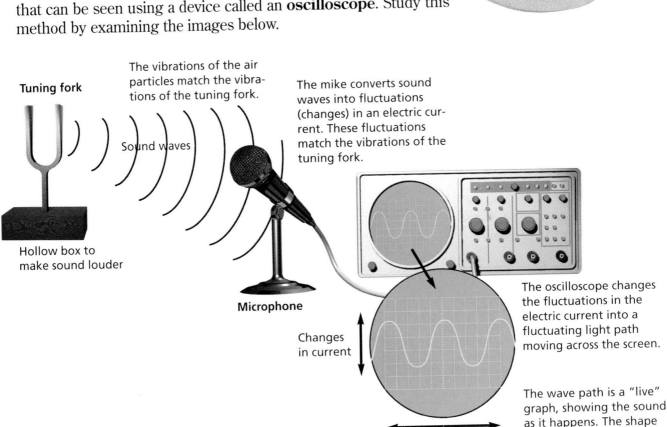

Tuning fork

The vibrations of the air particles match the vibrations of the tuning fork.

Sound waves

Hollow box to make sound louder

The mike converts sound waves into fluctuations (changes) in an electric current. These fluctuations match the vibrations of the tuning fork.

Microphone

Changes in current

The oscilloscope changes the fluctuations in the electric current into a fluctuating light path moving across the screen.

Time of sweep

The wave path is a "live" graph, showing the sound as it happens. The shape of the path matches the sound waves from the tuning fork.

How many vibrations, or complete waves, are represented in the wave picture on the oscilloscope at right?

As the tuning fork sounds, the light path sweeps from left to right. Suppose that the oscilloscope was set in such a way that one complete cycle of the light path crossed the screen in $\frac{1}{100}$ of a second. You wouldn't see the path as it was being traced out because it would happen too fast. Instead, you would see a steady wave picture; this picture would continue as long as the tuning fork was sounding.

If two complete waves appear on the screen in $\frac{1}{100}$ of a second, how many vibrations does the fork make in 1 second?

As you can see, the frequency of a sound can be determined by knowing how long it takes the light path to sweep across the oscilloscope screen and by knowing the number of waves appearing on the screen. How many complete waves would appear on the screen if the frequency of the sound were 300 vibrations per second (300 Hz)? 550 Hz?

If the time of sweep on an oscilloscope were set at $\frac{1}{50}$ of a second and the frequency of the sound were 200 Hz, how many complete waves would be pictured on the screen?

Exploration 2

Analyzing Wave Pictures on an Oscilloscope Screen

What to Do

Discuss the following questions with a partner. Be sure to write your conclusions in your ScienceLog.

1. This oscilloscope has its time of sweep set at $\frac{1}{100}$ of a second. What is the frequency of the pictured sound?

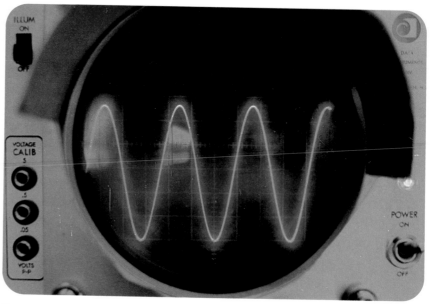

2. How does a wave picture on an oscilloscope show the loudness of sounds? Study these pictures and form your own conclusions.

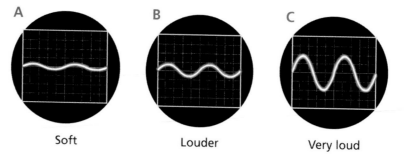

A	B	C
Soft	Louder	Very loud

3. How does a wave picture show the pitch of sounds?

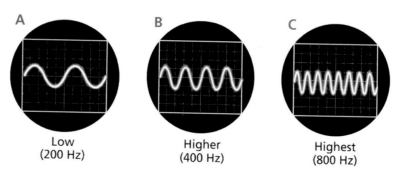

A	B	C
Low (200 Hz)	Higher (400 Hz)	Highest (800 Hz)

A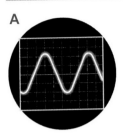

B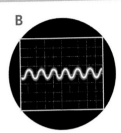

C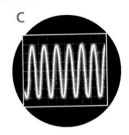

D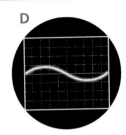

E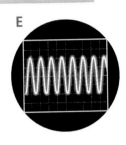

4. Check your conclusions to questions 1 and 2 by studying the five screen displays shown above and answering the following questions:

a. Match the screen displays above with each of the following:

- a sound that is both soft (quiet) and high-pitched
- a sound that is both loud and low-pitched
- a sound that is both loud and high-pitched
- a sound that is both soft and low-pitched

b. Which two screen displays show sounds of about the same loudness? of the same pitch?

c. Which screen display shows the lowest sound? the softest sound?

d. What are the frequencies of the sounds in D and E above if the time of sweep on the oscilloscope is $\frac{1}{100}$ of a second?

5. Below are the screen displays of five different sounds that you would probably recognize. The sounds differ primarily in quality.

A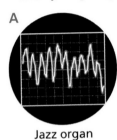

Jazz organ

B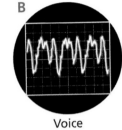

Voice

C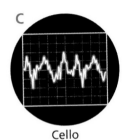

Cello

D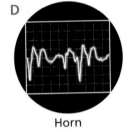

Horn

E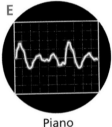

Piano

a. Do you recognize any similarities in the displays? What are they?

b. What differences are there among the wave patterns? Suggest some reasons for these differences.

6. Now consider the screen displays of two common noises. Note the haphazard displays of these nonmusical sounds.

Based on what you've observed, can you identify some of the differences between noise and music as shown by an oscilloscope? Describe these differences in your ScienceLog.

A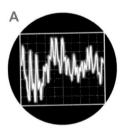

A sharp slap

B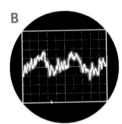

Office noise

Rubber tubing or stretched spring

How a String Vibrates

What accounts for the different qualities of sounds that you have seen pictured on the oscilloscope? The screen displays on the previous page suggest an answer. Also, recall that most things vibrate in a complex manner, causing a "mix" of many sounds. Do you see how the pitch and loudness of each of these component sounds could affect the quality of the sound you hear?

Examining how a string on a stringed instrument vibrates will make this clearer. First do the following activity with a partner:

Have one person hold steady one end of a long piece of rubber tubing or a stretched spring. The other person should slowly move the other end back and forth rather widely until the tubing or spring vibrates as a whole, as shown in (a) at right.

Now gradually add more energy; that is, make the tubing or spring move faster and faster until it assumes a more stable pattern of vibration. What happens? Were you able to produce the pattern shown in (b)? With even more energy, can you get other patterns of vibration, as shown in (c)? How many patterns did you get?

When you run a bow across a violin string, the string vibrates as a whole, just as the stretched tubing or spring did at first in the demonstration. However, at the same time, the violin string also vibrates slightly in various patterns, which consist of a varying number of parts. The second drawing below shows the string vibrating as a whole and in two parts at the same time. You know that the length of a vibrating string affects the frequency and pitch of the sound it makes. Therefore, when the violin string vibrates as a whole, it makes one sound. When it vibrates in two parts, it makes a higher sound. When it vibrates in three parts, it makes a still higher sound, and so on.

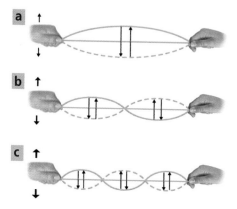

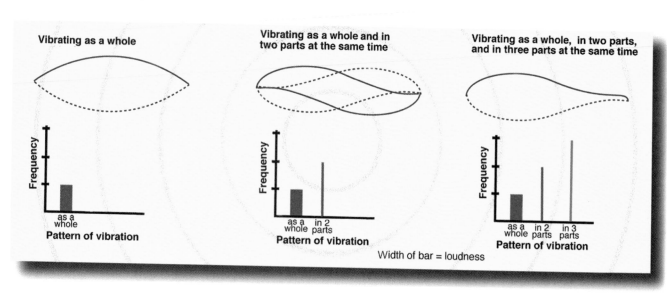

Vibrating as a whole

Vibrating as a whole and in two parts at the same time

Vibrating as a whole, in two parts, and in three parts at the same time

Frequency

Pattern of vibration

as a whole

Frequency

Pattern of vibration

as a whole | in 2 parts

Frequency

Pattern of vibration

as a whole | in 2 parts | in 3 parts

Width of bar = loudness

In reality, a violin string vibrates in many more than three parts. Generally, the lowest vibration as a whole is the most vigorous, so this sound is loudest. The higher sounds are softer. These higher sounds add the "wiggles" to the waves you see on the second and third oscilloscope screens shown at right. Put together all of the sounds made by the string's different vibrations, and you have the quality of the sound produced by the violin string.

The pleasant effect of certain vibrating strings has been known for centuries. Turn to page S116 of the SourceBook to find out more.

Wave Pictures of a Vibrating String

String vibrating as a whole

String vibrating as a whole and in two parts

String vibrating as a whole, in two parts, and in three parts

Locating the Higher Sounds

1. Pluck the lowest-pitched string of a violin (or guitar) (a).

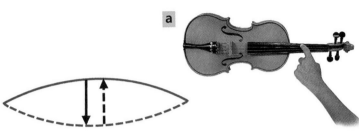

2. Then lightly touch the string exactly at its center to stop it from vibrating as a whole (b), and listen for a higher note. (This may take a little practice.) The note is the sound made by the string when it is vibrating in two parts. Does it sound one octave higher than the first note? Describe the quality of this sound.

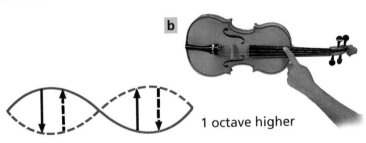

1 octave higher

3. Pluck the string again. Now touch it one-third of the way along its length. With a little skill, you can get the next highest **harmonic** (component tone) (c). This is the sound made by the string as it vibrates in thirds.

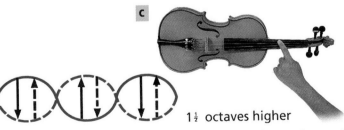

1½ octaves higher

4. Try to get even higher harmonics by touching the string closer and closer to one end. Could you get the string to vibrate in four parts? five parts? more parts?

CHALLENGE YOUR THINKING

1. Hear, There, and Everywhere

Consider each of the following situations:

a. What sounds, although they are loud enough to be heard, do you often miss? Why?

b. You can easily recognize a person just by the sound of his or her voice. Explain this in scientific terms.

c. What are some situations in which sounds are uncomfortably loud? Suggest ways to help yourself in these situations. Explain why each suggestion works.

2. Sound Words

tension, frequency, amplitude, energy, loudness, pitch, length, quality, resonance, distance, patterns of vibration

Use several appropriate words from the list above to write one or two sentences about each of the following:

a. distinguishing sounds of different musical instruments

b. sounds of vibrating strings

c. sound pollution

3. Steely Band

Pierre selected six different pieces of steel wire. He hung different numbers of identical masses from the wires, as shown at right. Then he plucked the strings and arranged them in order of pitch, from lowest to highest. Predict this order (and learn something about yourself)!

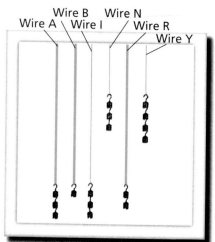

4. A Broad Scope of Sounds

Alejandro connected a microphone to an oscilloscope to see what sounds looked like. Here are some of the pictures that he saw.

a. Which screen shows the noisiest sound?

b. Which shows the purest musical sound?

c. Which shows the highest sound?

d. Which shows the loudest sound?

e. Which three sounds are the same pitch but were made by different instruments?

f. Which two sounds were made by the same instrument but are almost an octave apart?

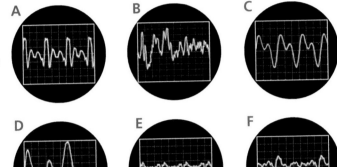

5. Instrumental Instruction

a. Marie blows into the trombone with the slide (1) all the way in and then (2) all the way out, with equal loudness both times. Which tone is lower? Why?

b. Joy plays the cello and Suzanne plays the upright bass. Which instrument is likely to produce higher sounds? Why?

c. Mrs. Cleaver's favorite music CD ends with a loud, sustained chord. She notices that when she plays the CD, her piano sounds the exact same chord for a few seconds after the music stops. Explain what is happening.

ScienceLog

Review your responses to the ScienceLog questions on page 399. Then revise your original ideas so that they reflect what you've learned.

Unit 6

SOUND

SourceBook

Without the air around us, there would be no sound waves produced by vibrating objects, and our ears would detect no sound at all. To find out more about the role of air in hearing and about how sound waves interact with each other, look in the SourceBook.

Here's what you'll find in the SourceBook:

The Big Ideas

In your ScienceLog, write a summary of this unit, using the following questions as a guide:

1. What causes sound?
2. What are some characteristics of sound?
3. How is sound transported from a sound source to a sound receiver?
4. Why can some animals hear sounds that people can't hear?
5. How can you calculate the speed of sound in air?
6. In what ways can you increase the loudness of a sound? the pitch of a sound?
7. How are the terms *frequency* and *amplitude* related to the pitch and loudness of sound?
8. How do microphones and oscilloscopes "hear" sounds?
9. How does an oscilloscope represent the pitch, loudness, and quality of different sounds?
10. How could you use a vibrating violin string to illustrate what gives rise to the distinctive quality of the sound it makes?

Checking Your Understanding

1. Because you have two ears, you can tell the direction from which sounds are coming. Your brain can detect the slight difference in arrival time between ears. Calculate the difference in the time taken by each ear to hear a sound, if the sound source is on your left.

2. What's wrong in the illustration shown at right?

3. A haiku is a Japanese nature poem consisting of three lines of five, seven, and five syllables, respectively. Read the examples that follow. They describe sounds, with a little science added too!

> *Stirrings in the night*
> *Quickly, quietly moving*
> *Tiny night noises*
>
> *Leaves lightly rustling*
> *Sound's volumes vary vastly*
> *Thunder deafening*

Write your own haiku about the sounds of nature. Be sure to include some of the scientific concepts you have learned.

4. 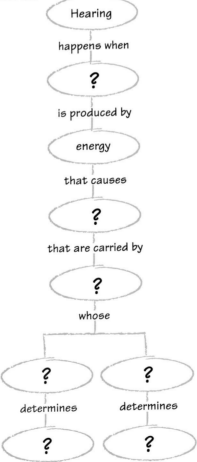 Copy the concept map into your ScienceLog. Then complete the map using the following words: sound, vibrations, frequency, amplitude, pitch, loudness, waves.

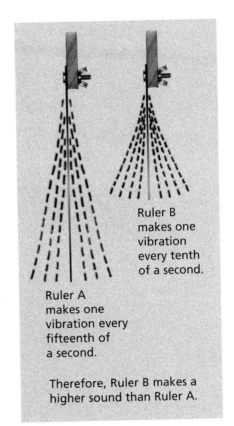

Ruler B makes one vibration every tenth of a second.

Ruler A makes one vibration every fifteenth of a second.

Therefore, Ruler B makes a higher sound than Ruler A.

Synthesizing Your Own Band

When you hear a musical instrument, your ear is responding to the specific frequencies, or vibrations, it creates. Most musical instruments create only a narrow range of frequencies, giving each instrument its distinctive sound. Not so for a synthesizer! A synthesizer can produce a range of frequencies so wide that it can it be used to imitate the instuments of an entire orchestra.

How Does It Work?

If you could look beneath the rows of keys, buttons, and knobs, you would find a generator, various filters, and an amplifier. The generator converts electricity from a wall plug into a vibrating electric signal. The rate of the vibration depends on which key on the keyboard is pressed. The electric signal then passes through a series of electronic filters that change the quality of the vibration according to the effect that is desired. Finally, the refined signal passes to an amplifier, where its strength is boosted to operate a speaker.

Any Sound at Your Fingertips

By adjusting the controls on the synthesizer, you can adjust the electronic filters to get just the sound you want. In this way, musicians can achieve the sounds of a piano, a violin, a trumpet, and even a snare drum. But synthesizers don't stop there. Some synthesizers can also duplicate natural sounds such as rainfall, wind, surf, and

▲ Musicians can use synthesizers to create special sounds that would be difficult to achieve any other way. The first portable synthesizer, shown above with inventor Robert Moog, went on the market in 1970.

thunder. The synthesizer can also be used to make entirely new sounds. In fact, many musicians play synthesizers to make unique sounds rather than to mimic conventional instruments.

Some synthesizers allow you to play many different parts of a song separately and then blend them together into a finished song. A microprocessor stores all the notes you played and even remembers how hard you pressed the keys. When you have finished recording each instrument's part, you can use the synthesizer to combine the parts and play the final composition.

Science Meets Music

The first commercially available synthesizer was created by an American physicist, Robert A. Moog, in 1964. Since then, synthesizers have become very sophisticated. If you were to go to a music store, you may even find synthesizers bearing Moog's name. Now you know where this name came from!

Find Out for Yourself

Go to a music store and try out a synthesizer. See for yourself the range of possibilities that this instrument provides. Try playing a violin sound. Do you think it is as rich sounding as a real violin? Why do you think this is so?

SCIENCE in ACTION

Designing Out the Noise

W ho hasn't covered their ears as a huge airliner passed overhead with an earth-shaking rumble? "Why can't they make those things quieter?" you ask. Richard Linn has done just that. As an aeronautical engineer, his job is to do something about that high-flying noise.

▲ Richard Linn reviewing an aerial photo of Dallas–Fort Worth International Airport

Turning Down the Volume

Aeronautical engineers specialize in the design, construction, or testing of aircraft. Also called aerospace engineers, they are responsible for designing anything that flies, from airplanes to spacecraft.

Richard has applied his 35 years of experience in aeronautical engineering to noise control. He analyzes and offers solutions for the noise created by the jet engines that send modern airplanes into the blue.

When asked what it was like when he got started, Richard replied, "Back in the 1960s the noise of airplanes really began bothering people. It became a social problem that needed attention."

The attention it received involved various kinds of research on how to reduce the noise from jet engines. At the same time, homeowners and political groups began putting pressure on Congress to do something about the problem. As a result, laws were passed that forced airplane manufacturers to make quieter jet engines.

As the years passed, however, air traffic increased, and the constant noise of airplanes flying overhead became just as bothersome as the louder jets had been. To attack this problem, aeronautical engineers turned to the high-tech world of computers. Richard explains the process: "We have new computer technology that allows us to use what we call *noise footprints*. These are airplane-simulation programs that help us determine the effect of engine noise on a community."

A Joint Effort

Richard feels that noise control should be a joint effort between the airlines and the community. He says, "It's frustrating to realize that in spite of the well-known noise problems, cities

▲ Airport employees who work on the runway must wear protective headgear to prevent hearing loss.

often zone or rezone the land around airports for homes, churches, and schools. The airlines can do only so much by themselves—without the community's help in zoning, we'll never completely solve the problem."

When asked how it feels to watch the first flight of a brand-new airplane soaring over a neighborhood, Richard responds, "It's great knowing that all your research and experimentation has paid off and you can demonstrate to the community that you have new, quieter technology."

Think About It

Do you think airports and airplane noise affect the animals and insects that live in the surrounding area? Discuss your thoughts with your classmates.

Listening for Fire

Many homes today are fitted with smoke detectors, but smoke detectors have their limitations. If a fire starts behind a wall, for example, by the time the smoke reaches a detector, the fire has already spread. However, it might not be long before there's a device that can "hear" a fire before it has had a chance to burst into flames.

▲ When the materials that make up this house burn, they produce high-frequency sound waves.

The Sounds of Heat

When a material heats up, it expands. If the material is made up of different components, each component may expand at a different rate. When a beam of wood is heated, for example, the cellulose fibers expand at a different rate than does the sap. When the wood fibers finally rupture from the heat, or when the sap boils, high-frequency sounds are emitted.

Piezo What?

William Grosshandler, a mechanical engineer at the National Institute of Standards and Technology, has developed an electronic system that "listens" for these high-frequency sounds. Grosshandler's system uses small sensors, each containing a piezoelectric crystal. These crystals are similar to the crystals used in quartz watches and clocks. Piezoelectric crystals produce an electric charge when they vibrate. In a watch, power from a small battery makes the crystal vibrate to produce a pulse of current at the precise rate of one pulse per second.

Sound Waves to Electric Charges

In Grosshandler's system, sound waves from a heated material cause the piezoelectric crystals to vibrate at a certain frequency, generating a specific electric signal. The electric signal is amplified, filtered, and routed to a microprocessor that is programmed to analyze the signal and determine whether or not it was caused by a heated material. For example, the microprocessor is programmed to differentiate between signals caused by wood fibers that are breaking and those caused by wood fibers that are about to burst into flames. Otherwise, the device would create false alarms.

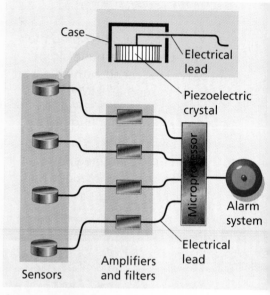

Case — Electrical lead — Piezoelectric crystal — Microprocessor — Alarm system — Electrical lead — Amplifiers and filters — Sensors

▲ In Grosshandler's system, sensors in walls send signals to a microprocessor before a fire gets out of control.

Grosshandler's system still needs some fine-tuning to improve its ability to recognize different signals in order to prevent false alarms. But Grosshandler believes it could be an effective way of detecting fires before they become dangerous.

Find Out for Yourself

You can see the vibrations caused by sound. Look at the surface of a speaker the next time you see one without its cover. Note how the speaker's surface vibrates to different frequencies of sound. Can you recognize a specific sound by the way the speaker vibrates? This is precisely the function of the microprocessor in Grosshandler's system!

A Hearing Aid for Fido

Normally, dogs have a keen sense of hearing. But many dogs, like humans, lose some of their hearing as they grow older. Thanks to dog's best friend, however, help may be on the way.

Testing a Dog's Hearing

Curtis Smith, an audiologist at Auburn University in Alabama, became interested in canine hearing loss when a friend's dog started going deaf. He contacted A. Edward Marshall, a researcher at Auburn's college of veterinary medicine, who had been working on a way to test dogs for deafness.

Marshall's hearing test involved fitting a dog with light-weight headphones. Electrodes were then attached to the dog's head to monitor its brain waves. Short sounds of different intensities were then played through the headphones. On a computer screen, Marshall could observe the response of the dog's brain to the sounds. By changing the frequency of the sounds, Marshall could determine the limit of the dog's auditory range and discover how much hearing the dog had lost.

A Canine Hearing Aid

How do you fit a dog with a hearing aid? Using a conventional hearing aid turned out to be impractical. The dog simply pawed its ear until the device

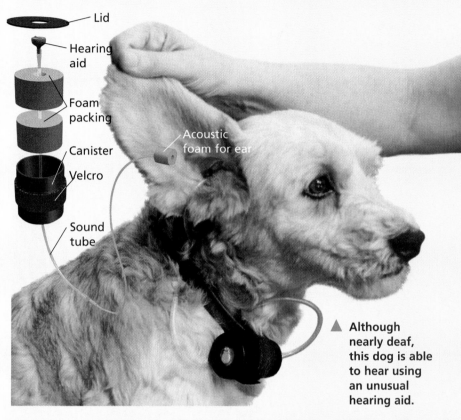

Lid

Hearing aid

Foam packing

Canister

Velcro

Sound tube

Acoustic foam for ear

▲ Although nearly deaf, this dog is able to hear using an unusual hearing aid.

fell out. The hearing aid that Marshall and Smith developed involves attaching the hearing aid hardware to the dog's collar. The amplified sound is channeled through hollow, light-weight tubes to the dog's ears. The tubes are held in place with plastic foam.

Still Room for Improvement

The hearing aid for dogs still has some problems. Although the collar system is tolerated by most dogs, they don't like it. Another problem is related to the dog's outer ear, or pinna. In humans, the pinna remains stationary. In dogs, however, the outer ear moves in many directions to pick up incoming sounds. This movement can cause the tubes of the hearing aid to work loose. Still another problem is that the device amplifies sounds only in the 500 to 6000 Hz range. These frequencies are much lower than a dog's normal hearing range.

Find Out for Yourself

Different animals hear sound in different frequency ranges. Do some research and make a chart to show what the range of sound is for several different animals. Don't forget to include yourself.

Unit 7

LIGHT

These natural, prism-like quartz crystals took thousands of years to form. From this you might infer that they are rare and valuable, like diamonds. But this is not true—quartz is one of the most common minerals on Earth. Nonetheless, quartz crystals are certainly beautiful and, due to their shape and composition, can create dazzling displays of light.

The large crystal shown here is actually colorless, so it is not the crystal itself that is beautiful, but rather the crystal's effect on the light that strikes it. If you could look straight through a section of this quartz, it would be similar to looking through thick window glass. However, a description of a crystal such as this one would likely include statements about color. How is it possible for a clear material like this quartz crystal to produce colors? You will learn more about this topic later in the unit.

You will also learn more about lenses and how they affect light. You will learn that lenses are really nothing more than transparent material shaped in specific ways for specific purposes. In fact, certain types of quartz crystals are ground into lenses for high-quality microscopes and telescopes. Although the quartz crystal shown here has not been shaped into a lens, let it focus your attention on the fascinating subject of light.

This transparent quartz crystal reflects light in such a way that it appears to have color.

The Nature of Light

1 Why do some stars appear reddish, some white, and some bluish white?

2 How are rainbows produced?

3 Is light matter or energy? Explain.

ScienceLog

Think about these questions for a moment, and answer them in your ScienceLog. When you've finished this chapter, you'll have the opportunity to revise your answers based on what you've learned.

Imagine that you've taken a shortcut home and you're walking down a back road on a moonless night. Looking around, you can see nothing but blackness. Without light to aid you, how would you find your way? What kind of items could you have carried to help you make your way through the darkness?

The items you named are probably objects that produce light. Light is something most of us take for granted. Indeed, you have probably never really had to walk home in total darkness because houses, apartments, and streetlights usually light up the night.

It would be hard to imagine life without light. But while you probably realize its importance, how much do you know about what light is? Here are some questions that may help direct your thinking about the nature of light.

1. When a flashlight is turned on, it gives off light. Does the flashlight lose mass as a result?

2. Why is a room with light-colored walls brighter than a room with dark walls?

3. Why is the sky blue when seen from the Earth but black when seen from space?

4. How do we see? Why can we see an object when a light is on but not when a light is off?

Make a list of other puzzling questions about light that you would like answered. By the time you have finished this unit, you'll be able to shed some light on the subject!

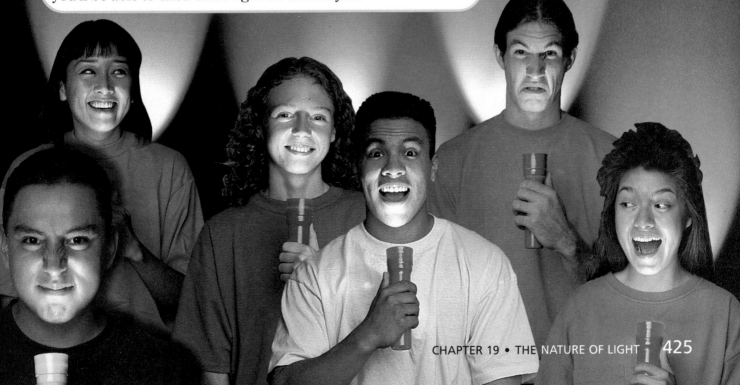

Light Brigade

Each of the Activities in this Exploration will give you a better understanding of what light is.

You Will Need

- a newspaper
- a light bulb or light source
- 2 thermometers
- white paper
- black paper
- tape
- scissors
- a radiometer
- blueprint paper
- a spring scale or balance
- copper wire
- a flashlight

What to Do

Form small groups. Perform at least three of the Activities, dividing them among the groups. While you're waiting for some of the light effects to occur in the longer Activities, you can start on a different Activity. Keep a record in your ScienceLog of what you find out about light. Later, you can report what you discovered to the other groups.

ACTIVITY 1

Light and Paper

Take a piece of newspaper 10 cm square or larger and put it in a window where the sun will shine on it. Leave it there for half a day. Compare its appearance with a piece of newspaper that has not been exposed to sunlight. Will a bright light bulb produce the same effect? Try it. What does this Activity tell you about light?

ACTIVITY 2

Light, Color, and Temperature

Tape a piece of white paper and a piece of black paper to the inside of a window that faces the sun. Then tape a thermometer to each piece, as shown below. Record the

temperature shown on both thermometers every minute for 5 minutes. What do your findings tell you about light and white paper? about light and black paper? What effect do you think light has on the temperature of different colored papers?

ACTIVITY 3

Flashlights and Mass

Find the mass of a flashlight. Then turn it on for 10 minutes. Now check its mass again. Is there any change? What does this tell you about light? Do you think light occupies space?

ACTIVITY 4

Lightly Done

Shape some copper wire into a flat design and place it on the surface of a piece of blueprint paper. Be sure to keep the paper covered and away from light until you are ready to perform the experiment. Expose the paper to direct sunlight or a sunlamp for 5 to 10 minutes. Then immerse the paper in water. Allow it to dry. What does your "photograph" look like? How was it made? What did light have to do with it?

ACTIVITY 5

The Light Windmill

Examine a light windmill, also called a *radiometer*. Put the radiometer in direct sunlight for a minute or so. What happens? Place it in the dark. What happens? Place it at different distances from a bright light bulb. What do you notice? In which direction does the radiometer rotate? What does the radiometer show you about light?

A radiometer

Light—A Quick-Change Artist!

Light obviously has many properties. Use what you learned about light in Exploration 1 to answer the following questions. Answer either *yes, no,* or *not certain.*

1. Can light make things move?
2. Can light cause the color of some things to change slowly?
3. Can light cause the color of some things to change rapidly?
4. Can light cause things to heat up?
5. Does light have mass?
6. Does light take up space?

What do you think? If you answered *yes* to the first four questions, you are correct. The Activities in Exploration 1 show that light can do two things: it can make things move, and it can cause changes in color and temperature. Isn't this what you would expect of *energy?* And would you expect light, as a form of energy, to have mass or occupy space as matter does? How did you answer questions 5 and 6 above? If you answered *no* to both questions, you are right again.

As the photographs on this page show, light is actually involved in many energy changes. All of the items shown are converting energy. That is, they are involved in energy changes. For each example, indicate the energy conversion taking place.

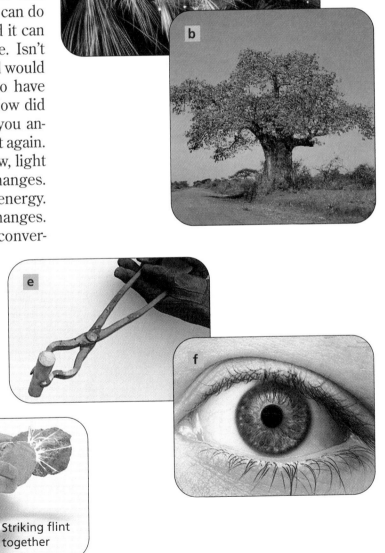

Striking flint together

LESSON 2

Light, Heat, and Color

Think of several things that produce light. What general observation could you make about the temperature of these things? Are they hot or cold? Is there a relationship between the light that an object gives off and the object's temperature? In the following Exploration you will take a closer look at this question.

EXPLORATION 2

Observing Hot Solids

You Will Need

- 3 D-cells—1 weak and 2 fully charged
- a copper wire (uninsulated, approximately 20 cm long)
- a flashlight bulb
- masking tape
- a clothespin
- a Bunsen burner
- a hot plate or electric stove

PART 1

Starting with setup (a), arrange the D-cells as shown below. Be sure to complete each circuit by touching the wire to the base

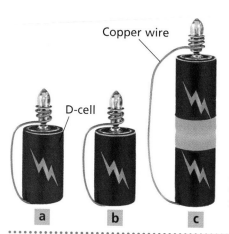

Copper wire

D-cell

a   b   c

of the appropriate cell. Feel the wire and bulb and observe the color in each case. Write down your observations in your ScienceLog.

Be Careful: The flashlight bulb may be hot.

a. A weak cell—just strong enough to make the filament start to glow
b. A strong cell
c. Two strong D-cells held together with masking tape

PART 2

Using a clothespin, hold a piece of copper wire in a flame until it glows (gives off light). What color do you see? If you were to place your hand close to the wire, what would you feel?

Be Careful: Do not touch the hot wire!

PART 3

Turn an electric stove element to low and gradually increase the heat to high. Observe the color of the burner and the heat it gives off in each case.

Be Careful: Keep your hands away from the heated surface!

PART 4

What is the color of the molten steel in the photo below? What do you think its temperature is?

As you probably realize, most things that produce light are hot. There are exceptions, however. Fluorescent lights and neon signs, for example, are relatively cool—even when lit for a long time. And fireflies don't get very hot either.

A Light Quiz

Consider what you learned in Exploration 2. Then complete the following statements by selecting the correct word or phrase from the choices provided. Record your choices in your ScienceLog.

1. Heat and light (rarely, often) occur together.

2. When electrical energy passes through a wire, the wire (gets cold, stays the same temperature, gets hot).

3. When a small amount of electrical energy passes through a light bulb, (white, reddish) light is produced.

4. When a large amount of electrical energy passes through a light bulb, (white, reddish-orange) light is produced.

5. When a small amount of electrical energy passes through an electric stove element, the element gets hot and (gives off, does not give off) light.

6. When a large amount of electrical energy passes through an electric stove element, the element gets hot and (gives off a reddish light, gives off a white light, does not give off light).

7. As an electric stove element becomes hotter, it gives off light that gets (brighter—turning from red to white, darker—turning from white to red).

8. Heating a wire in a hot flame (does not affect its appearance, causes it to glow).

9. When metal is heated to a very high temperature, it gives off (red, white) light.

10. The color of light given off by a substance often indicates (how large, how hot, what shape) the substance is.

How might you summarize your findings from Exploration 2? In general, is there a relationship between the temperature of an object and its color? Write down your ideas in your ScienceLog.

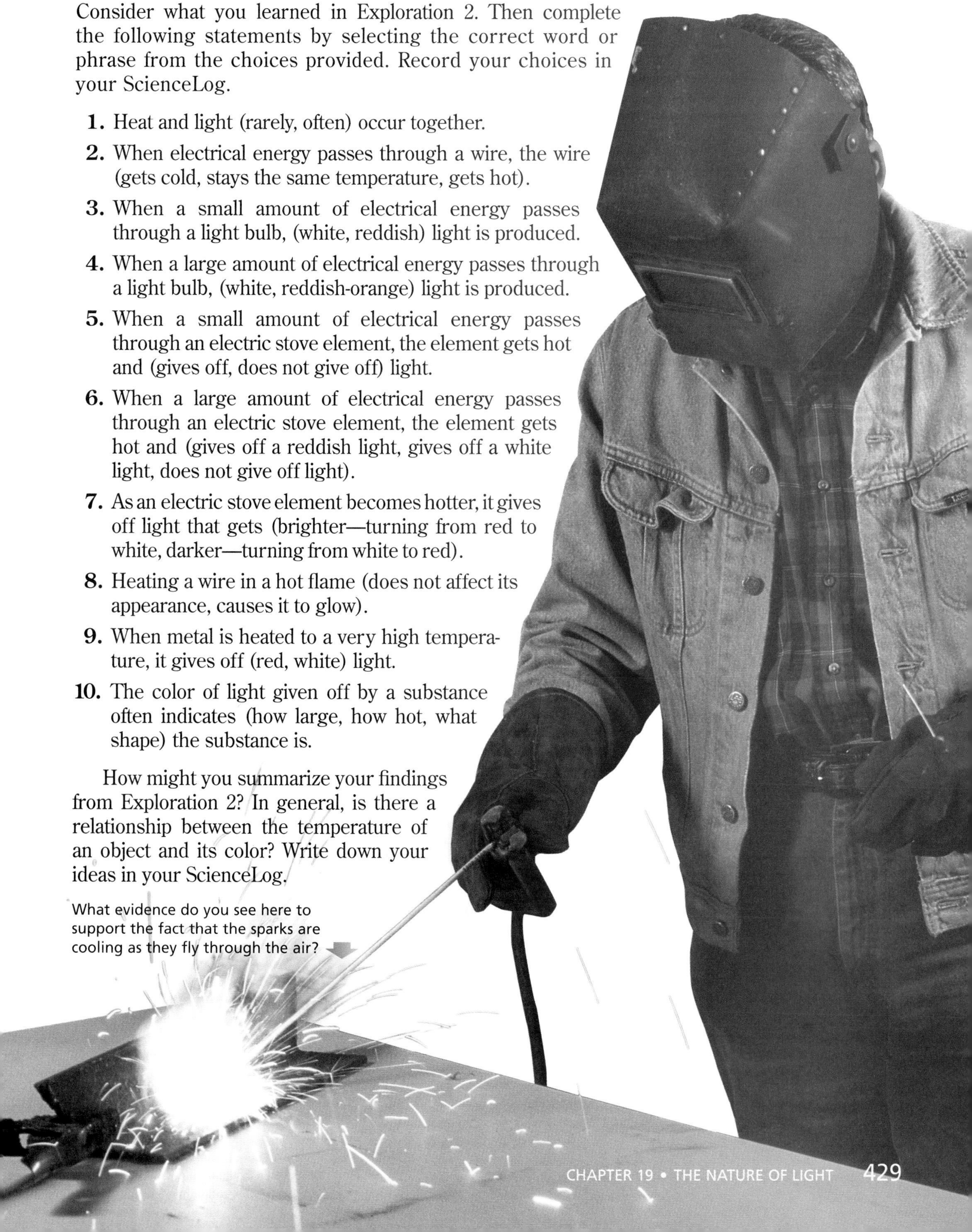

What evidence do you see here to support the fact that the sparks are cooling as they fly through the air? ➡

Light and Temperature

In your ScienceLog, place the items listed below in order of increasing temperature.

a. the surface of the sun

b. the human body

c. boiling water

d. the glowing filament in a 100 W bulb

e. red-hot iron

f. solid carbon dioxide (dry ice)

g. melting ice

h. white-hot iron

Now place the corresponding temperature next to each item on the list: $-80°C$, $0°C$, $37°C$, $100°C$, $500–650°C$, $1500°C$, $2200–2700°C$, $5000–6000°C$. Compare your lists with those of other students. Are you surprised at the results?

What Color Is Hot?

You have seen that the light given off by an object can indicate its temperature. Recall the color change in copper wire held over a flame. When the wire is red-hot, it is hotter than when it is dull red. When the wire is white-hot, it is hotter than when it is red-hot.

As an object becomes hot, it usually gives off light. The color of the emitted light changes as the temperature of the object rises. Certain colors generally represent certain temperatures. To understand the relationship between temperature and the colors of light, you need to consider the following:

• Where does color come from?

• Is white a color?

• Is black a color?

• What is the relationship between white light and other colors of light?

You may be surprised by how much you already know about color!

The surface of the sun is about four times as hot as white-hot iron.

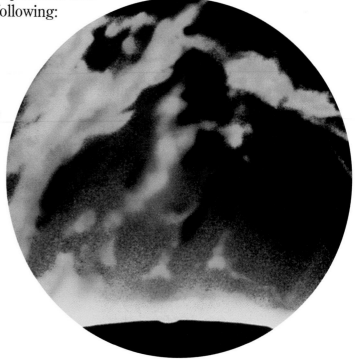

Answers to Light and Temperature:
(f) $-80°C$; (g) $0°C$; (b) $37°C$; (c) $100°C$; (e) $500–650°C$; (h) $1500°C$; (d) $2200–2700°C$; (a) $5000–6000°C$.

A Colorful Theory

In 1666 Isaac Newton, age 24, was experimenting with prisms. People had known about and used prisms since the time of Aristotle (about 350 B.C.), but Newton made some remarkable new discoveries. In Exploration 3, you can relive a great moment in science—one of Newton's first discoveries.

EXPLORATION 3

Lights, Prisms, and Filters

This series of experiments works best in a darkened room. The activities should be done in the order given. Use the questions provided to help you think about and develop your own "theory of color."

You Will Need

- 2 prisms
- a flashlight
- white cardboard
- 3–6 red filters (or colored cellophane)
- 3–6 blue filters
- 3–6 green filters
- a piece of white paper
- aluminum foil
- tape

EXPERIMENT 1

Light and Prisms

Following the setup shown below, shine a narrow beam of light on a prism in such a way that a rainbow of colors (called a **spectrum**) forms on the white cardboard. Notice how the colors blend into one another. How many different colors do you see on the screen? What are the colors at the ends of the spectrum? Which color seems to come through the thickest part of the prism?

Before Isaac Newton, people believed that the thickness of the prism determined the colors of the spectrum. According to this theory, white light passing through the thinnest part of the prism changes to red; white light passing through the thickest part of the prism changes to blue; and white light passing through a medium thickness of the prism changes to green. Do you agree with this explanation? Why or why not? How else could the rainbow of colors be produced from white light?

Isaac Newton didn't agree with the common explanation of his day. After completing the following experiments, you will understand Newton's reasoning.

Exploration 3 continued ▶

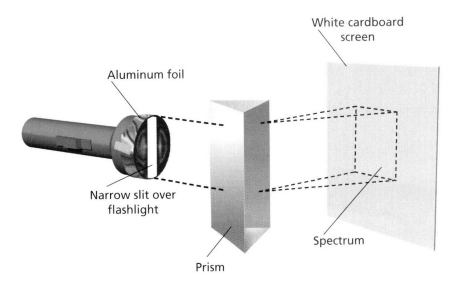

Aluminum foil

Narrow slit over flashlight

Prism

White cardboard screen

Spectrum

EXPERIMENT 2

Light and Filters

Return to the setup from Experiment 1. Place a red filter between the flashlight and the prism. Observe the screen.

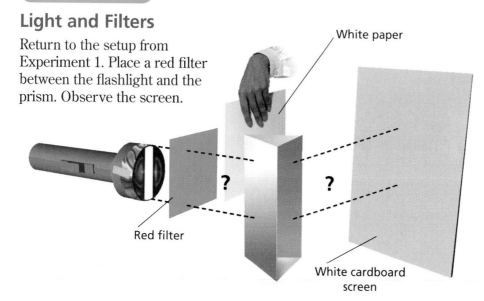

White paper

Red filter

White cardboard screen

How is the spectrum different from before?

Place a piece of white paper between the filter and the prism. What is the color of the light between the filter and the prism? Did the prism change this color of light? Shine the light through the thinnest part of the prism, the thickest part of the prism, and several areas in between. Did the various thicknesses of the prism alter the red light in any way?

Repeat this experiment, substituting a blue filter for the red one. Again, observe the color of the light. What is the color of the light between the blue filter and the prism? Did the prism cause any color change?

Now summarize the information you have acquired about light and color in your ScienceLog. State what you think white light may be made of, what you think prisms do to light, and what you think filters do.

EXPERIMENT 3

More About Filters

Prepare the setup shown in the figure below. What color of light comes through the filter? Does the red filter add something to white light, or does it take something away from it?

Repeat this experiment using a green filter in place of the red filter. Does the green filter add something to white light or take something away from it?

Consider This

Here is a good analogy for light passing through a filter: When a ball moving along the floor hits a puddle of water, does the ball gain or lose energy? Remember that light is energy. When light passes through the red filter, does it lose energy, gain energy, or retain its original energy?

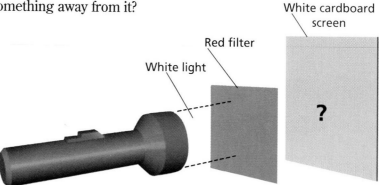

White cardboard screen

Red filter

White light

EXPERIMENT 4

A Combination of Filters

Return to the setup from Experiment 3. One by one, add more red filters, as shown below.

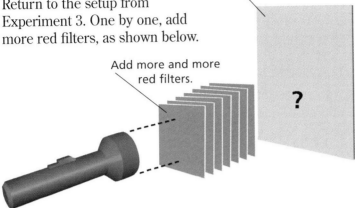

Add more and more red filters.

White cardboard screen

?

What happens to the white light as you add more red filters? Do the filters remove light energy? Could you add enough filters so that no light energy got through? How many red filters would it take? If no energy got through, what color would you see on the screen? In other words, what color is "no light"?

Repeat this activity using green filters, and consider the same questions.

Now shine the light through a single red filter, and then add either a dark blue or a dark green one, as shown below. Was the energy loss similar? Recall how many red filters you had to use so that no light energy reached the screen. With a single red filter, how many blue or green filters did you have to use to get the same effect? Why do you think this is so?

After considering the results of Experiments 3 and 4, would you agree that filters absorb some of the light energy in white light? In the following experiment, you will discover the nature of the energy loss caused by filters.

EXPERIMENT 5

Disappearing Colors

Shine white light through a prism and obtain a spectrum on the white screen.

Now place a red filter between the prism and the screen. What colors seem to have disappeared? In other words, what colors are absorbed by the filter? What color(s) pass(es) through the red filter? If only red light were shone on this filter, what would happen to the light? What effect would this filter have if green light were shone on it? If blue light were shone on it?

Repeat the experiment, this time using a dark blue filter between the prism and the screen. Consider the same questions.

Predict what would happen if you shined white light through two prisms (with the second prism facing in the opposite direction from the first one). Try it.

Your Theory

Think through all that you have discovered, and write down your own "theory of color." Then see how your theory compares with Isaac Newton's explanation, which is described next.

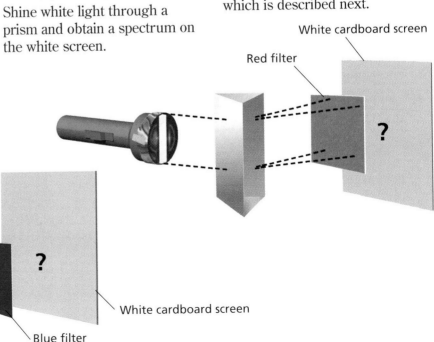

White cardboard screen

Red filter

?

Red filter

Blue filter

?

White cardboard screen

Newton's Bright Ideas

Consider Newton's theory about light and color. Newton said that light is not changed when it goes through a prism. Instead, he said, it is physically separated. Newton reasoned that, if the old theory were true—that is, if the thickness of a prism did change the color of white light—then another prism should change the resulting colors again. This did not happen when he tried it. What he found was that once the colors of white light were separated, they could not be changed any further. Newton called this the "critical experiment."

In addition, Newton noticed that the spectral colors always occurred in the same order. Thinking back to your investigations, can you recall what the order is? Some people remember the order of the colors of the spectrum by thinking of "Mr. Color," ROY G. BIV. What does each letter stand for?

Newton also noticed another interesting phenomenon about the spectral colors. He found that by using a second prism, the colors could be blended to form white light again.

Newton's theory of light and color was a totally new one. Many scientists of his day disputed it and attacked his ideas. Newton eventually wrote a letter saying that he was so harassed by arguments against his theory that he regretted losing his peace of mind in order to "run after a shadow" of an idea.

Isaac Newton
(1642–1727)

Things to Think About

1. How do red, green, and blue lights differ when passing through a prism?

2. How do you think Newton proved that any color produced by passing white light through a prism cannot be changed to any other color?

3. Why do you think other scientists disputed Newton's theory so strongly?

4. Using the illustration below and your own theory of light and color, predict the color of the light that enters the blue filter. Why do you think the screen appears black?

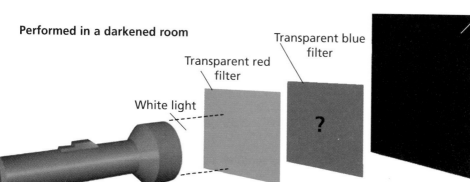

White screen appears black

Performed in a darkened room

Transparent blue filter

Transparent red filter

White light

?

5. Predict what you would see in the situation shown here, and explain your prediction.

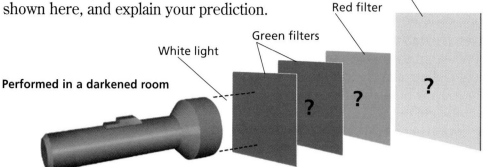

Performed in a darkened room

White light — Green filters — Red filter — White cardboard screen

? ? ?

6. Write a letter to a friend relating your personal discoveries about light. Be factual but expressive.

Activities to Try at Home

Each activity below shows a way of separating white light into its colors.

1. Dip the mouth of a jar into soapy water. Look at the light reflected by the soap film covering the mouth of the jar.

2. Observe the light reflected off of the surface of a compact disc.

3. With the sun behind you, look at a fine spray of water from the nozzle of a hose pointing toward the ground.

4. Look through a nylon stocking at the light coming through a small hole.

5. Observe the film that forms after a drop of oil hits the surface of a beaker of water.

Nylon stocking   Hole   Clear light bulb

Light

Light and Color

You now have a theory of light: *White light is composed of, and can be separated into, light of different colors.* The opposite of this is also true. Light of these different colors can be recombined to make white light. How does this theory explain the color of hot objects?

Which end of the spectrum do you think has the most "energetic" light? First recall the observations you made of the flashlight bulb in Exploration 2.

- A flashlight bulb attached to a nearly dead D-cell emits (gives off) reddish light. In this case the energy supplied to the bulb is small. The temperature of the bulb's filament is relatively low. What can you infer about the energy of red light?

- When the cell is stronger, the bulb emits a more yellowish light. The hotter filament gives off not only red light, but also orange and yellow light. What can you infer about the energy of orange light? of yellow light?

- If the bulb is connected to a new cell, even more energy is present. The filament is hotter. It emits not only red light, but the other colors of the spectrum as well, all of which combine to produce white light.

Based on these findings, how would you compare the energy of each color of the spectrum (ROYGBIV)?

Red light

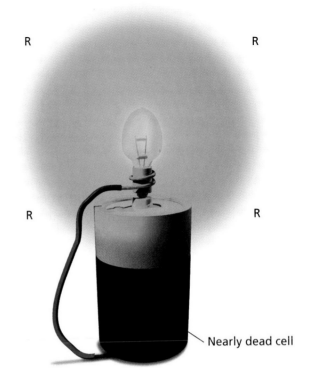

Nearly dead cell

My photo was taken with infrared light. To find out more about this form of light, read page S139 in the SourceBook.

White light

New cell

Adding and Subtracting Color

Adding Colored Lights

As you know, Newton was able to recombine the colors of the spectrum to produce white light. Did you observe a similar phenomenon in Exploration 3? (What happened when you shone white light through two prisms, one facing in the opposite direction from the other?)

Like Newton's experiment, the demonstration shown here involves mixing light of different colors. In this case, however, only red, green, and blue light are mixed. What do you predict will happen when these three colors of light are combined? when any two colors are combined?

In the pages ahead, you will have the chance to find out for yourself what happens when different colors of light are combined.

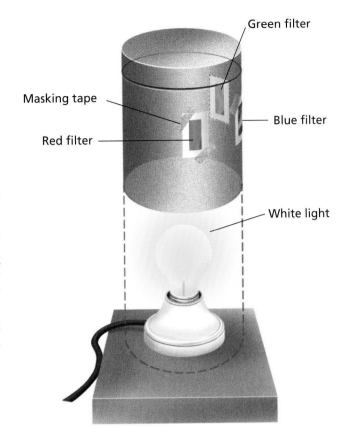

Performed in a darkened room

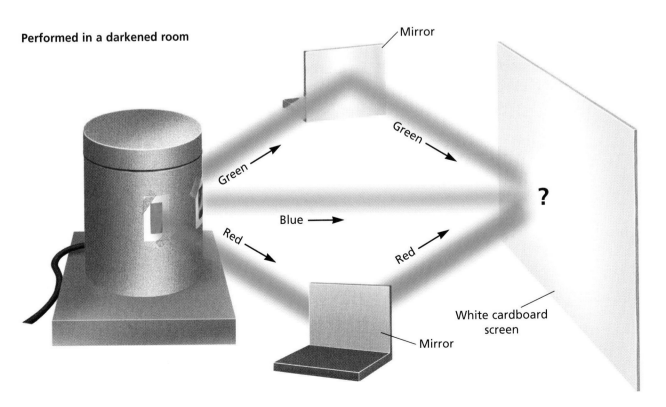

Color-Conscious Calculations

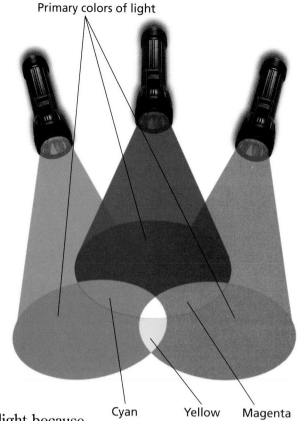

Primary colors of light

If you were to shine three flashlights covered with red, green, and blue plastic, respectively, on a white surface in a dark room, you would see the same results as if you carried out the demonstration on page 437.

With the help of the diagram at right and the results of the colored-light activity, you can do a little color math! What would you replace the question marks with in the following statements?

1. Red light + Blue light = ? light
2. R + ? = Y
3. B + G = ?
4. R + G + B = ?

Here are a couple of trickier equations:

5. Y + B = ?
6. Cyan + ? = White

Cyan   Yellow   Magenta

Secondary colors of light

Red, green, and blue are three special colors of light because they combine to make *white* light. They are called **primary colors of light**. When you mix any two primary colors of light together, you get **secondary colors of light**. How many secondary colors of light must there be? What are they?

In equations 5 and 6, did you observe that when a particular secondary color of light and a particular primary color of light are mixed, white light is produced? Pairs of colored lights that produce white light are called **complementary colors of light**. What is another pair of complementary colors of light other than those in 5 and 6?

Do you need all the colors (ROYGBIV) of the spectrum to produce white light? Explain.

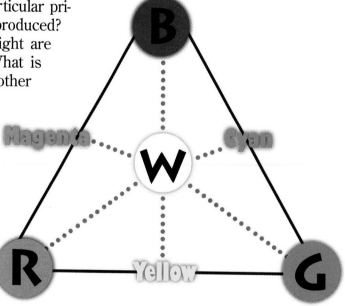

What information can you get from this light triangle?

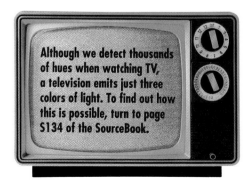

Although we detect thousands of hues when watching TV, a television emits just three colors of light. To find out how this is possible, turn to page S134 of the SourceBook.

Filter Fun

You have just been examining *additive* light phenomena. What do you think this term means? Review the results of Experiments 2, 3, and 4 of Exploration 3, in which you made use of filters. Are these experiments also additive light activities? Do filters "add" something to white light, or do they "subtract" from it? For example, when white light shines through a red filter, what color of light does the filter let through? Some colors apparently do not go through the filter; that is, they are subtracted from the white light.

Here is some "filter fun" you can do as a class experiment with an overhead projector. Place a colored filter or combination of filters on the projector and observe the light on the projection screen. Then deduce which colors are transmitted through the filter(s) and which are subtracted from the white light by the filter(s). In your ScienceLog, write what you think is happening in each case.

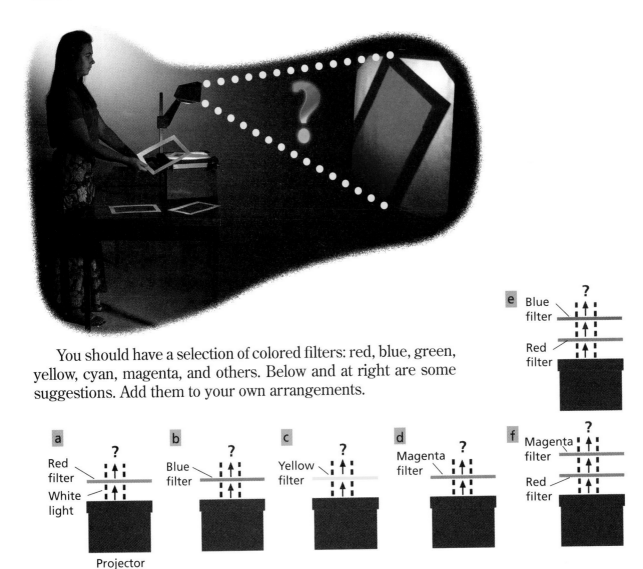

You should have a selection of colored filters: red, blue, green, yellow, cyan, magenta, and others. Below and at right are some suggestions. Add them to your own arrangements.

1. Lights Out!

It is nighttime and a storm has caused the electrical power in your house to fail. How many different ways can you find to light your room so that you can see your way around? For each way, identify the energy change involved.

2. Photo Facts

Photographers observe that when pictures are taken with the light from ordinary incandescent light bulbs, the developed photos have a reddish tint. Why? If a flash is used, the color is better. Why? Another way to improve the color of the photo is to place a blue filter over the camera lens. How does the blue filter help?

▲ Photo taken without flash

▲ Photo taken with flash

3. Color "Math"

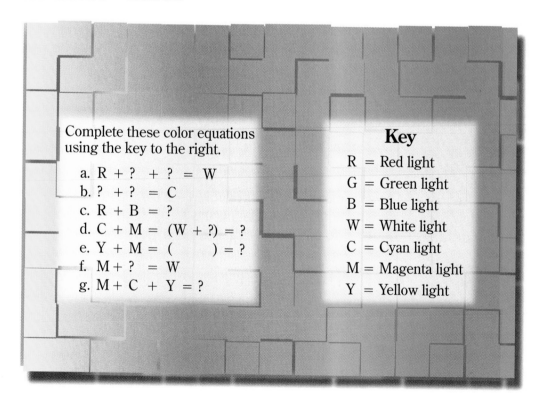

Complete these color equations using the key to the right.

a. R + ? + ? = W
b. ? + ? = C
c. R + B = ?
d. C + M = (W + ?) = ?
e. Y + M = (   ) = ?
f. M + ? = W
g. M + C + Y = ?

Key

R = Red light
G = Green light
B = Blue light
W = White light
C = Cyan light
M = Magenta light
Y = Yellow light

4. Color Commentary

Michelle asked, "What colors do you get when you mix different secondary colors of light together?" What do you predict? What color might each question mark represent in the illustration at right? Explain your prediction. You might test your prediction in an experiment.

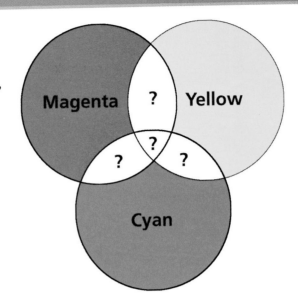

5. A Light Lunch

Millie says, "Plants change *light* into food. Therefore, light must be matter." Do you agree or disagree? Support your answer with a good argument.

6. Color Combo Quiz

Are you a color expert? Consider the illustrations shown below. Each one involves light of various colors. It's up to you to identify the colors you would expect to see at each labeled point, *A–F*. Remember to write in your ScienceLog, not in this book.

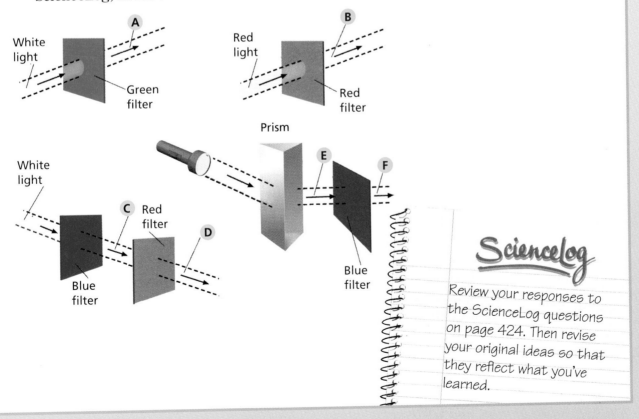

Review your responses to the ScienceLog questions on page 424. Then revise your original ideas so that they reflect what you've learned.

How Light Behaves

1 **Why is the sky blue?**

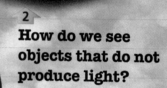

2
**How do we see
objects that do not
produce light?**

What determines 3
an object's color?

ScienceLog

Think about these questions
for a moment, and answer
them in your ScienceLog.
When you've finished this
chapter, you'll have the
opportunity to revise your
answers based on what
you've learned.

Light in Action

Perhaps you've heard the words *scattering, transmission,* and *absorption* before. These terms are commonly used by scientists when describing light. What do you think these words mean? Read the following statements, which use forms of these words. Rewrite the sentences and replace the words in boldface type with your own words to convey what you think these special light words mean.

- When a light beam shines on a mirror, it bounces off in a specific direction. However, when a light beam hits a white piece of paper or the white wall of a room, the light **scatters**.

- When a white light shines on a blue piece of paper, the blue color in the white light is **scattered** by the paper, while all the other colors are **absorbed**.

- When white light shines on a red filter, only the red color is **transmitted** through the filter. All the other colors are **absorbed** by the filter.

- When a beam of white light shines through smoky air, part of the light is **transmitted** through the smoke. The rest of the light is **scattered** by the smoke particles. Look at the photo at right.

A Light Box

By performing experiments with a *light box,* you can get a better understanding of scattering, transmission, and absorption. A light box shuts out most of the unwanted light in a room, making it easier for you to see the light used in the experiments. Follow the instructions on page 444 to build your own light box.

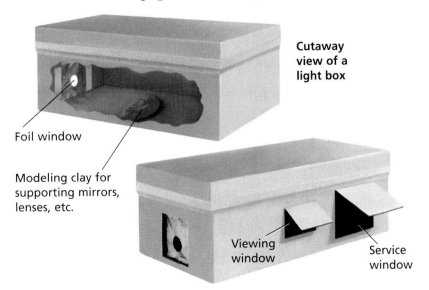

Cutaway view of a light box

Foil window

Modeling clay for supporting mirrors, lenses, etc.

Viewing window

Service window

Building a Light Box

Here is how to make a light box from simple, readily available materials. You will use this light box for many experiments in this unit.

Start with an empty cardboard box about 35 to 45 cm long, 15 to 20 cm wide, and 10 to 15 cm high. A large shoe box will work.

On one end, near the bottom, cut out a square window about 6 cm on each side. Then cut a square of aluminum foil about 8 cm on each side. About 3 cm from the bottom of the foil square, centered between the left and right edges, cut a round hole about 1.5 to 2.0 cm in diameter. Tape the square of foil over the window at the end of the box, as shown in the illustration on page 443. Be sure that you can see through the hole into the box's interior.

Cut two flaps in one side of the box—one near the center, the other one lower and to the right, at the end of the box away from the foil window. The flap near the center is for looking into the box. It should be about 6 cm × 4 cm. The flap to the right is for putting things into or removing things from the box. It should be about 8 cm on each side. You may find it easier to move objects into and out of the box by simply removing the lid. Be sure to replace the lid before you do any experiment.

Enlightening Experiences

You Will Need
- a light box
- a flashlight
- rough, black paper
- modeling clay
- white paper
- colored paper
- a wooden splint
- matches
- a piece of window glass
- a protractor
- a piece of paraffin wax
- a small beaker of water
- milk
- an eyedropper
- a stirring rod

PART 1

A Black-Surface Experiment

Set up the light box that you made. Position the flashlight so that its light shines through the foil window. Look through the viewing window.

What to Do

1. There is now some light in the box. Where does most of the light come from? Is it from the beam as it passes through the air or from light scattered when the beam hits the end of the box?

2. Put a piece of rough, black paper in the path of the beam near the end of the box. Prop it up with modeling clay, as shown below. What happens to the brightness of light in the box when the black surface is added? What happens to most of the light when it strikes the black surface?

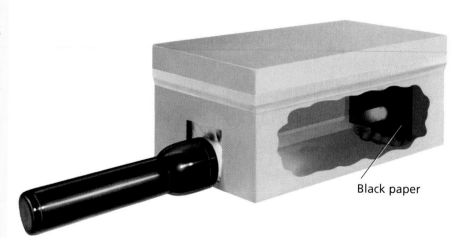

Black paper

3. Replace the black paper with a sheet of white paper. What does this do to the amount of light in the box? Which color of paper—black or white—absorbs more light? Which color reflects more light? Try other colors of paper as well.

4. Add some smoke from a smoldering wooden splint.

Caution: Beware of fire.

The particles of smoke scatter some of the light in all directions, allowing you to see the path it takes. Be careful not to use too much smoke. You must be able to see through the smoky air.

5. You should now be able to complete the following statements. Write them in your ScienceLog, and fill in the answers.

a. As a beam of light passes through the air, it (lights up, does not light up) the box.

b. When the beam of light hits the end of the box, it (lights up, does not light up) the box.

c. When the beam of light hits the black screen, the box is lit up (less than, more than, to the same extent as) in step (b).

d. White light falling on a black surface is absorbed (less than, more than, to the same extent as) when it falls on a white surface.

e. White light falling on a black surface is scattered (less than, more than, to the same extent as) it is scattered by a white surface.

f. A colored surface scatters (more light than, less light than, the same amount of light as) a white surface and (more light than, less light than, the same amount of light as) a black surface.

g. A green surface scatters (what color?) light from its surface. Therefore, it must absorb (what colors?) of light.

h. A black surface absorbs (what colors?) of light.

PART 2

Doing Windows

Replace the black paper that is in the path of the light beam with a square of window glass. Place the glass near the center of the box, just behind the viewing window. Add smoke so that you can see the light beam. Place the glass at various angles and observe the results.

What to Do

1. Place the glass at a 45° angle to the light beam, and then answer the following questions:

a. Does the beam of light scatter as it hits the glass, or does it bounce off (reflect) in a certain direction?

b. Can you see light pass through the glass? In other words, can you see any transmitted light?

c. Which seems brighter—the reflected beam or the transmitted beam? Why?

2. Repeat step 1 again, first placing the glass at a steep angle such as 80° (measured with respect to the horizontal) and then at a shallow angle such as 10°. For each angle, answer the questions in step 1.

3. Now write several statements like those you completed after Part 1 of this Exploration. They should describe what you have learned by doing this experiment.

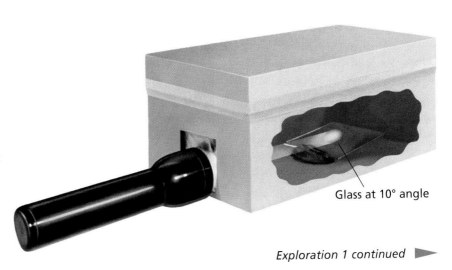

Glass at 10° angle

Exploration 1 continued ▶

PART 3

Wax Facts

What to Do

1. Replace the glass with a piece of paraffin wax. Hold the wax at various angles in the light beam. Be sure to use smoke to help you see the beam.

2. How does the appearance of the paraffin compare with that of the glass when each is in the beam? Is the box brighter when the paraffin is in place or when the glass is present?

3. Record your findings in a series of simple statements.

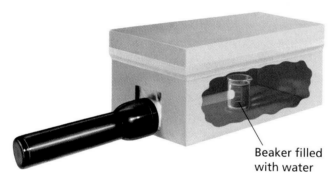

Beaker filled with water

PART 4

Light and Water

What to Do

1. Put a small beaker nearly filled with water in the light box. Turn on the flashlight and shine a light beam through the water. Can you see the beam in the water?

2. Add a drop or two of milk to the water and stir. Is the beam now visible in the water? What are the milk particles doing to the light? Is there evidence that light is transmitted through the slightly cloudy water? Is there evidence that light is scattered by the cloudy water?

3. Write a conclusion for this experiment using the words *transmitted* and *scattered*.

PART 5

Another Angle

What to Do

1. Take a large beaker or jar of tap water, and hold a small flashlight against one side of it. See the illustration below. Look at the light from the opposite side of the container as well as at right angles to the light beam. You can obtain the best results in a darkened room. What color is the light? What color, if any, is the water when viewed at right angles to the beam?

2. Add a few drops of milk to the water and stir. What color does the light seem to be now? What color, if any, is the water?

3. Repeat step 2, adding more milk until faint color effects are observed.

4. Describe your results in this experiment.

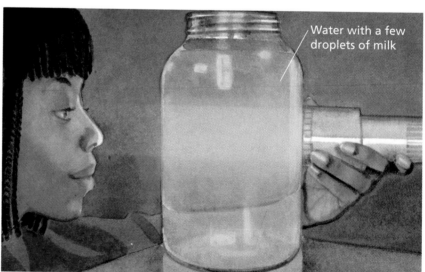

Water with a few droplets of milk

Interpreting Your Experiments

Recall the experiments you did in Exploration 1 of this chapter. Compare your concluding statements for each part of the Exploration with the explanations below.

Part 1

The inside of the light box was darker when the black paper was used than when the white paper was used. The reason for this has to do with how the light entering the light box interacts with each piece of paper. The black surface absorbs most of the light, turning this light energy into heat. The white surface, on the other hand, reflects most of the light and its energy. This is something to think about the next time you dress for a hot summer day. Should you wear light or dark colors? Why?

White surfaces reflect most of the light that strikes them.

Part 2

Glass is said to be **transparent**. How would you define this term? You might say that light can pass through a transparent material and that you can see objects on the other side of this material. However, not all of the light that strikes the glass passes through. Instead, a small amount of light is reflected at the surface of the glass. The amount of reflected light depends on the angle at which the light meets the glass. For example, when light strikes glass at a small angle such as 10°, more light is reflected than if the angle were increased to 70° or 80°. Taking this into consideration, how would you revise your definition of transparent?

How does the surface of a still pond or puddle resemble the window glass in Part 2 of Exploration 1?

Now consider this situation. Melissa sits in front of her bedroom window, and she can see the shrubs and trees outside. At the same time she can also see her own reflection. Why?

Under what conditions would Melissa be able to see her own reflection but not be able to see outside? Under what conditions would Melissa be able to see outside but not see her reflection?

Melissa is seeing reflected and transmitted images at the same time.

Part 3

How is the block of paraffin different from either the glass or the white paper? When you shine a light beam at glass, most of the beam passes through. When you use white paper, the light bounces off in all directions. But when you use paraffin, something different happens. Much of the light is scattered in all directions *within* the block itself, while a little of the light passes through.

Paraffin is **translucent**. It allows some light to pass through it, but scatters most of the light within it. You cannot see things clearly through a translucent material. Can you think of other materials that are translucent?

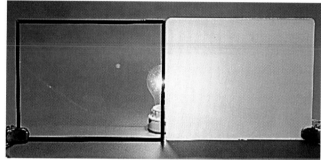

Glass and paraffin wax are positioned between you and a light source. Observe how only some of the light passes through the paraffin wax.

Parts 4 and 5

The experiments with the milk and water illustrate something else about how light behaves. In Part 4, the presence of a few milk particles in water caused light to scatter. What effect did this have on the light beam when viewed from a position at right angles to it?

When you added a little more milk in Part 5, you observed something different. Looking directly at the beam of light through the slightly milky water, you saw faint colors. The beam, viewed head-on, looked faintly yellowish red, but when viewed from the side, the surrounding water looked faintly blue. Why?

The reason is that the milk particles scatter mostly blue, violet, and indigo light. This is why the water around the beam looks bluish. The light coming out of the water looks yellowish red, however, because the beam's red, orange, and yellow light are not scattered.

The light that scatters from a beaker of milky water is of different colors. What color you see depends on your position around the beaker.

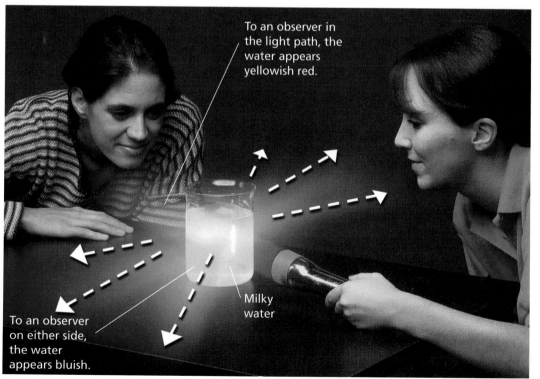

To an observer in the light path, the water appears yellowish red.

Milky water

To an observer on either side, the water appears bluish.

The results of these experiments will help answer some interesting questions. Examine the images at right. Can you provide answers to the questions accompanying them? Write your answers in your ScienceLog.

Why is space black?

Why is a sunset red?

Why does the ocean look blue?

Light's Path

Through your light-box experiments you were able to determine some characteristics about light. For example, you observed that light traveling through space is invisible, but you can see its path through air if smoke is present or through water if a bit of milk is added.

Now take a look at the diagram on page 444. Think about the path of the light beam you studied in Exploration 1. What did you observe about the edges of the beam? What was the beam's direction? When you observe a light beam passing through smoky air, what happens when the smoke completely disappears? Is the light still there? List some evidence in your ScienceLog to support your answer.

Examine the diagram below. What would you expect to observe on the white cardboard screen? Try setting up this experiment. Observe the screen. Then move the comb into and out of the path of the light beam, watching the screen as you do so.

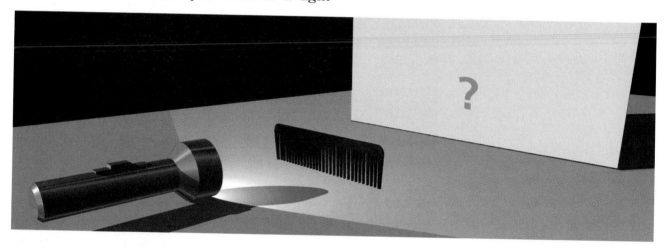

Based on the comb's shadow on the screen and the observations you made during the light-box experiments, what can you say about the way light travels? Does it move slowly or quickly? What evidence suggests that light moves in straight lines?

Light Lines

Study the shadows of a ball illuminated by two different-sized light sources. Examine the screens closely. How are the shadows different?

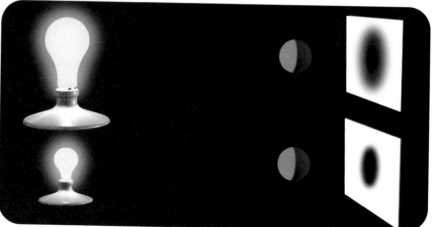

Sketch the light-bulb diagrams in your ScienceLog. Using a ruler, draw two straight lines from the top of the large light bulb—one to the top of the ball casting the shadow, the other to the bottom of the ball. Now draw two lines from the bottom of the large light bulb— one line to the top of the ball, the other to the bottom. Draw similar lines on the diagram with the small light bulb. Do the lines suggest an explanation for the two types of shadows cast on the screens? Write your explanation in your ScienceLog. Look again at the diagrams in your ScienceLog. Would shadows occur if light did not travel in straight lines?

In the next Exploration you will further investigate "light lines" by examining light as it passes through a pinhole. As you perform the experiment, look for evidence that light travels in straight lines.

EXPLORATION 2

Pinhole Images

You Will Need

- a lined white index card
- a metric ruler
- 2 clothespins
- a candle
- modeling clay
- matches
- several index cards with pinholes of different sizes
- a small jar lid

What to Do

1. Perform the experiment in a darkened room. Arrange the apparatus as shown in the diagram.

2. Start with the pinhole about 3 cm from the candle flame and the screen 3 cm from the pinhole. You should get a good image of the flickering flame. Describe the image.

Can you explain its appearance? (Hint: Make a drawing of the setup as shown below. Then draw thin "lines of light" from the top and bottom of the flame through the pinhole and to the screen.)

3. Observe the size of the image as the screen is moved (a) closer to the pinhole and (b) farther from the pinhole. Now examine the image as the pinhole is moved (c) closer to the flame and (d) farther from the flame. Make drawings of (a) through (d) using "light lines" to show how each image is formed.

4. Write a letter about light lines to a younger sibling or friend. Tell him or her how to obtain pinhole images and explain how these images suggest that light travels in straight lines.

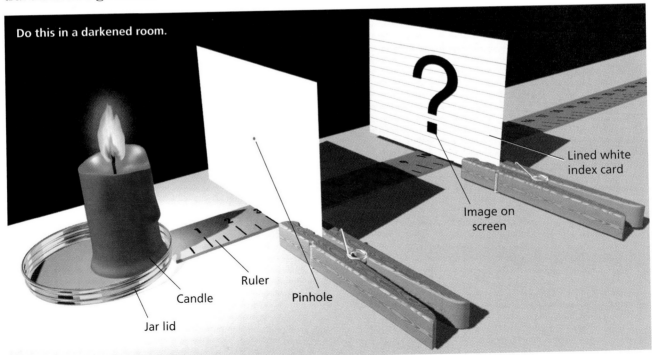

Do this in a darkened room.

Candle

Jar lid

Ruler

Pinhole

Image on screen

Lined white index card

LESSON 3 Reflection

Look at the symbols below. Are they hieroglyphics? some kind of code? Can you guess what they are? Try drawing the symbols that come next in the upper series. Having difficulty? Maybe a flat mirror will help. Cover up the right half of each of the symbols for a hint as to where to place the mirror, as well as to what the series of symbols is. Now draw the next two symbols in the series.

Can you use the mirror to crack the code in the lower series? You know that mirrors are good reflectors of light. But do you realize that *every* object you see reflects light? If it didn't, you would not be able to see the object. The activities in Exploration 5 will help you better understand the reflection of light.

EXPLORATION 3

Reflection Inspection

You Will Need

- a flashlight
- a light box
- modeling clay
- a flat mirror
- a protractor
- matches
- a wooden splint
- white cardboard (15 cm × 10 cm)
- green cardboard (15 cm × 10 cm)
- cardboard of various colors (15 cm × 10 cm)

PART 1

Light Reflected From a Flat Mirror

What to Do

1. Shine the flashlight into the light box. Place the mirror in the path of the light beam. Prop the mirror up using modeling clay. Use a smoldering splint to add smoke to the box, and then observe the light beam. On its way to the mirror, the light beam is called the **incident beam**. After it bounces off the mirror, it is called the **reflected beam**.

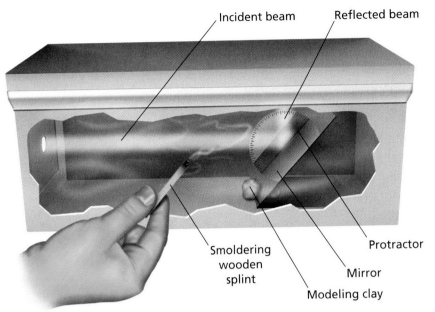

Incident beam
Reflected beam
Smoldering wooden splint
Protractor
Mirror
Modeling clay

Light Reflected From Cardboard

What to Do

1. In this activity, you will use white cardboard and green cardboard instead of a mirror. Use a smoldering splint to add some smoke to your light box. Place the white cardboard in the incident beam. Describe the appearance of the beam reflected by the white cardboard. How does this beam differ from the beam reflected by the mirror?

2. Rotate the white cardboard in the light beam. In which position does the cardboard reflect the brightest beam? the weakest beam?

3. Describe the appearance of the reflected beam at the point where it hits the top of the light box.

4. Repeat steps 1 through 3 using the green cardboard in place of the white cardboard. Answer the same questions as you go along.

5. How does the appearance of the beam reflected from the green cardboard differ from that of the beam reflected from the white cardboard?

6. Repeat the experiment using different colors of cardboard.

Exploration 3 continued

2. Place a protractor with its base against the mirror in such a way that it lies in both the incident beam and the reflected beam at the same time. The center of the two beams, where the light strikes the mirror, should be at the midpoint of the base of the protractor. What is the size of the angle between the incident beam and the mirror? between the reflected beam and the mirror?

3. Rotate the mirror to a different position. Again measure the angles between the incident beam and the mirror and between the reflected beam and the mirror.

4. What is the rule about the direction a reflected beam takes after leaving a flat mirror?

5. Suppose that a beam of light hits a mirror in a light box. The mirror is rotated backward by 20°. How many degrees will the reflected beam rotate? If necessary, use the light box and the figure below to help you answer this question.

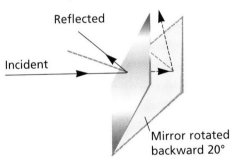

Reflected
Incident
Mirror rotated backward 20°

Reflection Reflections

PART 1

What type of reflection occurred when the mirror was used? The incident light beam was reflected from the mirror without being scattered. Reflection from very smooth surfaces is called **specular reflection.**

Here is a review to check what you discovered about light beams reflected from a mirror. Write your answers in your ScienceLog.

a How large is angle *X*?

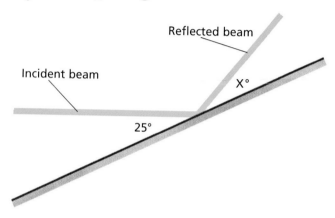

b Which is the correct reflected beam, *P*, *Q*, *R*, or *S*?

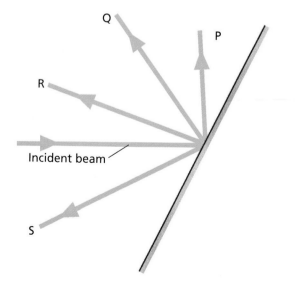

c Now that the incident beam has changed its position, which is the correct reflected beam?

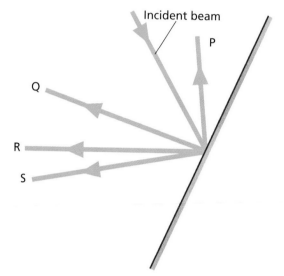

d Here the mirror is placed in a different position. In this case, which is the incident beam and which is the reflected beam?

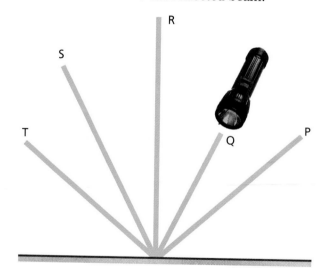

e Devise a rule for the reflection of light from a flat mirror. Here is a start: In specular reflection, the angle between a flat mirror and the incident beam is (equal to, greater than, less than) the angle between _____?_____.

f List some other surfaces on which specular reflection occurs.

How did the white cardboard reflect the beam of light? Was it specular, as in the case of the flat mirror, or more scattered—reflected in all directions?

The scientific term for "scattered reflection" is **diffuse reflection**. What causes diffuse reflection? Examine the two diagrams below to help you understand this question.

Specular Reflection

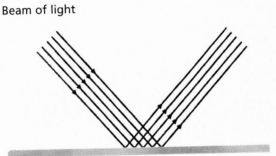

Beam of light

Surface of mirror is smooth (even when magnified)

Diffuse Reflection

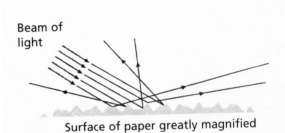

Beam of light

Surface of paper greatly magnified

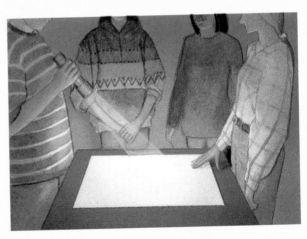

Diffuse reflection is more common than specular reflection. Why do you think this is so?

Was the reflection from the green cardboard specular or diffuse? What color was the light scattered from the green cardboard? What color was the smoke? When the light entered the light box, it was white. What colors found in white light were not visible? The following illustration suggests what happened.

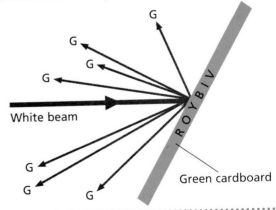

White beam

G

G

G

G

ROYBIV

Green cardboard

G

G

G

The Riddle of Color

When you shone light on the green cardboard in Exploration 3, what color did the cardboard reflect? This is how we perceive colors in objects—by the colored light that they reflect. What colors does a white surface reflect? a black surface? If you substituted red cardboard for the green cardboard in the light-box experiment on page 453 and shone white light on it, what do you think would happen? How would you label the colors in the diagram at right, which shows this experiment?

Look at the illustration below, and then explain why Lisa sees the color she does.

Do objects have the same color when they are viewed under white light as they do under light of another color? For example, if Lisa viewed her friend's sweater in the dim red light of a photographer's developing lab, would the sweater be the same color? Perhaps you have noticed that sometimes a sweater bought in a store illuminated by artificial light seems a different color when you look at it in sunlight or that the color of a jacket at night under certain street lights looks different from when the jacket is viewed in sunlight. Does the light shining on an object have an effect on the color you see? In the following Exploration you'll have the chance to find out.

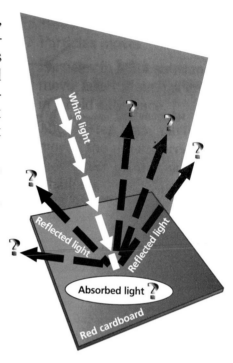

Changing Colors

You Will Need

- a cardboard tube
- a white piece of paper
- red, yellow, blue, and purple crayons
- tape
- 2 pieces of dark blue cellophane
- 2 pieces of dark red cellophane

What to Do

1. Cut out a small window at the bottom of the tube to construct a cardboard viewer like the one shown here.

2. Cut the paper into four rectangular pieces. Make a heavy mark on each with one of the colored crayons.

3. Tape the blue cellophane over the window of the tube, as shown below.

Now view each mark with this blue filter in place. What is the color of each mark? of the background? Cover the window with a second piece of blue cellophane. Is there any difference?

4. Suppose step 3 were repeated using red filters. Predict the colors of the crayon marks and the background. Check your predictions.

5. Use your viewer to look at the color photos in this unit under different lights. How good have you become at predicting the results of color mixing?

Did you know that visible light is part of a much bigger spectrum? To find out more, read pages S137–S140 in the SourceBook.

Mixing Paint

What happens when you mix blue, red, and green paint together? Do you get the same result as when you mix these colors of light? You can check the result for light on page 438. The diagram below shows what happens when you mix blue, red, and green paint together. Why the difference? Before reading further, discuss with a friend a possible reason for this difference.

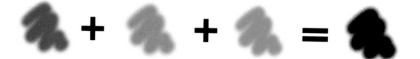

When you mix various colors of light, you are "building" white light. In other words, you are moving toward "lightness," or "whiteness." But colored paint, like colored objects, absorbs the white light that shines on it. If you mix paint of one color with paint of another color, the resulting blend absorbs more from white light than either paint would absorb separately. Less light means more "dark." Thus, mixing paint tends to increase "darkness," or "blackness."

How do the diagram on page 438 and the color-mixing schemes shown below support these ideas?

A Lesson in Art

If you have a set of watercolor paints, you can test other color-mixing experiences. Mix the dry paint with a little water. The more water you use, the "lighter" and more dilute the color will be. You won't need all the colors in your set—only three! Artists find that they can make all the colors they need if they start with *blue*, *yellow*, and *red*. In fact, artists call these the three **primary colors of paint**. How are these colors different from the three primary colors of light?

Here is an example. What color would you expect when blue and yellow paint are mixed together? Examine the diagram at right. How do you explain the color of this mix? Actually, the results of mixing paint lead us to more discoveries about the light reflected from objects. When light shines on colored paint, it reflects not only its own color, but small amounts of its neighboring colors in the color spectrum (ROYGBIV) as well.

Take a look at the diagrams at right. Notice that blue paint *absorbs* red, orange, yellow, and violet and *reflects* not only blue, but also a little green and indigo—blue's neighbors in the spectrum. Yellow paint absorbs red, blue, indigo, and violet and reflects yellow as well as its neighbors, green and orange. When blue and yellow paint are combined, together they absorb red, orange, yellow, blue, indigo, and violet. The only color reflected by the paint mixture is green, which is the color that you see.

Which two primary colors of paint do you think make the color orange? Sketch the situation in your ScienceLog, showing which colors of white light are absorbed and reflected by each primary color and by the mixture.

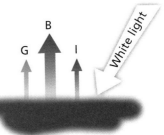

Blue absorbs ROYV.

What happens when you mix all three primary colors of paint together? Try this using equal amounts of paint and not too much water. How do you explain the result?

Examine the color triangle below. What information does it contain? Some hints are in the paint mixes to the right of the triangle. Add a few of your own paint-mix equations with the help of the color triangle. You might even want to test your paint-mix equations using real paint.

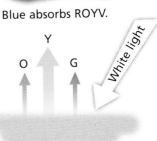

Yellow absorbs RBIV.

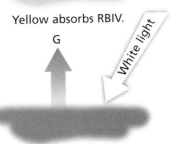

Blue and yellow mixed absorb ROYBIV.

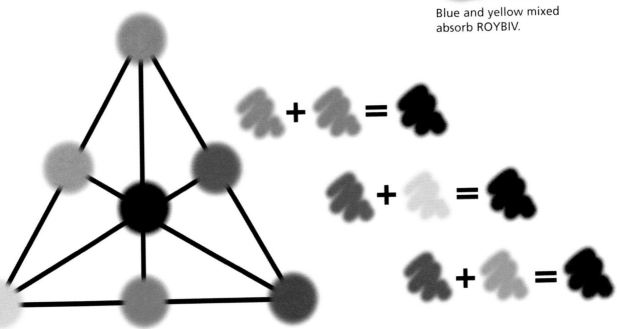

CHALLENGE YOUR THINKING

1. Colorful Moves

Brent made a color spinner like the one shown below. He wound the cord tightly and started the disc spinning by pulling back and forth on the cord. What did he see? (Both additive and subtractive color phenomena occur.) What would Brent see if the colored segments were yellow and blue only? red and blue? Explain.

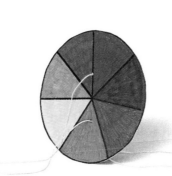

2. Straighten Her Out

Pam made this remark in class: "When you flip a light switch, light suddenly appears everywhere. How can you say that light travels in straight lines?"

How many different reasons could you give Pam to help her understand that light really does travel in straight lines? Describe them.

3. A Staged Question

A local production of Andrew Lloyd Webber's *Joseph and the Amazing Technicolor Dreamcoat* is in progress. At the front of the stage is Joseph in his brilliant coat of red, yellow, green, blue, orange, purple, and white stripes, all the brighter under the high-powered white stage lights. As the scene nears its end, the white lights are suddenly subdued and strong blue stage lighting predominates. Have the colors of Joseph's coat changed? Predict what you would see.

4. Time for Reflection

Devise more "reflection puzzles" like those on page 452, and then give them to a classmate to solve. In turn, solve some puzzles created by your classmate. Here is one to begin with. What is it? How does it differ from the first one on page 452?

5. Light Going

The amount of light that passes through a material depends on whether that material is transparent, translucent, or opaque (allows no light to pass through). Classify each of the following materials as one of the above types: wax paper, plastic wrap, brown paper, oiled brown paper, newspaper, a single ply of tissue paper, five sheets of tissue paper, an ice cube, sunglasses (not mirrored), salt water, a frosted glass.

6. Describe the Scene

Locate in the photo below where each of the following occur: specular reflection, diffuse reflection, transmission of light, reflection of blue light, scattering, and absorption of all colors.

ScienceLog

Review your responses to the ScienceLog questions on page 442. Then revise your original ideas so that they reflect what you've learned.

Light and Images

1 **How do you explain this?**

What is the purpose of
2 a lens on a camera?

3 **Why is this person's image so distorted?**

ScienceLog

Think about these questions for a moment, and answer them in your ScienceLog. When you've finished this chapter, you'll have the opportunity to revise your answers based on what you've learned.

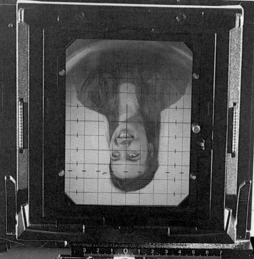

LESSON 1

Plane Mirrors

Consider the Plain Plane Mirror

Flat mirrors are also known as **plane mirrors**. Plane mirrors are what most people mean when they say "mirror." An image in a plane mirror obviously resembles the object that formed it, but is the image an exact duplicate? If not, how does it differ?

What's Your Image?

You read earlier that if an object did not reflect light, you could not see it. In fact, every object reflects at least some light. The fact that people standing at different places in a room can all see a particular object shows that light is being reflected in all directions from the object. Light bounces off of you in all directions.

If you stand in front of a plane mirror, some of the light reflecting off of you strikes the mirror and is reflected from it back to your eyes. What you see in the mirror is your image. Your body, standing in front of the mirror, is the *object;* the body that seems

to be behind the mirror is your *image*. Are the images in a plane mirror exactly the same as their objects? Hold a watch or a book in front of a mirror to find out. In fact, an object and its plane-mirror image are *not* identical. The following Exploration will help you learn more about images in a plane mirror.

EXPLORATION 1

Plane-Mirror Insights

You Will Need

- graph paper
- a plane mirror
- a pencil
- a flat piece of colored glass
- modeling clay
- a metric ruler
- an index card

What to Do

1. Make a dark line over one of the horizontal lines on a piece of graph paper. Stand the plane mirror straight up along the line you made. Use modeling clay to stand the mirror up along that line.

2. Put your pencil in front of the mirror. Where in the mirror is the image of the pencil?

3. Now place the pencil six squares in front of the mirror. Where is the image now? How many squares is it behind the mirror? Place the pencil in another position, and check the location of the image again.

4. Replace the plane mirror with a piece of colored glass. The glass reflects light just as a plane mirror does. However, you can also see through the glass. Both of these properties will help you make your plane-mirror insights.

5. Draw a straight line across the center of a new sheet of graph paper. Place the glass along the line. Use modeling clay to hold the glass upright. Draw a triangle in front of the glass, near the middle of that half of the page. Look at the image of the triangle. While looking

Exploration 1 continued ▶

through the glass, trace the lines of the image on the paper behind the glass. Now remove the glass and modeling clay.

6. Draw a line from one point on the triangle to the corresponding point on its image. Measure the distance along this line

   a. from the glass to the image (called the *image distance*) and

   b. from the glass to the object *(object distance)*.

   How do the two distances compare? What is the angle between the glass and the line joining the points on the object and image? What conclusion can you make about where you will see the image of an object that is placed in front of a plane mirror?

7. Measure the sides of the object triangle and those of the image triangle. How do the sizes of the object and the image compare?

8. Draw a center line on a new piece of graph paper. Then set the glass vertically on the line as shown below. Put your pencil on the paper and slowly move it toward the glass, drawing a line as you go. Imagine that your pencil point is a car on a road driving toward the glass.

Observe the image of the pencil and line. Now, gradually curve your pencil line to the left. What direction does the image turn? Now trace over the image line that you see on the paper behind the glass. As you follow the line toward the glass and around the curve, in which direction is the imaginary car turning?

9. Make a card with the words KID, OXO, POP, and WOW printed on it. Predict what will happen to the words when you lay the card down in front of the glass. Be sure the words are facing you right-side up. What are the images of the words like?

10. Next, predict what will happen when you hold the card upright in front of the glass. How are the images of the words different? Turn the card so that the words are upside down. What does the image in the glass look like now? Where have you seen that image before?

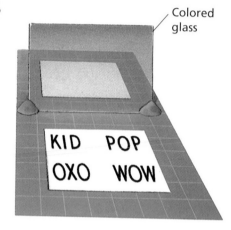

Colored glass

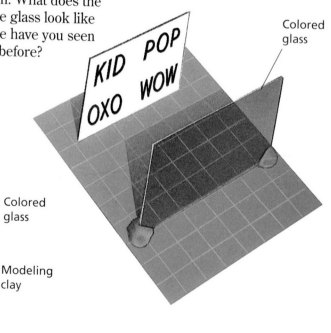

Colored glass

Colored glass

Modeling clay

Looks Like Leonardo's

Try writing your name on a card in such a way that it can be read correctly in a mirror. Interestingly, this is how the famous Renaissance scientist Leonardo da Vinci habitually wrote his notes. He did this so that other people couldn't read them.

Checking the Facts

You have seen lots of plane mirrors, including the ones you used in Exploration 1 of this chapter. Some of the following statements are true; others are false. Decide which statements you agree with, disagree with, or are uncertain about.

1. The image in a plane mirror is the same size as the object.

2. Images of objects in plane mirrors appear to be on the surface of the mirror.

3. If an object moves farther away from a mirror, the image seems to move backward in the mirror.

4. If an object moves farther away from a mirror, its image appears to grow smaller.

5. The perpendicular distance from the object to the mirror is the same as the apparent perpendicular distance from the image to the mirror.

6. If a point on an object and the corresponding point on the image are joined by a straight line, the line intersects the mirror at 90°.

7. You need a plane mirror half your size to see all of yourself at one time.

8. If a moving object in front of a plane mirror turns right, the image seems to turn left.

9. The images of objects held at a right angle (90°) to a mirror are upside down.

10. The images of objects held in front of a plane mirror are reversed left to right.

11. The images of letters *H, O,* and *X* look the same as the objects in a plane mirror, no matter how the letters are held in front of it.

12. The images in a plane mirror aren't where they seem to be.

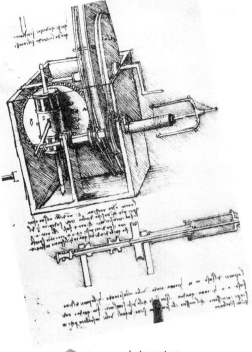

Leonardo's notes for a yarn-spinning machine

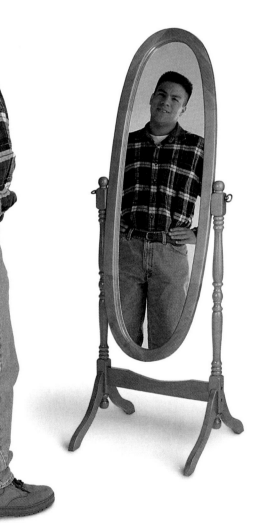

Real or Virtual?

Images are representations of objects. Your bathroom mirror *reflects* light to produce an image of your face. A movie screen *projects* light to produce an image on a screen from a frame of film. Obviously, both the movie projector and the bathroom mirror give you images. But one of these images is **real** while the other image is **virtual**. Why? (Look up *virtual* if you are uncertain about its meaning.) What is the difference between a real image and a virtual one?

The image on a movie theater screen is made by light that is projected onto the screen. If you stood in front of the screen, the image would be projected onto you. Look at the photo to the right. You could place a light meter at the location of the screen image and measure its brightness. For these reasons, the movie-projector image is a real image.

Now think about a mirror image. Where does the image seem to be located? If you answered, "the same distance behind the mirror as the object is in front of the mirror," you are correct. Imagine putting a screen at the location behind the mirror where the image appears to be. Would you get an image? Would a light meter record a reading for the brightness of the image?

The answer to these questions is no. This is why the image in a mirror is virtual, not real. Virtual images cannot be formed on a screen. They are a sort of optical illusion having an apparent location, at which there is no actual light present. On the other hand, real images are formed on a screen by the actual presence of light.

Look at the examples of images provided on this page. Which are real? Which are virtual? How do you know?

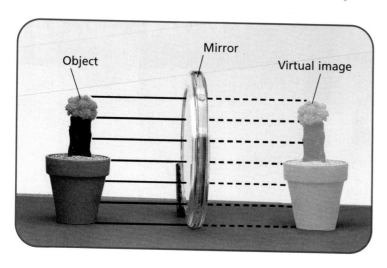

Object   Mirror   Virtual image

Convex Mirrors

Not all mirrors are flat. Sometimes the reflective surface is curved in some way. How does this affect the image? You have probably seen each of the items at right behaving as a mirror. What do the images reflected in these mirrors look like? What is similar about each of these mirrors?

Each object's surface is a **convex mirror**. Convex mirrors curve outward toward the viewer. How do you think convex mirrors differ from plane mirrors? What are objects and images like in convex mirrors? Consider the following questions about convex mirrors:

- How do the images compare in size with their objects?
- Are the images *erect* (right-side up) or *inverted* (upside down)?
- Are the images behind or in front of the mirror?
- Are the images reversed from left to right?
- Are the images real or virtual?

First, try answering these same questions with respect to plane mirrors. If you can do so, you will have a head start on understanding convex mirrors. But to really understand convex mirrors, do the following Explorations.

EXPLORATION 2

Exploring Convex Mirrors

You Will Need

- a large, shiny spoon
- a light box
- mirrors (1 plane, 1 convex)
- a wooden splint
- matches
- a flashlight
- a pencil
- a candle
- a jar lid
- modeling clay
- a sheet of white paper

PART 1

Comparing Mirror Images

Examine your image on the back of a spoon where the curve is greatest.

a. How do the size of the image and the size of the object compare?

b. Is the image right-side up or upside down?

c. Is the image reversed?

d. How does the field of view in this convex mirror compare with that in a plane mirror?

Exploration 2 continued

PART 2

How Convex Mirrors Affect Light

Use modeling clay to support a convex mirror in your light box. Insert a smoking splint, and then shine a flashlight or projector beam through the hole in the foil window. Draw a diagram to show what happens to the light that strikes the convex mirror.

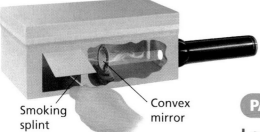

Smoking splint

Convex mirror

PART 3

Classifying Convex-Mirror Images

Recall that plane mirrors produce virtual images. Which type of images do convex mirrors produce—real or virtual? The following activity will help you answer this question.

In a darkened room, place a lit candle near one end of a table.

Place a convex mirror supported with modeling clay at the other end of the table, facing the candle. Hold a piece of paper between the candle and the mirror, but slightly to one side so as not to block the light from the candle. Move the paper slowly from side to side and back and forth to see whether you can focus an image of the candle on it, that is, obtain a real image. Are you able to obtain a real image?

PART 4

Locating the Image

Hold a pencil midway between your face and a convex mirror. Now move the pencil toward the mirror. How does the image seem to move? Move the pencil away from the mirror. How does the image seem to move? Where does the pencil seem to be?

Summary

Based on the observations you made in this Exploration, are images in a convex mirror behind or in front of the mirror? real or virtual? smaller, larger, or the same size as the object?

?

Convex mirror

Uses of Convex Mirrors

Consider the following applications of convex mirrors, and think of the characteristics that make them so useful.

• A car's passenger-side rear-view mirror is usually convex. Why is this so? Such mirrors usually have a warning attached: "Objects may be closer than they appear." Why?

• Large convex mirrors are found in many stores. Why?

How many more uses for convex mirrors can you find?

Converging Lenses and Real Images

What do a camera, binoculars, eyeglasses, and an animal's eyes have in common?

If you said, "They all contain lenses," you are correct. A **lens** is a transparent solid with curved surfaces. Lenses that are thicker in the center than at the edges are called converging lenses. What do you think a **converging lens** does to light?

As the name implies, converging lenses cause light rays to come together. Examine the cross sections of lenses shown below.

Which would you classify as converging lenses? *Farsightedness* is a condition in which things far away are in focus but things nearby are not. One of these converging lenses is used to correct far-sightedness. Which do you think it is and why?

The lens shown at the far left of the diagram is called a double-convex lens. What do you think it could be used for?

Converging Lens Experiences

You Will Need

- a book or piece of paper with writing on it
- a double-convex lens
- a light box
- modeling clay
- a candle
- a jar lid
- a screen
- a wooden splint
- matches
- a flashlight
- a blank index card

PART 1

Strange Sights

1. Place a book or a piece of paper with large print on it about 1 m from your face. Hold a double-convex lens about halfway between you and the book. Slowly move both the lens and your eyes closer to the book. Continue until the lens touches the book.

2. Describe the appearance of the print when you first looked at it through the lens. Was the print erect or inverted? Determine whether the image of the print was larger than, smaller than, or the same as the print's actual size.

3. Describe the appearance of the print when the lens was only a few centimeters away from it.

4. Repeat the procedure slowly, making notes in your Science-Log of every change that takes place in the "image" of the print as you bring the lens closer and closer to the print.

5. On the basis of your observations, choose the word that best fits in each of the following items. You may have to double-check your observations.

   a. When the converging lens is far away from the object, the image of the object formed by the lens is (erect, inverted).

   b. When the converging lens is moved slowly toward the object, at first the image of the object gets (smaller, larger).

   c. As the converging lens moves closer to the object, at one point the image becomes (clear, distorted).

   d. When the converging lens gets even closer to the object, the image is changed. It is now (inverted, erect). It is also (smaller, larger) than the object.

   e. As the converging lens moves still closer to the object, the image (increases, decreases) in size.

How Converging Lenses Affect Light

1. Set up your light box, and insert a smoking wooden splint for a few seconds. Mount a double-convex lens on some modeling clay so that the lens lies in the beam of light from the flashlight.

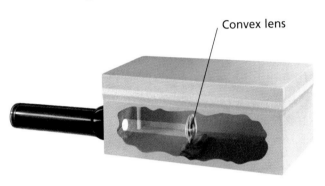

Convex lens

2. Draw a diagram of the beam of light before and after it reaches the lens. Is there any difference in the behavior of a beam of light passing through a double-convex lens and a beam bouncing off a convex mirror? If so, describe the difference.

3. When the beam of light passes through a converging lens, it comes to a point at a position known as the *focal point* of the lens. Label the focal point in your diagram.

   Estimate how far, in centimeters, the focal point is from the lens. This distance is called the *focal length* of the lens.

PART 3

Converging-Lens Images

1. Using a double-convex lens, find the image of a lighted candle on a blank white index card. The card should be on the side of the lens opposite the candle. Discover in what positions the lens, candle, and screen should be located in order to obtain the following:

   a. an image of the object larger than the object itself

   b. an image the same size as the object

   c. a smaller image

   It is sometimes easier to leave the lens in one position. Start with the object fairly far from the lens. Try to locate its image on the screen. Then systematically change the position of the object.

2. For each situation in step 1, draw a diagram in your ScienceLog.

3. You have been looking at *real images* made by a double-convex lens. Why are these images real and not virtual? Do converging lenses ever form virtual, erect images? (See Part 1.) How can you make this happen? Position the candle and lens in such a way that you get this type of image. Where does the image seem to be? How does it compare in size with the object? Why couldn't you use the screen to locate the virtual image?

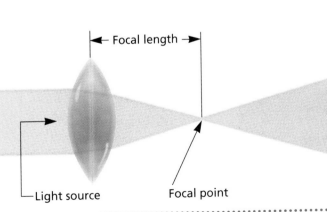

← Focal length →

Light source   Focal point

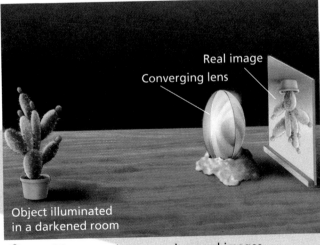

Real image
Converging lens

Object illuminated in a darkened room

⬆ Double-convex lenses produce real images.

Picture This

As you might imagine, converging lenses have many uses. One common device that uses a converging lens is a camera. Can you name some others?

Study illustrations *A* and *B* below, which show how a converging lens helps a camera produce a photograph. Imagine you have been asked to tell a fifth-grade class how the camera produces an image. Make a lesson plan, as if you were their teacher. Write what you would say or do to help the class understand how the device works.

A Special Converging Lens

You have probably heard the human eye compared to a camera. How accurate is this analogy? Study the diagram of the human eye shown below. In your ScienceLog, match the labeled parts of the eye to the descriptions below. Then answer the questions that follow.

- Transparent front part of the eye; refracts (bends) the light entering the eye
- Layer made of tough tissue that forms the protective outer layer of the eye
- Disk-shaped soft, transparent structure; attached to muscles that can change its shape
- Layer containing millions of light-sensitive cells
- Adjustable diaphragm that alters the size of the eye's light-admitting aperture

Eye Exam

1. Which part of the eye is analogous to the shutter in a camera? Which part is analogous to the film?

2. Which part of the eye is analogous to the camera lens?

3. In what, if any, ways is the eye *unlike* a camera?

4. Does the eye form real or virtual images? Explain.

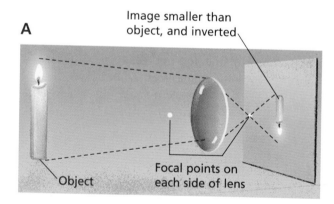

A

Image smaller than object, and inverted

Object

Focal points on each side of lens

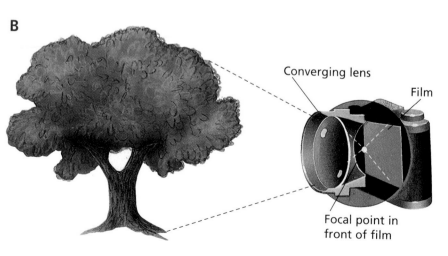

B

Converging lens

Film

Focal point in front of film

Photo after development

Lens

Pupil

Vitreous body

Iris

Cornea

Optic nerve

Retina

Choroid

Sclera

Real Images and Concave Mirrors

At right are two candles and their images. Which image do you think was produced by a plane mirror and which was produced by a convex mirror?

You have studied mirrors that curve outward. You are now going to take a look at mirrors that curve inward. Such mirrors are called **concave mirrors**. A spoon is a simple example of both a concave mirror and a convex mirror. Which side is which?

EXPLORATION 4

Discovering Concave Mirrors

You Will Need

- a large, shiny spoon
- a concave mirror
- a metric ruler
- a pencil
- a light box
- a wooden splint
- matches
- modeling clay
- a flashlight
- a candle
- a jar lid
- a sheet of paper

PART 1

A Concave-Spoon Mirror

Look at your image in the concave side of the spoon, holding it 40 cm or so from your eye. What do you see? Now move the spoon very close to you. How has your image changed?

Now, with the other hand, bring a pencil 5–10 cm in front of the spoon. Move it sideways and up and down. What do you observe? Now move the pencil slowly toward the spoon, noting every change in its image. Continue until the pencil touches the spoon.

Exploration 4 continued ▶

The following activities will help you to understand why concave mirrors form the images they do.

  PART 2

Light and Concave Mirrors

Set up the light box as before, but this time include a concave mirror supported by a piece of modeling clay. Use a smoldering wooden splint to put smoke into the light box. Aim the flashlight through the hole and at the concave mirror.

a. Draw a diagram in your ScienceLog to show what happens to the light when it strikes the concave mirror.

b. Label the focal point on your diagram. Label the focal length (distance between the focal point and the mirror) on your diagram.

Concave mirror

PART 3

Concave-Mirror Images

Hold a sheet of paper between a concave mirror and a candle. The paper should be slightly off to one side, so that light from the flame can reach the mirror. Move the mirror and its base slowly back and forth toward the paper. Look for an image on the paper.

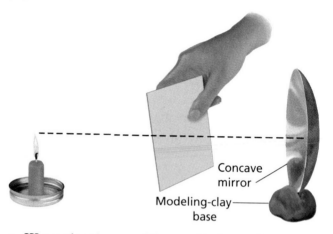

Concave mirror

Modeling-clay base

a. Was a clear image of the candle formed? If so, draw a diagram to show the locations of the candle, the paper, and the concave mirror when the image was clearest.

b. Are the images formed by concave mirrors real or virtual? How can you tell?

c. Do the images appear to be behind the mirror or in front of it?

d. Gradually move the candle toward the mirror. Keep the paper between the candle and the mirror, but adjust the paper to get a sharp image. What change takes place in the location of the image? Is the size of the image always the same?

e. Continue the experiment until you can construct three diagrams. The first should show where the candle, paper, and mirror are placed if a very small image is desired. The second should show the setup that produces an image the same size as the object. The third diagram should show how the candle, paper, and mirror should be arranged to produce an image larger than the object.

f. Now place the candle very close to the mirror. Look into the mirror. What do you see? What kind of image is it—virtual or real? How can you tell?

Comparing Lenses and Mirrors

1. Using a candle as an object, you were able to get images with a concave mirror that looked like those at right. Which images were real? Which were virtual? What other optical device gave you the same kinds of images?

Candle images

2. When an image is real, what characteristic do you expect the image to have? When an image is virtual, what other characteristic do you expect it to have?

3. If a real image is smaller than the object, which is located closer to the device forming the image—the object or the image?

4. All of the mirrors shown below form virtual images.

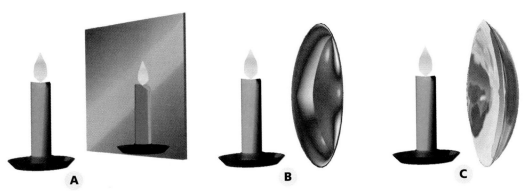

Which mirror(s) give(s) only virtual images? Which mirror(s) can give both virtual and real images? Do virtual images appear in front of the mirror, on the mirror, or behind the mirror? Do real images appear in front of the mirror, on the mirror, or behind the mirror?

5. Which mirror or lens would you use to
   a. magnify your finger?
   b. examine the tip of your nose?
   c. see yourself life-size and right-side up?
   d. see yourself smaller than you are, but right-side up?
   e. see yourself smaller than you are and upside down?

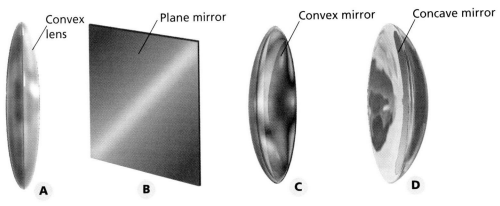

Uses of Concave Mirrors

If you look around, you'll probably notice that most of the curved reflecting surfaces you see are convex. But there are surprisingly many uses for concave mirrors. For example, what is the purpose of the concave mirror in the device at right?

Recall that light passing through a converging lens comes together at a focal point. Similarly, when a beam of light is aimed at a concave mirror, almost all of the light goes to the focal point.

The opposite is also true; if you place the filament of a light bulb at the focal point of a concave mirror, the light reflects away from the mirror as a beam. Suggest other examples of concave mirrors that work on the same principle as a flashlight reflector.

Study the diagrams below of a solar cooker and a searchlight. How do the two devices differ from each other? How are they the same?

Some reflecting telescopes have concave mirrors that measure 5 or 6 m in diameter. What purpose could such a large mirror serve? Why would a telescope with a large main mirror work better than a telescope with a small main mirror?

As you probably guessed, the bigger the concave mirror in a telescope, the more light it can collect from very faint stars. The best telescopes can intensify the available light a billionfold. Study the photograph and the diagram shown below to determine how a reflecting telescope works.

Most flashlights have a concave mirror. Do you know why?

A solar cooker. Where is the focal point?

A searchlight

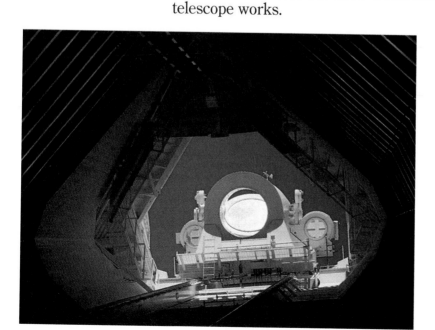

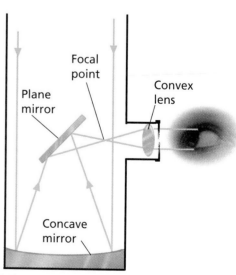

When you look into a reflecting telescope, you see the image at the focal point of the large concave mirror.

Plane mirror

Focal point

Convex lens

Concave mirror

Refraction

The Bending of Light Beams

Light travels through empty space at a speed of about 300,000 km/s. Moving at this speed, someone could travel around the world about $7\frac{1}{2}$ times in 1 second! Light travels slightly slower in air than through empty space. In water, light travels even slower—at about 225,000 km/s. Light's speed in glass is only about 200,000 km/s. That's still pretty fast, though!

When light passes through one material into another, its speed changes and some unusual things happen. Complete the following experiments to find out more about this phenomenon.

EXPLORATION 5

Light-Bending Experiences

You Will Need

- 2 pennies
- water
- a plastic cup
- a pencil
- a felt-tip marker
- rubbing alcohol and vegetable oil
- a protractor
- a 600 mL or 1 L beaker
- milk
- a flashlight
- chalk dust
- a small piece of cardboard
- aluminum foil
- a rubber band

PART 1

Which Penny Is Closer?

Fill a plastic cup with water until nearly full. Place a penny inside the container, up against the side. Put another penny on the table outside the cup, close to the penny inside the cup. Look directly down at the pennies.

a. Describe what you see.

b. Which penny seems closer?

c. Move your head slowly from side to side while observing the two pennies from above. Does either penny appear to move across the background of the table? If so, which one?

Water

Remember, reflected light from the outside penny comes from the penny, through the air, and to your eye. Reflected light from the penny in the water first passes through the water and then through the air to reach your eye. What effect occurs when light must pass through two materials to reach your eye?

PART 2

Measuring Apparent Depths

Place a penny in a cup filled with water. Look down at the penny. Indicate where the penny appears to be by making a mark on the side of the cup while still looking down. Compare the apparent depth of the penny with the real depth.

Substitute alcohol and then vegetable oil for the water in the cup. Compare the depths at which you see the pennies.

Water, alcohol, or vegetable oil

Exploration 5 continued ▶

PART 3

Bent Objects

Fill a large container, such as a 600 mL or 1 L beaker, two-thirds full of water. Put a long pencil into the container. Observe the pencil from all sides and from above. Save this setup for Parts 4 and 5.

a. Describe the appearance of the pencil from different angles.

b. When seen from above, is the part of the pencil that is below the surface apparently bent upward or downward?

PART 4

Bent-Light Paths

Use the setup from Part 3 (minus the pencil). Stir in 2 or 3 drops of milk—just enough to give the water a faint milky appearance. Cover the top of the setup with a piece of cardboard that has a hole about 0.5 cm in diameter at the center. Darken the room. Shine a flashlight through the hole, at an angle, into the water. Observe the light from the side. Shake a small amount of chalk dust into the air above the water so that the beam of light is visible above the water, as well as in the water.

a. How does the beam appear to bend when it goes into the water? Look at the beam through the side of the container. Draw a diagram showing how the beam bends.

b. Is the beam of light going into or coming out of the water? When you observed the underwater part of the pencil in Part 3, did you see it as a result of light going *from* your eyes to the pencil or light coming *to* your eyes from the pencil?

c. Why did the part of the pencil that is underwater appear to bend upward, while the light beam seems to bend downward where it enters the water?

PART 5

Total Internal Reflection

Darken the room. Move the container near the edge of the table and direct the flashlight beam upward from below the water level through the side of the container. Use the flashlight setup shown in the diagram below. Again shake a bit of chalk dust over the water to make the beam visible as it comes out of the water.

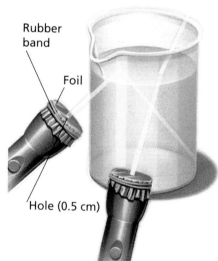

Rubber band

Foil

Hole (0.5 cm)

a. Do you observe bending? Make a sketch showing how the beam bends. Slowly raise the flashlight so that the angle between its beam and the surface becomes smaller. What happens to the beam in the air? At one particular angle the beam stops coming out of the surface. At all smaller angles the beam is reflected back into the water as though there were a mirror at the surface. This phenomenon is called *total internal reflection*. Total internal reflection occurs only when light reaches a boundary between one material and a second material through which light can travel at a faster speed. Thus, light traveling in water can be totally internally reflected when it comes to a boundary between the water and air—provided the light hits the boundary at a sufficiently small angle.

b. Using a protractor, measure the angle between the light beam and the water surface when total internal reflection is first observed.

Light has some peculiar characteristics. To learn more about light, turn to pages S142–S143 of the SourceBook.

Interpreting Your Findings

Check your understanding of light bending by answering the following questions in your ScienceLog.

1. Does a penny viewed in water appear deeper, shallower, or at the same depth as it really is?

2. In the case of question 1, is light going by the path eye → air → water → penny or by the path penny → water → air → eye?

3. In situation *A*, which path will the light beam most likely follow—*1, 2,* or *3*?

4. a. In situation *B*, light is traveling from the water to the air to the person's eye. Which beam of light—*1* or *2*—will most likely reach the person's eye?

   b. Which beam of light—*1* or *2*—will probably not exit the water, but will instead be reflected as *R*?

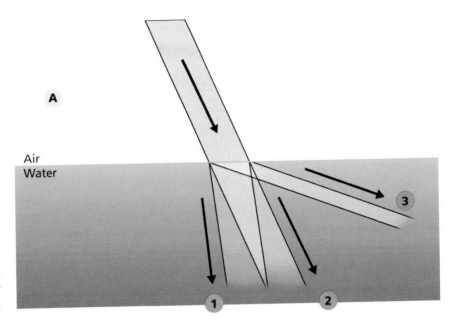

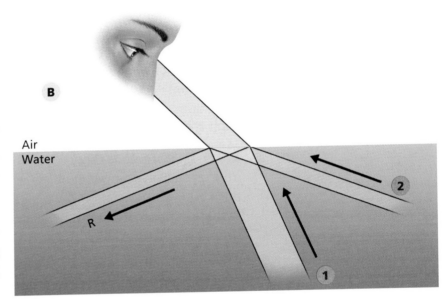

In Conclusion

As you have seen, light bends when it passes from one medium to another. The scientific term for this phenomenon is **refraction.** What other examples of refraction can you think of?

1. Help the Artist

Explain what's wrong in each of these pictures. Then redraw each one so that it's correct. The lens has a focal length of 10 cm.

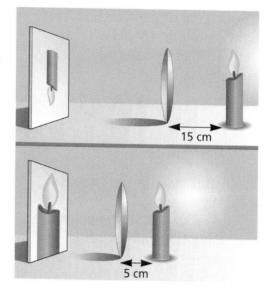

2. The Right Mirror for the Job

Dr. Hughes uses a small, slightly concave mirror when examining her patients' teeth. Her motorcycle has a convex rearview mirror. Why does Dr. Hughes use these different types of mirrors? How is each type of mirror appropriate for its situation?

3. Eye Fooled You!

Your eye always views an object as if its light path never changed direction. This explains why a pencil immersed in water appears to bend upward. Write a brief explanation of the accompanying illustration based on what you know about light paths.

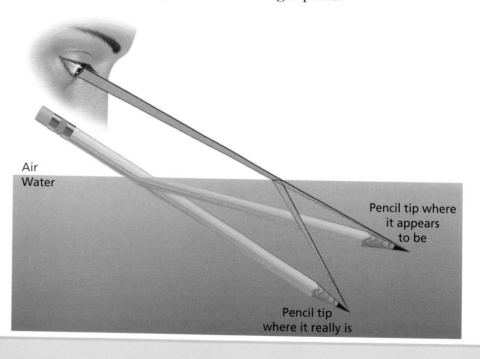

Air
Water

Pencil tip where it appears to be

Pencil tip where it really is

4. Mirror, Mirror

As the photos show, a reflective cylinder produces a different image depending on whether it is held vertically or horizontally. Explain how the cylinder is able to produce such different images. What kind of mirror is the cylinder—convex, plane, concave—or something else? Explain.

5. You Make the Call

You know that optical devices such as plane mirrors, convex mirrors, and converging lenses produce images. You also know that some of the images produced by these devices are real, and some are virtual. The illustration below shows an object and several images of it that were produced by the optical devices just mentioned. Which are real and which are virtual? Which optical device could have produced each image? Where would the object have to be placed in relation to the optical device to get a given result?

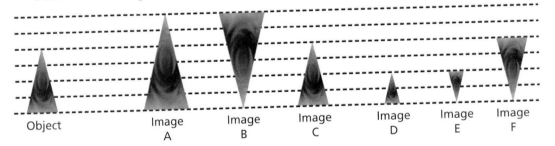

Object   Image A   Image B   Image C   Image D   Image E   Image F

6. Ponder This

Use the principles of light refraction to explain why a clear, still pond seems to be shallower than it really is.

ScienceLog

Review your responses to the ScienceLog questions on page 462. Then revise your original ideas so that they reflect what you've learned.

Making Connections

Unit 7

LiGHT

The Big Ideas

In your ScienceLog, write a summary of this unit, using the following questions as a guide:

1. How can you tell whether light is matter or energy?
2. How can colored light be produced from white light?
3. What is the relationship between the temperature of an object and the color of the light it emits?
4. What are filters, and how do they work?
5. What gives an object its color?
6. Along what kind of path does light travel?
7. How is mixing colored lights different from mixing colored paints?
8. What is reflection? refraction?
9. How do the various types of lenses and mirrors differ in terms of the images they produce?

Checking Your Understanding

1. Objects lit by a strong, nearby light source cast shadows that expand very quickly, unlike objects that are lit by a faraway light source. Use your understanding of light to explain this.

2. Why do athletes, such as baseball players, often smear a dull, black substance below their eyes on sunny days?

3. All lenses cause a prismatic effect; that is, they bend each component color of light by a slightly different amount, causing a distorted image. For this and other reasons, the best telescopes always use mirrors rather than lenses.
   a. Why do mirrors not exhibit the problem of the prismatic effect?
   b. In what other ways are mirrors advantageous?
   c. How might you correct the prismatic effect using lenses alone?

4. How could you use two mirrors to see the top of your head?

5. What's inside each mystery box? The only clues you have are the light beams that enter and leave each box.

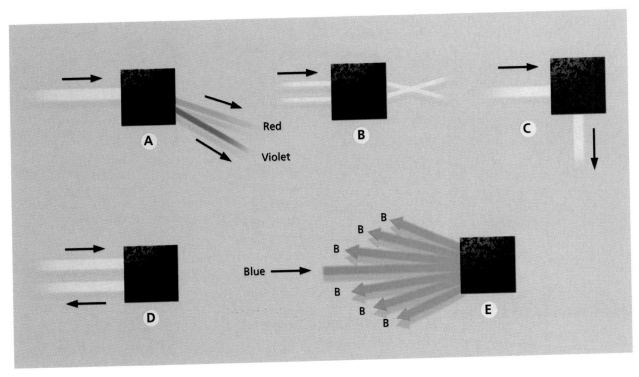

6. concept map Draw a concept map showing the relationship between these words or phrases: converging lens, reflected, real image, refracted, virtual image, light, and plane mirror.

Mirrors in Space

How would you like to live in continuous daylight? Would you enjoy the extra time to garden or to play outdoors, or would it disrupt your normal cycle of work, play, and sleep? These questions are not as strange as they may seem now that Russian scientists have put a mirror in space.

Shedding New Light

Usually, only the moon reflects a noticeable amount of the sun's light to our planet's dark side.

▲ The moon acts much like a mirror, reflecting light from the sun back to the Earth's surface.

But on February 4, 1995, the Russian space mirror, named *Banner*, illuminated a small area of predawn Europe to about the brightness of a full moon. *Banner* was released from the Russian space station, *Mir*, 362 km above the Earth's surface. The cosmonauts unfolded the 4 kg space mirror like a fan. Its very thin aluminum-coated plastic created a disk 19.5 m in diameter. For 8 minutes the mirror was able to reflect a stream of sunlight 4 km wide! The beam of light moved across the Atlantic and parts of Europe before disappearing into the sunlit side of the globe.

Commercial Possibilities

The success of this experiment convinced some Russian scientists that a series of space mirrors could illuminate large areas of the Earth at night. They speculate that such a system could save billions of dollars in electrical lighting costs, could reduce the use of natural resources, and could allow outdoor projects such as planting and harvesting crops to continue around the clock. However, these predictions may be optimistic considering the dimness of the reflection. Also, to accomplish any of this, scientists must first overcome a technological challenge: the mirrors must be able to focus light on a specific location while orbiting at thousands of kilometers per hour. Presently, the reflected beam of light moves as *Banner* moves. However, the scientists are hopeful

▲ The space mirror *Banner*, or *Znamya* in Russian, unfolds like a huge umbrella. An artist's conception (inset) shows two space mirrors in action.

that they can create space mirrors that will continuously project a beam of light onto one area.

Enlightened Predictions

Many scientists in Russia and around the world are concerned about the idea of perpetual sunlight. How would animals that depend on darkness for hunting and protection survive? How would the growth cycle of plants be affected? How would humans be affected physically and emotionally? Choose one of these questions, and make predictions about how our world could change.

Neon Art

In 1898 two chemists made an exciting discovery. They passed an electric current through a gas they had collected while experimenting with liquid air. The gas began to glow with a fiery, reddish orange light. The chemists, Sir William Ramsay and Morris Travers, had discovered a new element! They named the element *neon*, from the Greek word meaning "new."

A New Art Form

Not long after the discovery of neon, people began to think of artistic and decorative uses for it. By the 1920s, neon signs were in wide use. Today neon signs are so common that we are likely to take them for granted. But have you ever wondered how they are built or how they work?

Tubes of Gas

The first step in creating a neon sign or sculpture is to shape the glass. To do this, a glass tube is heated until it becomes flexible, and then it is gently bent into shape. Next, electrodes are installed in each end, and a vacuum pump is used to pump the air out of the tubing. Finally, the tube is filled with invisible neon gas.

Color Magic

When an electric current is applied to neon gas, a reddish orange light is produced. But if a few drops of mercury are added, the light changes to a brilliant blue! The color changes because the atoms of different gases give off different-colored light when they are energized. However, this is not the only

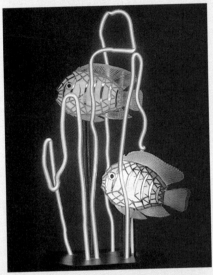

▲ The discovery of neon led to new art forms.

way to change the color of a neon light. Green, yellow, and orange light can be created by coating the inside of the glass tubes with powdered phosphors. When ultraviolet rays from the mercury vapor strikes the phosphors, they glow, giving off the desired color. Still other colors can be achieved by replacing the neon gas with argon gas or by using different colors of glass. With all of these techniques at their disposal, neon artists have a complete palette of glowing colors to work with!

▼ Making neon signs requires skill and patience. The artisan must blow air through the tube as it is heated to prevent the tube from collapsing as it is bent.

Think About It!

When you see a green neon sign or a red stoplight, the color you see is the color of the light itself. How is this different from what you see when you look at a green leaf or a bright red T-shirt?

Light Pollution

Have you ever seen a large city at night from a distance? Soft light from windows forms the outlines of office buildings. Bright lights from stadiums and parking lots shine like beacons. Scattered house lights twinkle like sparkling jewels. The sight can be stunning!

Unfortunately, these lights are considered by astronomers to be a form of pollution. This pollution, called light pollution, is reducing our ability to see beyond our own cosmic neighborhood. In other words, the light pollution from cities and other light sources is making it difficult to see the night sky. Astronomers around the world are losing their ability to see through our atmosphere into space.

Sky Glow

Twenty years ago, stars were very visible above even large cities. The stars are still there, but now they are obscured by the glow of city lights. This glow, called sky glow, is created when light reflects off of dust and other particulates suspended in the atmosphere. Even remote locations around the globe are affected by the cumulative effects of light pollution. The sky glow affects the entire atmosphere to some degree.

The majority of light pollution comes from outdoor lights such as automobile headlights,

▲ Lights from cities can be seen from space, as shown in this photograph taken from the space shuttle *Columbia*. Bright, uncovered lights (inset) create a glowing haze in the night sky above most cities in the United States.

street lights, porch lights, and the bright lights illuminating parking lots and stadiums. Other sources include forest fires and gas burn-offs in oil fields. Air pollution makes the situation even worse, adding more particulates to the air so that there is even greater reflection.

A Light of Hope

Unlike other kinds of pollution, light pollution has some simple solutions. In fact, light pollution can be cleaned up in as little time as it takes to turn off a light! While turning off most city lights is impractical, several simple strategies can make a surprising difference. For example, using covered outdoor lights instead of uncovered ones keeps the light angled downward, preventing most of the light from reaching particulates in the sky. Also, using motion-sensitive lights and timed lights helps eliminate unnecessary use of

light. Many of these strategies also save money by saving energy.

Astronomers hope that public awareness will help improve the dwindling visibility in and around major cities. Some cities, including Boston and Tucson, have already made some progress in reducing light pollution. It has been projected that if left unchecked, light pollution will affect every observatory on Earth within the next decade.

See for Yourself

With your parents' permission, go outside at night and find a place where you can see the night sky. Can you count the number of stars that you see? Now turn on a flashlight or porch light. How many stars can you see now? Compare your statistics. By what percentage was your visibility reduced?

Science Snapshot

Garrett Morgan (1877–1963)

One day in the 1920s an automobile collided with a horse and carriage. The riders were thrown from their carriage, the driver of the car was knocked unconscious, and the horse was fatally injured. One witness of this scene was an inventor named Garrett Morgan, and the accident gave him an idea.

A Bright Idea

Morgan came up with an idea for an electrical and mechanical signal to direct traffic at busy intersections. Morgan's idea for the signal included electric lights of different colors. These lights ensured that the signal could be seen from a distance and could be clearly understood.

Morgan patented the first traffic light in 1923. This traffic light looked very different from those used today. Unlike the small, three-bulb signal boxes that now hang over most busy intersections, the early versions were T-shaped and had the words *stop* and *go* printed on them.

Morgan's traffic light was operated by a preset timing system. An electric motor turned a system of gears, which operated a timing dial. As the timing dial turned, it caused switches to turn on and off. The result was a repeating pattern of green, yellow, and red lights.

Morgan's traffic light was an immediate success, and he sold the patent to General Electric

Nov. 20, 1923.

G. A. MORGAN

TRAFFIC SIGNAL

Filed Feb. 27. 1922

1,475,024

2 Sheets-Sheet 1

▲ **Morgan's patent for the first traffic light**

Corporation for $40,000, a large sum in those days. Since then, the traffic light has been the mainstay of urban traffic control.

The technology of traffic lights, however, continues to improve. Some newer models, for example, can change their timing based on the traffic needs for a particular time of day. Some models can even use sensors installed in the street to monitor traffic flow. Other models have

sensors that can be triggered by an ambulance so that they automatically turn green, allowing the ambulance to pass.

The Man

Born in Paris, Kentucky, Garrett Morgan was one of 11 children. Having received only an elementary school education, Morgan educated himself. When he was 14, he left home to seek work in Cleveland, Ohio. He was broke and had few skills, but he was mechanically inclined and quickly taught himself enough about sewing machines to get a job repairing them. He worked very hard, and in 1907 he opened his own sewing-machine repair shop. Just two years later, he bought a house and started a tailoring business.

One night when Morgan was experimenting with a lubricant for the sewing needle, he noticed that the lubricant straightened the fuzz on a fur cloth he had been using. He had accidentally invented hair straightener, the first of many products that Morgan invented.

Think About It

Traffic control is not the only system that uses light as a signal. What are some of these other systems, and what makes light such a good medium for communication?

CONTINUITY OF LIFE

As though responding to some invisible signal, newly hatched leatherback turtles head straight for the surf. Unfortunately, many will be snatched up by waiting predators before ever reaching the water. And for those that survive the gauntlet of jaws and beaks, life will continue to be perilous for many months to come. Perhaps one turtle in fifty survives to adulthood.

The tiny turtles would face much greater odds against survival if they did not head immediately to the water. But how do they know to do this? Encoded in the cells of each turtle is a set of behavioral instructions, including the order to make this instinctual trek. These behavioral instructions are inherited by the turtles from their parents, in the same way that their physical characteristics are inherited.

Although these turtles appear identical, each is, in fact, unique. The mechanism of heredity is able to recombine the traits of the parents in so many ways that each offspring gets a slightly different set of life instructions. These instructions, responsible for the turtles' similar-yet-different form, will also one day direct surviving females back to the shore where their lives began. There they will lay eggs, passing life instructions to a new generation.

Newly hatched leatherback turtles, Les Hattes Beach, French Guiana

489

Discoveries

1 **Why don't we all look alike?**

How is your body **2** **able to mend a cut?**

ScienceLog

Think about these questions for a moment, and answer them in your ScienceLog. When you've finished this chapter, you'll have the opportunity to revise your answers based on what you've learned.

How old do you think **3** **you were when your heart began to beat?**

A Family Likeness

"You got your mom's chin." "You look so much like your aunt." You might have heard your relatives say things like this. Such similarities are called family "likenesses."

Look at yourself carefully in a mirror. Notice details such as the color and shape of your eyes and the shape and size of your nose, eyebrows, mouth, ears, and hands. Then look at other family members (your parents, sisters, brothers, cousins, aunts, uncles, and grandparents) and at family photos for resemblances.

You're the spitting image of your dad!

Your nose is just like Grandpa's.

How do these siblings resemble one another?

Family Relations

The diagram at the right shows how everyone in Juan's immediate family is related. Use it to answer the following questions:

1. How is Juan related to each of the other people in the diagram?

2. How are Alicia and Laura related?

3. What is meant by a **generation** of a family?

4. How many generations are included in the diagram?

5. If Laura and Nathan have children, how will their children be related to Juan and Alicia?

6. From whom might Alicia get her looks? If she "has her father's ears," from whom did he get them?

7. Who has more *ancestors*, Juan or his father? Explain.

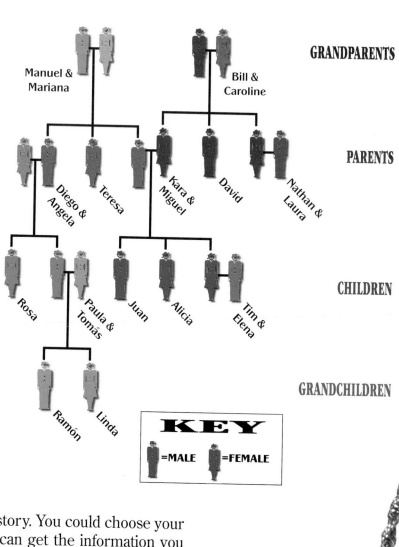

FOUR GENERATIONS OF A FAMILY

GRANDPARENTS

Manuel & Mariana

Bill & Caroline

PARENTS

Diego & Angela

Teresa

Kara & Miguel

David

Nathan & Laura

CHILDREN

Rosa

Paula & Tomás

Juan

Alicia

Tim & Elena

GRANDCHILDREN

Ramón

Linda

KEY

= MALE   = FEMALE

A Family-Tree Project

Now is the time to begin a family history. You could choose your own family or one from which you can get the information you need. The family history will feature a diagram like the one drawn for Juan's family. Include as many generations as possible. Some families are fortunate enough to have four or five generations living at the same time. You may also include relatives who are no longer alive. Go back as many generations as you can by asking questions of family members and by looking through family records. The resulting **family tree,** as this type of diagram is often called, could become quite large.

Preparing a family tree can take quite a bit of time. You can add to it as you proceed through this unit. If you'd like, you could collect photographs to include on the family tree. The photos may help you observe family members' characteristics, or **traits,** more carefully than you could by relying on memory alone. The following Exploration will introduce you to some traits that you know about and others that are probably unfamiliar to you.

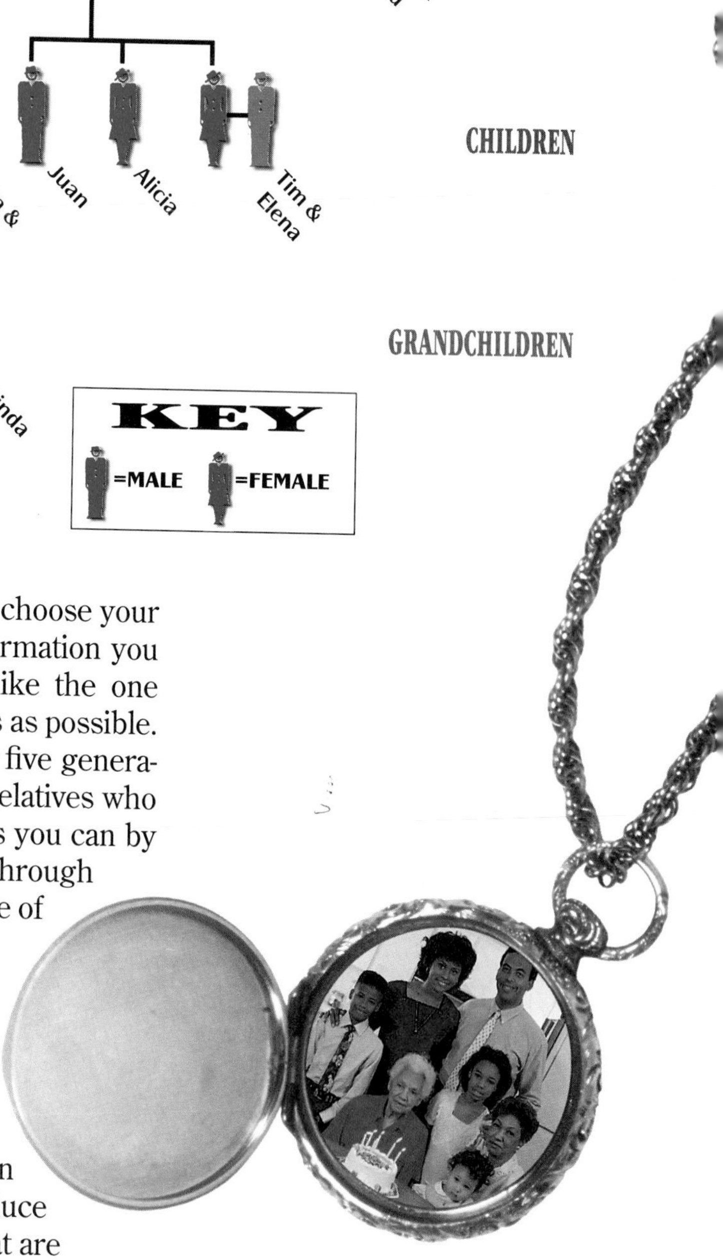

A Traits Test

You Will Need

- an index card
- a paper punch
- scissors
- a knitting needle

What to Do

1. Prepare the index card as shown in the diagram.

2. Use the illustrations to help you answer the questions below. For each question, if your answer is *yes,* darken the area by the number, between the hole and the edge of the card. If your answer is *no,* leave the card unchanged. See the example shown.

3. When you have answered all the questions, use scissors to cut out the areas you darkened for step 2. This will open the holes for all the *yes* answers, as shown. Pass in your card.

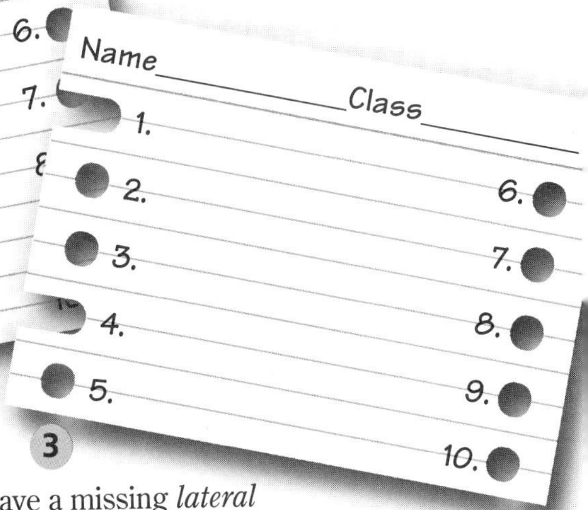

Determining Your Traits

1. Can you tell the difference between red and green? If you do *not* see a number in the circle, then you have red-green colorblindness.

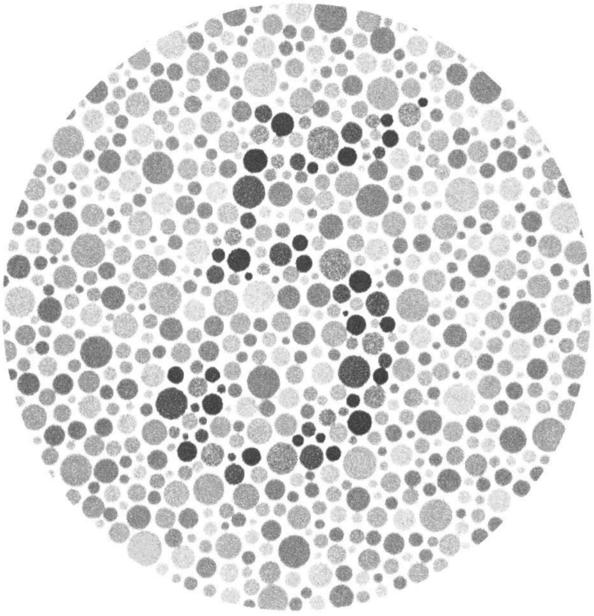

2. Do you have a missing *lateral incisor*? (Lateral incisors are the teeth on either side of your two front teeth. You should have a total of four lateral incisors—two in your upper jaw and two in your lower jaw.)

Exploration 1 continued ▶

3. Do you have free ear lobes?

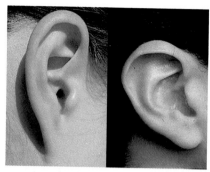

Attached   Free

4. Is the last segment of your thumb straight?

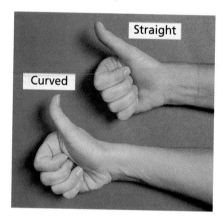

Straight

Curved

5. Do you have dimples?

6. Do you have mid-digit hair?

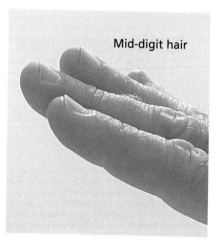

Mid-digit hair

7. Do you have a white forelock—strands of white hair growing just above your forehead?

8. Is your fifth finger (pinky) straight?

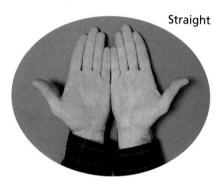

Straight

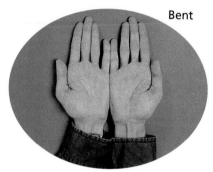

Bent

9. Do you have an indentation in the middle of your chin (cleft chin)?

10. Do you have a widow's peak?

Widow's peak

The Results

When all the cards have been collected, stack them together. Insert the knitting needle in each hole, in turn, and then lift the pile of cards. Gently shake the cards until some fall. Which ones will fall? For example, if you answered *yes* to question 1, will your card stay in or fall out of the group? Count how many cards fall and how many remain. Record the numbers in a chart. Repeat this process for each question.

Analyzing the Data

1. Looking at the first four characteristics, how many people in your class answered *yes* for the same two characteristics? for three? for more?

2. Do any two people in your class have the same responses for all of these traits?

3. What do these results tell you about the characteristics of people?

4. What are some characteristics that all people have in common?

What Is Life?

Families share likenesses. So do all humans, all mammals, all vertebrates, and all animals. Some characteristics are shared by *all* living things. If you were asked to describe the characteristics of living things, what things would you include? The list of questions below suggests four essential properties that all living things exhibit. See whether you can determine the four properties.

1. When you cut your finger, what does your body do?

2. Why does an earthworm avoid daylight?

3. Cats don't live forever, so why are there so many of them around?

4. Why do we perspire?

Biologists have identified four essential properties of life.

- Self-preservation—staying alive; obtaining energy to live and grow

- Self-regulation—the control of life processes, such as respiration

- Self-organization—forming, grouping, and repairing body cells

- Self-reproduction—making new life of the same kind (such that the offspring of monkeys are monkeys and the offspring of whales are whales)

Now try matching the four questions with the four properties.

This unit is involved with the fourth essential property of life—self-reproduction—and deals with many interesting questions.

- Why can two relatives resemble each other in some ways but still look very different?

- Can two people look exactly alike?

- Who did you get your looks from? Where did they get their looks?

- If someone loses a toe in an accident, is it likely that his or her children will be born with a missing toe?

- How is life passed on?

- How does a human baby develop before birth? Is it possible to predict what a baby will look like before it's born?

- What is a "test-tube" baby?

- Could there ever be another "you"?

Under which property of life would you classify *perspiring*?

Recipes for Life

Throughout history, people have given many different explanations for how living things reproduce. As you read the following statements made centuries ago, suggest why each of the explanations might have seemed reasonable at the time it was offered.

Hairs from a horse's tail become horsehair worms, which are found in pools.

Flies come from rotting meat.

The barnacle goose grows from goose barnacles found on rocks beside the ocean.

Put one dirty shirt and some grains of corn into an old pot. In 21 days there will be a lively crop of mice.

The emanations rising from the bottom of the marshes bring forth frogs, snails, leeches, herbs, and a good many other things.

Some insects are born on dewy leaves; some are born from the hair and flesh of animals.

What do you think about these explanations? Choose any one of the statements listed above, attributed to seventeenth-century scientist Jan Baptista van Helmont. What could you do to investigate that claim scientifically?

Well into the 1500s, ideas about the reproduction of living things were still being influenced by Aristotle, who lived in Greece 2000 years before van Helmont. Aristotle said that living things could spring from nonliving things by a process called *spontaneous generation*. For example, Aristotle said that eels grow from mud and slime at the bottoms of rivers and oceans.

What happened to change people's ideas about how new generations of living things come about? Read the following story about Francesco Redi, whose experiment eventually changed people's ideas about reproduction.

Redi's Experiment

In 1628 an English doctor named William Harvey published a book in which he suggested that some living things might come from tiny eggs and seeds instead of from nonliving things (as had been thought previously).

When Francesco Redi, an Italian doctor, read Harvey's book, he decided to investigate whether flies grow spontaneously from rotting meat. His experiment is pictured below.

Redi's Experiment   **Results**

A Meat in open jars

Maggots on the meat

B Meat in tightly closed jars

No maggots on the meat

Think About It

1. What must have happened in part *A*? in part *B*?

2. Where did the maggots in part *A* come from?

3. Redi repeated his experiment, making one important change: he substituted a gauze cloth for the solid lid he had used in part *B*. Why do you think he did this?

4. Why was part *A* so important? What name is given to that part of the experiment?

5. What conclusion could be drawn about spontaneous generation from this experiment?

Although people no longer believe in spontaneous generation, they remain deeply interested in the development of new living things. For example, when you were very young, you probably asked, "Where do babies come from?" In the next section, you'll begin your exploration of this and other questions about reproduction by studying single cells, the basic units of life.

Cells—The Basic Units of Life

You've probably used a microscope to look at single-celled living things like those shown below. They can be found in pond water.

In the following Exploration, you'll look at *Protococcus* —an alga that forms the greenish stain on tree trunks, wooden fences, flowerpots, and buildings.

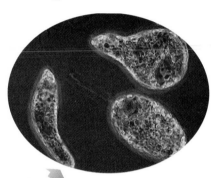
Euglena

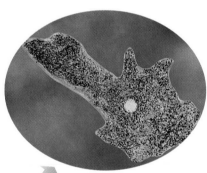
Amoeba

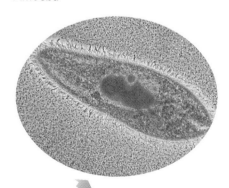
Paramecium

EXPLORATION 2

A Cell Has a Nucleus

You Will Need

- *Protococcus* (or other alga)
- a microscope
- water
- a microscope slide and coverslip
- an eyedropper
- a knife
- a plastic container with lid

What to Do

1. Locate some *Protococcus,* the green "moss" that grows in the places mentioned above. Scrape a small sample into a container. Bring it to the classroom and make a wet mount of it. If you can't find *Protococcus* outdoors, look for algal growth on the glass in an aquarium. Such algae may not be *Protococcus* but will serve your purposes here.

2. Use low power and then high power to examine the algal cells.

Making a wet mount

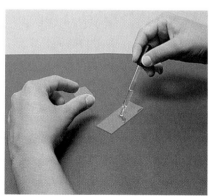

3. Draw both a single cell and a group of cells.

4. You'll probably notice that each cell contains several chloroplasts—the parts of the cell that are responsible for photosynthesis. One other structure that should be clearly visible in all of the algal cells is the nucleus. The **nucleus** of a cell controls most of the activities that take place in that cell. Find the nucleus in one of your cells and label it on your drawing. As you will soon discover, the nucleus contains the "recipe" for life. In what process, then, must the nucleus play an important role?

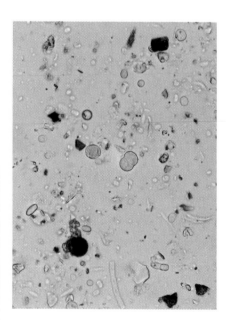
Microscopic view of *Protococcus*

Trouble in the Toy Factory!

Things are not going well at Joy's Toy Factory. It seems that things have gotten totally chaotic. Nobody knows what to do because nobody seems to be in charge. In the old days when the company was small this wasn't a problem. But the company has grown so much recently that it needs some kind of structure.

The toy makers drew up a statement to give to the owner of the factory, asking that a head office be set up. At right is the beginning of the diagram they prepared to describe what a head office could do for them. Copy the diagram into your ScienceLog and complete it to show what the responsibilities

of a head office could include. Take into consideration the problems illustrated in the cartoon, and add any other ideas you have.

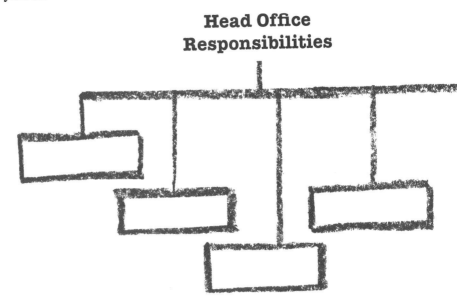

Head Office Responsibilities

The Cell Is Like a Factory

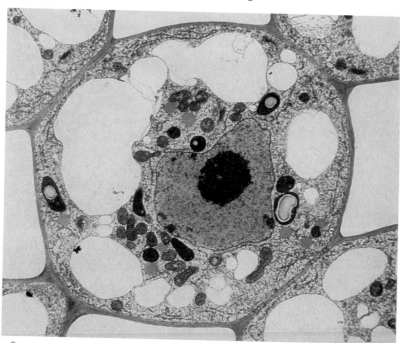

A root cell of a plant

The photograph above shows that a cell has many parts. Identify as many parts as you can. As in a factory, a lot of complicated action takes place within a cell. It needs direction from a kind of "head office." The head office of a cell is its nucleus. The nucleus in the cell shown above is certainly conspicuous. Just as a factory's head office holds the plans to build a new factory if necessary, the nucleus contains the blueprints for creating new cells.

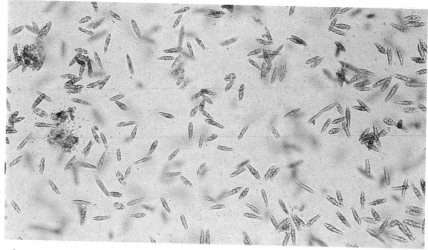

Euglenas

Consider an organism made of a single cell, such as a *Euglena*. A single-celled organism actually multiplies by dividing! When the organism is ready, its nucleus directs it to split into two identical cells, each of which contains all it requires for all of life's activities. This process is called **cell division**.

Look at the diagram of cell division below.

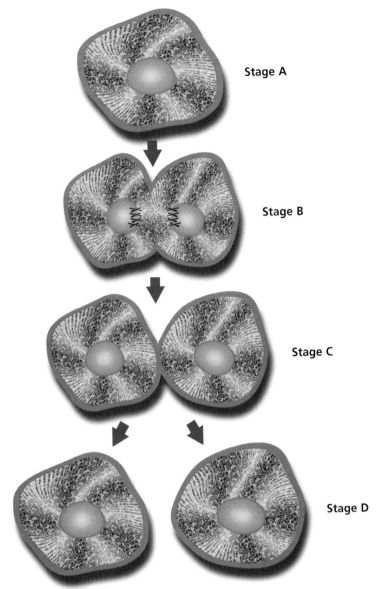

Stage A

Stage B

Stage C

Stage D

1. What happened to the "parent" of the daughter cells?

2. Draw a diagram to show the next cell division. (Draw stage *D* from the diagram, and add further stages as needed.)

3. Did you see any sign of cell division when you did Exploration 2? Look at your drawings for evidence.

Cell division also takes place in multicellular bodies such as yours. It happens when you need new cells for growth, repair, or other purposes. For example, skin cells are dividing continually to make new skin and to heal cuts.

Next, you will have a closer look at what happens inside a cell during cell division.

Not So Simple

Gloria and Yoon visited their town's new high-tech Science Discovery Center. Listen in as they talk to their guide, a microbiologist.

Guide: *During cell division, the materials within the cell, including the nucleus, organize themselves into two equal portions. As a result, the two new daughter cells are identical—and just like the "parent" cell.*

Gloria: *But wait! If you divide one apple into two equal parts, you end up with two half-apples. How can one cell produce two whole cells simply by dividing?*

Guide: *Good point! The split isn't at all simple. The two new cells are actually each a bit smaller than the original one. They'll gradually grow larger until it's time for each of them, in turn, to divide. Each of the resulting cells is a complete living thing that can carry out all of the activities necessary for life. To carry out these activities, they must take in some substances, such as food, and release others, such as wastes.*

Yoon: *So what does that have to do with cells dividing?*

Guide: *The new, smaller cells have a surface area that is large for their volume. This enables them to take in and release substances very well. When the cells grow in size, the increase in volume is greater than the increase in surface area. So their capability to take in and release substances lessens. Eventually, each cell must divide. But before the cell divides, the material in the nucleus doubles!*

Gloria: *No way! You're saying matter can appear from nowhere. That's impossible!*

Guide: *You're right. To see how the material doubles, let's use some special miniaturizing equipment that will make you small enough to enter a cell. So let's strap on the equipment and get going, okay?*

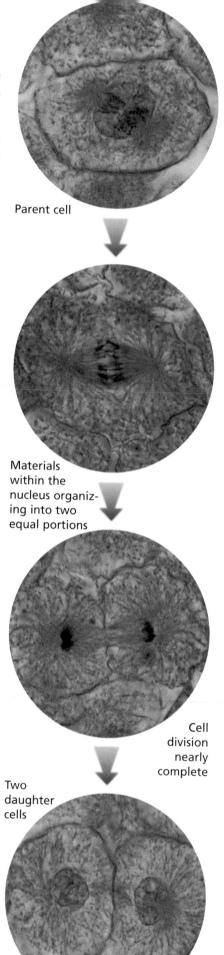

Parent cell

Materials within the nucleus organizing into two equal portions

Cell division nearly complete

Two daughter cells

Yoon: *You mean we are going to be cell explorers?*

Guide: *Right! You'll see what it's like to be a cytonaut. (*Cyto *means "cell.")*

The miniaturizing equipment worked fast. Gloria and Yoon soon felt like they were underwater, moving through a liquid containing many objects they'd never seen before.

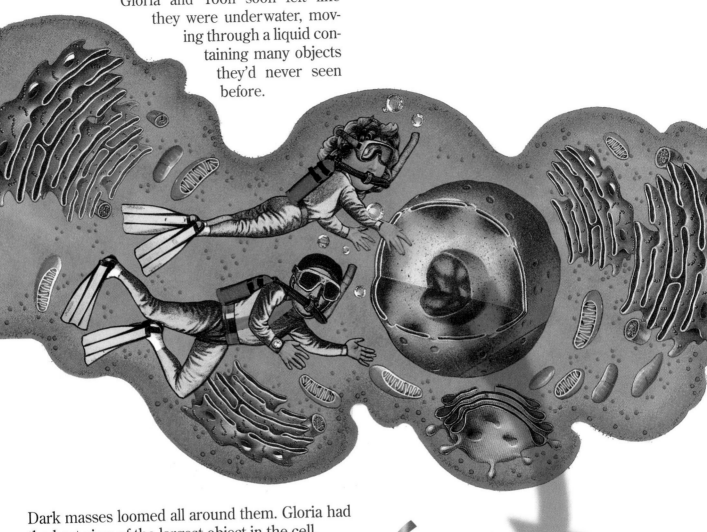

Dark masses loomed all around them. Gloria had the best view of the largest object in the cell.

Gloria: *What's that?*

Guide: *That's the nucleus. Watch it carefully. The cell is about to divide!*

Gradually, several long ladderlike objects came into view inside the nucleus. Each "ladder" was twisted in such a way that it looked something like a spiral staircase. Gloria and Yoon gasped as each ladder rung began to break apart near its middle (a). Right away, separate components that had been floating in the cell fluid moved toward the broken rungs and attached themselves (b).

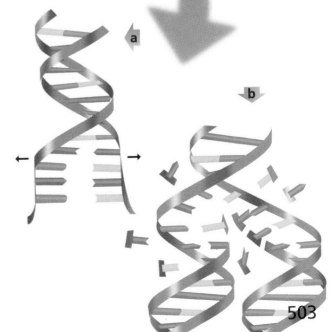

503

New, complete ladders took shape before their eyes. Now they saw twice as many ladders, but each spiral ladder was joined to its duplicate (c). "Like Siamese twins," Gloria thought.

Guide: *Look! The membrane around the nucleus is dissolving.*

Gloria and Yoon then watched as the spiral ladders lined up across the center of the cell. Suddenly, each "Siamese twin" separated, forming two sets of spiral ladders (d). Everyone watched as one set of spirals moved toward them, while the other set faded from view behind a film.

Guide: *Guess what? You're in one of the new nuclei now. A membrane has enclosed you in this new nucleus, keeping you separate from the other new nucleus, which is now in its own cell. You just witnessed cell division from inside the nucleus! Pretty cool, huh? Let's return to normal size now.*

The photos below represent stages (the formation of new nuclei that Yoon and Gloria saw) of a process called **mitosis**. You'll notice that there aren't any spiral ladders here. That's because the photos show the way you'd see cell division through a microscope. Microscopes simply aren't powerful enough to show these tiny ladders.

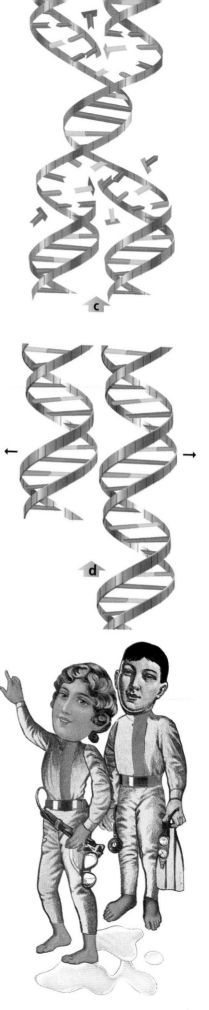

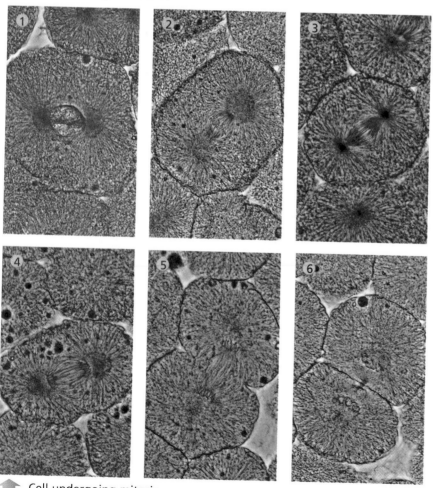

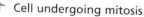
Cell undergoing mitosis

A Cytologist's Expertise

A **cytologist** studies cells and their structures. How well informed a cytologist are you right now? Check your expertise by doing the following matching exercise. Write eight sensible statements about cells in your ScienceLog by matching the words in Column I with the most appropriate words in Column II.

Column I

a. Cells are

b. Euglenas, paramecia, and amoebas are

c. Cells can

d. Nuclei are

e. Cell division

f. Before a cell divides, material in the nucleus

g. Our skin cells undergo cell division for

h. During cell division,

Column II

growth and repair.

produces new cells like the old.

control centers of life processes.

the material in the nucleus is equally divided.

the basic units of life.

reproduce.

single-celled living things.

makes a copy of itself.

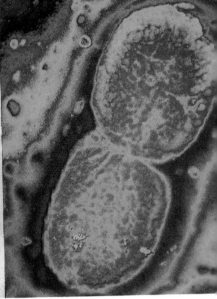

These *E. coli* cells are dividing. *E. coli* is a type of bacteria found in your digestive tract.

Twosomes

Cell division is performed by individual cells. It is a successful form of reproduction for many single-celled organisms, such as bacteria. Multicelled organisms (such as plants and animals) grow and repair damaged tissues by cell division, but they usually reproduce by a more complicated method. This method of reproduction requires two individuals, one male and one female. It is called **sexual reproduction**. In sexual reproduction, two cells—one from the male and one from the female—unite. The offspring, which forms from the union of these cells, inherits a mixture of traits from both parents. It's through this mixing of traits that diversity arises.

By contrast, the kind of reproduction that involves only a single parent is called **asexual reproduction**. Many single-celled organisms reproduce asexually by dividing into two cells. Each cell is then a new individual. Many plants also can reproduce asexually. For example, a new plant can be produced by planting a small stem that has been cut from another plant.

Going Further

Many single-celled organisms can reproduce sexually, but the process differs from sexual reproduction among multicelled organisms. How do single-celled organisms reproduce sexually?

Mendel's Factors

"What would you get if you crossed a snake plant with a trumpet vine?"

If you've ever heard jokes like this, you probably already know what "crossed" means. In the case of plants, it means that the pollen from one plant is used to pollinate another plant. When pollen from a flower's *stamens* (a plant's male organs) contacts the ovules in a flower's *pistil* (the female organ), parts of the pollen and ovules unite. Sexual reproduction has occurred, and eventually seeds are formed. We cross plants to produce seeds that will grow into plants with a desirable mixture of characteristics inherited from the parent plants.

By the way, the answer to the joke above is, *a snake in the brass*!

Far-Reaching Labors

What would you get if you crossed a tall pea plant with a short pea plant? (This time, the question is real—not a joke.) The answer is that you would get a second generation of tall pea plants. Why not short plants or plants that fall somewhere in between? In the 1850s, Gregor Mendel, an Austrian monk with a passion for gardening, became intrigued by such questions. He had become curious about pea plants because he'd noticed that in some patches of his pea plants, all the plants grew tall. In other patches, there were only short plants, and in a third type of patch, there were mixtures of tall and short plants. He spent the next 7 years growing pea plants and looking for answers to satisfy his curiosity. In doing so, he made some very important discoveries.

Gregor Mendel

Tall

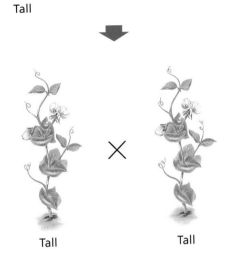

Tall   ×   Tall

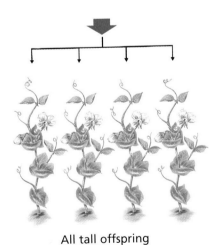

All tall offspring

As you read about the stages of Mendel's investigations, search for clues to explain the results.

1. Mendel planted the seeds of tall pea plants in one garden area. He planted the seeds of short pea plants in another area. Why didn't he plant all the seeds in the same area of the garden?

2. He wrapped the flowers of each plant with pieces of cloth. Why?

3. He collected the seeds of each group of plants and planted them the next spring in separate beds. What was he trying to find out?

4. Mendel repeated this procedure many times until he was satisfied that the seeds from tall plants produced only tall plants, and the seeds from short plants produced only short plants. Why did this matter?

5. Mendel then planted the seeds from the pure tall plants and the seeds from the pure short plants. When they bloomed, he transferred pollen from tall plants to the flowers of short plants and later collected the seeds. From which plants (tall or short) did he collect the new seeds?

Short

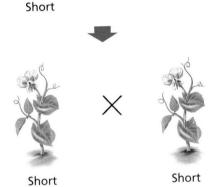

Short   ×   Short

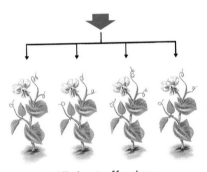

All short offspring

Pure tall   Pure short

6. The next year, the seeds produced by **cross-pollination** were planted. All the resulting plants were tall. What might be inferred from this? How could it be tested?

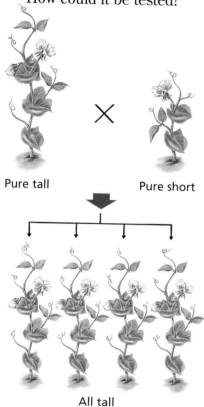

Pure tall × Pure short

All tall

7. Next, Mendel tried transferring pollen from short plants to tall plants, but the results were always the same—all the offspring were tall. How would you explain this?

Pure short   Pure tall

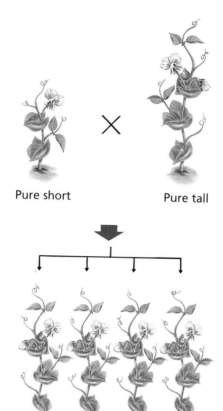

Pure short   Pure tall

All tall

To Mendel, it seemed that each male and female gave its own "instructions" for height to the seeds (and therefore to the next generation of peas). Why do you think he inferred this? He called the "instructions" *factors* and concluded that factors are inherited in pairs. For example, his pure tall parent plants contributed the factor for tallness, and his pure short parent plants contributed the factor for shortness. When two such plants were crossed, somehow tallness always "won." It seemed stronger, so Mendel called it **dominant**. He called the apparently weaker factor for shortness **recessive** because it seemed to recede into the background. Even though the recessive factor didn't show up in the offspring, Mendel suspected that it was still present, just hidden.

8. Finally, Mendel proceeded one generation further. He crossed the offspring that resulted from crossing a pure tall plant with a pure short plant. The results were surprising. From a total of 1064 plants, Mendel counted 787 tall plants and 277 short plants. He noticed a definite pattern among the test plants. The number of tall plants versus short plants can be expressed as a fraction. What is this fraction (rounded off)? Why was this such an important result?

Tall   Tall
(From cross between pure tall and pure short)

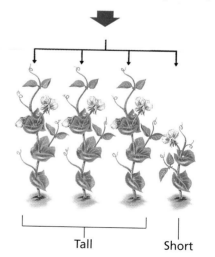

Tall   Short

Putting It All Together

Mendel's hunch was right. The recessive factor was still there among the offspring that had come from crossing pure tall plants with pure short plants. Mendel gave the name **hybrid** to offspring that contained one dominant factor and one recessive factor. The diagram on the right can be used to show what Mendel did. Write a one-sentence description for each section of the diagram.

Mendel went on to investigate other characteristics of pea plants, such as seed color. He found that the same pattern prevailed. Again and again, three-fourths of the second generation of plants (produced from the hybrids) showed the dominant trait, and one-fourth showed the recessive trait. Mendel was right. The recessive factor was still there in the hybrids, but it was hidden. It reappeared in the next generation.

Mendel experimented for many years. He wrote in one of his reports, "It indeed required some courage to undertake such far-reaching labors."

Mendel's work did not get immediate notice. However, his results eventually became the basis for the science of **genetics,** which is the study of how traits are passed from one generation to the next.

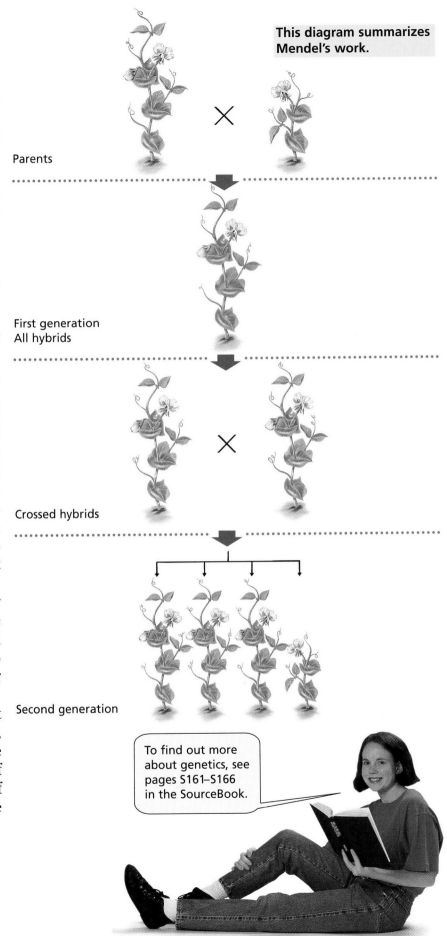

This diagram summarizes Mendel's work.

Parents

First generation
All hybrids

Crossed hybrids

Second generation

To find out more about genetics, see pages S161–S166 in the SourceBook.

Dominant to Recessive: Why 3 to 1?

Mendel observed that when hybrids are crossed, the next generation shows the traits in about the same mathematical relationship each time: three-fourths show the dominant trait, while one-fourth show the recessive trait. This 3-to-1 ratio occurred again and again in Mendel's studies of traits other than plant height. It didn't seem to appear by chance. The larger the number of plants involved in each test, the more clearly defined the 3-to-1 ratio became. (Why?)

Using pieces of paper to represent Mendel's factors, the following Exploration shows how likely the 3-to-1 ratio is. As you do the Exploration, remember that Mendel was working with living things made of cells. In Exploration 3, what part of a living cell might the bag represent?

EXPLORATION 3

Crossing Two Factors

You Will Need

- 200 squares of paper (about 2 cm × 2 cm)
- 2 paper bags or other opaque containers

What to Do

1. Mark 100 squares with a capital *T*. Let this stand for the dominant factor for height—tallness.

2. Mark 100 squares with a small *t*. Let this stand for the recessive factor for height—shortness.

3. Drop 50 squares with a *T* and 50 squares with a *t* into a paper bag labeled "Male Hybrid." Then drop 50 squares with a *T* and 50 squares with a *t* into a paper bag labeled "Female Hybrid."

Combinations	Number drawn
TT	
Tt	
tt	

4. Without looking into the bags, select one square from the Male Hybrid bag and one from the Female Hybrid bag. Place them together on a table.

5. Repeat step 4 until all the squares have been used up and all possible combinations of the male and female factors have been made. Set each possible combination *(TT, Tt,* or *tt)* in its own area of the table.

6. Count the number of pairs of *TT* combinations, *Tt* combinations, and *tt* combinations. Record the number of each in your ScienceLog, in a table containing the information shown at left.

Looking for Meaning

1. How many combinations of squares did you have?

2. In how many pairs did you have at least one *T*? What fraction is this of the total number of squares?

3. In how many pairs did you have only *t*'s? What fraction is this of the total number of pairs?

4. What is the ratio of pairs with at least one *T* to pairs with two *t*'s?

More Peas, Please

Mendel thought that each parent contributed, by chance, one of its factors for height to the peas. Suppose both parents supplied a *T*. The combination of *TT* would result in a tall plant in the next generation. The combination of a *T* from one parent and a *t* from the other parent would also result in a tall plant in the next generation, since *T* is dominant and *t* is recessive. If a plant inherited a *t* from each parent, however, the plant would be short. The diagram below illustrates this.

1. How many different combinations of factors are possible?

2. How does the number of combinations with at least one *T* compare with the number that have only *t*'s?

3. For the cross shown in the diagram below, how many tall plants would you expect for every short plant?

4. What did Mendel actually find in his investigations for this particular cross?

5. Refer again to the diagram outlining Mendel's investigations (page 509). How would you label each plant shown, using combinations of *T* and *t*?

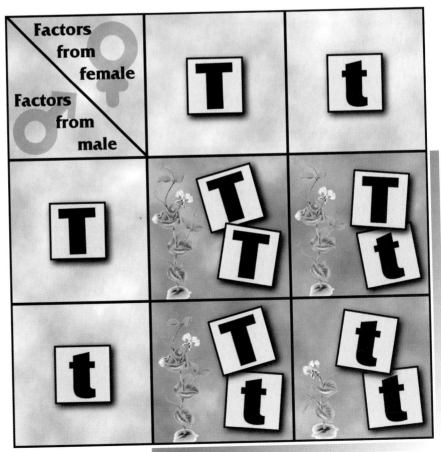

Possible Combinations in the Offspring

Making Punnett Squares

The diagram on page 511 shows all the possible combinations that can be passed on to the offspring of hybrid parent pea plants. Such diagrams are called **Punnett squares**. The passing on of traits from one generation to the next is called **heredity**. Punnett squares are commonly used in studying heredity.

Now it's your turn to practice working with Punnett squares. In your ScienceLog, draw and complete a Punnett square that shows the possible results of mating a black guinea pig with a brown one. Black is the dominant color, while brown is the recessive color. The black guinea pig in this cross is purebred for a black coat. To indicate the dominant factor for coat color, use the capital letter *B*. For the recessive factor, use the lowercase letter *b*.

What would be the results of crossing two hybrid guinea pigs? In your ScienceLog, draw and complete a Punnett square for such a cross. What ratio of black to brown coat color would you expect the offspring of these hybrids to show?

Brown

Black

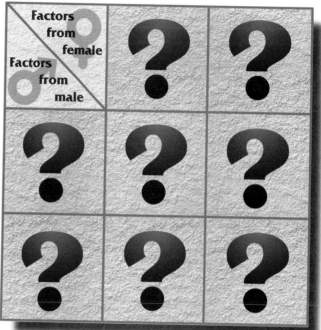

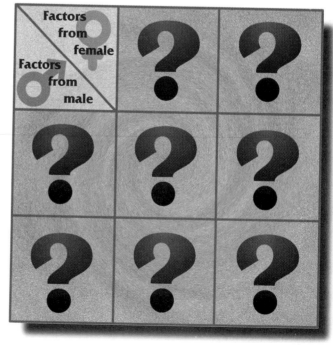

Predictable Ratios

In Exploration 3, you observed that there were three pairs with at least one *T* for every one pair that consisted only of *t*'s. This is a ratio of 3 to 1. Had you chosen only a few pairs instead of 100, would you have gotten this 3-to-1 relationship? Why or why not?

To arrive at his results, Mendel experimented with many different inherited characteristics of pea plants, such as seed shape and color, pod shape and color, stem length, and flower position. He kept careful records of his experiments and used these records to calculate the ratios in which the traits appeared. In every case, the offspring of hybrid crosses produced the 3-to-1 ratio. One of Mendel's chief contributions was showing that many inherited traits would appear in predictable mathematical patterns.

Thinking Back

Now look over the data you recorded for Exploration 1. All the *yes* answers indicate dominant traits. That is, every *yes* answer you gave shows that you have the dominant expression of that characteristic. Every *no* answer shows that you have the recessive expression of that characteristic. If you show the recessive expression of a characteristic, what kind of factor did each of your parents contribute? Would the same be true if you show the dominant expression of a characteristic? Explain.

Going Further

1. Should you expect there to be a 3-to-1 ratio of dominant to recessive traits for the characteristics you observed in Exploration 1? Why or why not? Suppose everyone in your school completed an index card. For each trait assessed, how do you think the number of dominant to recessive traits would compare?

2. Having a cleft chin is a dominant trait. So is having six fingers. Do you think that three-fourths of the students in your school have cleft chins? How about six fingers? How would you explain your observations?

3. Mendel's pea plants were either tall or short. How does this compare to the heights of your classmates? How might you explain this?

As you can see, studying human inheritance factors is much more complex than studying Mendel's pea-plant factors.

Michael Douglas (top) and his father Kirk show off their famous cleft chins. A cleft chin is a dominant trait. Why, then, don't most people have it?

And Where Did You Get That Nose?

Now use what you've learned to add to your family-tree project. Using the questions you answered in Exploration 1, interview or research as many members of the family you chose as possible. Keep a careful record of the information you find.

The diagram on the right is similar to the family tree you drew earlier, but this one traces the inheritance of a particular trait in a family—dimples. Using a separate diagram for each trait, note as many different traits as you can for the family you are studying. Each of these diagrams is called a *pedigree*.

Are there any family members who have characteristics that are not possessed by either of their parents? What can pedigrees tell you about traits that are passed on from generation to generation? Pedigrees are not usually used for studying the inheritance of traits in plants. Why are pedigrees used for studying the inheritance of human traits?

A Modern Look at Mendel's Factors

You know a lot now about dominant and recessive factors. By now, too, you've seen how certain physical traits are *inherited,* or passed from one generation to the next. What are the mysterious *factors* that determine these traits? It took scientists quite a while before they discovered an answer to this question.

Toward the end of the nineteenth century, certain structures in the nucleus of a dividing cell were seen through a microscope. Because these structures easily absorbed dye or stain, they were called **chromosomes** (meaning "colored bodies"). Each type of living thing has a specific number of chromosomes in its cells. The chromosomes are in pairs, like Mendel's factors. Human body cells, for example, contain 23 pairs, or a total of 46 chromosomes.

By 1910, Mendel's factors were identified as individual sections of chromosomes. These components were named **genes**, after a Greek word relating to "birth." Each chromosome in a pair carries one of the two genes that determine a trait. Altogether, human chromosomes carry at least 100,000 genes. Genes not only determine an organism's traits, but also provide instructions to the organism's cells as it grows.

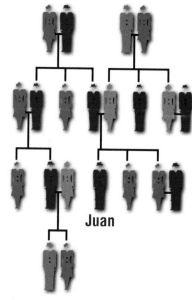

Juan

KEY

	Female with dimples		Female no dimples
	Male with dimples		Male no dimples

A pedigree of the dimples trait in Juan's family

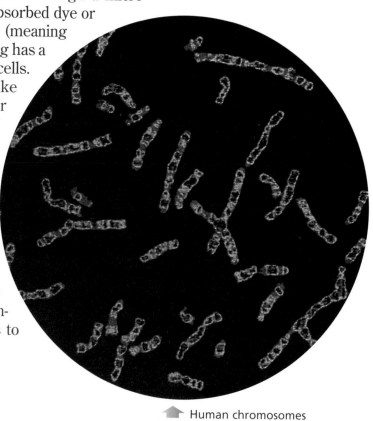

Human chromosomes

Sex Cells

When sexual reproduction takes place, a cell from one parent unites with a cell from the other parent to form a new cell. This process is called *fertilization*. In flowering plants, pollen contains a sperm nucleus (the male contribution), which unites with an ovum, or egg cell (the female contribution), inside an ovule. In the animal kingdom, a sperm from a male unites with an ovum, or egg, from a female. Eggs and sperm are also called **sex cells**.

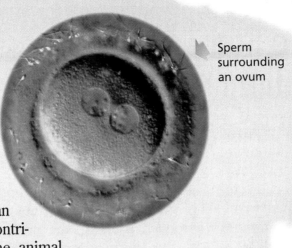

Sperm surrounding an ovum

Electron micrograph of sperm penetrating ovum (background)

Now here is a problem! If each human sex cell supplied 46 chromosomes, the offspring would develop from a total of 92 chromosomes. But human cells have only 46 chromosomes. What happens to make this number remain constant, generation after generation? Through a complicated process, sex cells with only 23 *single* chromosomes (one chromosome of each pair) are produced. This way, when the two sex cells unite, the resulting cell has 23 pairs of chromosomes, or a total of 46 chromosomes—the correct number.

The production of sex cells begins like ordinary cell division: the material in the nucleus doubles. The later stages cause the chromosomes to be distributed among four cells, instead of between two cells.

At right is a simplified view of what happens when sex cells form.

If this were a human cell, it would contain 23 pairs of chromosomes. Only three pairs are shown here so that you can more easily see what is happening. Look at the illustration for each step. What is happening to the number of chromosome strands at each step?

In a male, these four cells all become sperm. In a female, only one of the four cells becomes an egg cell. The other three disintegrate.

You have just observed the process that results in the formation of sex cells. This process is called **meiosis**.

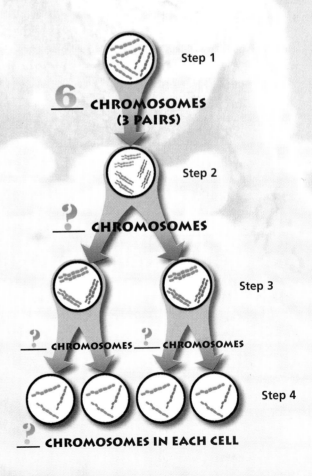

Step 1

__6__ CHROMOSOMES (3 PAIRS)

Step 2

__?__ CHROMOSOMES

Step 3

__?__ CHROMOSOMES __?__ CHROMOSOMES

Step 4

__?__ CHROMOSOMES IN EACH CELL

Life Story of the Unborn

The development of a new human life is a remarkable process. At five different points in this lesson, you will pause to observe the development of an *embryo*. It all began with the fertilization of an egg cell by a sperm cell. Cell division then began, and from there, all kinds of things started happening.

But first things first. It really begins with two sex cells, each formed through the process of meiosis. Female sex cells (egg cells) are produced in the female's ovaries, while male sex cells (sperm cells) are produced in the male's testes. Each sex cell has only 23 single chromosomes (as opposed to the 23 pairs, or 46 chromosomes, in a body cell). Do you remember why there's a difference? To examine the events that come before fertilization and the development of an embryo, read Before You Were You, and examine the diagrams on this page. For each numbered item, write a brief description of the item's function in your Sciencelog.

Before You Were You

The sperm cells are placed in the vagina (1) by the male reproductive organ, the penis (2). The sperm move upward through the uterus into the Fallopian tube (3), where one sperm cell (4) may fertilize an egg cell (5) that has been released from the ovary (6). The fertilized egg cell begins to divide (7) as it moves into the uterus (8). There, perhaps 1 week after the egg has been fertilized, the embryo embeds itself (9) in the wall of the uterus.

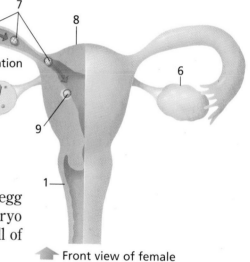

Front view of female reproductive system

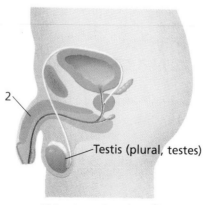

Side view of male reproductive system

Testis (plural, testes)

Side view of female reproductive system

This diagram shows some details of the embryo in the uterus wall.

1. What do you think happened at (10)?

2. What role might the amniotic cavity (11) play?

3. What observation can you make about the cells of the embryo (12) at this stage?

4. What part do you think the yolk sac (13) plays?

5. Why is it important that the mother's blood vessels (14) be near the embryo?

The life-support system for the embryo (15) is called the **placenta** (16). It is an organ that develops around the embryo and that attaches the embryo to the wall of the uterus. The placenta is rich with blood vessels that come from both the mother and the embryo. However, the blood of the mother does not mix directly with the blood of the embryo. Instead, substances are exchanged between the mother's blood and the embryo's blood across the walls of the blood vessels in the placenta.

As the embryo develops, it grows an umbilical cord (17) that connects the embryo to the placenta. Substances enter and leave the embryo through this cord. What substances are needed by the embryo? What substances must leave the embryo?

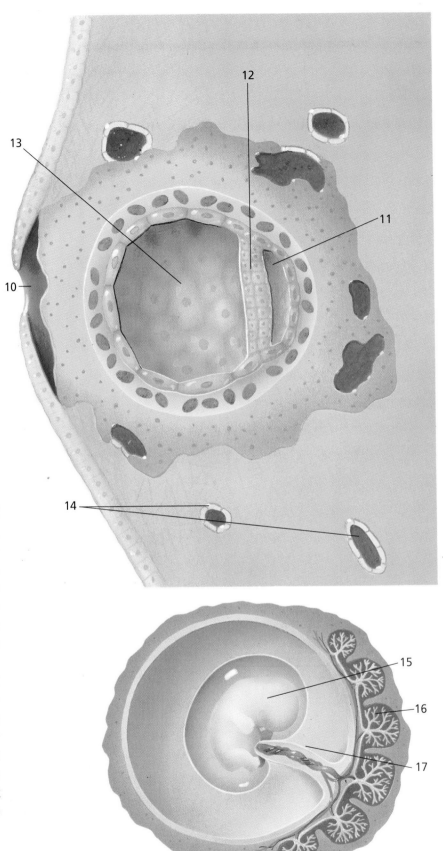

Compare these two diagrams. What necessary part of the life-support system has not yet developed in the upper diagram?

Your Early Development

Now consider what you already know about your own development. In your ScienceLog, place the following happenings in the order you think they occurred. Decide, for example, what might have happened during the first month, second month, and so on. The list includes abilities and activities.

a. Your fingers and toes took shape.

b. Your heart began to beat.

c. Your memory began.

d. Leg and arm stumps appeared.

e. Your eyes could respond to light.

f. Your mother was aware of your movement.

g. Your baby teeth started to take shape, and tiny buds for your permanent teeth began to develop.

h. You began the cycle of sleeping and waking, and you developed the ability to dream.

i. Your sex was clearly indicated by internal sex organs.

j. You developed fingerprints.

k. Your face and neck began to develop.

As you follow each episode in Adventures of a Life in Progress, refer to your list to see how well you imagined your own beginning!

⬆ A human being in the making

Adventures of a Life in Progress

Episode 1—The First Month

Your life started with the fertilization of an egg cell by a sperm cell. The egg was about 0.1 mm in diameter—even smaller than the period at the end of this sentence. The sperm that fertilized the egg was only one-thirtieth the size of the egg. After fertilization, you began to divide, and things began to happen like clockwork. Later, cells with a specific shape and function began to appear.

By the age of 17 days, blood cells had formed, and shortly thereafter, you had a heart tube. Your heart began to twitch and began the rhythmic beating that must continue until your life's end. By the end of the third week, the cells that would eventually form your sperm or eggs were set aside. In the fourth week, a tube was formed in what became your mid-back region. At the front end, your brain later developed. The back part of the tube later formed your spine. Next came your food canal. By the 25th day, you had a head end and a tail end (and a bit of a tail also). You had no face or neck, and your heart lay close beside where your brain would eventually be.

By the end of the first month, you had tiny bumps where arms and legs would later develop. Your lungs, liver, kidneys, and most other organs had begun to form. Your head had two pouches that would become eyes, two sunken patches of tissue where the nose would form, and a sensitive area behind each eye where the inner ear would form. A lot happened in 1 month. Can you imagine what size and shape your embryo was by then?

b

a

d

c

g

e

f

⬆ In what order would you place these drawings to illustrate the beginning of a new human life?

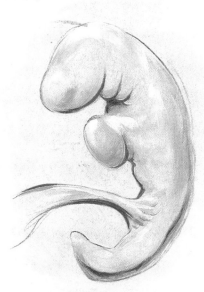

⬆ This is what you looked like at 1 month of growth in the womb, shown about 10 times actual size.

Jean Hegland recorded in a journal the thoughts she and her husband had as they awaited the birth of their child. She published her thoughts in a book called *The Life Within*. Here is a part of her journal for the first month.

> Then comes heart, and spinal cord, and gut, all primitive, all bulging and shifting like a rose opening in a time-lapsed film, all arising out of that speck of protein that was two cells, that became one, that doubled and divided, transforming from one thing to another as wildly and exactly as objects in a magician's show, where rabbits become scarves that become tulips and then coins, in a perfect, dizzying succession of change. And so these organs, these basic bits of human being, appear like rabbits out of nowhere—out of the intention of the universe—and a new creature takes shape.

What sort of instructions are being followed to make all of the embryo's body parts form at a certain time, in a certain way? Compare the events of the first month with the list of developments you wrote in your ScienceLog. Were your predictions close?

Helping the Natural Process

In 1978 a baby's birth made history. That baby's name was Louise Brown, and she was the first "test-tube" baby. Did she really grow in a test tube? What does the term mean? Why was this test-tube procedure done?

Sometimes, certain problems interfere with the natural process of fertilization. **In vitro fertilization** is a procedure that allows certain people to have children. Find out what this term means and why the procedure is performed. As you investigate this form of reproductive technology, as it is called, ask yourself some basic questions, such as the following:

1. Why would someone choose to try (or not to try) this procedure?

2. What are the costs, in terms of
   a. risks to parent or baby?
   b. money?
   c. time?

3. How successful is in vitro fertilization likely to be?

Louise Brown—the first "test-tube" baby—with her younger brother, also conceived by in vitro fertilization.

Outline a brief presentation that you could give to someone interested in the procedure. Include the pros and cons of in vitro fertilization and back up your statements with information gathered in your research. End by giving your recommendation for guidelines as to when in vitro fertilization is appropriate and when it is not.

Twin Heartbeats

Sometimes two embryos develop at the same time. The resulting children may be so much alike that even their parents have trouble telling them apart. These are called *identical twins*. Sometimes, though, twins are no more alike than any other siblings (children of the same parents). These are *fraternal twins*. Why are twins sometimes identical and sometimes fraternal? Consider two possibilities.

A The mass of cells from a single fertilized egg separates into two halves early in development. Two babies result.

B Two eggs are released by an ovary. They are fertilized by two different sperm cells. Each is implanted in the uterus and continues to grow. Two babies result.

A

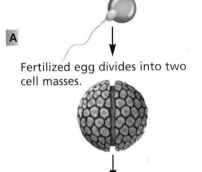

Fertilized egg divides into two cell masses.

Two cell masses develop in one sac.

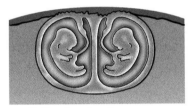

Twins develop in same sac.

B Two eggs are fertilized by two sperm and divide.

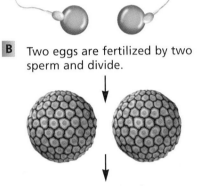

Two masses of cells develop separately, each in its own sac.

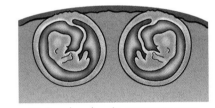

Twins develop in separate sacs.

What Do You Think?

Think about the two possibilities above. Do they tell you why twins are sometimes identical and sometimes fraternal?

1. Which instance, *A* or *B*, would produce two identical babies?

2. It is possible at birth to tell whether twins are identical. How?

3. Could identical twins be (a) both boys, (b) both girls, or (c) one boy and one girl? Provide an explanation for your answers.

4. Could fraternal twins be (a) both boys, (b) both girls, or (c) one girl and one boy?

   Provide an explanation for your answers.

Episode 2—The Second Month

This month was a time of rapid growth. You grew about 5 times longer and about 500 times heavier. Your face and neck developed. Your brain developed at a faster pace than any other part of your body. Why do you think this happened?

By the end of the second month, your sex was clearly indicated by internal sex organs, which could probably have been seen externally too. Bones and muscles began to form. The arm and leg buds that formed last month began to grow into limbs, and your fingers and toes took shape. Gradually, constrictions formed to mark off elbows and wrists, knees and ankles.

Your bones first consisted of a soft substance called cartilage. Your bone structure changed throughout your prenatal (before birth) life, and it will continue to change in your postnatal (after birth) life. In fact, not until you become a mature adult will your skeleton be fully formed.

How big were you by the end of the second month? How many events did you predict correctly?

A 5-week-old embryo

At 6 weeks, you were little more than 1 cm long. The dark masses are this embryo's heart and liver.

As an 8-week-old embryo, you were only the size of a walnut, but you had all of your basic organs and systems. Your head was almost half your total size.

Sometime during this month, this bit of a child, this baby grown to the size of a peanut, will begin to move, to flail its stubby hands and stretch its delicate legs . . . I wonder what it must be like to bend a brand new elbow, to wiggle freshly made toes . . .

And then there are the thousands of things that could go wrong . . .

Each dividing cell, each growing organ must follow the 4-million-year-old plan exactly—the first time. There are no second chances. There is no going back, no catching up, no fixing mistakes. . . .

. . . we have been warned about Thalidomide, radiation, rubella. We have been told about nutrition and exercise, and what the Chinese knew a thousand years ago, that the child of a tranquil mother is calmer, brighter, and weighs more at birth. And so I drink milk for the sake of this creature's bones, eat protein for its mushrooming brain. I avoid alcohol, aspirin, and stress . . .

Opinions and Viewpoints

As you have seen, the developing embryo is very delicate. What are some things that can affect its development? What are the parents' responsibilities to the developing embryo? This subject has prompted quite a bit of research and thought. Use the following questions to help you form your own informed opinions and viewpoints.

1. How dangerous is it for pregnant women to drink alcohol, to smoke, or to abuse drugs? What can happen to the developing embryo as a result of these activities? Find out all you can about this topic by consulting your library for books, magazines, and newspaper articles on the subject. For additional assistance, ask a nurse or doctor.

2. What is rubella? Investigate how a pregnant woman's exposure to this can affect the developing embryo.

3. Thalidomide was a drug used in some countries for a short time to help pregnant women who had nausea during the first months of pregnancy. It had disastrous results.

   a. Do some research about this drug and the tragedy of its use.

   b. What evidence indicates that it was used during the second month of pregnancy?

4. How does stress experienced by the mother affect the developing embryo?

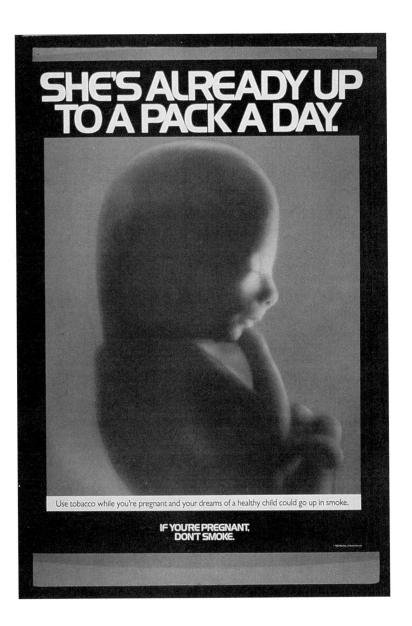

SHE'S ALREADY UP TO A PACK A DAY.

Use tobacco while you're pregnant and your dreams of a healthy child could go up in smoke.

IF YOU'RE PREGNANT, DON'T SMOKE.

Episode 3—The Third Month

From the third month on, the developing child is called a **fetus**. A male fetus undergoes a lot of sexual development during the third month. This could also be called the "tooth month" because your first teeth started to take shape, and tiny buds for your permanent teeth began to develop. Your vocal cords formed during this month. Your digestive system, liver, and kidneys all began to function. Bones and muscles developed, and your face and body began to take on their distinctly human form.

After 3 months of growing, you were about 7 cm long.

. . . its umbilical cord connects our Riddle to its placenta like an astronaut to his spacecraft or a deep-sea diver to her boat. A placenta is a weird, dense pudding of an organ, just thicker than a beefsteak and roughly round as a dinner plate. On one side, the pale cable of umbilical cord rises from it like a tree trunk from a rooty tangle of blood vessels, and on the underside, those veins have thinned to tiny capillaries that entwine with the capillaries in the wall of my uterus. There, this fetus and I communicate in . . . the language of diffusion, so that its needs are met, molecule by molecule, before it knows them. There, its blood rushes constantly to meet mine, to strip it of its lode of oxygen and nutrients, and give me in exchange carbon dioxide, uric acid, wastes.

. . . After it is born, and the cut stump of its umbilical cord sloughs off, this baby will be scarred forever with the pretty pucker of its navel, a remembrance of the communion it once took from my blood.

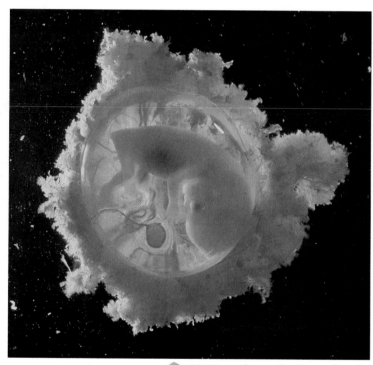

At 11 weeks, each of your hands was about the size of a teardrop, and your body weighed about as much as an envelope and a sheet of paper.

You have now reached the end of the first **trimester** of the developing life. Review what has happened. How does the order of events compare with what you predicted on page 518?

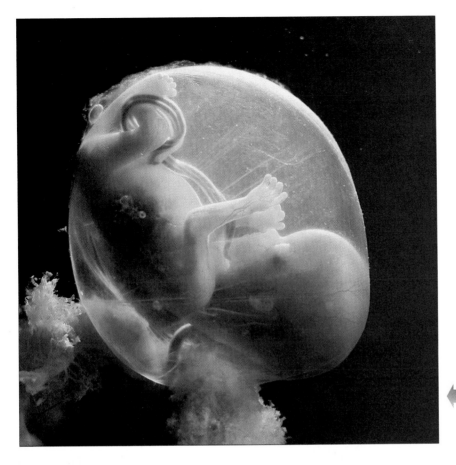

By the 12th week, your eyelids closed and remained shut for about 3 more months.

Analysis and Reflection

1. a. The writer of *The Life Within* calls her unborn child a "Riddle." Why do you suppose she does that?

   b. Explain "the language of diffusion." Where are the membranes that permit diffusion? What substances diffuse? How is this like the process in a hen's egg?

   c. You may have seen the "cut stump" of an umbilical cord on a newborn. What does "sloughs off" refer to?

2. If a pregnant woman smokes cigarettes, do any of the substances from the smoke pass through the placenta into the fetus? Find out about this.

3. Suppose someone said, "A pregnant woman can do whatever she wants to do. She can smoke, drink, take drugs, and not take care of herself, if that's what she decides to do." Would you agree or disagree with the statement? Gather information to support your opinion.

Episode 4—The Second Trimester

A 4-month-old fetus

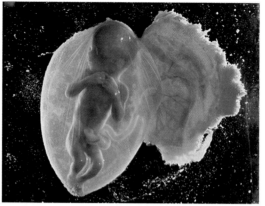

Fetus at $4\frac{1}{2}$ months

Your eyes (sealed at the age of 3 months) reopened during the sixth month. You had more taste buds on your tongue and in your mouth than you have now.

In the second trimester of development, your fingerprints and toeprints formed (month 4), followed by oil and sweat glands, scalp hair, nails, and tooth enamel (month 5). Your eyebrows and lashes continued to grow (month 6).

Your mother first became aware of your movement during the fourth month. You were half your birth height (a third of that was the length of your head), and you were wrinkled and dark red in color.

During the fifth month, some of your skin cells died and fell off. They were replaced by new ones. A mixture of the dead cells and a substance from the skin glands formed a protective layer over your skin. Your mass was about half a kilogram. Your head was well balanced, and your back grew straighter.

They say that already your face, with its newly finished lips, is like no one else's face . . . a twenty-week-old fetus is a porcelain doll of a child, complete with fingernails and eyebrows and a tender fringe of lashes sprouting from the lids of its still-sealed eyes. Fragile, luminous, it seems an unearthly being—more sprite than hearty human—through whose skin shines an intricate lace of veins.

. . . Always when it moves I know it now. It pummels my kidneys, kicks my stomach, punches my lungs. I know when it is resting, and when it wakes and stretches. I know it well by now . . . It knows me too. It dangles . . . inside me . . . listening to all my doings. It hears all that I hear, the sounds intensified and distorted like the . . . racket one hears under the water in a swimming pool.

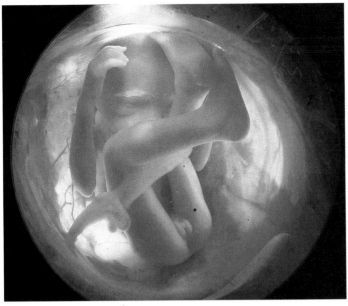

By 22 weeks, your mother could feel your movements, and she may have even been aware of your sleeping and waking cycles.

Thought-Provoking Questions

1. Why do you suppose the features of the fetus are already unique at such an early age?

2. Why do some mothers-to-be read and talk to their growing fetuses?

Episode 5—The Third Trimester

In the seventh month, your eyes could respond to light, and the palms of your hands could respond to a light touch. Fat began to fill in the wrinkles during the eighth and ninth months. By the end of the seventh month, your mass was about 1.4 kg. If you had been born at this point, you could have survived as a premature baby. Your activity increased, and you might have even changed position in the uterus. During this trimester, you were "practice-breathing," even though your lungs were filled with fluid. If your mother smoked just *one* cigarette, the practice-breathing stopped for up to half an hour.

Medical researchers say that memory begins during the seventh month . . . I wonder what this child will remember . . .

. . . neurologists claim that by the eighth month this fetus will have begun a cycle of waking and sleeping whose rhythms are already influenced by the sun. And . . . when it sleeps, it dreams—or at least its brain waves resemble those of a dreamer.

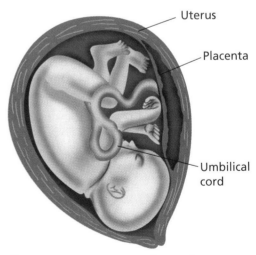

By the end of the ninth month, a fetus turns itself upside down, or into a head-down position. The umbilical cord supplies the fetus with food and oxygen and removes waste until after birth, when the cord is tied and cut.

Birth

At the end of pregnancy, the combined effects of several hormones and the stretching of the uterus cause the uterus to begin to contract with strong muscular movements. With these movements, the baby is pushed out of the uterus. Jean Hegland suggests, "For a baby, contractions may be caresses that express the fluid from its lungs and stimulate its skin in the same way that other mammals . . . do by licking their newborns."

At birth, you gasped, filled your lungs with air, and cried. The placenta and umbilical cord you no longer needed were cut away. When your wound healed, you were left with the scar of the navel. The navel has been called a "permanent reminder of our once-parasitic mode of living."

People have very different life experiences, but everyone shares the prenatal phase, which forms the first chapter in the personal history of each human being.

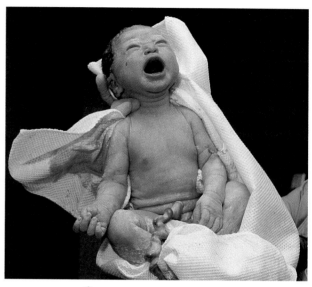

Within seconds after birth, a baby takes its first breaths.

Going Further

1. When a newborn baby cries, are there tears? How do you explain this?

2. How old must a baby be before starch can be added to its diet? Why?

3. What eye development has to take place after birth?

4. Explain what you think Jean Hegland meant when she wrote, "In the years to come, when we celebrate this Riddle's birthday . . . we will be commemorating little more than a change in its means of respiration and nutrition."

5. A sizable percentage of babies in the United States are delivered by *Caesarean section*. Find out what you can about this procedure. How is it performed? What are the advantages of delivery by Caesarean section? the disadvantages?

During pregnancy, the mother's breasts prepare for the production of milk. After the baby is born, breast milk can provide all the nutrients needed for almost the entire first year of a baby's life.

Be a Witness

Find out more about mammal birth by watching it or by talking to someone who has seen a mammal birth. Here are some suggestions for what you might do.

1. Plan a visit to a farm, an agricultural fair, or a kennel to watch the birth of a mammal, such as a colt, a calf, or a puppy.

2. Interview a pregnant woman during her pregnancy and later, after her child's birth.

3. Interview a father, both before and after his child's birth. Many fathers are present during the birth of their children.

4. Watch a film of a human birth.

I Want to Know!

Patrick's class arranged for a visit from an obstetrician (a doctor specializing in human birth). Everyone was supposed to do some reading in advance and come up with some good questions. How might the obstetrician have answered them?

1. I read that a newborn has no bacteria living in its digestive system. Is that true? How do we get the bacteria?

2. One book said the fetus lives the life of a parasite. Do you agree? Explain.

3. Are the mother and fetus really separated, such that they don't touch? Would the mother's body really reject the fetus if they were touching?

4. If the mother's blood does not pass through the developing fetus, how does the fetus get what it needs?

5. I read that the fetus is floating and weightless, like an astronaut in space. Can that be so? Is there any advantage to that?

6. Someone told me that viruses and poisons can get into the fetus from the mother. Is that true? Can the AIDS virus do that?

What other questions would you like to ask? Make a list. Arrange to invite a doctor or another health professional to your class. You will likely make a lot of new discoveries.

CHALLENGE YOUR THINKING

1. Family Reunion

Could you draw a family tree for single-celled organisms? Could such organisms hold a family reunion? Why or why not?

2. A Contradiction in Terms

Explain to a 9-year-old how this statement can be true: Some living things multiply by division.

3. Spontaneous Generation Revisited

Micki told her mother this morning, "There are a bunch of fruit flies around the bananas in the cupboard, Mom. They just appeared—out of nowhere!" Is Micki right? Did the fruit flies grow from the bananas? Explain what must be happening, and suggest how she can keep fruit flies from becoming more numerous.

4. It's Alive!

What life-sustaining characteristics are shown when

a. your body digests food?

b. a fox chases a rabbit?

c. a dog pants?

d. a salamander grows a new tail?

e. a fly lays eggs on food?

f. a wound heals?

5. Sex Cells and Body Cells

a. How do sex cells differ from the rest of the body's cells?

b. Why are the following processes necessary?

- mitosis
- meiosis
- fertilization

6. Playing Catch-Up

A baby born after only 8 months of development has about a 70 percent chance of survival. It has to "catch up" to be as strong as a 9-month-old fetus. What are two improvements in its condition that must be accomplished by a premature baby?

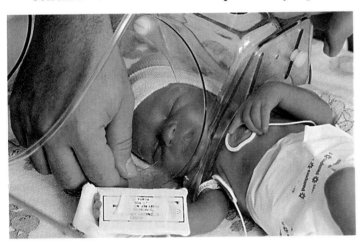

ScienceLog

Review your responses to the ScienceLog questions on page 490. Then revise your original ideas so that they reflect what you've learned.

Instructions for Life

1 Write down as many ways of giving instructions as you can think of.

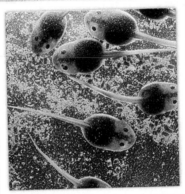

2 What kind of cell instructions are necessary when a new life begins?

ScienceLog

Which has a greater effect on a person, inherited characteristics or the circumstances of the person's life?

3

Nature or Nurture?

Think about these questions for a moment, and answer them in your ScienceLog. When you've finished this chapter, you'll have the opportunity to revise your answers based on what you've learned.

Living With Instructions

Miguel arrived home looking quite pleased.

"I got a model kit on sale—75 percent off the regular price," he announced, as he hurried to open the cardboard carton. "Look at this," he said, as dozens of parts fell out onto the table.

"Oh no! There are no instructions—not even a diagram," he moaned. "I know it's a model of an airplane, but I'm not sure what it's supposed to look like. The illustration and model number on the outside of the box don't help me much."

Miguel called the store for help. The store manager said, "You'll have to send away for the instructions—to the head office."

What sort of instruction sheets should Miguel expect to get?

Instruction sheets for model kits have illustrations that show what the pieces look like, diagrams and text that explain how to assemble them, and pictures that show what the completed model should look like.

All Kinds of Instructions

Not all instructions include illustrations and diagrams. Some instructions use only words, or a combination of words and numbers. Computers, for example, receive instructions in the form of an alphanumeric code. Earlier, you used holes in cards as a source of information about traits. The holes and cut-away sections did not resemble any of your traits, but they did pass on information. What mechanical means did you use to retrieve, or *translate,* the information? What information did you get? What other examples of instructions can you think of that do not use pictures?

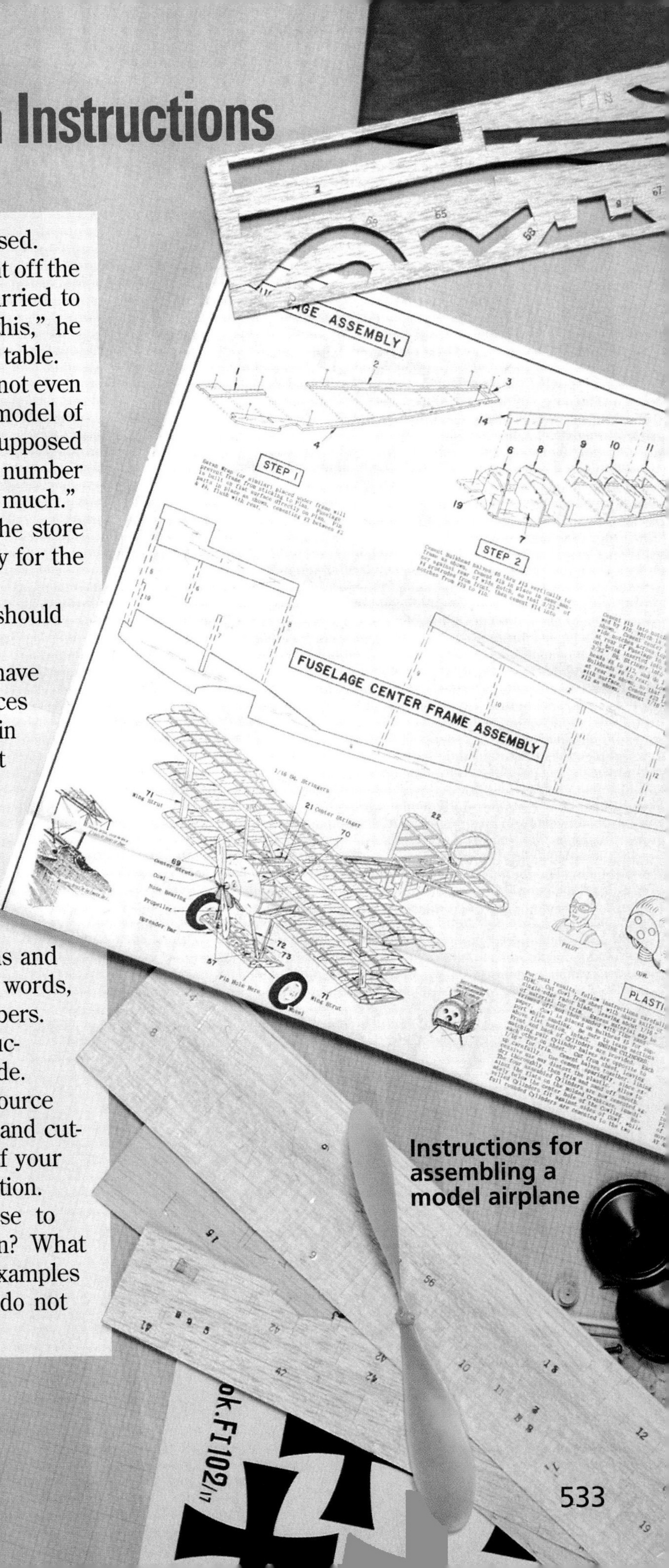

Instructions for assembling a model airplane

Look at the following examples of sets of instructions. For each set of instructions, consider these questions:

1. What do you think the instructions are about? Do the instructions look like the end result?

2. What notation (set of characters) does each use? How is the notation arranged?

3. How do you think each set is translated so that its instructions can be followed or its information acted upon?

In the examples below, information is provided in a series of symbols or characters arranged in rows or lines. These sets of instructions are forms of *linear programming*. Which sets of instructions were developed for recent technology? Which have been used for centuries?

The opening bars of Beethoven's Sonatina in F major

Sonatina in F Major

Allegro assai

Ludwig van Beethoven (1770–1827)

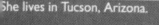
She lives in Tucson, Arizona.

ISBN 0-06-092253-2
90000

Barcodes contain coded information that can be read by a computerized scanner.

dc in next sc, repeat from * to end of rnd. Then ch 1, and join with sl st to 3rd st of ch-4 first made. 9th rnd: Ch 10, and complete cross st as before, skipping 2 dc between each leg of cross st and inserting hook under ch-1 sp. Ch 3 between each cross st. Skip 2 dc between each cross st. Repeat around (20 cross sts) and join last ch-3 with sl st to 7th st of ch-10 first made. 10th rnd: Sl st in 1st 2 sts of 1st sp, ch 5, * dc in same sp, ch 3, sc in next sp, ch 3, dc in next sp, ch 2, repeat from * to end of rnd. Join last ch-3 with sl st to 3rd st of ch-5 first made. 11th rnd: Ch 6, dc in same sp, * ch 4,

A few lines from some crochet instructions

```
IF UC = 4 THEN GOTO 220
IF UC = 7 THEN GOTO 250
IF UC < 4 OR UC > 4 OR OR UC < 7 OR UC > 7 THEN
GOTO 165
UL$ = "CORMORANTS & SEAGULLS": CL$ =
"UPPER":TC$ = "WATERBIRDS":U = 5:AR = 36521
X$ = "CORMORANTS SEAGULLS, BLUEFISH HERRING
SARDINES SQUID, MOLLUSKS URCHINS STARFISH
WORMS, CRUSTACEANS INSECTS, BACTERIA ZOO-
PLANKTON, ALGAE PHYTOPLANKTON PROTO-
ZOANS,"
GOTO 280
UL$ = "WHALES & TUNA":CL$ = "FOURTH":TC$ = "FISH
& MAMMALS":U = 4:AR = 3521
```

This shows part of a computer program that keeps track of the food chains in a particular habitat.

A passage in Braille. Blind people read Braille by running their fingertips over the dots.

The Instructions That Shape Life

Like Miguel and his model airplane, scientists often have to work without instructions. For example, biologists spent many years looking for the instructions that guide the development of new living things. They needed to find out whether there was a "head office" where the instructions were kept, what notation was used, how the notation was arranged, and how it was translated. In the next Exploration, you will follow the story of how they found answers to some of these questions.

EXPLORATION 1

Cell Search

What to Do

Read the descriptions below of some important findings made by scientists. Notice how technology helped them as they looked for answers to their questions about reproduction.

A. Body Fabrics

The tissues of the body (such as skin, bone, and muscle) were first described as woven materials, such as linen or rope. During the nineteenth century, improved microscopes allowed biologists to see the cells of living things for the first time.

B. Colored Ribbons

A special dye enabled scientists to observe the nucleus with much greater clarity. The scientists saw certain colored structures in the nucleus that looked like fine, ribbonlike strands. These structures were named *chromosomes,* meaning "colored bodies." The function of the chromosomes was not known at first.

C. Fertile Ground

Scientific techniques allowed biologists to observe living cells and some of their life processes. Using petri dishes and microscopes, they watched fertilization of egg cells (ova) by sperm cells.

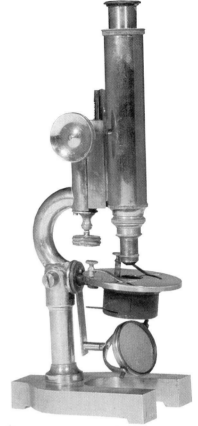

In the nineteenth century, improved microscopes advanced the study of living things.

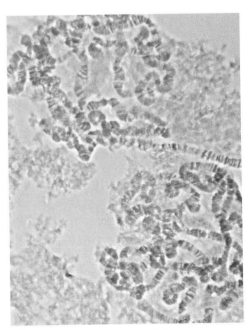

A microscopic view of chromosomes

Exploration 1 continued

D. Inner Division

Biologists were keenly interested in the behavior of the chromosomes. Chromosomes seemed to appear right before cell division, and then a complex series of movements caused the chromosomes to be distributed equally between the two new cells. Because the same number of chromosomes appeared each time, they concluded that the chromosomes must be duplicated between each division. The careful way in which they were distributed to each new cell led biologists to believe that the chromosomes carried the hereditary instructions for the cells.

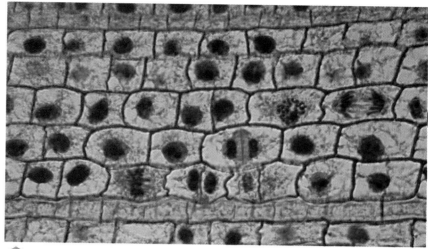

A section of plant tissue containing dividing cells

E. Shadowy Shapes

By the 1950s, scientists were studying the structure of crystals with a photographic technique that uses X rays instead of light. With this technique, Rosalind Franklin exposed crystallized molecules from the nucleus to X rays and then made photographs of the "shadows" cast by those molecules. James Watson and Francis Crick used her photographs to determine the structure of those molecules.

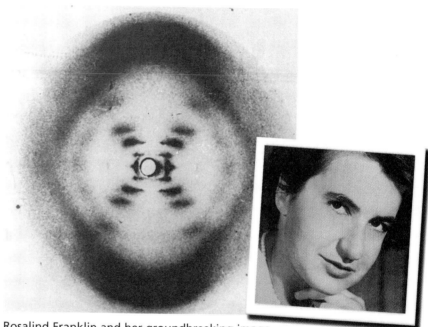

Rosalind Franklin and her groundbreaking image

Inferences

Match each of the following inferences to observations A–E on this and the preceding page.

   a. Two parents each contribute part of the instructions that lead to the development of offspring.

   b. A model could be constructed to resemble a molecule of material in the nucleus.

   c. The chromosomes contain instructions for offspring. These instructions enable living offspring to be constructed from nonliving materials.

   d. The nucleus is the "head office" of a cell—the location of the instructions for the development of new life.

   e. Body tissues are communities of similar cells joined together to form a structure.

Thinking It Over

What answers did biologists have, so far, about the mysteries of reproduction? As you read the following paragraph, think about the words that complete the sentences.

Biologists knew that two sex cells, ____1____ and ____2____, unite to start a new life. They called the combining of these cells ____3____. The center of activity in the formation of the sex cells was always the ____4____, where long, ribbonlike objects called ____5____ took part in the process of cell division. The biologists believed these objects to be the sets of ____6____ for the development of a new individual. They believed these sets were a form of ____7____ programming. They also were able to identify the ____8____ of a molecule of chromosome material.

Watson and Crick, with their model of DNA. Their model provided scientists with a valuable tool in the study of heredity.

Substance of Life

What was the material in the nucleus that biologists were viewing? Its shape and structure were identified in 1953 by James Watson and Francis Crick, who received a Nobel Prize for their discovery. The material is called **DNA**—an abbreviation for the chemical that contains the instructions for cells, *deoxyribonucleic acid.* It is found in the chromosomes and is essential for all forms of life.

The models of DNA at right may remind you of the spiral-ladder-shaped structures that Gloria and Yoon saw in the nucleus at the Science Discovery Center. The spiral ladders they watched were molecules of DNA.

Further investigation showed how the DNA molecule stores the instructions for life. There are four chemicals that bond together, two at a time, to form the rungs of the DNA ladder. The instructions for life are based on specific sequences of these bonded pairs. Much is known about how the chemical code is translated into action as new living things develop. However, much also remains to be discovered.

Model of a DNA molecule

The structure of DNA is much like a coiled ladder or a spiral staircase.

The structure of DNA, uncoiled so that you can see the "ladder rungs" more clearly. The letters C, G, T, and A stand for the chemicals that combine to form the rungs of the "ladder."

Mind-Benders

Consider these facts about DNA—the code of life.

- DNA has an alphabet of four chemical "letters" that form the rungs of the spiral ladders. The DNA in a human contains more than a billion of these letters, which would fill about 3000 novels of average length.

- If you could completely uncoil a DNA molecule in one of your cells, it would measure almost 2 m. Each cell contains many strands of DNA, and you have more than 10 trillion cells in your body. All the DNA they contain could stretch to the sun and back.

- Chromosomes contain long strands of DNA, which are "read" in segments called genes. A section of the DNA spiral ladder with 2000 rungs might make up only one gene. Each gene is like a "sentence" in the language of life. Each of your cells contains about 100,000 genes, or "sentences" of instructions.

- A full set of DNA for all 6 billion people alive now could fit on 1 teaspoon and would have a mass of about 1 g.

- The long series of letters shown below is part of the instructions for a protein in human blood. You can see that it is in the form of linear programming. The whole code for the protein would fill up this page. To print the genetic code for one single cell from your body, however, you would need about 1 million pages the size of this page.

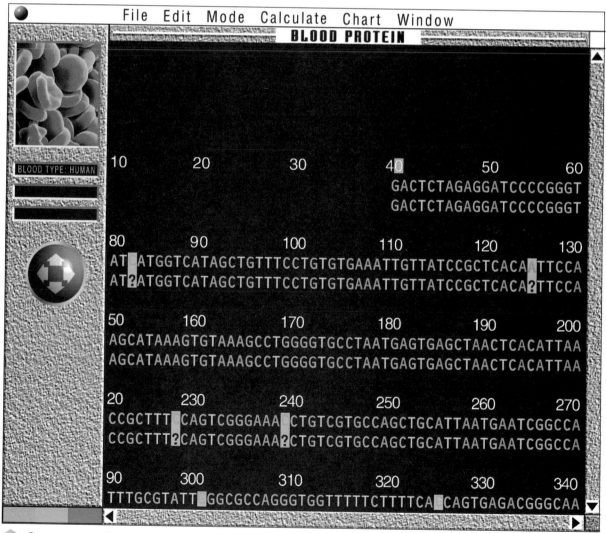

File   Edit   Mode   Calculate   Chart   Window

BLOOD PROTEIN

BLOOD TYPE: HUMAN

```
       10           20           30          40          50          60
                                                     GACTCTAGAGGATCCCCGGGT
                                                     GACTCTAGAGGATCCCCGGGT

       80           90          100          110         120          130
AT ATGGTCATAGCTGTTTCCTGTGTGAAATTGTTATCCGCTCACA TTCCA
AT?ATGGTCATAGCTGTTTCCTGTGTGAAATTGTTATCCGCTCACA?TTCCA

       50          160          170          180         190          200
AGCATAAAGTGTAAAGCCTGGGGTGCCTAATGAGTGAGCTAACTCACATTAA
AGCATAAAGTGTAAAGCCTGGGGTGCCTAATGAGTGAGCTAACTCACATTAA

       20          230          240          250         260          270
CCGCTTT CAGTCGGGAAA CTGTCGTGCCAGCTGCATTAATGAATCGGCCA
CCGCTTT?CAGTCGGGAAA?CTGTCGTGCCAGCTGCATTAATGAATCGGCCA

       90          300          310          320         330          340
TTTGCGTATT GGCGCCAGGGTGGTTTTTCTTTTCA CAGTGAGACGGGCAA
```

Computer-rendered DNA sequence

Stop and Think

Read the dialogue below and analyze what the characters are saying. Sort out and summarize the thoughts expressed about genes, chromosomes, DNA, and genetic instructions.

When you think you understand the structure and organization of genetic material, reconsider the function of DNA. The instructions of the genetic code control the formation of a new life—from the first cell division to the formation of cells of various sorts (such as nerve and muscle cells) to the organization of the cells into an organism. How do you think the instructions account for all the characteristic traits of an organism? Does this mean that every organism is completely programmed by its genes?

What other factors, besides genes, affect who you are? In the following lesson, you'll consider the importance of environment on the formation of life.

Environment or Heredity?

The instructions have been given and all the pieces have been assembled properly. As a result, a baby has been born. The new human being has 100,000 gene pairs inherited in two sets—one from each parent. The huge number of traits that the gene pairs code for represent a unique combination. With 100,000 gene pairs to juggle, you can see how no one else is—or ever will be—exactly like you. But genes are not the only reason you are who you are. Factors from your environment also greatly affect your unique combination of characteristics.

Family-Tree Time

At this point you can add to your family-tree project. You will need to trace the family's traits back several generations. Talk to the family members you are studying to learn about those traits. In the project, include lists of the following traits:

1. observable traits
2. traits that may be inherited but that will not show up until later in life
3. recessive traits that may have "skipped a generation"

Geneticists call the description of an organism's traits its **phenotype**. Your phenotype is the physical expression of your genes. For example, your phenotype may be for brown eyes.

The actual pair of genes you have inherited for each of your traits is called your **genotype**. If your phenotype is for brown eyes, your genotype may actually include a "brown eye" gene and a "blue eye" gene.

For each individual, the traits you listed in 1, 2, and 3 above will fall under one or both of these categories—*phenotype* and *genotype*. See whether you can identify which traits are part of your phenotype and which are part of your genotype. Can a trait be part of your genotype but not part of your phenotype? Can the reverse be true?

▲ What are the phenotypes for several traits of the horses shown here? Can you tell what their genotypes are for these traits?

There's just one more piece of information to add to your family-tree project. Read the passage below to discover what that information is, and then complete your family tree. How could the information you gathered for this family tree be valuable to a family? In what ways do you feel more informed about heredity after doing this project? Record your thoughts about these questions in your ScienceLog.

How Important Is Your Environment?

Which has a greater effect on what you are—the genes you inherit, or your environment? What does environment include? Food and nutrition are major factors. Consider a neglected, stray dog that is fortunate enough to be taken in by someone who provides food and good living conditions. The results of good nutrition and care are often dramatic. Would this dog's genotype be the same in each environment? What about its phenotype?

Investigate, as far back as possible, the length of life of people in the family you're studying. Look for patterns. For example, many members of the father's family may have lived to be 80 or 90 years old. Many people believe that longevity, the length of a person's life, is strongly influenced by heredity. In what ways could your environment affect your life span? Identify as many environmental factors as possible.

Add the information about the longevity of your ancestors (or the ancestors of the family you chose) to your family-tree project. How can people alter their environment to try to extend life?

"Environment can affect a person as much as heredity does."

Do you agree or disagree? Begin collecting information that illustrates why this statement is actively debated. Think about the dog described above. Ask people you know about their opinions.

A Field Trip

Visit an experimental nursery and explore the environment. Find out from someone there how they make use of heredity to produce healthy, productive animals or plants. Ask about environmental effects on living things. In your ScienceLog, restate the previous quote for the plants or animals you investigated: Environment can (cannot) affect animals (plants) as much as heredity does.

After being taken in and properly fed, this stray dog looked quite a bit different. Did its genotype or phenotype change?

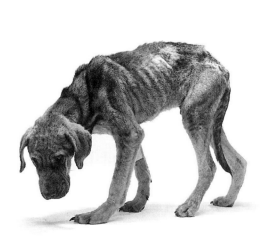

Surprising Evidence or Coincidence?

The twin brothers shown at right were separated at the age of 4 weeks and were not in touch with each other for 40 years. They grew up in different areas with different families. They were brought together again by a group doing an investigation about heredity and environment. The results surprised everyone. The two men had much in common beyond appearance. They both liked math and hated spelling in school, had similar hobbies, went to the same beach areas in Florida each year, drove the same kind of cars, married women named Betty, named their firstborn sons James Alan and James Allen, had dogs named Toy, had high blood pressure, gained and lost weight at the same stages in their lives, and had the same type of recurring headaches. When reunited they were even wearing similar clothes. How does this study add to your understanding of the role of genes and environment in development?

The twins, moments after their reunion

EXPLORATION 2

How Alike?

1. If you know any identical twins, arrange to interview each one separately. Make up a set of questions in advance about their likes, dislikes, habits, and abilities. Ask each person the same questions, and then compare their answers.

2. If you know any fraternal twins, arrange to ask them the same questions. Their inheritance is not identical, but they have been living in the same environment. What environmental factors affect fraternal twins that do not affect siblings who are not twins?

3. If you don't know any twins, you can do research to discover more about them. Many studies have been done on twins and other multiple births. What are some unusual documented similarities among twins, triplets, or quadruplets? What is the likelihood that a woman will have a multiple birth?

4. What do you think now about the effects of heredity and environment? Which one matters more?

When Heredity Goes Wrong

Most of the time, the instructions for life are precisely followed. But suppose something goes wrong. Although 99 percent of the babies born are the result of normal development, genetic errors can cause problems during development. Some embryos are so damaged that they cannot develop beyond the first stages and are rejected by the mother's body—often before she even knows she is pregnant. It has been estimated that this happens 500,000 times every year in the United States alone.

Why do these things happen? What goes wrong? Could it be that sometimes, in the process of cell division, instructions get improperly duplicated? Perhaps sections of instructions in the sex cells are lost, altered, turned upside down, or placed out of order. Perhaps the instructions are perfect, but there is an error in their translation in the developing embryo.

Study the flowchart, beginning at the top and working down. What is it saying? Write a script to accompany this chart so that someone else will understand it. What would be a good title for the flowchart?

Disaster or Lucky Mistake?

A **mutation** is a permanent change in the genetic code. Mutations may produce changes that are disastrous to living things. On the other hand, without mutations, living things would not have changed, or evolved, to their present form. The remarkable diversity of the living things around you would not exist if mutations had not occurred over time.

What are some examples of nature's "lucky mistakes"? Perhaps this is a subject you would like to investigate.

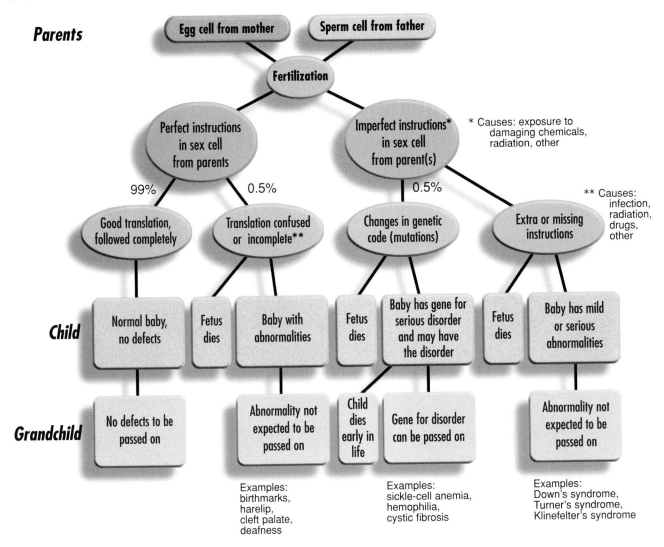

What the Future Holds

While you were still very young, you began to examine the life all around you. You have learned a lot about yourself, your environment, and other living things. You have learned something about the structure of living things and the uses or functions of their parts. You have become especially well acquainted with the plants and animals you use for food. Your eyes alone have provided you with a lot of information about the world around you.

Your experience has mirrored that of scientists through history. Their examination of living things was limited, for a long time, to what their unaided eyes could tell them. After the invention and perfection of the microscope, their main approach changed from studying the whole animal to examining the animal's tiny details. Now biologists are able to study life at the level of the molecule of life—DNA. When you hear the term *molecular biology,* you will know what it means.

Technology has advanced to the point that biologists can now work within the nucleus of a cell. They can manipulate, or move, DNA from one cell to another. This manipulation of DNA is called **genetic engineering**. It has become possible to perform some remarkable procedures. However, the greatest realization seems to be that biologists know enough now to recognize how much more they have to learn about living cells. Molecular biologists and other scientists whose work deals with heredity are called *geneticists.*

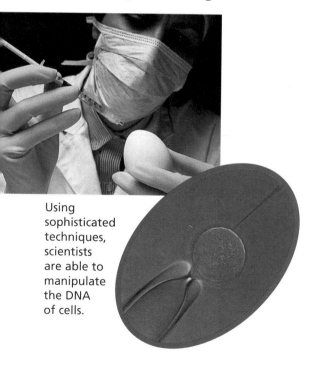

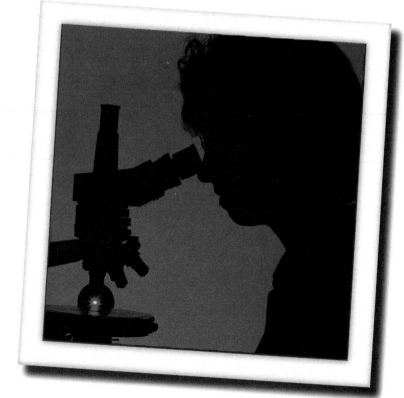

Using sophisticated techniques, scientists are able to manipulate the DNA of cells.

It is important to note here that there are regulations to control what researchers do with the genetic structure of cells. For example, there are limits to what can be attempted with human fertilized eggs. Biologists themselves want the regulations. Form small groups to discuss why people might fear the results of genetic engineering that involves humans. Do you think the possible dangers posed by genetic engineering outweigh its benefits to human life?

On the Horizon

What should you expect to hear about in the future? Several possibilities are listed below.

• *Improved ways of detecting genetic disorders early in pregnancy.* Amniocentesis is a procedure used to obtain fetal cells for examination. A long needle is used to withdraw some of the fluid surrounding the fetus. This fluid contains skin cells that have been shed by the fetus. Testing of the fetal cells can detect certain genetic disorders. Unfortunately, there is a slight risk of death to the fetus in performing this procedure. A new technique is being developed that requires only a sample of the mother's blood. The technique involves finding the few fetal blood cells that enter the mother's bloodstream through placental leaks. What do you think would be some advantages of using the new technology?

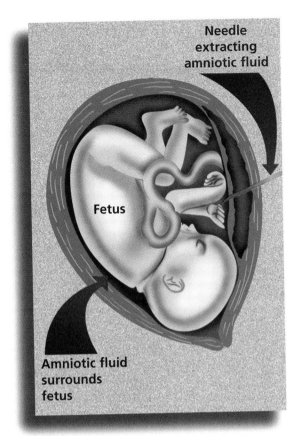

Amniocentesis involves withdrawing some of the amniotic fluid that surrounds a fetus and then analyzing cells from that fluid.

A genetic counselor advises couples about the probability of passing on genes for a genetic disorder to their offspring.

- *Discoveries about the location of genes that cause birth defects and diseases.* The genes associated with cystic fibrosis and a form of muscular dystrophy have already been located.

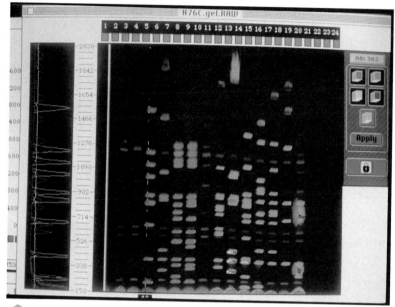

⬆ Researchers are gradually putting together a map showing every human gene.

- *Gene therapy.* There is hope that many genetic disorders can be corrected through gene therapy. In gene therapy, some cells are removed from the patient, healthy genes are inserted into them, and the cells are returned to the patient's body. The substance lacking in the person's body is then produced by the cells with the new genes. Would this procedure have any effect on the descendants of a person who had undergone gene therapy?

- *Procedures for changing genes in a fertilized egg.* Why might this procedure have more far-reaching effects than the gene therapy described above? Explain.

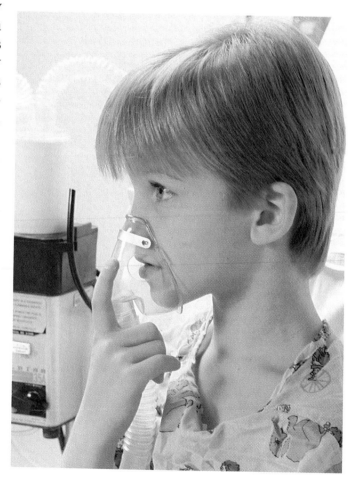

With gene therapy, genetic disorders such as cystic fibrosis could be cured. ➡

- *Clones.* When you cut off parts of a mature plant (such as leaves or stems) and grow new plants from them, you are producing **clones**. Clones have the same genetic makeup as the mature plant. They are new plants produced without sexual reproduction.

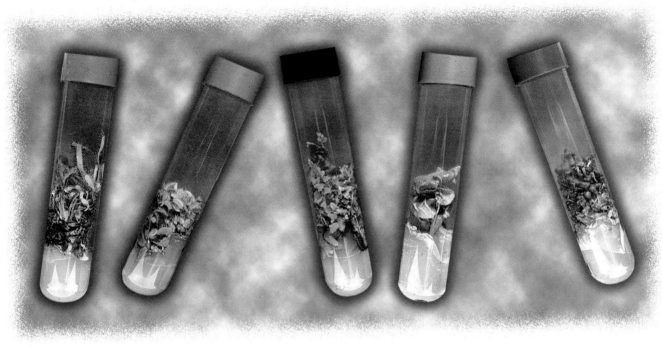

These plants, which were each grown from a single cell, are all clones of other plants.

It is now possible to clone certain simple animals. Are human clones possible? This issue is being debated right now.

Keep in mind that human clones would not first appear as full-grown adults. Instead, they would have to mature as every other embryo does—and you have already seen the kind of effect that environment can have on development.

What lies ahead is unknown. The noted scientist Lewis Thomas has said that in the future, people will travel through a "wilderness of mystery." He says, "Science will not be able to explain the full meaning of all that it makes possible." As a result, Thomas believes that every variety of talent will be needed for the future, especially the contributions of poets, artists, musicians, philosophers, historians, and writers. What talents do *you* think will be needed? How could such talents help explain the process of life and how to make use of technology?

Genetic engineering may be very important in the future. To find out more about genetic engineering, read page S172 in your SourceBook.

1. Coming Soon, to a Theater Near You

The 16-year-old clone was nervous. She had just been reassigned as COPI-864 (Clone, On Parts Inventory), and she was one of the next to be used if transplants were needed—a leg, a kidney, or perhaps her brain. Although most clones had been brainwashed into accepting their roles, COPI-864 plotted her escape.

At home that evening, Diego and Tracy talked about the movie and the whole idea of human clones. They wanted to find out all they could about the possible results of cloning. They were interested not just in humans, but in all sorts of living things.

They started writing down a list of experts who could contribute knowledge and valuable viewpoints to a discussion about cloning. Complete the list they started (shown below) by adding at least five more people.

- Owner of prize cattle or other farm animals
- Director of a zoo for endangered wildlife

What might each expert have to say?

2. Let's See the Evidence!

Are there inherited characteristics that are not affected by the environment? Are there human characteristics that are not affected by genes? Support your answers with evidence.

3. If X Is to Y . . .

Explain in your own words how the items in each group are related to each other.

a. linear programming, DNA, trait

b. double helix, Watson and Crick, X rays

c. heredity, genotype, phenotype

4. 100 Radish Seeds

Outline how you could use 100 radish seeds to investigate the effects of environment on plants.

5. From Beads to Genes

Below is a design for beading on a costume.

a. Translate it into a form of linear programming.

b. How is your example of linear programming similar to the instructions for life in the DNA molecule?

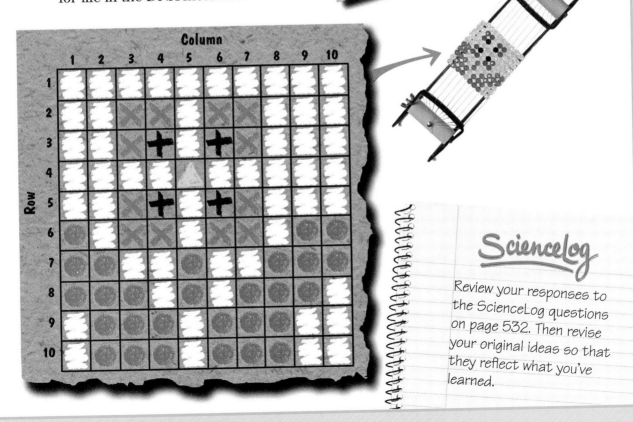

ScienceLog

Review your responses to the ScienceLog questions on page 532. Then revise your original ideas so that they reflect what you've learned.

Unit

CONTINUITY OF LIFE

The Big Ideas

In your ScienceLog, write a summary of this unit, using the following questions as a guide:

1. What does it mean to "inherit" traits?
2. What are four abilities that all living things share?
3. What part of a cell controls the formation of new cells? How is this accomplished?
4. How do you distinguish asexual reproduction from sexual reproduction?
5. What patterns did Mendel discover in his research?
6. What parts of an embryo develop early in pregnancy? midway through? late in pregnancy?
7. How is genetic information organized?
8. What is meant by the statement, "Chromosomes are a form of linear programming"?
9. How can the environment affect development (both before and after birth)?

Checking Your Understanding

1. Think about the situations presented below. What would be the possible effects on a couple's descendants if
   a. the father had lost a finger due to an accident at the plant where he worked?
   b. the mother (unvaccinated) was exposed to rubella during the early weeks of pregnancy?
   c. the mother abused drugs during pregnancy?
   d. the mother went to a scary movie about vampires during the early weeks of pregnancy?

SOURCEBOOK

If this unit got you interested in genetics, you can find out more about the subject in the SourceBook. Read about how a person's sex is determined by genes, how DNA copies itself, and more.

Here's what you'll find in the SourceBook:

UNIT 8

e. the mother was exposed to radiation during the early weeks of pregnancy?

f. the father was exposed to rubella during the early weeks of the mother's pregnancy?

g. a woman was vaccinated against rubella before she became pregnant?

h. a woman who abused drugs (of any sort, including alcohol and tobacco) gave them up before she decided to have a child?

i. the father had abused drugs known to cause chromosome damage?

j. the father worked around insecticides suspected of causing genetic mutations?

2. Straight or Curved Thumbs?

a. Draw a Punnett square to show the possible combinations among the children shown in the pedigree at right.

b. Fill in the children's genotypes on the pedigree.

c. Who has straight thumbs and who does not in this family?

d. Use Punnett squares to show the possibility of straight thumbs among the grandchildren.

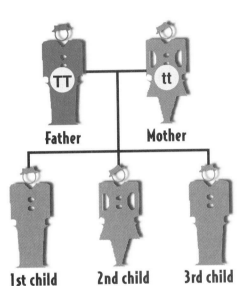

Father   **Mother**

1st child   **2nd child**   **3rd child**

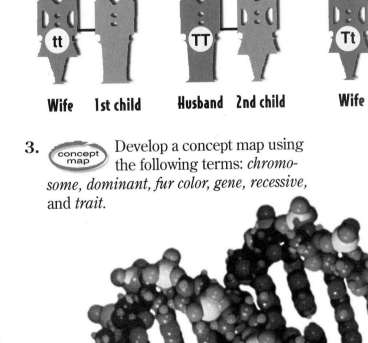

Wife   **1st child**   **Husband**   **2nd child**   **Wife**   **3rd child**

3. *concept map* Develop a concept map using the following terms: *chromosome, dominant, fur color, gene, recessive,* and *trait.*

Searching for Prehistoric Genes

In the movie *Jurassic Park*, scientists not only duplicated the genetic material of dinosaurs but also re-created the dinosaur itself. A little far-fetched? Maybe, but in fact scientists have retrieved and duplicated prehistoric DNA, and they've done it with the help of an "old sap" called amber.

Trapped in Sap

Golden in color and as clear as glass, amber looks like a gem, but it's really fossilized tree sap.

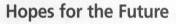

▲ Many different species of prehistoric plants and animals have been found trapped inside lumps of amber.

Amber has been used in jewelry making for hundreds of years, but today it is popular for another reason. Millions of years ago, microscopic organisms, pollens, insects, and other very small animals were trapped in the sap that oozed from prehistoric trees. As the sap hardened to amber, it mummified its prehistoric victims so well that millions of years later fragments of their DNA, or genetic material, are still preserved.

Extracting the DNA

Genetic researchers are now extracting fragments of prehistoric DNA from the mummified bodies of organisms found in amber. After carefully cleaning the amber to prevent contamination, scientists extract the organic material. In one method, thin slices are cut from the amber until the material can be scooped out. In another method, holes are drilled in the amber, and the material is sucked out with a needle.

The DNA retrieved from prehistoric tissue is almost always badly damaged. Most DNA from living organisms is made up of millions of molecules joined in unbroken strands. The DNA recovered from fossil samples is much shorter—only a few hundred molecules long—and has many missing pieces. However, with new techniques, the extinct DNA can actually be duplicated. The DNA can then be compared with the DNA of a living organism thought to be related to the extinct one.

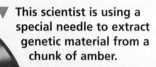

▼ This scientist is using a special needle to extract genetic material from a chunk of amber.

Hopes for the Future

The search for prehistoric genes is already leading scientists to a better understanding of how living organisms evolved. Some researchers also hope to clone DNA and revive prehistoric species of bacteria, viruses, fungi, and pollen. They believe this will lead to discoveries of new medical drugs, industrial compounds, and natural pesticides. And what about re-creating a dinosaur? Not very likely. The DNA fragments are simply too damaged and incomplete.

Find Out for Yourself

Why is amber such a good preserver of animal and plant tissue? Do some research to find out.

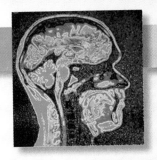

Goats to the Rescue

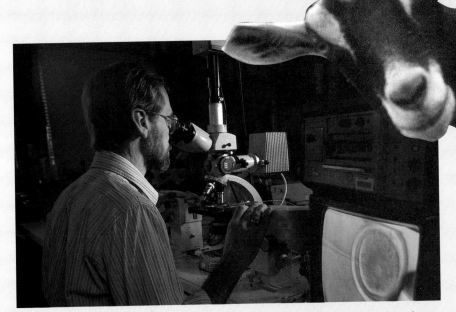

They're called transgenic (tranz JEHN ik) goats because their cells contain a human gene. They look just like any other goats, but because of their human gene they produce a drug that can save lives.

Life-Saving Genes

Heart attacks are the number one cause of death in the United States. Many heart attacks are triggered by large blood clots that interfere with the flow of blood to the heart. Human blood cells produce a chemical called TPA (tissue plasminogen activator) that dissolves small blood clots. If TPA is given to a person having a heart attack, it can often dissolve the blood clot, stopping the attack and saving the person's life. But TPA is in short supply because it is difficult to produce large quantities in the laboratory. This is where the goats come in. Researchers at Tufts University in Grafton, Massachusetts, have genetically engineered goats to produce this life-saving drug.

Drug-Producing Goats

Producing transgenic goats is a complicated process. First, fertilized eggs are surgically removed from normal female goats. The eggs are then injected with hybrid genes that consist of human TPA genes "spliced" into genes from the mammary glands of a goat. Finally, the altered eggs

▲ A scientist at Tufts University injects human TPA genes into fertilized goat eggs. Above, the first transgenic goat

are surgically implanted into female goats, where they develop into young goats, or kids. Some of the kids that are born actually carry the hybrid gene. When the hybrid kids mature, the females produce milk that contains TPA. Technicians then separate the TPA from the goat's milk for use in heart-attack victims.

The Research Continues

Transgenic research in farm animals such as goats, sheep, cows, and pigs may someday produce drugs more rapidly, cheaply, and in greater quantities than current methods. The way we view the barnyard may never be the same.

◄ The TPA in this milk might help save the life of a heart-attack victim.

Find Out for Yourself

Using chemicals produced by transgenic animals is just one of many gene therapies. Do some research to find out more about gene therapy, how it is used, and how it may be used in the future.

Science Snapshot

Barbara McClintock
(1902–1992)

In 1918, Barbara McClintock's mother was deeply concerned that her daughter would have trouble fitting into society. McClintock had unusual interests and aspirations for a woman of her era. She was about to enter Cornell University, where she was going to become a scientist!

A Relish for Research

In college, McClintock studied genetics, botany, cytology, and zoology. A perfectionist in her research, she soon was recognized as an outstanding researcher in the field of plant genetics. McClintock began working with a type of corn called maize. Her work resulted in some ground-breaking discoveries about genes.

Jumping Genes

During her research with maize, McClintock noticed that the color of kernels changed unexpectedly from one generation to the next. The responsible genes seemed to move, or "jump," from place to place on the chromosomes. This observation led McClintock to develop a radically new model for gene behavior. The idea that genes could change locations on the chromosome contradicted the prevailing view that genes and chromosomes were stable parts of a cell nucleus.

Scientists were shocked by McClintock's findings, and most of them discounted her ideas for more than 20 years. Over time, however, as more and more research supported her hypothesis, McClintock's model gradually gained acceptance. Finally, in 1983, 65 years after her mother had agonized over McClintock's college plans, McClintock was awarded a Nobel Prize for her work in genetics.

In 1983, Barbara McClintock won the Nobel Prize for her work in genetics.

The genes that McClintock discovered are called transposons because they transpose, or shift, on the chromosomes. When such a gene changes location, it often causes a mutation or causes a gene to become inactive. Such genetic changes can lead to new or different characteristics in the offspring of the affected organism.

▶ Transposons cause surprising changes in the patterns of colored corn kernels.

Fieldwork

Many scientists doubted McClintock's findings because a great deal of evidence indicated that genes maintained their positions on the chromosome. Scientists had even constructed maps that showed the relative positions of genes on certain chromosomes. How is this similar to the debate over the shape of the Earth?

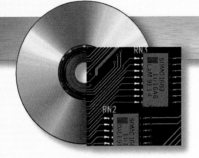
Supersquash or Frankenfruit?

*T*he fruits and vegetables you buy at the supermarket may not be exactly what they seem. Scientists may have genetically altered these foods to make them look and taste better, contain more nutrients, and have a longer shelf life. So make way for genetically engineered foods because they may already be on the shelves of your local grocery store.

From Bullets to Bacteria

Through genetic engineering, scientists are now able to duplicate one organism's genes and place them into the cells of another species of plant or animal. This technology enables scientists to give plants and animals a variety of new traits. The new traits are then passed along to the organism's offspring and future generations.

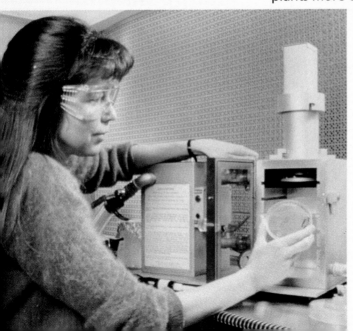

▼ A scientist using a "gene gun" to insert DNA into plant cells

Scientists alter plants by inserting DNA with new properties into the plant's cells. The DNA is usually inserted by one of two methods. In the agrobacterial method, the new DNA is placed inside a special bacterium. The bacterium carries the DNA to the plant cell and transfers it into the cell. The DNA particle-gun method works a little differently. In this method, microscopic particles of metal coated with the new DNA are actually fired into the plant cells by way of a special "gene gun."

High-Tech Food

During the past decade, scientists have inserted genes into more than 50 different kinds of plants. Most of the new traits from these genes make the plants more disease resistant or more marketable in some way. For example, scientists have added genes from a caterpillar-attacking bacteria to cotton, tomato, and potato plants. The altered plants produce proteins that kill the caterpillar's crop-eating larvae. Scientists are also trying to

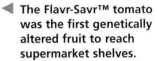

◄ The Flavr-Savr™ tomato was the first genetically altered fruit to reach supermarket shelves.

develop genetically altered peas and red peppers that stay sweeter longer. A genetically altered tomato that lasts longer and tastes better is already in many supermarkets. One day it may even be possible to create a caffeine-free coffee bean.

Are We Ready?

As promising as these genetically engineered foods seem to be, they are not without controversy. Some people are afraid that new, harmful genes could be released into the environment or that foods may be changed in ways that endanger human health. For example, could the nutritional value of some foods be inadvertently changed or reduced in an attempt to make the food look better and last longer? All of these concerns will have to be addressed before the genetically altered food products are widely accepted.

Find Out for Yourself

Are genetically altered foods controversial in your area? Survey a few people to get their opinions about genetically altered foods. Do they think grocery stores should carry these foods? Why or why not?

Life Processes

In This Unit

Now that you have been introduced to two of the major life processes, photosynthesis and respiration, consider these questions.

1. What are the basic chemicals of life, and what are their primary functions?

2. In what ways are photosynthesis and cellular respiration complementary processes?

3. What is metabolism and how does it relate to digestion?

In this unit you'll explore the chemical basis of life and then see how plants make the sugar that almost all other organisms depend on. You'll also learn more about basic life processes that are characteristic of all living things.

THE CHEMISTRY OF LIFE

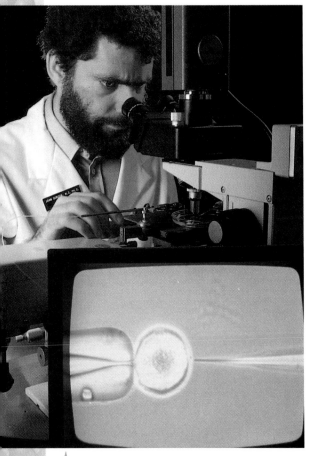

Using an extremely fine syringe, a biologist injects nucleic acid into this mouse cell. This procedure allows scientists to study the biochemistry of the nucleus.

A carbon atom readily bonds with other carbon atoms in various formations. Because carbon forms chemical bonds so readily, numerous carbon-containing compounds exist. Over 4 million compounds of carbon are now known, and the list continues to grow.

Organic compound

A compound that contains carbon and usually is produced by living things

Everything that happens in a living thing is the result of chemical reactions. Just flexing your leg, for example, requires many different chemical changes. Think about this as you bend your leg. Chemical changes in your brain cause nerve cells to send the "bend your leg" message from your brain to your leg. Then another series of chemical changes in your leg causes some of the muscle cells in your leg to contract and others to relax. As a result, your leg bends.

A better understanding of the chemical processes of life may be the key to solving many human problems, from producing enough food for the people of the world to curing diseases such as cancer, heart disease, and AIDS. Currently, scientists are doing extensive research to help them further understand the chemical basis of life as well as the processes that occur in all living things.

The Elements of Life

As you know, all matter—living and nonliving—is made up of particles called *atoms*. Atoms are the smallest particles of elements. But of the 91 naturally occurring elements, just four of them—carbon, hydrogen, oxygen, and nitrogen—make up more than 99 percent of the living matter on Earth. Atoms of these four elements combine with each other by forming very strong and stable chemical bonds.

Carbon atoms form the "backbone" of most biological molecules. Two special abilities of carbon make life as we know it possible. First, carbon atoms can chemically bond with up to four different atoms. Second, carbon atoms can bond to other carbon atoms, making long chains, rings, or branched structures.

Chemists use the term **organic compounds** when referring to most carbon-containing compounds, since they are normally formed by living organisms. Indeed, *organic* means "coming from life." Compounds that do not contain carbon, like water (H_2O)—and a few that do, like carbon dioxide (CO_2)—are called *inorganic compounds*.

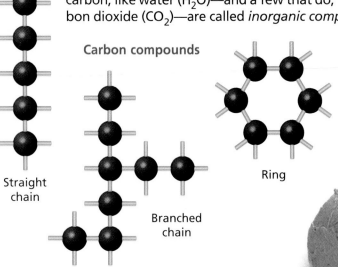

Carbon compounds

Straight chain

Branched chain

Ring

Most organic molecules are very large. Some organic molecules contain thousands or even millions of individual atoms. These atoms are arranged in chemical units called *monomers* that repeat over and over throughout the molecule. Molecules that are made of many monomers are called *polymers*. The prefix *mono* means "one," and the prefix *poly* means "many."

Basic Biochemicals

The organic compounds made by living things are often called *biochemicals* because they are literally the "chemicals of life." But each of the many kinds of biochemicals can be classified into one of four major groups: *carbohydrates*, *lipids*, *proteins*, and *nucleic acids*. As you read about each kind of organic compound, keep in mind that plants synthesize all of these compounds from the inorganic materials carbon dioxide, water, and a variety of minerals. Animals, on the other hand, must eat foods that contain the monomers they use to build the organic compounds that their bodies need.

Carbohydrates   Sugars, starch, and cellulose are examples of the organic compounds known as **carbohydrates**. The photo below shows some of the sources of these substances in your diet. Carbohydrates are made entirely of carbon, hydrogen, and oxygen. Almost all carbohydrates contain stored energy that, as you will see later, can be made available to living cells. In fact, most cells prefer to use carbohydrates as their primary source of energy.

Many carbohydrates are polymers of smaller, repeating monomers called *monosaccharides,* or simple sugars. Glucose, a product of photosynthesis, is one type of monosaccharide. When two monosaccharides join together chemically, they form a *disaccharide*, or double sugar. Common table sugar is a disaccharide made of glucose and fructose. Many glucose molecules join together to form the *polysaccharides* starch and cellulose.

▲ Starch is a polysaccharide, a polymer formed from many glucose units, or monomers.

Fruits and honey contain the simple sugars glucose and fructose. Complex carbohydrates, such as starches, are found in bread, cereal, and pasta.

Carbohydrates

Organic molecules composed of one or more monosaccharides that are often used for energy by cells

Lipids Fats, oils, waxes, and steroids are examples of the group of organic compounds called **lipids**. Most lipids consist of three long *fatty-acid molecules* chemically bonded to a *glycerol molecule*. These kinds of lipids are often called *triglycerides* because of their three-chain structure. Like carbohydrates, lipids consist of carbon, hydrogen, and oxygen. But there are usually many more hydrogen atoms in lipids because of their very long carbon chains. Lipids contain energy that can be used by cells, but more often lipids serve as excess-energy storage molecules. Animals store excess lipids primarily as *fat*; plants store excess lipids as *oil*.

In addition to their energy-storage function, lipids play other important roles in living things. Cell membranes, for example, are made of two layers of lipids. *Waxes* are made by many organisms. Waxes on the leaves of plants protect the tissues they cover from dehydration, and those that line the inside of your ears protect your ear canal and eardrum from dirt and microorganisms.

Cholesterol is another type of lipid that you have probably heard much about. Although excess cholesterol can lead to heart disease, a small amount of it is necessary in animals. Cholesterol is used for making cell membranes, nerve and brain tissues, and certain *hormones*. Hormones are important chemicals that regulate life processes such as growth and reproduction. The photo below shows some of the sources of lipids in your diet.

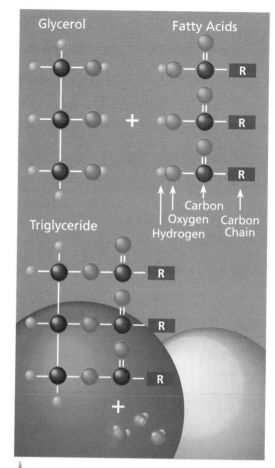

▲ Most lipids are made by joining three fatty-acid molecules to a glycerol molecule.

Lipids

Organic molecules composed of fatty acids and glycerol that store energy and make up cell membranes in living things

▼ Lipids are found in each of the items shown here.

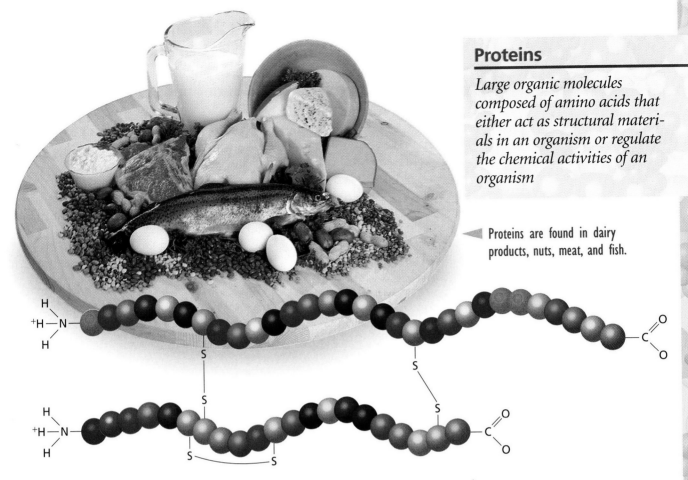

Proteins

Large organic molecules composed of amino acids that either act as structural materials in an organism or regulate the chemical activities of an organism

Proteins are found in dairy products, nuts, meat, and fish.

Proteins Most of the different kinds of organic molecules found within living things are **proteins**. The photo above shows some of the sources of protein in your diet. All proteins contain carbon, hydrogen, oxygen, and nitrogen, and many also contain sulfur. Proteins are complex polymers of smaller molecules called *amino acids*.

There are 20 different kinds of amino acids. These amino acids make up all the different proteins in living things in much the same way that the 26 letters of the alphabet make up all the different words in the English language. In the same way that words with different meanings are spelled with different letters, proteins (with different chemical properties and functions) are made by varying the kinds, the number, and the sequence of amino acids. For example, two proteins that differ by only one amino acid can have completely different properties, just as the words *live* and *love* have completely different meanings. The longest words in the English language contain only a few dozen letters, but the longest proteins have over 1000 amino acids. Just think how many different words you could make if words had more than 1000 letters! Because of their great size, many different kinds of proteins are possible.

Proteins are very important to the structure and function of living things. Some proteins, such as *collagen* in skin and *myosin* in muscle cells, form strong, elastic fibers. Other proteins, called *enzymes*, are important to the chemical activities of cells. Enzymes are involved in nearly every chemical reaction in a living cell. Still other proteins, such as hemoglobin in the blood, carry substances to places in the body where they are needed. Cell membranes also contain many *carrier proteins* that transport particles into and out of cells.

Insulin, with only 51 amino acids, is one of the smallest proteins in humans. Each color symbolizes a different amino acid contained in the insulin molecule. Two "sulfur bridges" hold the two strings of amino acids together.

Nucleic acids

Large organic molecules composed of nucleotides that store the information for building proteins

Nucleic Acids   The largest and most complex organic molecules in living things are the nucleic acids—DNA (deoxyribonucleic acid) and RNA (ribonucleic acid). Nucleic acids are very long chains of monomers called *nucleotides*. Nucleotides are composed of the elements carbon, hydrogen, oxygen, nitrogen, and phosphorus. To give you an idea of how long some nucleic acids are, your body contains about 25 billion kilometers (more than four times the distance to Pluto) of DNA molecules alone!

DNA is located in the nucleus of each cell. It is a very important molecule because it contains the instructions for putting the amino acids of proteins together. These instructions are in the form of a code called the *genetic code*. While DNA functions as the "recipe book" for proteins, RNA does the actual physical work of synthesizing the proteins.

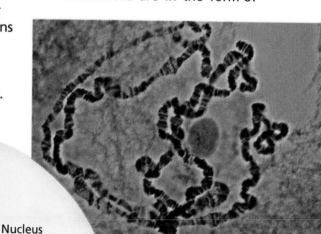

Each group of three nucleotides in a DNA molecule is a "code word" for a specific amino acid.

RNA copies the protein-building information from DNA and moves out of the nucleus and into the cytoplasm of the cell.

RNA nucleotides act as a pattern for lining up the amino acids required for building proteins.

Nucleus

DNA

Protein

RNA

▲ This is a close-up of DNA from a fruit fly. DNA contains information needed to make proteins.

Water

You could live about 5 weeks without eating a bite of food. However, you could live only about 5 days without drinking water. Water is the most important inorganic substance in living things, making up from 50 to 98 percent of the mass of a living cell. About 65 percent of the total volume of your body is water. If you were to lose only 15 percent of your body's normal amount of water, it could be fatal.

We share our need for water with all living things. For one thing, water dissolves the chemicals that participate in the many chemical reactions that occur inside organisms. Among its many important properties, water has a high capacity to absorb and give off heat without great changes in its temperature. Water's *heat capacity*, therefore, keeps living things from both freezing and overheating. As the only common substance that is a liquid at most of the temperatures found on Earth, water is also the habitat of thousands of living species.

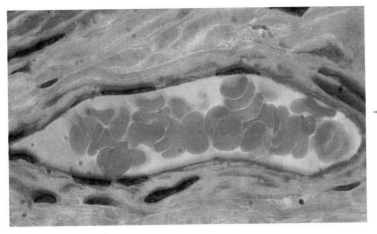

◄ Without water, there can be no life. These blood and muscle cells exist in solutions that are mostly water.

Polar molecule

A molecule that carries unevenly distributed electrical charges

A Polar Substance Many of the properties of water are due to its chemical structure. A water molecule forms when two hydrogen atoms chemically bond with an oxygen atom. The negatively charged electrons of the molecule spend more time near the oxygen nucleus than they spend near the hydrogen nuclei. As a result of this arrangement of electrons, the oxygen side of a water molecule has a slight negative charge, and the hydrogen side has a slight positive charge. Molecules, such as water, that have positive and negative poles are said to be **polar molecules**.

Polar molecules are attracted to other molecules that have electric charges. Because water is polar, it makes an excellent *solvent*. A solvent is a substance that can dissolve another substance.

Adhesion and Cohesion Water molecules tend to stick to things they cannot dissolve, such as glass, soil particles, and cells. This property of water is called *adhesion*. As you know, adhesion is partly responsible for water's ability to rise inside small tubes such as the vessels in plants that conduct water from the roots to the leaves. Water molecules are attracted to each other as well, due to their polarity. This property, called *cohesion*, keeps water columns from breaking as they rise inside plants.

◄ The unequal distribution of positive and negative charges in a water molecule results in its bent shape.

When crystals of common table salt (NaCl) are dropped into water, the water molecules are attracted to the individual sodium and chloride ions that make up the salt. The attraction of the water molecules pulls the ions away from the salt crystals, and each ion becomes surrounded by water molecules. As a result, the salt disappears into the water, or dissolves.

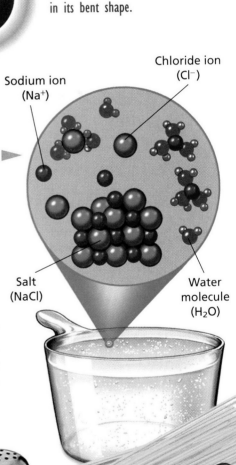

Sodium ion (Na⁺)

Chloride ion (Cl⁻)

Salt (NaCl)

Water molecule (H₂O)

Water's Chemical Role   Water plays an extremely important part in the chemical reactions that occur in living things. First of all, dissolved substances react much more rapidly than undissolved substances. In fact, many substances will not react at all unless they are dissolved in water. Water dissolves most of the substances that participate in the chemical reactions of life. Second, water is often a participant in chemical reactions. For example, splitting apart large organic polymers requires the addition of water. This type of reaction is called *hydrolysis*, which means "splitting with water." On the other hand, large organic molecules are made by a reaction known as a *condensation* reaction. In a condensation reaction, water is produced when monomers are joined to form polymers.

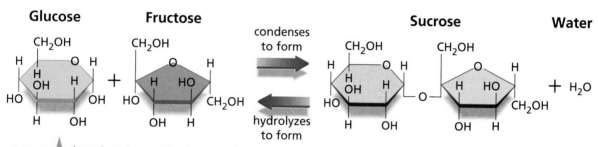

▲ A condensation reaction between glucose and fructose produces sucrose (table sugar) and water. This reaction is reversible by hydrolysis, which uses water to break sucrose down into glucose and fructose.

SUMMARY

Carbon is the element that forms the backbone of most of the important compounds of living things. These organic compounds include carbohydrates, lipids, proteins, and nucleic acids. Carbohydrates are made from monosaccharides, and lipids are made from fatty acids and glycerol molecules. Both are used as energy-storage compounds. Proteins are made by joining amino acids together. Some proteins are structural materials, while others, called enzymes, make possible the chemical reactions necessary for life. Nucleic acids are made of long chains of nucleotides. DNA is a nucleic acid that contains the instructions for making proteins. RNA copies information from DNA and uses it to build proteins. Water is the most abundant inorganic substance in living things. It dissolves many materials and participates in chemical reactions.

BUILDING BIOCHEMICALS WITH LIGHT

All of the organisms on Earth are ultimately dependent on the sun. Without the sun's warming rays, the Earth would soon become a frozen wasteland, a sphere of ice on which no living thing could survive. But the sun brings more than just warmth to the Earth. The sun also supplies the energy that plants need to build the organic molecules other living things depend on.

Chlorophyll

A green chemical that absorbs energy from the sun during photosynthesis

Photosynthesis

During *photosynthesis*, plants use sunlight, water (H_2O), and carbon dioxide (CO_2) to make organic molecules and oxygen (O_2). Plants use some of these organic molecules as building blocks for making lipids, nucleic acids, proteins, and carbohydrates. Remember that carbohydrates are a major source of energy for most living things. A basic formula used to represent any kind of carbohydrate produced by a plant is CH_2O.

The Place   Photosynthesis begins when the sun's energy falls on plant organelles called *chloroplasts*. Chloroplasts are found in cells inside the leaves of plants. A chemical called **chlorophyll** inside the chloroplasts absorbs the sunlight.

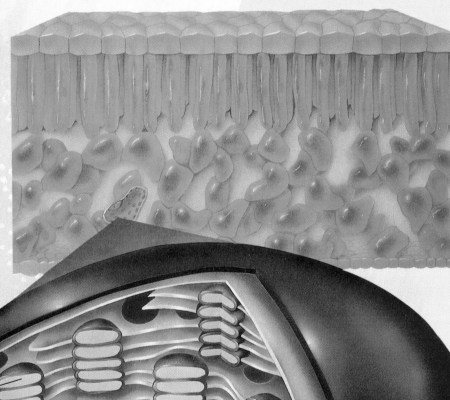

Inside the chloroplasts of ▷ plant cells are structures called grana, which contain chlorophyll molecules.

Many of the reactions of photosynthesis occur in the grana.

Other reactions of photosynthesis occur in the cellular substance that surrounds the grana. Sugar is made during these reactions.

The Process The overall chemical equation that describes photosynthesis is shown below. The equation shows that a basic carbohydrate is made from a molecule of carbon dioxide and a molecule of water, with a molecule of oxygen being released as a byproduct. During some of the reactions of photosynthesis, the water molecules are split. As a result, oxygen is produced. This oxygen eventually leaves the plant and enters the atmosphere by passing through the *stomata* in the leaves. The carbon dioxide and the hydrogen from water are used in a separate series of reactions that occur in the cellular material that surrounds the grana inside chloroplasts. These reactions lead to the formation of carbohydrates and other organic molecules.

The Products Some of the organic compounds that plants produce, such as carbohydrates, contain energy that cells use. The carbohydrates made by plants include glucose, sucrose, and starch. Some plants have specialized parts in which starch is stored. Potatoes are one example of such a part. One polysaccharide that plants make, called *cellulose*, is not used by plants for energy. Plants use cellulose to make their cell walls.

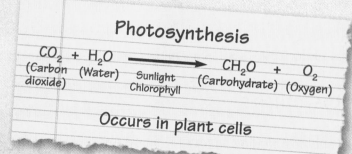

Photosynthesis

$$CO_2 + H_2O \xrightarrow[\text{Chlorophyll}]{\text{Sunlight}} CH_2O + O_2$$
(Carbon dioxide) (Water) (Carbohydrate) (Oxygen)

Occurs in plant cells

O_2

CH_2O

CO_2

Plants take in carbon dioxide through their leaves and water through their roots. These two chemicals, along with chlorophyll and the energy in sunlight, are used by plants to form carbohydrates and oxygen.

H_2O

Reaping the Benefits of Photosynthesis

The carbohydrates that a plant makes—sugar and starch—are used as energy sources by the plant. An animal that eats the plant takes in these molecules and also uses them for energy to carry out its own life processes. Any extra energy is stored in the animal's body as fat or as animal starch, which is also called *glycogen*. As one animal eats another animal, the energy that came to the plants from the sun is passed on again and again. In this way, the energy that is used by almost all living things on Earth can be traced back to sunlight. But how do plants and the animals that eat plants actually get energy from carbohydrates? The answer lies in a chemical process called *cellular respiration*.

Skin cells

Mitochondria

Some of the reactions of cellular respiration occur in the cytoplasm of cells. Other reactions of cellular respiration occur inside organelles called mitochondria.

Getting Energy From Glucose As you know, plant and animal cells rely on organic molecules, primarily glucose, for energy. When an animal ingests plant material, it digests the starch in the plant, breaking it down into glucose molecules. Plants must also convert their own starch into glucose to get energy out of it. The glucose then fuels cellular respiration, the chemical process that transforms the energy in glucose into energy that can be used by the organism. As in photosynthesis, many different chemical reactions occur during cellular respiration. The overall process of cellular respiration can be represented by the chemical equation shown above.

During cellular respiration, the energy contained in an organic compound, such as glucose, is released. But this released energy cannot be used directly by cells. Instead, the energy is used to form another molecule called **ATP,** which, chemically speaking, is a type of nucleotide. As you know, most nucleotides are used to build nucleic acids. But the function of ATP is different. ATP molecules are "packets of energy" that cells use to live. The energy contained in ATP is used by organisms in a variety of life processes, including movement, building organic polymers, and cell division.

Cellular Respiration

$$CH_2O + O_2 \longrightarrow CO_2 + H_2O + Energy$$

(Carbohydrate) (Oxygen) (Carbon dioxide) (Water) (ATP)

Occurs in both plant and animal cells

ATP

*An organic molecule (**adenosine triphosphate**) used to deliver energy for life processes*

A Precarious Balance

If you compare the equation for photosynthesis with the equation for cellular respiration, you will see that one is basically the reverse of the other. Products of photosynthesis (carbohydrates and oxygen) are the reactants of cellular respiration, and products of cellular respiration (carbon dioxide and water) are the reactants of photosynthesis. Because of this relationship between photosynthesis and cellular respiration, it has been hypothesized that plants and animals could coexist in a sealed environment such as a glass globe. In fact, artificial ecosystems have been created and do last quite a long time. Perhaps the most famous of these was the Biosphere II project, shown to the right. This project was an attempt to demonstrate that plants and animals (including humans) could live together in a closed environment for an extended period of time.

From 1991 to 1993, eight "biospherians" lived under an air-tight glass dome that contained a variety of plant and animal species. During their stay in Biosphere II, they raised their own livestock, grew their own crops, and collected scientific data. Today, the facility is being used as a research center. In it, scientists conduct experiments to further refine our understanding of the intricate balance between plant and animal life and the transfer of energy within an ecosystem.

Atmospheric oxygen

Atmospheric carbon dioxide

Energy

Energy

▲ Life on Earth is possible because of the interaction between photosynthesis and cellular respiration. If plant life disappeared from the Earth, most other organisms would soon cease to exist due to a lack of food and oxygen.

Summary

Plants make organic molecules such as carbohydrates by carrying out photosynthesis. Photosynthesis occurs inside plant organelles called chloroplasts. Photosynthesis uses water, carbon dioxide, and energy from the sun to produce organic compounds and oxygen. Plants and animals use glucose in cellular respiration. During cellular respiration, glucose and oxygen are used to produce energy in the form of ATP. Carbon dioxide and water are produced as byproducts. Photosynthesis and cellular respiration are interdependent processes because the products of photosynthesis are the reactants of cellular respiration, and the products of cellular respiration are the reactants of photosynthesis.

BASIC LIFE PROCESSES

The many chemical activities of living things can be grouped according to several major processes. Among these processes are *obtaining raw materials*, *metabolism*, *excretion*, and *regulation*. You may have observed processes that resemble these in nonliving things. For example, streams gather and transport sediments, while an air conditioner adjusts the air temperature in a room. However, all organisms—and the cells of which they are made—perform not just one, but all, of these processes. As you read about them, notice how each process is related to and depends on the others.

▲ Even though a computer has complex organization, responds to stimuli, and has moving parts, it is not alive. What can the mouse do that the computer cannot do?

Obtaining Raw Materials

To remain alive, living things must take in raw materials such as food, water, and mineral nutrients. *Food* is any substance that provides organisms with organic compounds. When you eat a snack or a meal, you supply your body with new organic molecules. Some organisms, such as most plants, algae, and certain bacteria, take in only inorganic substances. From these, they make the organic substances they need by carrying out photosynthesis. Organisms that make their own food, called *producers*, form the basis of all food chains. Organisms called *consumers* must eat other organisms to obtain organic substances.

◀ The movement of materials into and out of cells is regulated by the cell membrane, which allows certain substances to enter and keeps others out.

The materials obtained by both producers and consumers must be transported to the individual cells of the organism, where they are used. However, in order for cells to take in materials, the materials must first pass through cell membranes. Movement through a cell membrane is accomplished by two processes: diffusion and active transport.

DID YOU KNOW...

that if diffusion suddenly ceased, you could not live more than a few minutes? The oxygen that you inhale moves into your blood by diffusion, and the oxygen in your blood moves into all your cells by diffusion. Brain cells require a lot of oxygen and begin dying after 5 or 6 minutes without it.

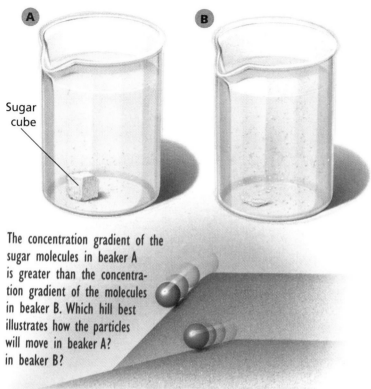

Sugar cube

The concentration gradient of the sugar molecules in beaker A is greater than the concentration gradient of the molecules in beaker B. Which hill best illustrates how the particles will move in beaker A? in beaker B?

Concentration gradient

The difference between the concentrations of a substance in two areas

Diffusion and Osmosis The movement of particles that occurs when the concentration of a dissolved substance differs in two neighboring areas is called *diffusion*. A difference in concentration can be pictured as a hill. Just as a ball would roll down the slope of a hill, from high to low, the movement of diffusing molecules and ions also occurs from high to low—high concentration to low concentration. In other words, the particles move down (or with) the concentration gradient.

The **concentration gradient** is simply the difference between the concentrations of a substance in two areas. The steeper the slope of the hill is, the faster the ball rolls. The same is true for diffusion—the greater the concentration gradient is, the faster the process of diffusion occurs.

Of course, molecules and ions do not really roll downhill. However, they are constantly moving back and forth between areas of different concentration. Since there are more particles to move from high to low, and fewer to move in the opposite direction, an area that is less concentrated gains more particles than it loses. As a result, the less concentrated area experiences an overall increase in molecules or ions of that kind.

Osmosis follows the same principle as diffusion but refers only to the diffusion of water through semipermeable membranes, such as those surrounding cells. The concentration of water molecules depends on the number of dissolved particles contained in the water. There are three types of environments that can exist around cells—hypotonic, hypertonic, and isotonic. The number of dissolved particles and the concentration of water molecules is different in each environment.

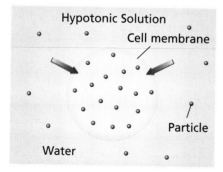

In a hypotonic environment there are more water molecules (fewer dissolved particles) outside the cell than inside. Therefore, the water molecules tend to move into the cell in order to equalize the concentration.

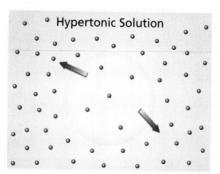

In a hypertonic environment there are fewer water molecules (more dissolved particles) outside the cell than inside. Therefore, water tends to leave the cell.

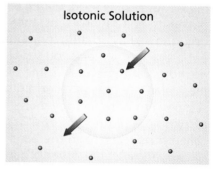

In an isotonic environment the concentration of water molecules is equal on both sides of the membrane. Therefore, there is no net movement of water in either direction.

Diffusion and osmosis are things that you might say "just happen" to a cell. The cells themselves do not take an active part in obtaining materials by either process. However, only certain materials can enter a cell by these means. First of all, the material entering a cell must be in higher concentration outside the cell. Second, the materials must be able to dissolve in water and must have molecules or ions small enough to pass through the tiny openings in a cell membrane.

Active Transport Sometimes a cell may need a substance that is already more concentrated inside the cell than it is outside the cell. Such substances cannot get in by diffusion. However, there *is* a method by which these substances can enter. This method is called **active transport**. Unlike diffusion, active transport requires energy. Where do you think this energy comes from? If you said from the ATP molecules made during cellular respiration, you are absolutely correct.

To understand why active transport uses energy but diffusion and osmosis do not, imagine that you are riding your bicycle and you come to a hill. If the hill goes down, you can relax because all you have to do is steer. You don't get tired because you don't use energy to roll downhill. But if the hill goes up, you have to use energy to pedal up the hill. As long as the hill goes up, you must continue to use energy.

Active transport

The movement of materials into and out of cells against a concentration gradient, requiring the use of energy

▲ Pedaling uphill is like active transport.

◀ Coasting downhill is similar to diffusion and osmosis.

Substances entering a cell by diffusion or osmosis are like your bicycle rolling downhill. Because both diffusion and osmosis occur with (or down) the concentration gradient, no energy is needed. On the other hand, active transport moves materials *against* (or up) the concentration gradient. Just as you must expend energy to ride your bicycle uphill, a cell must use energy to move materials against a concentration gradient.

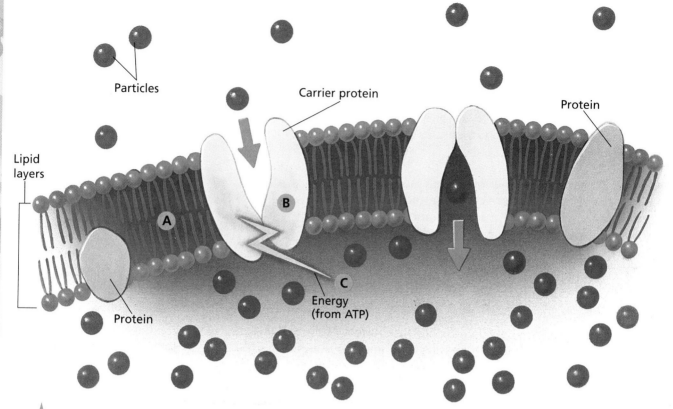

Particles

Carrier protein

Protein

Lipid layers

A

B

C

Energy (from ATP)

Protein

▲ The cell membrane (A) is composed of a double layer of lipids in which many kinds of proteins are embedded. Carrier proteins (B) act like gates. When a particle has to go against its concentration gradient through one of these gates, a chemical reaction inside the cell breaks down ATP to supply the needed energy. The energy released from the ATP (C) is used to change the shape of the carrier protein and move the particle through the membrane.

The secret to active transport is in the structure of the cell membrane. Large protein molecules called *carrier molecules* are positioned in the lipid layers of the membrane. These molecules change their shape and move ions or molecules into and out of the cell. Changing the shape of the carrier protein is what requires energy.

Metabolism

Living things are able to organize and reorganize the substances they take in for specific

functions. This involves breaking down large organic molecules into smaller molecules, as well as joining small molecules together to make larger ones. Together, all the chemical breaking-down and building-up processes of organisms are referred to as **metabolism**.

There are three basic chemical processes that make up metabolism: digestion, cellular respiration, and synthesis. In *digestion*, large food molecules are broken down into simpler compounds. Some of these compounds are then used in the process of *cellular respiration*, which supplies the energy for cell activities. Although the simple sugar glucose is the primary fuel for cellular respiration, fatty acids and amino acids can be used as well. In fact, if an

organism is suffering from starvation, it will break down any available organic compound to obtain energy.

Some of the products of digestion, such as sugars, amino acids, and nucleotides, are combined in new ways to make the specific carbohydrates, lipids, proteins, and nucleic acids that an organism needs for its activities. This process of putting smaller molecules together to make larger molecules is called *synthesis*. Photosynthesis is a great example of synthesis. As the name implies, it is a process that uses light to synthesize a product—organic molecules. From the raw materials obtained from its environment, an organism can synthesize many of the substances that are necessary for life.

Excretion

Building things up and breaking things down usually produces some materials that are unusable. These unusable materials are called *wastes*. Wastes that are produced by the chemical building-up and breaking-down processes of metabolism are called *metabolic wastes*. They include carbon dioxide, water, several nitrogen-containing compounds, and inorganic salts. If these wastes build up in a cell or an organism, they may become harmful, or toxic.

The removal of metabolic wastes is called **excretion**. Individual cells and unicellular organisms eliminate these wastes primarily by diffusion. In large multicellular animals, wastes are usually gathered from the area around individual cells by the fluids of the *circulatory system*. These wastes are then filtered out of the fluid for removal from the body. In humans, organs such as the lungs, skin, and kidneys play an important role in excretion. In plants, the primary metabolic waste product is oxygen, which diffuses out of the leaves through the stomata.

Excretion

The removal of metabolic wastes

In simple one-celled organisms, wastes simply diffuse into the surrounding environment.

In complex animals, such as humans, special organs rid the body of metabolic wastes. Lungs remove carbon dioxide and water from the body. Skin excretes water, nitrogen-containing compounds, and salts in the form of perspiration. Kidneys remove nitrogen-containing compounds, salts, and water, which are excreted in the form of urine.

Waste particles

Regulation

To remain alive, all organisms must continually monitor the activities of their cells, tissues, organs, and organ systems and keep them operating in a coordinated manner. This life process is called **regulation**. In order to regulate its activities, the individual parts of an organism must be able to send and receive messages. Plants, for example, regulate their activities by producing chemical substances called *growth regulators*. These chemicals initiate such changes as cell specialization, growth, tropisms (responses to light, gravity, or touch), and flower formation. Animals, on the other hand, have two highly developed systems that work individually and together to regulate the activities of their bodies.

Regulation

The coordination of the internal activities of an organism

One of these systems, the *nervous system,* acts very quickly, controlling adjustments that must be made immediately. For example, fleeing a predator is a response that requires quick action. The messages of the nervous system are passed along by transporting ions into and out of *nerve cells.* Such *electro-chemical messages* are transmitted from nerve cell to nerve cell at a speed of 100 m/s.

The other regulatory system in animals is called the *endocrine system.* The endocrine system acts more slowly than the nervous system. It co-ordinates body activities and controls gradual changes such as growth and development. The organs of the endocrine system, called *endocrine glands,* produce a variety of chemicals. These chemicals, called *hormones,* act as "chemical messengers" that travel through the bloodstream to affect certain tissues and organs.

There are cases in which the nervous system and the endocrine system work together to regulate activities in the body. For example, the nervous system monitors the level of a hormone called *thyroxine* in the bloodstream. Thyroxine, which is produced by the thyroid gland, increases the rate of cellular respiration in cells. The thyroid gland is signaled to produce more thyroxine when the body's cells are producing too little ATP and to produce less thyroxine when ATP levels rise too high.

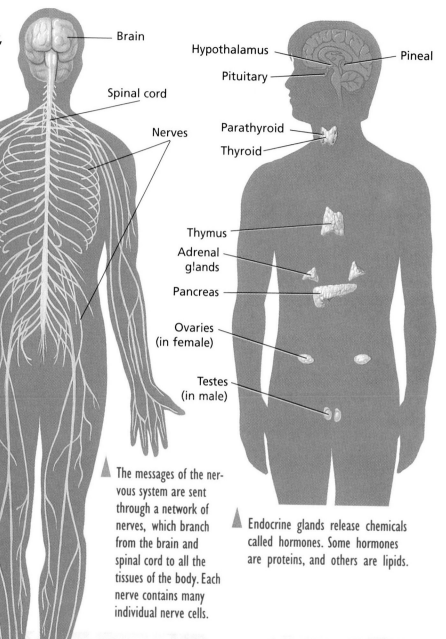

Brain

Spinal cord

Nerves

Hypothalamus

Pituitary

Pineal

Parathyroid

Thyroid

Thymus

Adrenal glands

Pancreas

Ovaries (in female)

Testes (in male)

▲ The messages of the nervous system are sent through a network of nerves, which branch from the brain and spinal cord to all the tissues of the body. Each nerve contains many individual nerve cells.

▲ Endocrine glands release chemicals called hormones. Some hormones are proteins, and others are lipids.

SUMMARY

Living things carry out certain processes that are necessary for maintaining life. First, organisms must take in raw materials. The materials are then transported to the cells, where they pass through membranes by diffusion or active transport. These materials are processed by the various chemical reactions of metabolism to produce the energy and the many different molecules that organisms need to live. Then, by the process of excretion, metabolic wastes and excess water are removed from cells and organisms. The various functions of an organism are regulated by slow-acting chemical messengers and (in animals) a quick-acting nervous system.

Concept Mapping

The concept map shown here illustrates major ideas in this unit. Complete the map by supplying the missing terms. Then extend your map by answering the additional question below. Write your answers in your ScienceLog. **Do not write in this textbook.**

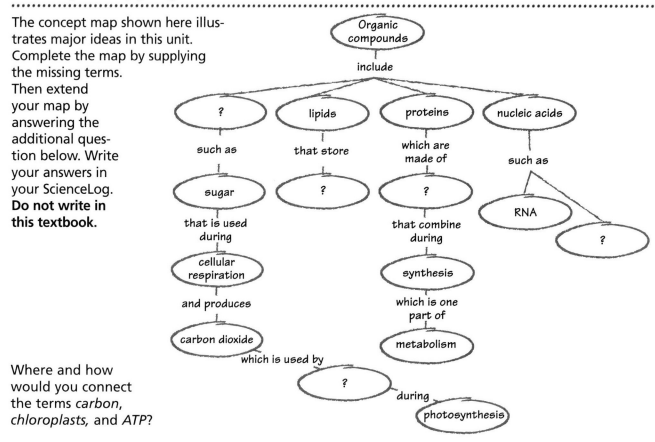

Where and how would you connect the terms *carbon*, *chloroplasts,* and *ATP*?

Checking Your Understanding

Select the choice that most completely and correctly answers each of the following questions.

1. Which biochemical is most readily used as an energy source by cells?
   a. protein
   b. amino acid
   c. carbohydrate
   d. glucose

2. Which is NOT an organic compound?
   a. ATP
   b. DNA
   c. H_2O
   d. $C_6H_{12}O_6$

3. Water is important to living things because
   a. it participates in hydrolysis reactions.
   b. it contains carbon atoms.
   c. it is an uncharged molecule.
   d. it is made of amino acids.

4. Energy from the sun reaches humans by way of
   a. starch.
   b. glucose.
   c. ATP.
   d. All of the above

5. Which is NOT true of metabolic wastes?
   a. They are excreted.
   b. They are produced only by animals.
   c. They may result from cellular respiration.
   d. They may be used by other organisms.

Interpreting Graphs

A cell is placed in a salt solution. Which graph (*A* or *B*) indicates that the solution is hypertonic? Explain your answer.

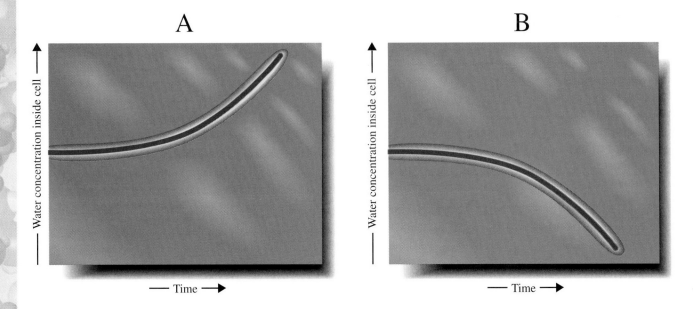

A

B

Water concentration inside cell → Time →

Water concentration inside cell → Time →

Critical Thinking

Carefully consider the following questions, and write a response in your ScienceLog that indicates your understanding of science.

1. While doing research, an organic chemist determines that a certain polymer contains carbon, hydrogen, oxygen, nitrogen, and phosphorus. The chemist then hydrolyzes a fresh sample of the compound. Name the monomers that result from the hydrolysis.

2. Let's say that a scientist is able to make water that has cohesive, but not adhesive, properties. Why would using this special water to water the plants in Biosphere II eventually have an effect on *all* of the life in this facility?

3. One theory suggests that the extinction of the dinosaurs resulted from a giant meteorite that crashed into the Earth about 65 million years ago. Scientists think that the tremendous impact could have thrown a giant cloud of dust into the atmosphere, preventing sunlight from reaching Earth's surface for many months. Explain how this could have caused the extinction of the dinosaurs.

4. Explain the following statement: A cell low in ATP is probably starving for raw materials.

5. Does your body's cellular respiration produce metabolic wastes at the same rate during the day as it does at night when you are sleeping? Explain your answer.

Portfolio Idea

Imagine that you are a reporter who is interviewing a water molecule and a glucose molecule. During the interview, each molecule boasts that *it* is the most important molecule on Earth. Use all the information that you gain from your interview to write a story that supports one of the molecules as being the most important. Be creative, but use facts and logic in reaching your conclusion.

Particles

IN THIS UNIT

Now *that you have been introduced to the concept of particles, consider the following questions.*

1. What is a *mole*? How is it used in measuring matter?

2. How can the kinetic molecular theory of matter be used to explain certain physical properties of matter?

3. Are there particles of matter that are smaller than atoms? What are these particles called?

4. In what way does the particle model of matter fail? How could you account for this failure?

In this unit, you will take a closer look at particles of matter and theories about the nature of particles.

PARTICLES OF MATTER

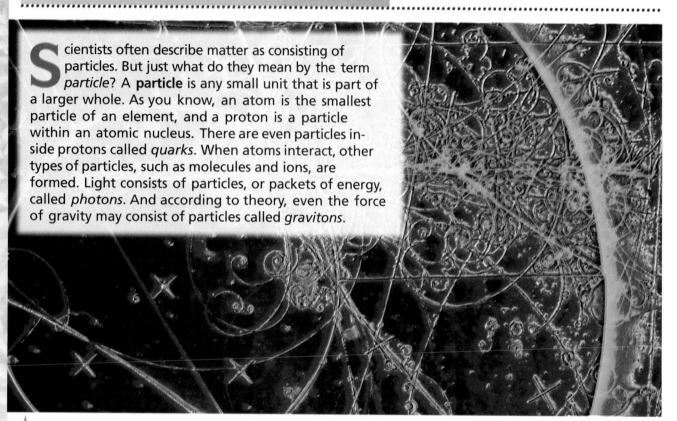

Scientists often describe matter as consisting of particles. But just what do they mean by the term *particle*? A **particle** is any small unit that is part of a larger whole. As you know, an atom is the smallest particle of an element, and a proton is a particle within an atomic nucleus. There are even particles inside protons called *quarks*. When atoms interact, other types of particles, such as molecules and ions, are formed. Light consists of particles, or packets of energy, called *photons*. And according to theory, even the force of gravity may consist of particles called *gravitons*.

▲ This artificially colored photograph shows the tracks of many subatomic particles in a bubble chamber. These particles appear when atoms collide and break apart.

Particle

Any small unit (of matter or energy) that, with others, forms a larger whole

▲ These models represent molecules of different chemical compounds.

Molecules and Ions

You frequently hear the word *molecule* used to describe any particle composed of two or more atoms. We often say that all compounds—substances made from two or more elements—are made of molecules. This, however, is not quite accurate. In the scientific sense, the term *molecule* refers only to neutral groups of atoms that are joined by a particular type of chemical bond.

You may recall that there are two distinct types of chemical bonds—ionic and covalent. An ionic bond forms between atoms that have opposite electrical charges. A covalent bond forms when two or more atoms share electrons. The particles we call *molecules* consist of atoms that are joined by covalent bonds and have no net electric charge. For example, water consists of water molecules in which two hydrogen atoms share electrons with one oxygen atom. Thus, the chemical formula for a molecule of water is H_2O.

On the other hand, a substance such as sodium chloride (table salt) does not consist of molecules. Rather, it contains particles called *ions*, which form when electrons move from one atom to another. Atoms that lose electrons take on a positive charge, and those that gain electrons take on a negative charge. Two ions with opposite charges attract each other, forming an ionic bond.

By magnifying grains of table salt, you can see that salt crystals are shaped like cubes.

In a crystal of table salt, opposing electric charges hold positive sodium ions (Na^+) and negative chloride ions (Cl^-) together. Chloride ions surround each sodium ion, and sodium ions surround each chloride ion, as shown in the illustration to the right. For every Na^+ ion in a salt crystal there is one Cl^- ion. However, since the ions in a salt crystal do not *share* electrons, their groupings are not considered molecules. The designation NaCl is simply a way of writing formulas and equations in which sodium chloride takes part. Therefore, the symbol NaCl is called a *formula unit.*

Can you see how useful the term *particle* might be? You can talk about particles of sugar (sugar molecules) and particles of salt (Na^+ and Cl^- ions), as well as the properties they share, without having to consider the details. This comes in handy in a number of areas of study.

Na^+

Cl^-

The shape of a salt crystal is determined by the arrangement of its ions.

Counting Particles

All types of particles, no matter what they are, share certain characteristics. One of these is the ability to be counted. Particles can be counted either individually or in groups. But since particles of matter are so small, we need a special way to count them.

The Mole   When you go to the store to buy eggs, you usually buy them by the dozen. When you buy a dozen eggs, you know that you are getting 12 eggs. The word *dozen* is a counting term for "groups of 12." Another counting term you may be familiar with is the *gross,* which is 12 dozen. When you buy a gross of oranges you get 144 oranges. It does not matter whether you buy eggs or oranges, because the number of items in a dozen or a gross is always the same. If you buy a dozen, you get 12. If you buy a gross, you get 144.

In science, a very useful counting term for particles of matter is called the *mole* (mol). No, we are not talking about little furry animals! A **mole** is a unit for measuring the *amount of a substance.* Just as there are always 12 items in a dozen—no matter what you are counting—a mole of any substance always contains the same number of particles.

Mole

The SI unit for the amount of a substance

Avogadro's number

The number of particles in 1 mol of a substance (6.022137 × 10²³)

Avogadro's Number   Because the basic particles of matter are so small, the number of particles in a mole is very large. But just how many particles are there in a mole? It has been determined that one mole of any substance contains 6.022137×10^{23} particles. This number is called **Avogadro's number** in honor of the Italian scientist Amedeo Avogadro, who first suggested the idea of molecules. Written out, Avogadro's number is 602,213,700,000,000,000,000,000.

Imagine getting the help of all 6 billion people on Earth to count the number of particles in only 1 mol of a substance. If each person counted 1 particle per second, it would take over 3 million years to count all the particles in a mole! To put it another way, if you made $40,000 every *second* at your job and you had been working since Earth formed 4.5 billion years ago, you would not yet have earned Avogadro's number of pennies.

Using the Mole   Using the mole is like using any other standard quantity, such as a dozen or a gross. Suppose you were in charge of preparing breakfast for a group of 36 students and you wanted to serve breakfasts with 2 eggs and 1 sausage patty on each plate. You could order, say, 6 dozen eggs and 3 dozen sausage patties. This way, you would have just enough eggs and sausage patties so that none would be left over after 36 breakfast plates were filled.

In a similar way, chemists usually determine the amounts of the chemicals they will need for a particular chemical reaction in terms of moles. For example, to make water molecules, you need twice as many hydrogen atoms as oxygen atoms. Therefore, you could use 2 mol of hydrogen atoms and 1 mol of oxygen atoms. Just as you were sure you could serve 36 identical breakfast plates by buying eggs and sausage patties in dozens, chemists ensure that they get the correct ratio of atoms by using moles. But how can moles be translated into units you are already familiar with? To understand the answer to this question, you must look at the periodic table.

◄ If you could stack 6.02×10^{23} pennies on top of one another, the stack would reach to the other side of the Milky Way galaxy.

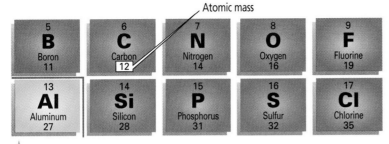

Atomic mass

| 5 **B** Boron 11 | 6 **C** Carbon 12 | 7 **N** Nitrogen 14 | 8 **O** Oxygen 16 | 9 **F** Fluorine 19 |
| 13 **Al** Aluminum 27 | 14 **Si** Silicon 28 | 15 **P** Phosphorus 31 | 16 **S** Sulfur 32 | 17 **Cl** Chlorine 35 |

By finding the atomic mass of an element on the periodic table, you can determine the number of grams in 1 mol of that element.

Moles to Grams   If you look at the periodic table, you will find that hydrogen (H) has an atomic mass of 1. This value tells you the number of grams in 1 mol of that element. In other words, there is 1 g of hydrogen in 1 mol of hydrogen atoms. Likewise, the atomic mass of oxygen (O) is 16. Thus, 1 mol of oxygen has a mass of 16 g. As you might expect, 2 mol of hydrogen atoms would be twice as much as 1 mol, or 2 g of hydrogen. *To find the number of grams in 1 mol of any particular element, just look up its atomic mass and use it as the number of grams.*

Let's again use eggs to illustrate how moles relate to grams. Suppose that you have three sizes of eggs—small (30 g), medium (40 g), and large (50 g). If you have a dozen of the small eggs, you would have a total of 360 g of eggs. A dozen medium eggs would be 480 g, and a dozen large eggs would be 600 g. You could order 360 g of small eggs or 600 g of large eggs and expect to get 1 dozen in either case.

In the same way, scientists use grams to determine the number of moles that they want to use. If you know the atomic mass of the atoms of an element, you can determine how many grams of that element you need to get 1 mol (or 6.02×10^{23} atoms), as the photos on this page illustrate.

Moles of Compounds   Atoms of elements are *not* the only particles that you can count with moles, or Avogadro's number. You can also have a mole of ions or a mole of molecules. Therefore, you can measure amounts of *compounds* in moles. Recall that a water molecule (H_2O) has two atoms of hydrogen and one atom of oxygen. Therefore, 1 mol of water contains 2 mol of hydrogen and 1 mol of oxygen. Still, there are only 6.02×10^{23} *molecules* in 1 mol of water.

To determine the number of grams in 1 mol of water, you would first find the atomic masses of the hydrogen and oxygen atoms. Then you would add the atomic masses of each of the atoms in one molecule. For a water molecule (H_2O), your calculation would be as follows: $1 + 1 + 16 = 18$. Therefore, by applying the same rule that you used for moles of elements, you would need 18 g of water to get 1 mol of water molecules.

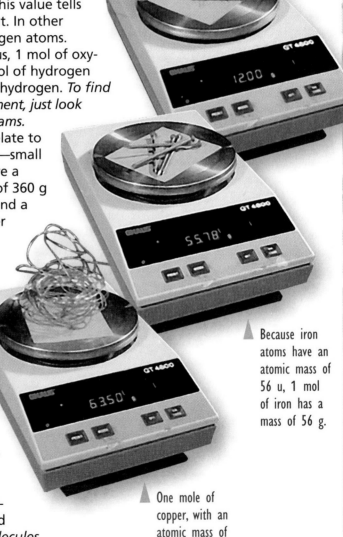

Carbon atoms have an atomic mass of 12 u. Therefore, 1 mol of carbon has a mass of 12 g.

Because iron atoms have an atomic mass of 56 u, 1 mol of iron has a mass of 56 g.

One mole of copper, with an atomic mass of 64 u, has a mass of 64 g.

The Mole in Chemistry Knowing the mass of a mole of any substance is very important to chemists. Again, suppose that you wanted to combine hydrogen with oxygen to get water. In nature, hydrogen gas occurs as H_2 molecules, and oxygen gas occurs as O_2 molecules. Therefore, to make water, you must combine hydrogen and oxygen according to the following balanced chemical equation:

$$2H_2 + O_2 \rightarrow 2H_2O$$

The equation states that two *molecules* of H_2 react with each *molecule* of O_2. That is, you will need twice as many H_2 molecules as O_2 molecules for your reaction. This is where moles come in handy. If you use twice as many moles of H_2 as you do of O_2, you will have the right proportions of the two chemicals to make water. Suppose that you use 10 mol of H_2 and 5 mol of O_2; how many grams of each would you need?

Two molecules (or moles) of H_2 plus one molecule (or mole) of O_2 yields two molecules (or moles) of water.

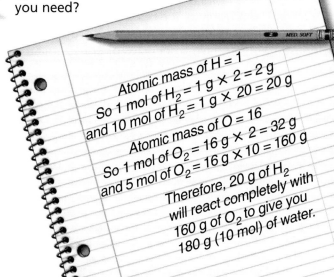

Atomic mass of H = 1
So 1 mol of H_2 = 1 g × 2 = 2 g
and 10 mol of H_2 = 1 g × 20 = 20 g

Atomic mass of O = 16
So 1 mol of O_2 = 16 g × 2 = 32 g
and 5 mol of O_2 = 16 g × 10 = 160 g

Therefore, 20 g of H_2 will react completely with 160 g of O_2 to give you 180 g (10 mol) of water.

SUMMARY

Scientists often use the term *particle* when discussing matter and energy. Molecules are particles made up of atoms held together by covalent bonds. Ions are particles formed when atoms gain or lose electrons. The simplest grouping of ions in a compound formed by ionic bonds is called a formula unit. *Mole* is a term used for counting the basic particles of matter, whether they be atoms, molecules, or ions. Avogadro's number is the number of particles in a mole of a substance. A mole of any one substance has exactly the same number of particles as a mole of any other substance. The mass of a mole of any substance is the atomic mass of its particles given in grams.

PARTICLES IN MOTION

Joseph Black, the Scottish chemist who in the 1750s first identified carbon dioxide gas, devoted much of his life to investigating the effects of heat on matter. Among the things he studied were the freezing, melting, and boiling points of different substances. In one experiment, Black observed how the temperature of water changes as it is heated to boiling. Of this experiment he wrote:

The liquid gradually warms, and at last attains the temperature, which it cannot pass without assuming the form of vapor . . . However long and violently we boil a liquid, we cannot make it hotter than when it began to boil. The thermometer always points to the same degree, namely, the vapor point of that liquid. Hence the vapor point of a liquid is often called its boiling point.

Why is it impossible, under ordinary conditions, to raise the temperature of boiling water above 100°C? How do particles of liquid water differ from particles of water vapor? What holds particles of boiling water together? Forces of attraction between particles of matter and the *kinetic molecular theory of matter* can be used to answer these and other questions about the behavior of matter.

Forces of Attraction

Have you ever wondered what holds a piece of paper, a steel beam, or a raindrop together? As you know, many different forces draw particles of matter together. Electrons stay in orbit about an atom's nucleus because the negatively charged electrons are attracted by the positively charged nucleus. Water from a faucet falls toward the Earth because of the pull of gravity. Yet when you pour water into a clean glass, some of it creeps up the inside of the glass, *against* the pull of gravity. Two forces of attraction—cohesion and adhesion—are responsible for the behavior of the water in a glass.

1 Because water tends to adhere (stick) to glass, it moves up the side of the cylinder.

2 The surface of the water forms a curve called a meniscus.

3 Because water molecules cohere (stick to each other), the water's surface remains unbroken.

If you look closely, you will ▶ see that the surface of the water in a clean glass graduated cylinder is not level.

Cohesion

The force of attraction between like particles

Adhesion

The force of attraction between unlike particles

Cohesion Particles of matter that are alike, such as water molecules, tend to stick together. This force of attraction between like particles is called **cohesion.** In solids, cohesion may sometimes be very strong. Thus, cohesion holds solids, such as sheets of paper and steel beams, in a definite shape. Cohesion also holds together the particles of liquids. However, the cohesion between the particles of liquids is not usually as strong as the cohesion between the particles of solids. Imagine trying to hold up a sheet of water in the same way you hold up a sheet of paper. Because there is less cohesion between water molecules than there is between paper particles, a sheet of water would not hold together.

Adhesion Particles of different kinds of matter may also stick to each other. For example, when water touches clean glass, the water and glass particles attract one another much more stongly than water particles attract each other. The force of attraction between the particles of two different substances is called **adhesion.** There is a great deal of adhesion between glues and the substances to which they stick.

Surface Tension Have you ever watched water dripping from a leaky faucet? It does not drip molecule by molecule. Rather, the water molecules stay together until a large drop forms. As the drop falls, it becomes a sphere. But without a container, what keeps drops of liquid water in this shape? The spherical shape of a falling water drop results from a property of liquids called *surface tension.* Surface tension is the tendency of the particles at the surface of a liquid to pull together. As the diagram to the left shows, *surface tension* results from cohesion between a liquid's molecules. Because of surface tension, some insects are able to walk across the surface of a pond without sinking into the water.

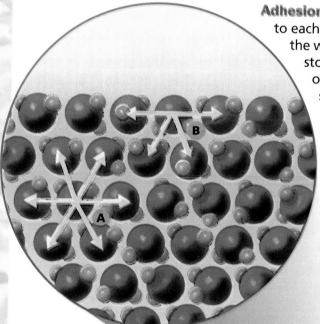

1 A molecule beneath the surface of the water (position **A**) is equally attracted to other water molecules on all sides.

2 A molecule on the surface of the water (position **B**) is attracted to other water molecules beneath it as well as next to it, but there are no water molecules to attract it upward.

3 Thus, the molecules at the surface are pulled closer together, causing them to act like a thin elastic film.

▼ Surface tension allows this water strider to walk on water.

DID YOU KNOW...

that surface tension is what makes raindrops round? In a drop of water, all surface particles are pulled toward the center of the drop, giving it a spherical shape.

S28

The Kinetic Molecular Theory of Matter

One of the most important theories of modern science is the *kinetic molecular theory of matter*, which states that *matter is made of very tiny particles that are in constant motion*. The particles that make up matter are, of course, atoms, molecules and ions. Because they are constantly moving, these particles have kinetic energy—the energy of motion. The faster the particles move, the more kinetic energy they have.

States of Matter   The kinetic molecular theory can be used to explain the difference between the three states of matter—solids, liquids, and gases. In a solid, the atoms or molecules do not have enough kinetic energy to overcome the cohesion that holds them together. Therefore, the particles of a solid cannot move about freely; they simply vibrate back and forth about fixed positions. For this reason, solids have a definite shape and a definite volume.

The atoms or molecules in a liquid have more kinetic energy than they do when in a solid state. Although the particles of a liquid still cling together because of cohesion, they are free to move around or slide past each other. The motion of these particles resembles the way that marbles in a plastic bag would roll around if you squeezed the bag. Because the particles of a liquid are held together by cohesion, liquids have a definite volume. But because the particles of a liquid can change position, liquids have no definite shape. Liquid water, for example, can be poured from a tall, narrow glass into a short, wide glass, but it will occupy the same volume in both glasses.

Gas particles have enough kinetic energy to overcome the forces of cohesion altogether. The atoms or molecules of gases fly off in straight lines until they collide with other particles or the walls of a container. For this reason, gases expand until they fill the container they occupy. Thus, they do not have definite shapes or volumes.

▲ The atoms or molecules in a solid vibrate in fixed positions. In many solids, the particles have a very orderly arrangement that gives the solid a crystal shape.

◀ Particles in a gas have a lot of kinetic energy and little cohesion, and they can move freely from place to place. On average, the distance between the particles in a gas is very great compared with the size of the particles.

◀ Liquid molecules can move around but are always in contact with other molecules. The shape of the container in which a liquid is held determines the shape of the liquid.

Diffusion

The mixing of the particles of one substance with the particles of another substance because of the motions of their particles

Diffusion We can see that atoms and molecules are in motion when we observe **diffusion**, which is the mixing of the particles of two or more substances as a result of their motion. You have already learned that diffusion is an important way that living systems distribute materials. Although you cannot see diffusion happening inside your body, you experience it every time you smell something. Because gas particles travel much farther between collisions than do particles of solids or liquids, diffusion is more rapid in gases. For example, if you open a bottle of strong perfume inside a room, the perfume molecules will quickly diffuse through the air. Air molecules help to distribute the perfume molecules, and you can soon smell the perfume all over the room.

Another way to observe diffusion is to place a drop of ink into a glass of water. The ink particles move among the water molecules until they are equally distributed throughout the glass. As you might predict by using the kinetic molecular theory, the temperature of the water influences the rate of diffusion. Because particles move faster at higher temperatures, diffusion is more rapid at higher temperatures.

Diffusion even occurs between the particles of solids. However, because the particles of solids are not nearly as free to move around, diffusion occurs very slowly in solids. For example, if you were to stack a lead brick on top of a gold brick, a very slow diffusion would occur between these two solids. After several months, particles of one solid could be detected in the other.

Over time, ink gradually mixes with water to form a uniform mixture. This is called *diffusion*.

Changes of State We can also use the kinetic molecular theory of matter to explain how matter changes states. Recall Joseph Black's discovery that the temperature of a liquid at its boiling point remains constant even when it is heated further. As you know, temperature is a measure of the average amount of kinetic energy of the particles in a substance. Thus, when the temperature of a substance increases, it gains kinetic energy and its particles move faster. However, the molecules in a boiling liquid *cannot* move faster unless they break free of the cohesion that holds them together. Therefore, the temperature of boiling water stays the same even when you add more heat. Additional heat provides the energy needed to overcome the cohesion of water molecules. These molecules escape the liquid as a gas. Until all molecules have escaped, the temperature of the water stays at its boiling point.

Likewise, adding heat to a solid increases the kinetic energy of its particles. When the particles acquire enough energy to move around each other freely, the solid melts. Once at its melting point, however, the temperature will remain constant until all the solid has melted. After melting, the particles of the solid can move even faster when more heat is added. As a result, the kinetic energy of the particles and the temperature of the substance can increase.

What happens when gases and liquids cool? As heat leaves a substance, its particles lose kinetic energy and slow down. Once the particles lose enough kinetic energy, cohesion begins to draw the particles together. When the particles of a gas cohere (stick together), they get much closer together, and we say that the gas *condenses*. Heat is given off in the process. When the particles of a liquid lose their ability to move past each other freely, we say that the liquid freezes.

Changing States of Water

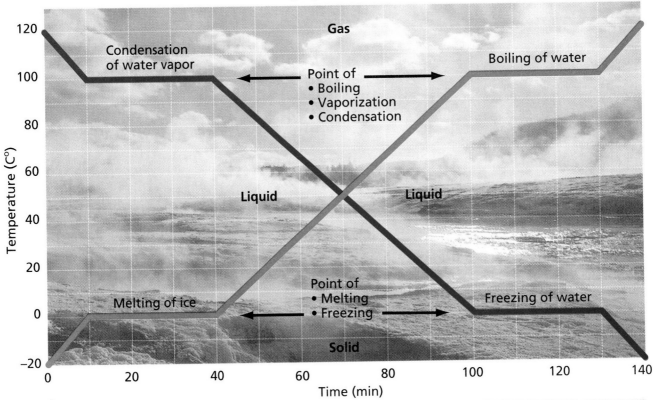

This graph compares the changes in temperature that occur as ice changes from a solid to a liquid and then to a gas, and as water vapor changes from a gas to a liquid to a solid. Notice that the boiling, vaporization, and condensation points of a substance are the same, as are the melting and freezing points.

S31

Applications of the Kinetic Molecular Theory of Matter

Conditions that affect the way particles move, such as temperature and pressure, may change the volume occupied by matter. Because the particles in solids and liquids are already quite close together, the volumes of solids and liquids change very little as temperature changes. In most cases, pressure has even less effect on the volume of solids and liquids. However, any change in temperature or pressure greatly affects the volume of a gas. The particles of a gas will readily spread out or move together to fit the conditions. For example, if you leave a balloon full of air in bright sunlight, the balloon expands as the air inside warms up. If you push down on the handle of a bicycle pump, the volume of the air inside the pump decreases and the pressure increases. Several scientific laws describe the effect of temperature and pressure on matter. Let's see how the kinetic molecular theory of matter can be used to explain these laws.

1 atm = *atmospheric pressure at sea level (101.3 kPa)*

Boyle's Law   Robert Boyle was an Irish scientist who lived during the seventeenth century, around the same time as Isaac Newton. He studied how gases behave when the pressure on them changes. Robert Boyle experimented with several gases, such as oxygen and carbon dioxide. In each case, he found that *doubling* the pressure always reduced the volume of a gas by *one-half,* if its temperature remained constant. His observations led to what is now called **Boyle's law,** which can be stated as follows: *The original volume (V_1) occupied by a certain amount of gas multiplied by its pressure (p_1) is equal to its new volume (V_2) multiplied by its new pressure (p_2),* or

$$V_1 \times p_1 = V_2 \times p_2$$

For example, suppose that the air trapped by the piston seen on this page has a volume of 200 mL (V_1) when the pressure is equal to 1 atm (p_1). If the total pressure increases to 2 atm (p_2), its new volume (V_2) will be 100 mL.

$$200 \text{ mL} \times 1 \text{ atm} = V_2 \times 2 \text{ atm}$$

$$V_2 = \frac{200 \text{ mL} \cdot 1 \text{ atm}}{2 \text{ atm}}$$

$$V_2 = 100 \text{ mL}$$

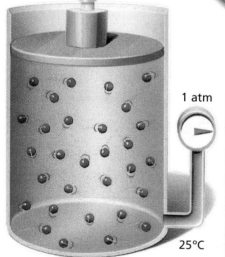

1 atm

25°C

A gauge measures the pressure exerted on a gas by a piston. As pressure increases, volume decreases in proportion to the increase in pressure.

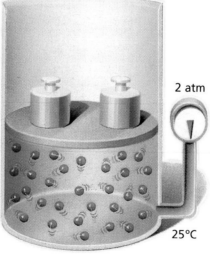

2 atm

25°C

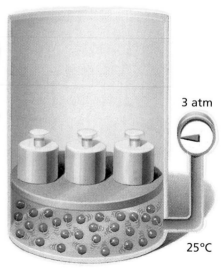

3 atm

25°C

The kinetic molecular theory of matter can be used to explain Boyle's law in the following way: As the force that causes the pressure pushes down on the air inside the cylinder, the air inside the cylinder pushes upward against the piston with an equal force. Otherwise, the piston would go all the way down to the bottom of the cylinder. The pressure exerted on the bottom of the piston to hold it up is a result of the constant bombardment by the trapped air particles inside. Each time the volume occupied by the air inside the cylinder is reduced by *half*, the number of particles per unit of volume *doubles*. Thus, twice as many particles strike the surface of the piston per second and therefore exert twice the force.

Pascal's Law Blaise Pascal was a French mathematician, scientist, and philosopher who also lived during the seventeenth century. Among other things, he is known for his work on the effect of pressure on liquids. Pascal observed that pressure applied to a liquid in a closed container is felt throughout the container. This observation is described in *Pascal's law*: *Whenever pressure is increased at any point, an equal change in pressure occurs throughout the liquid.*

Hydraulic systems, such as the hydraulic lift shown below, work on the principle of Pascal's law. A hydraulic lift has an *input piston*, which is used to apply pressure to a liquid in a closed container, and an *output piston*, on which the pressure can act. Because the output piston has a larger surface area than the input piston, a small force exerted on the input piston creates a large force on the output piston. The brake system of an automobile uses the same principle. A small amount of pressure applied to the brake pedal is transmitted by the brake fluid to the brake pads, where the force is multiplied many times—enough to bring the car to a stop.

Pascal's law can also be explained by the kinetic molecular theory of matter. Liquids, like gases, are fluids. A *fluid* is any substance that can flow and change shape. Fluids have these characteristics because their particles can move about when pressure is applied to them. Unlike the particles of gases, however, the particles of liquids cannot be compressed (pushed closer together) because they are already close together. Thus, when pressure is applied to a liquid in a closed container, it cannot be compressed and it cannot change shape. As a result, the pressure applied at one point causes the liquid to push on all sides of its container.

$\dfrac{100 \text{ N}}{\text{cm}^2}$

$\dfrac{100 \text{ N}}{\text{cm}^2}$

According to Pascal's law, pressure applied to the cork in a bottle full of liquid is dispersed equally throughout the liquid in all directions.

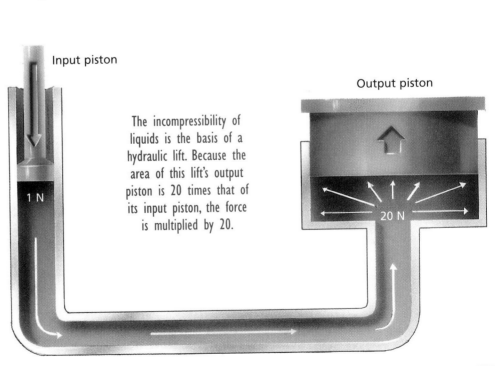

Input piston

Output piston

The incompressibility of liquids is the basis of a hydraulic lift. Because the area of this lift's output piston is 20 times that of its input piston, the force is multiplied by 20.

1 N

20 N

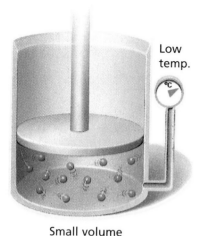

Low temp.

Small volume

High temp.

Large volume

▲ According to Charles' law, when pressure remains constant, the volume of a gas increases as the temperature increases.

◀ Charles' law explains why the hot air inside this balloon is less dense than the surrounding air.

Charles' Law In the late eighteenth century, the French scientist Jacques Charles observed that the volume of air increases steadily as its temperature increases. This observation is described by *Charles' law*: *The volume of a gas increases as the temperature increases, if the pressure remains the same.*

When using the Kelvin temperature scale, this relationship can be expressed as a direct proportion, which is written as

$$\frac{V_1}{T_1} = \frac{V_2}{T_2}$$

In other words, if you double the Kelvin temperature of a gas, its volume doubles. As is the case when you use Boyle's law, if you know three values, you can calculate the fourth.

How does the kinetic molecular theory apply to Charles' law? If the temperature of a gas increases, the gas particles move faster and collide with the walls of their container more often. Because of this, they exert more force. If pressure is to remain constant, the particles must travel greater distances, requiring the container to expand and the volume to increase.

SUMMARY

The kinetic molecular theory states that matter is composed of very tiny particles that are in constant motion. This scientific model helps explain many of the physical properties of matter, as well as diffusion, changes of state, and the effects of changes in temperature and pressure on volume. Boyle's law describes how the volume of a gas changes as the pressure changes. Pascal's law describes the effects of applying pressure on a liquid in a closed container. Charles' law describes how the volume of a gas changes as the temperature changes.

DID YOU KNOW...

that gases expand about 1/273 (0.37%) in volume when heated by 1°C? This degree of expansion is the basis for the Kelvin, or absolute, temperature scale. At absolute zero, which is −273°C or 0 K, the motion of the particles of a substance reaches its minimum.

PARTICLES OF PARTICLES

Each element consists of its own unique atoms—the smallest particles of matter that we usually deal with in the field of chemistry. But are atoms the smallest particles found in nature? The answer is no. Among the other particles that you have come across in your investigations of matter are protons, neutrons, and electrons. These particles are studied by physicists.

The current view of the universe considers *matter* to be the most basic entity (thing). This means that every substance that exists must be composed of fundamental particles of matter. Until the late nineteenth century, matter was thought to be formed of indivisible atoms. Then, due to the work of J. J. Thomson, Ernest Rutherford, and James Chadwick, new models of the atom—composed of protons, neutrons, and electrons— were proposed.

More recently, even smaller particles have been identified. Protons and neutrons, for example, are now thought to be made of three smaller particles called **quarks.** Although quarks have never been isolated, physicists have inferred that they exist from the interactions of atoms in huge machines called particle accelerators.

▲ Atoms consist of smaller pieces called subatomic particles.

▲ This model of a helium nucleus shows that protons and neutrons are made of three smaller particles called quarks.

Quark

A theoretical particle that is thought to make up protons and neutrons

Particle Interactions

Scientists have used their knowledge of subatomic particles and the forces involved in their interactions to explain the behavior of matter and the particles of which it is composed. According to present theories, four fundamental forces help to hold everything together.

Physicists use this particle ▶ accelerator, at Fermilab in Batavia, Illinois, to explore the interior of atoms.

Gravitational Force The primary interaction that affects stars, planets, and most of the visible objects around you is *gravitational force*. Gravity pulls everything near the Earth toward the center of the Earth and keeps you from flying off into space. The strength of this pull is directly proportional to the masses of the objects that are attracted and is inversely proportional to the square of the distance between them—the greater the distance, the less the attractive force.

Electromagnetic Force At the level of atoms and molecules, interactions that involve *electromagnetic force* have a greater influence than does gravity. Electromagnetic interactions occur between particles that are either electrically charged or magnetic. Remember that like electric charges repel each other, and opposite electric charges attract each other. Magnetic fields work in a similar way. Electrons are held in orbit about atomic nuclei because of the electrical attraction between negatively charged electrons and positively charged nuclei. Atoms stick together to form molecules for the same reason. The theory that explains electromagnetic force is called *quantum electrodynamics*, or QED.

Strong Force The *strong force* is what holds an atomic nucleus together. As you know, atomic nuclei contain the small particles called protons and neutrons. Each of these particles is thought to be made of still smaller particles called quarks. The strong force holds the three quarks inside each proton and neutron together. The strong force also holds the nucleus itself together. This means that the strong force is stronger than the electromagnetic force that tends to repel protons, which have like electrical charges.

Weak Force The fourth fundamental force is called the *weak force*. The weak force is responsible for several types of radioactive decay. This force also helps to hold particles such as neutrons together. In weak-force interactions, particles called *neutrinos* are either produced or absorbed. Neutrinos are very small. In fact, physicists are still trying to determine whether they have any mass at all. As indicated by their name, neutrinos do not have any charge.

Some scientists think that there are four forces that hold everything together. Each force operates at a different level of organization, as shown here.

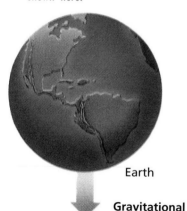

Earth

Gravitational force

Moon

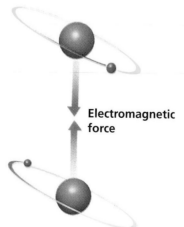

Electromagnetic force

Atoms and molecules

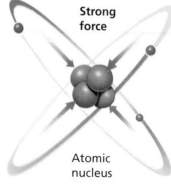

Strong force

Atomic nucleus

Weak force

Neutron

Quantum Theories

The theories explaining the electromagnetic, strong, and weak forces are called *quantum* theories. These theories assume that only discrete packages, or *quanta*, of energy can be transferred when particles interact. For example, photons are quanta of light energy. The transfer of quanta of energy can be compared to buying eggs at a grocery store. You can buy 1 egg or a dozen eggs, but you cannot buy 2.7 eggs. The same is true of photons of light. An object might absorb 2 or 3 photons, but it cannot absorb 2.7 photons.

Because electromagnetic, strong, and weak forces can all be described by quantum theories, many scientists have tried to combine these forces into a theory that explains them all. One theory, called the *electroweak theory*, combines the electromagnetic and weak forces. The electroweak theory explains everything that can be explained by considering electromagnetic force and weak force as separate forces. This theory has even successfully predicted the existence of some previously unknown particles.

An egg can be divided only by breaking the shell. Quanta, however, are indivisible.

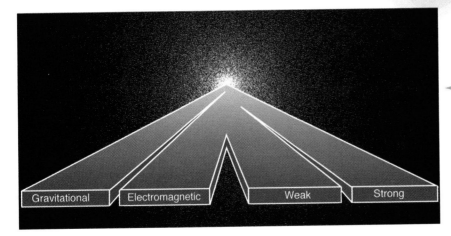

Gravitational   Electromagnetic   Weak   Strong

Scientists theorize that in the first few seconds after the big bang, the four fundamental forces separated from one original unifying force. Gravitational force was the first to separate from the other three. Next, the strong force separated from the electroweak force. Finally, the electroweak force separated into the electromagnetic and weak forces.

Since the electroweak theory was proposed, a number of theories that describe the strong force and the electroweak force in a single theory have also appeared. As a group, these are called *grand unified theories*, or *GUTs*. All GUTs share some features, including the prediction that electroweak and strong forces become a single force at extremely high energies when particles are very close together. Eventually, physicists may be able to explain all four fundamental forces with a single "theory of everything."

Hydrogen atoms combine by nuclear fusion to form helium. Smaller particles and a great deal of energy are released by this reaction.

Where Do We Go From Here?

Before the mid-twentieth century, the *law of conservation of mass* (matter cannot be created or destroyed) and the *law of conservation of energy* (the amount of energy in the universe is constant) were thought to always hold true. But we now know that matter can be transformed into energy. Every second, for example, 657 million tons of the sun's hydrogen is converted into 653 million tons of helium. The "missing" 4 million tons of matter is converted into radiant energy, some of which we see as sunlight.

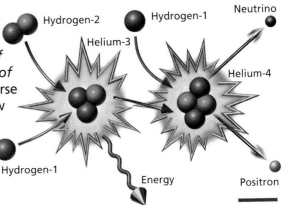

Hydrogen-2   Hydrogen-1   Neutrino
Helium-3   Helium-4
Hydrogen-1   Energy   Positron

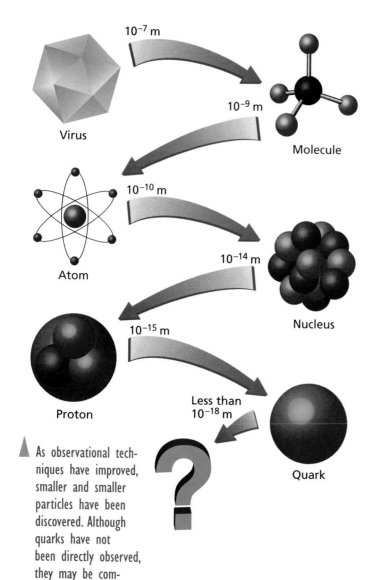

10⁻⁷ m

Virus

10⁻⁹ m

Molecule

10⁻¹⁰ m

Atom

10⁻¹⁴ m

Nucleus

10⁻¹⁵ m

Proton

Less than
10⁻¹⁸ m

Quark

▲ As observational techniques have improved, smaller and smaller particles have been discovered. Although quarks have not been directly observed, they may be composed of something even smaller.

In 1905, Albert Einstein proposed a mathematical relationship between mass and energy that he expressed with his equation $E = mc^2$. This famous equation states that *energy* is equal to *mass* multiplied by the *speed of light* squared. To account for the creation of energy from mass, a new law combining the two previous laws was needed. This law has become known as the **law of conservation of mass and energy.** It states: *The total amount of matter* and *energy in the universe does not change.* Whether mass is converted into energy or energy is converted into mass, the total amount of mass and energy remains the same.

Our idea of the atom has also changed. The "indivisible" atom has been subdivided far beyond protons, neutrons, and electrons. It has been broken into quarks, leptons, bosons, and many other pieces. In fact, over 200 subatomic particles have been cataloged. Just how far can this go? If matter can become energy, and energy can become matter, can we consider any particle of matter as fundamental? Perhaps there is a "particle" common to both matter and energy that can answer such questions.

DID YOU KNOW...

that the top quark is nearly as massive as a gold atom?
After 20 years of searching, scientists were surprised to find that this quark, whose discovery was confirmed in 1995, has so large a mass.

S U M M A R Y

According to current theories, matter consists of particles that are affected by four forces—gravitational, electromagnetic, strong, and weak—that hold everything in the universe together. Each force operates at a different level of matter, from galaxies to subatomic particles. Quantum theories, which include all but the force of gravity, assume that quanta of energy are transferred when particles interact. Grand unified theories attempt to combine three fundamental forces, while a "theory of everything" would combine all four. The relationship between mass and energy is $E = mc^2$. According to the law of conservation of mass and energy, the total amount of matter and energy in the universe does not change.

Unit CheckUp

Concept Mapping

The concept map shown here can be used to illustrate major ideas in this unit. Complete the map by placing terms from the list in the appropriate positions. Then extend your map by answering the additional question below. Use your ScienceLog. **Do not write in this textbook.**

Where and how would you connect the terms *formula unit, electromagnetic force,* and *the kinetic molecular theory of matter*?

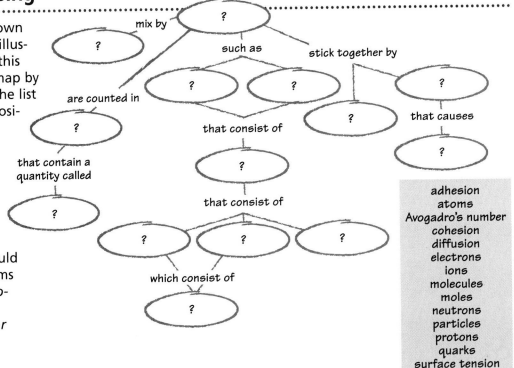

adhesion
atoms
Avogadro's number
cohesion
diffusion
electrons
ions
molecules
moles
neutrons
particles
protons
quarks
surface tension

Checking Your Understanding

Select the choice that most completely and correctly answers each of the following questions.

1. All of the following are particles of matter except
   a. molecules.
   b. ions.
   c. atoms.
   d. photons.

2. If you wanted to indicate the number of atoms in a certain amount of a substance, you could use the SI unit called a
   a. mole.
   b. formula unit.
   c. gram.
   d. liter.

3. Adhesion
   a. is the force of attraction between unlike particles.
   b. increases as the kinetic energy of molecules increases.
   c. is greatest in solids.
   d. decreases as the temperature of a substance decreases.

4. According to Boyle's law,
   a. the volume of a gas increases as its temperature increases.
   b. the volume of a gas decreases as the pressure increases.
   c. the pressure applied to a liquid in a closed container is evenly distributed.
   d. the temperature of a substance increases as the pressure on it increases.

5. The force between particles that acts over the greatest distance is
   a. weak force.
   b. strong force.
   c. gravitational force.
   d. electromagnetic force.

Interpreting Photos

Explain how mass and energy are conserved in the process seen in this photo.

Critical Thinking

Carefully consider the following questions, and write a response in your ScienceLog that indicates your understanding of science.

1. Suppose that your teacher asks you to bring 5 mol of sodium bicarbonate ($NaHCO_3$) to your lab table. What information would you need to know in order to determine how much sodium bicarbonate to obtain? Look up this information, and calculate the amount you will need. Show your work.

2. The force of cohesion between atoms of liquid mercury is greater than the force of cohesion between water molecules. Which substance, mercury or water, has the greatest surface tension? Explain your answer.

3. The condensation point of a gas is the same as its boiling point. Likewise, the freezing point of a liquid is the same as its melting point. How would you explain these observations using the kinetic molecular theory of matter?

4. Suppose that 100 N of force are required to raise an object with a hydraulic lift, but that you can apply only 0.1 N of force. How could you alter the hydraulic lift shown on page S33 to accomplish this task? Explain.

5. What will the volume of a 2 L balloon be after the temperature falls from 300 K to 270 K? Show your work.

Portfolio Idea

Imagine that you could be shrunk to the size of the smallest subatomic particle and go on a voyage inside an atom. Write a short story describing your adventure into the interior of an atom. Be sure to describe all of the particles and forces that you would encounter on your journey and to explain how these particles and forces interact to produce what we know as matter.

MACHINES, WORK & ENERGY

IN THIS UNIT

Now that you have been introduced to machines and how they relate to work and energy, consider the following questions.

● **1.** What simple machines make up the mechanical systems of your home?

2. How is the input energy supplied to each system? What does the output energy do?

3. What is the mechanical advantage and efficiency of each mechanical system?

In this unit, you will take a closer look at the physics behind machines and at how machines affect our daily lives.

FORCE, WORK, AND POWER

When the forces applied by each of these arm wrestlers are equal, neither arm wrestler can move the arm of the other.

Do you realize that you are constantly under the influence of a far-reaching force? Even though you might not feel it, this force is pulling on you right now. The force is, of course, gravity. As you have learned, the size of the gravitational force pulling on you is equal to your weight. Fortunately, another force usually balances your weight. When you are standing or sitting, for example, an equal force pushes up on you from below. This force, supplied by the ground or the seat of the chair, acts in the opposite direction of the gravitational force. In situations involving balanced forces, there is no change in motion.

When a force is not balanced, however, something will change. For example, if you push hard enough on the back of a chair, the chair will probably move. The chair starts to move because the force you supply is *unbalanced* (not opposed by an equal force).

Work

The result of using a force to move an object over a distance

Work: Force Over a Distance

In everyday language, you might say that sitting at your desk reading this book is "work." But in a scientific sense, reading is not work. To a scientist, **work** involves using a *force* to move an *object* over a *distance*. Thus, work is related to motion. In order for work to be done, an object must be moved over some distance.

Think about the last time you played tug of war. When you win a tug-of-war contest, you do work in the scientific sense. The purpose of such a contest is to move an object (the other team) over a distance (across the center line). Your team must apply a force (pull on the rope) to move the other team. From your team's perspective, this force is the *effort*. As your team pulls on the rope, the opposing team resists your team's force by pulling in the opposite direction.

If neither team moves in a tug-of-war contest, nobody is doing any work. But if your team's effort is greater than the

In the scientific sense, climbing stairs is a lot more work than holding a barbell above your head.

1 As long as the efforts of both teams are equal, all forces are balanced and nobody moves. These people might be pulling and straining, yet neither team is doing *work* in the scientific sense unless one team moves some distance.

2 By using more force than the opposing team, your team can move the opposing team across the line. Now your team has done *work* in the scientific sense.

other team's effort and causes them to *move*, your team has done work. To find out how much work your team did in winning the contest, you must know how much force your team used and how far the other team moved.

To find how much work your team did, multiply the *force* that your team applied by the *distance* over which the force was applied. In other words, find the work that your team did by using the following equation:

$$\text{Work} = \text{Force} \times \text{Distance,}$$
$$\text{or}$$
$$W = F \times d$$

Because we measure force in newtons and distance in meters,

the unit of work is the *newton-meter* (N·m). This unit is also called a *joule* (J).

Let's assume that your team applied a force of 10,000 N more

than the other team and that your team pulled the other team a distance of 3 m. The work your team did can be calculated as follows:

$$W = F \times d$$
$$W = 10,000 \text{ N} \times 3 \text{ m}$$
$$W = 30,000 \text{ N·m,}$$
$$\text{or}$$
$$30,000 \text{ J}$$

Therefore, your team did a total of 30,000 J of work in winning the tug of war.

◀ By definition, a joule is a force of 1 N acting over a distance of 1 m, or 1 N·m. This person is doing 1 J of work. As you can see, 1 J is not a very large amount of work.

S43

In the previous example, we assumed that the force stayed the same throughout the tug-of-war contest. In most situations, however, the applied force is not constant. For example, suppose you are pulling on a spring. The farther the spring stretches, the greater the force you must apply. In cases where the amount of applied force changes, *average force* is used to calculate the work that is done.

Power

The rate at which work is done

As you pull back on the string of an archer's bow, you must apply different amounts of force. You apply a small force when you start pulling back on the bowstring, but you must apply more force as you continue to pull. In this situation, you would use average force to calculate work.

Power: Work Over Time

Suppose your team is challenged to a rematch of the tug-of-war contest. This time you beat the other team even faster than you did the first time. In the first contest, it took your team 5 minutes to pull the other team 3 m. Now it takes your team only 2 minutes. Assuming that the force your team used was the same in both games, how do you think this time difference affected the amount of work your team did? Actually, the work was exactly the same in both cases.

The amount of work that you do does not depend on how fast you do it. Notice that in the equation for work, there is no variable for time. To describe how fast a certain amount of work is done, we use the term **power**. The faster an amount of work is done, the greater the power that was produced. Power equals *work* divided by *time* (in seconds). This relationship is given in the following equation:

$$P = \frac{W}{t}$$

These runners will do about the same amount of work, but they do not have the same power. Which runner has more power? Why?

James Watt defined power in terms of a workhorse. One *horsepower* (hp) was the power of a horse lifting 550 lb. 1 ft. in 1 second. One horsepower = 746 W.

Because work equals force times distance, the equation for calculating power can also be written as follows:

$$P = \frac{(F \times d)}{t}$$

If you exert a force of 1 N over a distance of 1 m for 1 second, the power would be calculated as follows:

$$P = \frac{(1 \text{ N} \times 1 \text{ m})}{1 \text{ s}}$$

$$P = 1 \frac{\text{N·m}}{\text{s}} = 1 \frac{\text{J}}{\text{s}} = 1 \text{ W}$$

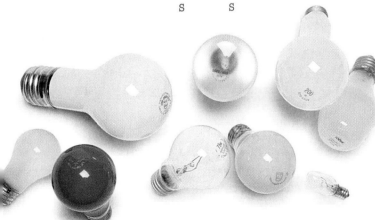

You can see from the equation that power is measured in *joules per second* (J/s), or *watts* (W). The watt is the SI unit of power. It was named after James Watt, the inventor of the sliding-valve steam engine. To generate 1 W of power, you must do 1 J of work in 1 second. Because a watt is rather small, however, the *kilowatt* is often used to indicate power. One kilowatt of power equals 1000 W. To generate 1000 W, or 1 kW, of power, you must do 1000 J of work in 1 second.

How much power did your tug-of-war team have in each of the two matches? Recall that your team did 30,000 J of work in both of the matches. Therefore, you would calculate the power of your team in each match (P_1 and P_2) as follows:

$$P_1 = \frac{W}{t} = \frac{30,000 \text{ J}}{5 \text{ min}} \times \frac{\text{min}}{60 \text{ s}}$$

$$P_1 = 100 \text{ J/s, or } 100 \text{ W}$$

$$P_2 = \frac{30,000 \text{ J}}{2 \text{ min}} \times \frac{\text{min}}{60 \text{ s}}$$

$$P_2 = 250 \text{ J/s, or } 250 \text{ W}$$

Note: The term min/60 s is a conversion factor for changing minutes into seconds.

What common household appliances have power ratings similar to these?

Electricity and Energy Use

Work is done by electrical appliances as they transform the electrical energy in electricity into other forms of energy. Because the watt is the unit used to measure the rate at which objects do work, the rate at which electrical appliances, such as light bulbs and hair dryers, transform energy is expressed in watts.

The total amount of energy an appliance uses depends on how long it operates. The units for total energy consumption are *watt-hours* (Wh) or *kilowatt-hours* (kWh). For example, a 60 W light bulb that burns for 1 hour uses 60 W × 1 h, or 60 Wh, of electricity. When several electrical appliances operate at the same time, a great deal of electricity may be used. Therefore, the amount of electrical energy used in a home is usually measured in kilowatt-hours (where 1 kWh = 1000 Wh).

As the following calculation shows, a kilowatt-hour is actually a unit of energy, not a unit of power:

$$1 \text{ kWh} = 1000 \text{ W} \times 1 \text{ h}$$

$$1 \text{ kWh} = \frac{1000 \text{ J}}{\text{s}} \times 3600 \text{ s}$$

$$1 \text{ kWh} = 3{,}600{,}000 \text{ J}$$

When you pay your electric bill, you are really paying for energy. For example, a 60 W light bulb consumes 216,000 J of energy in 1 hour:

$$60 \text{ J/s} \times 3600 \text{ s} = 216{,}000 \text{ J}$$

▲ An electric meter, such as this one, measures the number of kilowatt-hours of electrical energy that you use. On an average day, a house may use from 60 to 100 kWh of electrical energy.

S U M M A R Y

When a force causes an object to move over a distance, work is done. To calculate work, which is measured in joules, multiply force by distance. Power is the rate at which work is done. To calculate power, which is measured in joules per second, or watts, divide work by time. Electrical energy is usually measured in kilowatt-hours, which are units of energy, not power.

SIMPLE MACHINES

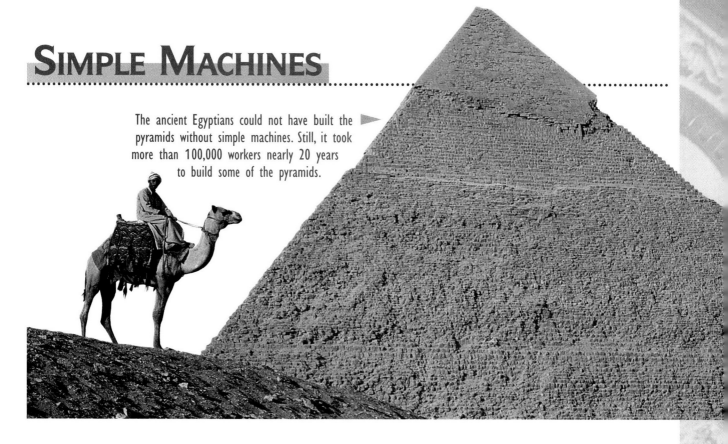

The ancient Egyptians could not have built the pyramids without simple machines. Still, it took more than 100,000 workers nearly 20 years to build some of the pyramids.

Can you imagine how much work was done in building the ancient Egyptian pyramids? Each of the stone blocks used to build the pyramids has a mass of thousands of kilograms. Yet because it took many years to build a pyramid, the power involved was not very great. You have already seen that people generate very little power in a tug-of-war contest. In the two previous examples, your team generated only 100 W and 250 W—about the same amount of power as a light bulb.

As you can see, humans cannot apply much force, and human power is not very effective in doing large amounts of work quickly. This is why humans invented machines. For example, moving a heavy boulder probably requires more force than you alone could normally supply. You might not be able to move it at all by simply pushing on it. But by applying the same force to a metal bar placed under the boulder, you can produce a force large enough to move the heavy boulder. The bar is a type of *simple machine*. You do work on the bar by applying a force to one end and moving that end. The other end of the bar, in turn, does work on the boulder by lifting it some distance.

A **simple machine** is a device that changes the size or direction of a force. While a simple machine can make your task easier, it cannot do more work than is put into it. Remember, the amount of work depends on the force applied to an object and the distance the object moves. If you want to decrease the force needed to do work, you must increase the distance over which the force is applied. On the other hand, if you want to decrease the distance over which work is done, you must increase the force applied. These relationships are true for all machines. Among the devices that we think of as simple machines are the screw, wedge, wheel and axle, and pulley. However, these simple machines are forms of just two basic machines—the *inclined plane* and the *lever*.

This metal bar is a simple machine that enables you to apply more force to an object.

Simple machine

A device that changes the size or direction of a force

S47

The Inclined Plane

An *inclined plane* is a simple machine with no moving parts. All simple machines, including the inclined plane, work on the same basic principle. An inclined plane, such as the ramp shown in the photograph to the left, simply increases the distance over which an applied force acts. As the length of an inclined plane increases, the force needed to move a load to the same height decreases. If you shorten the length of an inclined plane (make it steeper), more force is required. Again, the amount of work done is the same in both cases. It is only a matter of which is more important in getting the job done— the amount of force you must use or the distance you must travel.

Ramps are examples of inclined planes connecting one level to another.

Because an inclined plane forms a sloping surface, a load must move through a greater distance to rise to a certain height. Therefore, you can move a load up an inclined plane with less force than would be required to lift it straight up to the same height.

The Screw   Though they are simple machines in their own right, screws are really forms of the inclined plane. By looking at the photograph below, you can see that a screw is an inclined plane wrapped around a cylinder. Screws are very useful because the length of the inclined plane can be spread over a large distance. The closer together the threads on a screw are, the longer the inclined plane is. It will take more turns to fully tighten such a screw, but it will take less force to do it.

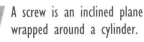

A screw is an inclined plane wrapped around a cylinder.

The Wedge Two inclined planes placed back to back make a wedge. When a wedge is used to split a log, as shown in the photo to the right, the two inclined planes move the resistance (the wood) on either side apart. As you might imagine, a short, wide wedge would split a log without being driven very far. However, the wedge would have to be pounded with a great deal of force. A long, narrow wedge would be a lot easier to pound but would have to be driven farther. There are probably many wedges on your kitchen counter. A knife is really a very narrow wedge.

A wedge is two inclined planes placed back to back. Wedges are used to split an object in two.

The Lever

A pole or bar used to move a boulder is an example of a *lever*. The diagram below shows the basic parts of a lever. Although a lever might seem to do more work than is put into it, this is not the case. A lever only *transfers* effort. When the effort arm is longer than the resistance arm, applying a small force to the end of the effort arm results in a larger force at the end of the resistance arm. This is because the effort force acts over a greater distance than the resistance force. As in all machines, the work done by a lever can never exceed the work put into the lever. A lever merely changes the way that work is done.

DID YOU KNOW...

that principles of simple machines were developed by Greek mathematician Archimedes in about 250 B.C.? Archimedes once suggested that given a place to stand, he could move the Earth with a lever.

A typical lever

A A lever always turns on a fixed point called a *fulcrum*. The fulcrum is the point that supports the lever.

B The *effort arm* is the part of a lever between the fulcrum and the point where force is applied.

C The *resistance arm* is the part of a lever between the fulcrum and the load.

As seen below, there are three classes of levers, based on the location of the fulcrum. Notice that each class of lever changes the way work is done in a different way. A first-class lever reverses the direction of the applied force. If the effort and resistance arms are not equal, a first-class lever will also change the size and distance of an applied force. Second- and third-class levers change the size and distance but not the direction of an applied force. You probably have seen and used many examples of all three kinds of levers.

Three Classes of Levers

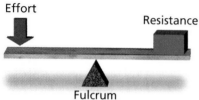

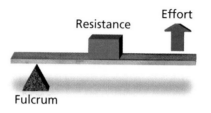

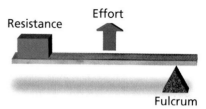

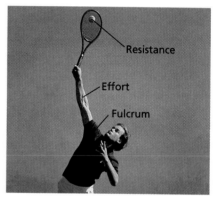

1 A first-class lever has the fulcrum between the resistance and the effort and *reverses* the direction of the applied force. Because the effort and resistance arms of this particular lever are equal, the effort and resistance are equal and so are the distances that each moves.

2 A second-class lever has the resistance between the fulcrum and the effort and *does not change* the direction of the applied force. A small amount of effort applied over a long distance moves a large resistance over a short distance.

3 A third-class lever has the effort between the fulcrum and the resistance and *does not change* the direction of the applied force. A large amount of effort applied over a short distance moves a small resistance over a long distance.

A screwdriver is a wheel and axle. This person is using a screwdriver to increase the amount of force applied to a screw.

The Wheel and Axle Have you ever tried to turn the shaft of a doorknob when the knob was missing? If so, you know how much harder it was to open the door. Imagine trying to make a car turn a corner without a wheel on the steering column. A doorknob and a steering wheel are common examples of the *wheel and axle*, a machine that is similar to a lever. A wheel and axle has a fulcrum at the center of the axle. The resistance arm of a wheel and axle is the radius of the axle, and the effort arm is the radius of the wheel. Effort applied to the wheel is transferred to the axle.

Have you noticed that large trucks and tractors have very large steering wheels? In a wheel and axle, the size of the wheel determines how much force acts on the axle. The larger the wheel, the more distance it covers in turning the axle. Therefore, a small force used to turn a large steering wheel will cause a large force to act on the axle of the steering wheel. This large force is transferred to the turning mechanism, making it much easier to turn a heavy truck.

The Pulley A pulley is similar to a wheel and axle. However, the axle on a pulley does not turn. Instead, the wheel of a pulley turns as a rope runs over it. A pulley can be used in two differ-ent ways—it can be *fixed* to something or it can *move* along the rope. With a *fixed pulley*, the load (resistance) hangs down on one side, and the effort is applied to the other side. A fixed pulley does not change the effort force or the distance it moves; it simply changes the direction of the force. With a *movable pulley*, the resistance hangs from the axle of the pulley, and the pulley hangs on a rope. A movable pulley does not change the direc-tion of the effort force, but it changes both the size of the force and the distance over which it acts.

Ⓐ A fixed pulley changes the direction of the effort force but does not multiply effort force. Such pulleys can be useful because it is often easier to pull downward than to lift upward. When you pull down on the rope of a fixed pulley, the load rises. Notice that one length of rope supports the entire load (resistance). Because the effort and resistance distances are the same, there is no gain in force.

Ⓑ A movable pulley reduces the effort force but does not change the direction of effort force. Notice that two lengths of rope support the load (resistance). Because the fixed rope supports half the weight of the load, it takes half as much effort to lift the load. Since force decreases, dis-tance increases. The effort distance is twice the resistance distance.

Sᴜᴍᴍᴀʀʏ

Machines help people overcome the limitations of human strength. The inclined plane and the lever are the most basic machines. All other simple and complex machines are forms of these two machines. Simple machines change the size or direction of a force. They do not increase the amount of work done. To decrease the force needed to accomplish work, the distance over which the force is applied must increase. Con-versely, if distance decreases, force must increase.

S51

MACHINES AT WORK

A ny tool that helps us do work is a machine. Pencil sharpeners, drills, screwdrivers, and crowbars are machines. Even a baseball bat is a machine. Machines help do work in three ways. First, a machine can change the force applied to an object. Second, a machine can change the direction of a force. Third, a machine can change the distance through which a force moves. You already know that work equals force multiplied by distance. Because of this relationship, machines can reduce the force needed to do a given amount of work, thus making work easier.

Work Is Work

Although machines can make it easier to perform a task, they do not reduce the amount of work needed to do the job. For example, it takes the same amount of work to lift a heavy crate regardless of how the crate is lifted. You might lift the crate without the aid of a machine, or you might use a simple machine such as an inclined plane or a pulley to lift it. If you use a machine, you may exert less force, but you must exert that force over a longer distance. No matter how you do it, the amount of work required to lift the crate to a certain height is always the same. In other words, the amount of work you put into a machine (work input) can never be less than the work that comes out (work output).

In an **ideal machine**, *work input* always equals *work output*. Since work equals force multiplied by distance, we can set up the following equations:

$$\text{work input} = \text{work output}$$
$$\text{(effort force)} \times \text{(effort distance)} = \text{(resistance force)} \times \text{(resistance distance)}$$
$$F_e \times d_e = F_r \times d_r$$

Now suppose you are faced with a task in which it is important to minimize the force you must apply. For example, suppose you want to lift a 360 N load 0.5 m off the ground with a lever. If you can only apply a force of 90 N, how far would the opposite end of the lever have to move?

$$F_e \times d_e = F_r \times d_r$$
$$90 \text{ N} \times d_e = 360 \text{ N} \times 0.5 \text{ m}$$
$$d_e = \frac{360 \text{ N} \times 0.5 \text{ m}}{90 \text{ N}}$$
$$d_e = 2 \text{ m}$$

▲ We even use machines to play baseball. What type of machine is a baseball bat?

Ideal machine

An imaginary machine in which there is no friction

S52

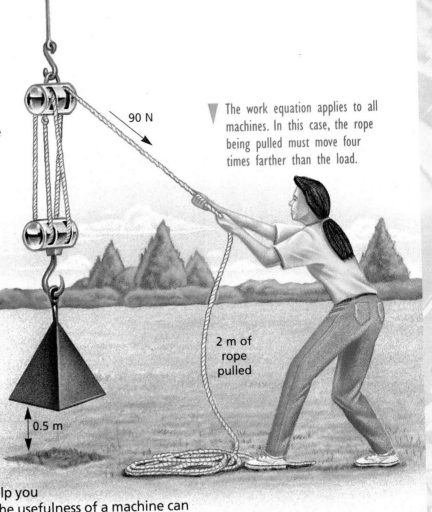

You can see from this example that in order for the *effort force* to be less than the *resistance force,* the *effort distance* must be greater than the *resistance distance.* In this case, the resistance force (F_r) is four times greater than the effort force (F_e) you applied. However, your effort force moved four times farther than the load was lifted. As the figure to the right illustrates, a machine can change the size of a force, but it cannot supply more work than is put into it.

90 N

▼ The work equation applies to all machines. In this case, the rope being pulled must move four times farther than the load.

2 m of rope pulled

0.5 m

Machines and Mechanical Advantage

If machines do not increase the amount of work you can do, what would make one machine better than another? Besides the fact that you can choose whether to increase force or distance, what else might help you decide to use a particular machine? The usefulness of a machine can be rated by comparing the amount of force you get out with the amount of force you put in.

You can find the amount by which an ideal machine increases force by simply dividing the output (resistance) force by the input (effort) force. The resulting number is called the **mechanical advantage** (MA) of the machine. Mechanical advantage is a ratio of the output force of a machine compared with the input force. For example, suppose an automobile mechanic uses an engine hoist to pull an engine from a car. If the engine weighs 4000 N and the mechanic uses a force of only 50 N to lift it, the mechanical advantage of the hoist can be figured as follows:

$$\text{MA} = \frac{\text{Output force}}{\text{Input force}} = \frac{4000 \text{ N}}{50 \text{ N}} = 80$$

The mechanical advantage of the hoist is 80. This means that while raising the engine, the machine applied 80 times more force to the engine than the mechanic applied to the machine.

Mechanical advantage can also be *less* than one. By using a machine with a mechanical advantage of less than 1, you exchange distance or speed for force. For example, when you turn the handle of an eggbeater, the blades turn much faster than the handle.

Mechanical advantage

The factor by which a machine multiplies effort in order to equal resistance

The mechanical advantage of an eggbeater is less than 1. The blades turn faster than the handle, increasing speed (or distance over time) rather than force. ▶

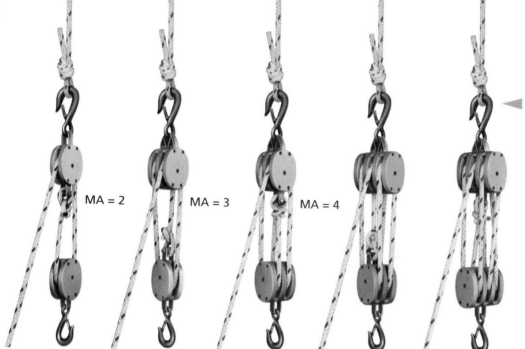

MA = 2 MA = 3 MA = 4

The easiest way to determine the mechanical advantage of a pulley system is to count the ropes that support the resistance force, or the load. What is the mechanical advantage of the two pulley systems on the right?

Computing Mechanical Advantage Mechanical advantage can sometimes be calculated without knowing the input and output forces. Levers and their related machines, the wheel and axle and the pulley, increase effort force because their effort arms are longer than their resistance arms. Thus, you can determine a lever's mechanical advantage by simply relating effort distance to resistance distance. This is done by dividing the length of the effort arm by the length of the resistance arm. For example, if the length of the effort arm of the lever you plan to use is 2.0 m and the length of the resistance arm is 0.2 m, you would find the mechanical advantage as follows:

$$MA = \frac{\text{Effort arm}}{\text{Resistance arm}} = \frac{2.0 \text{ m}}{0.2 \text{ m}} = 10$$

Your lever has a mechanical advantage of 10. In other words, this lever multiplies your effort force by 10.

The mechanical advantage of an inclined plane can be found by dividing its length by its height. For example, if the length of an inclined plane is 3.0 m and its height is 0.6 m, you would find its mechanical advantage as follows:

$$MA = \frac{\text{Length}}{\text{Height}} = \frac{3.0 \text{ m}}{0.6 \text{ m}} = 5$$

So the force needed to push a box up this inclined plane would ideally be one-fifth of the force you would need to lift the box directly. The same equation can be used to calculate the mechanical advantage of a wedge or screw.

The Effect of Friction Friction, as you have learned, is a force that results when two surfaces rub together. It opposes the motion of two objects in contact with each other. For example, suppose you use an inclined plane to lift a heavy box. Dividing the length of the inclined plane by its height will give its *ideal mechanical advantage*. However, because there is friction between the box and the surface of the inclined plane, some of the force you use is needed to overcome the friction. Therefore, the *actual mechanical advantage* of the inclined plane is less than what you calculated.

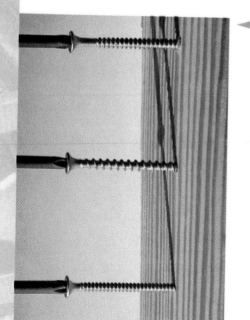

The closer together the threads on a screw are, the greater its mechanical advantage is. Which of these screws has the greatest mechanical advantage?

The actual mechanical advantage of this inclined plane is less than its ideal mechanical advantage because some force is used to overcome friction.

Imagine that you are moving a 500 N box up an inclined plane, or ramp, into a truck. Suppose the ideal mechanical advantage of the ramp is 5. This means that you should only have to apply a 100 N force to move the box up the ramp. Unfortunately, friction between the box and the ramp increases the input force you must exert to move the box. Therefore, the actual mechanical advantage of the ramp is less than 5. This is true in every real situation when a machine does work. The greater the friction is, the smaller the actual mechanical advantage will be.

The Efficiency of Machines

Because all machines have at least two surfaces in contact, friction is a major consideration in the design and application of machines. And since all machines have some friction, the actual mechanical advantage of a machine is always less than its ideal mechanical advantage. In other words, a real machine always puts out less work than is put into it. The difference between the work that comes out of a machine and the work put into it is a measure of the machine's **efficiency**, which is expressed as a percentage. The smaller the difference, the greater the efficiency.

One way to determine the efficiency of a mechanical system is to divide the work output of the system by the work input. For example, suppose you use 16 kJ (kilojoules) of energy to pedal a bicycle. Of this amount, 5 kJ is used to overcome friction. The efficiency of the bicycle would be calculated as follows:

$$\text{Efficiency} = \frac{\text{Work output}}{\text{Work input}} = \frac{16 \text{ kJ} - 5 \text{ kJ}}{16 \text{ kJ}} = \frac{11 \text{ kJ}}{16 \text{ kJ}} = 0.69 = 69\%$$

Therefore, your bicycle has an efficiency of 69 percent. This means that only 69 percent of the work you do in pedaling your bicycle is used by the bicycle to move forward. Efficiency, then, is a measure of how well machines overcome friction.

Every new appliance sold in the United States now carries a notice indicating the efficiency of the machine. This translates into savings—the higher the efficiency of the appliance, the lower your utility bill.

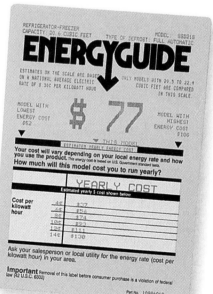

Efficiency

The ratio of the work done by a machine to the work put into it

Increasing Efficiency One way to improve the efficiency of a machine is to reduce friction. Friction can be reduced in many ways. For example, oil or grease makes the surfaces of rotating parts turn more easily. Bearings, such as the ball bearings inside bicycle wheels, are also used to reduce friction between rotating parts.

Many people throughout history have tried to invent machines that would be 100 percent efficient. These machines were called perpetual-motion machines. They were designed to produce at least as much energy as they consumed, or to accomplish as much work as was put into them. In other words, a perpetual-motion machine is one that would continue operating forever with no additional input of energy. However, no one has ever succeeded in inventing a perpetual-motion machine because no one has ever been able to totally eliminate the loss of energy or work due to friction.

The oil on this rotating spindle helps reduce friction, thereby increasing efficiency.

S U M M A R Y

Machines make work easier, but they do not decrease the amount of work needed for any task. The mechanical advantage of a machine is the relationship between its output force and its input force, or between its effort distance and resistance distance. Because of friction, the work that comes out of a machine can never equal the work that is put into it. Efficiency is the ratio of work output to work input. Despite many efforts to build a perpetual-motion machine, no one has been able to make output work equal input work because of the presence of friction.

Concept Mapping

Starting with the terms supplied below, construct a concept map that illustrates major ideas from this unit. Arrange the terms in an appropriate manner and connect them with linking words. Then extend your concept map by adding as many additional terms from the unit as you can. Use your ScienceLog. **Do not write in this textbook.**

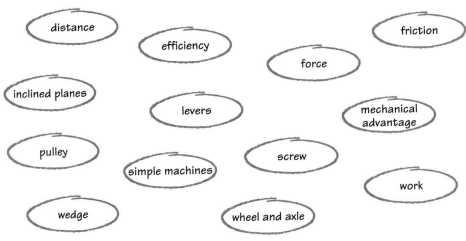

Checking Your Understanding

Select the choice that most completely and correctly answers each of the following questions.

1. The quantity that is determined by the distance moved and the force used is called
   a. power.
   b. work.
   c. friction.
   d. newtons.

2. Which of the following would constitute a power rating?
   a. meters per second
   b. joules per second
   c. watts per hour
   d. watts per meter

3. Which of the following simple machines would be considered a type of lever?
   a. ramp
   b. screw
   c. wedge
   d. wheel and axle

4. The screw is a simple machine that works like the
   a. lever.
   b. inclined plane.
   c. wheel and axle.
   d. pulley.

5. With an ideal machine, the work you get out of it equals
   a. the distance it must move.
   b. the force you use on it.
   c. the work you put into it.
   d. the resistance it has.

Interpreting Photos

What type of simple machine is shown in the photo below? How would the mechanical advantage of this machine be calculated? Identify any sources of friction that could decrease the efficiency of this machine.

Critical Thinking

Carefully consider the following questions, and write a response in your ScienceLog that indicates your understanding of science.

1. Describe how you would determine the amount of power involved in walking up a flight of stairs. Use actual values in your description.

2. How much energy does a 100 W light bulb use in 24 hours? Show all your work.

3. How does a knife function as a double inclined plane? Why does sharpening a knife make it cut better?

4. Why can you use either force or distance to calculate ideal mechanical advantage? Use formulas to support your answer.

5. Some people think that levers are more efficient than inclined planes. Do you think they are right? Explain your answer.

Portfolio Idea

Locate several examples of non-electric mechanical devices in and around your home. Then produce a technical brochure for one or more of these devices. Each brochure should have a photograph or drawing of the device; a schematic diagram showing the types of simple machines that make up the device (levers, pulleys, wheels and axles, inclined planes, screws, and wedges); and a written description of the device, how it works, and how it makes your task easier. Also try to determine the mechanical advantage for each simple machine that you identify in your mechanical devices.

OCEANS AND CLIMATES

Unit 4

IN THIS UNIT

Now that you have been introduced to oceans and climates, consider the following questions.

1. How does the composition of the atmosphere affect life on Earth as we know it? How has the atmosphere changed over time?

2. In what ways do heat and moisture control the weather patterns in the atmosphere?

3. How are the resources of the ocean utilized, and in what ways must these resources be protected?

In this unit, we will take a closer look at the atmosphere and its structure, and at the ocean's resources.

THE ATMOSPHERE

Unlike Earth, Mercury has no atmosphere.

You could, perhaps, live about 5 weeks without food and about 5 days without water. But you would last less than 5 minutes without air. In a sense, then, Earth's blanket of air, called the **atmosphere**, could be considered its most important resource.

Earth's atmosphere is one of its most distinguishing features. Life could not exist without it. In addition to supplying the gases required for life processes, the atmosphere protects organisms from exposure to harmful radiation and from bombardment by meteors. The atmosphere also gives our planet a moderate climate. Without an atmosphere, the Earth would have a climate like Mercury, with extremely hot days and terribly cold nights. There would be no wind, no clouds, and no liquid water. Without an atmosphere, our planet would be nothing like it is today. However, the atmosphere was once very different from the way it is now. In fact, you would not be able to breathe the air that was present in the atmosphere during most of Earth's history. How, then, did the present atmosphere change to become capable of supporting life as we know it?

Atmosphere

The layer of gases that surrounds the Earth

Evolution of the Atmosphere

Like living things and the physical features of the Earth, the atmosphere is thought to have evolved over several billion years to reach its present form. The Earth's earliest atmosphere probably formed shortly after the planet condensed from a swirling cloud of gas and

According to one scientific model, the Earth's present atmosphere took billions of years to develop. Its first atmosphere, consisting of gases left over from the planet's formation, was scattered by the solar wind.

A second atmosphere then formed from gases released by volcanoes. The action of lightning or ultraviolet radiation on atmospheric gases formed complex organic compounds that eventually developed into primitive living things.

Organisms capable of photosynthesis eventually evolved. Oxygen, a byproduct of photosynthesis, began to accumulate in the atmosphere.

dust about 4.5 billion years ago. This early atmosphere most likely consisted of gases left over from the Earth's formation: hydrogen, ammonia, methane, and water. But once the sun (which probably formed about the same time) began to shine, the high-energy radiation streaming from it most likely scattered these gases into space.

The Primitive Atmosphere   Evidence suggests that after the original atmosphere was scattered, a new atmosphere formed from gases that were trapped inside the rocks of Earth's crust when it first solidified. Most of these gases—which included hydrogen, water vapor, carbon dioxide, and nitrogen—were released during the volcanic eruptions that were common during Earth's early history. But as this new atmosphere was forming, it was also evolving. Hydrogen, which is too light to be held by gravity, escaped into space. And as Earth and its atmosphere cooled, much of the water vapor condensed and fell as rain. As a result, the first oceans formed when rainwater collected in low places on the Earth's surface.

Conditions in this primitive atmosphere may have allowed certain important chemical reactions to occur. For example, energy from ultraviolet solar radiation and the electrical discharges of lightning could have caused some of the gases in the atmosphere to combine into a variety of complex organic compounds. These compounds collected in the oceans. Scientists think that the first life-forms on Earth developed from these organic compounds more than 3.5 billion years ago.

Oxygen Enters the Picture   The early atmosphere contained almost no free (uncombined) oxygen. Oxygen (O_2) was introduced to the atmosphere by two processes. A small amount of oxygen entered the atmosphere when water molecules were split by lightning and solar radiation. However, most of the oxygen on Earth today is the result of photosynthesis. During photosynthesis, carbon dioxide and water are used to make sugar. Oxygen is a byproduct of that process. Evidence suggests that primitive organisms evolved the ability to carry out photosynthesis fairly early in the Earth's history.

Some of the free oxygen in the atmosphere formed the gas ozone (O_3), which collected in a layer high above the Earth. This was an important development because this layer absorbed most of the sun's high-energy ultraviolet radiation, which is harmful to living things. Once the layer stabilized, the rapid evolution of early life-forms began. Increasing numbers of photosynthetic bacteria and algae began to release a steady supply of oxygen gas into the atmosphere and the ocean.

At some point, oxygen became common enough that an ozone shield formed. Life evolved rapidly.

When the oxygen level in the atmosphere reached a certain point, larger life-forms became possible. On a huge scale, organisms used up carbon dioxide to form calcium carbonate shells.

Oxygen continued to accumulate. Today, it makes up more than 20 percent of the atmosphere.

Early corals and other organisms used much of the carbon dioxide dissolved in ocean water to make their skeletons or shells.

Over hundreds of millions of years, oxygen accumulated in the atmosphere and oceans. By the beginning of the Cambrian geologic period, about 570 million years ago, enough oxygen had become dissolved in the oceans to support larger life-forms. An evolutionary explosion resulted. Most of these primitive animals had shells made of *calcium carbonate*—a compound formed by the reaction of carbon dioxide and the element calcium, which dissolves in water. Much of the carbon dioxide that had been part of Earth's early atmosphere was used by these animals to make their shells. Over millions of years, the shells of dead organisms collected on the ocean floor and were compressed and cemented together to form limestone. Today, most of the carbon dioxide from the primitive atmosphere remains chemically bound in massive deposits of limestone.

The Atmosphere Today

For some time after the Cambrian period, oxygen continued to accumulate in the atmosphere. Eventually, the concentration of oxygen in the atmosphere stabilized near today's levels. Today, the proportions of oxygen, carbon dioxide, and nitrogen in the atmosphere are fairly stable, kept in equilibrium by cycles that exchange gases among the living and nonliving parts of the environment. Currently, the atmosphere consists of about 78 percent nitrogen, 21 percent oxygen, and 0.9 percent argon, with traces of water vapor, carbon dioxide, and other gases. This mixture of atmospheric gases is called **air**.

Air

The mixture of gases in the atmosphere

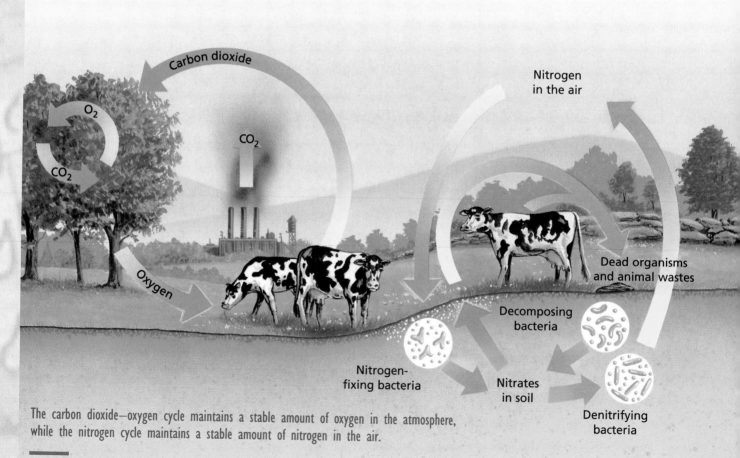

The carbon dioxide–oxygen cycle maintains a stable amount of oxygen in the atmosphere, while the nitrogen cycle maintains a stable amount of nitrogen in the air.

Pressure Changes If you have ever traveled up a mountain, you might have noticed that your ears "popped." This happens because the air pressure outside your ears gradually decreases as you climb, while the air pressure inside your ears remains the same. Your ears pop when the pressure difference between the outside and inside of your ears suddenly equalizes. But why does air pressure change as you climb? Examine the diagram on this page for some clues.

Temperature Changes The temperature also changes as you ascend. If you measured the change, you would find that the temperature falls about 6.5°C for each 1000 m of altitude gained. For example, if you ascended in a balloon to an altitude of 3 km, you would find the temperature about 20°C lower than at sea level.

Layers of the Atmosphere
The Earth's atmosphere consists of several layers. Each layer is characterized by either a rise or a fall in air temperature. A steady decrease in pressure also takes place the farther up you go. The lowest and densest layer, the *troposphere*, lies next to the surface. By mass, the troposphere contains about 90 percent of the atmosphere. It is also the layer in which almost all of Earth's weather occurs. The sun's energy heats the Earth's surface, which in turn heats the air next to it. Warm air near the surface becomes less dense and rises. This is balanced by cool air sinking to the surface somewhere else. The constant movement of air in the troposphere drives the Earth's wind and weather systems.

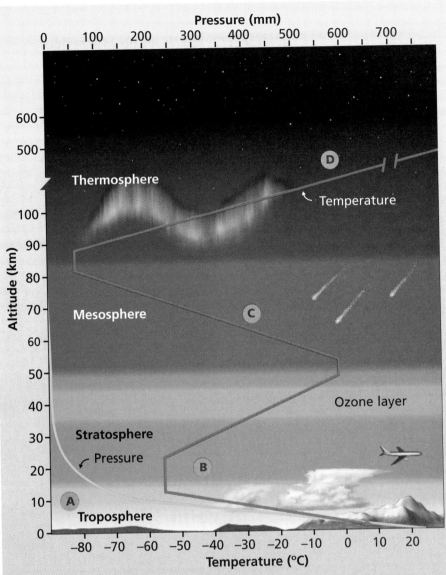

The layers of the atmosphere are marked by changes in temperature and pressure. The temperature of the air changes with altitude. At first it drops, but at higher altitudes, the absorption of high-energy solar radiation by ozone molecules causes the temperature to rise. Air pressure is simply the weight of all the air overlying a given area, and it is caused by gravity pulling the atmosphere toward the surface. At sea level, the weight of the air around you exerts a pressure of about 100,000 N/m²! At higher altitudes, less of the atmosphere is overhead, so air pressure is less.

A The troposphere is the lowest layer of the atmosphere. Virtually all weather takes place there. Atmospheric pressure is greatest in the troposphere and declines with altitude.

B Temperature in the stratosphere remains constant for the first 10 km or so. It then begins to rise due to the presence of ozone, most of which is located in the middle to upper stratosphere.

C In the mesosphere, the temperature drops steadily with increasing altitude as the ozone concentration and the atmospheric pressure decrease.

D The thermosphere blends gradually into the vacuum of space. The absorption of high-energy solar radiation, however, causes temperatures in the ultra-thin air of the thermosphere to soar.

Jet stream

A belt of high-speed wind circling the Earth between the troposphere and the stratosphere

The *stratosphere* lies above the troposphere. In the stratosphere, the air is very thin and contains little moisture or dust. As a result, practically no weather occurs there. At the base of the stratosphere are broad, fast-moving "rivers" of air, called **jet streams.** The jet streams circle the planet in the mid- and upper latitudes, affecting weather patterns in the troposphere below. The stratosphere also contains most of the ozone in the atmosphere. Ozone molecules absorb some of the sun's energy, causing the temperature in the stratosphere to increase.

Above the stratosphere is the *mesosphere.* Once again, throughout this layer the temperature drops steadily with increasing altitude.

The uppermost layer of the atmosphere is the *thermosphere.* The "thermos" in thermosphere is a reference to the high temperatures found there. Special instruments have recorded temperatures in the thermosphere as high as 2000°C. In the upper part of the thermosphere, or *exosphere,* gas molecules are so far apart that little interaction occurs among them, and the Earth's atmosphere gradually fades into the vacuum of space.

Fast-moving jet streams form at the boundary between polar and tropical air masses.

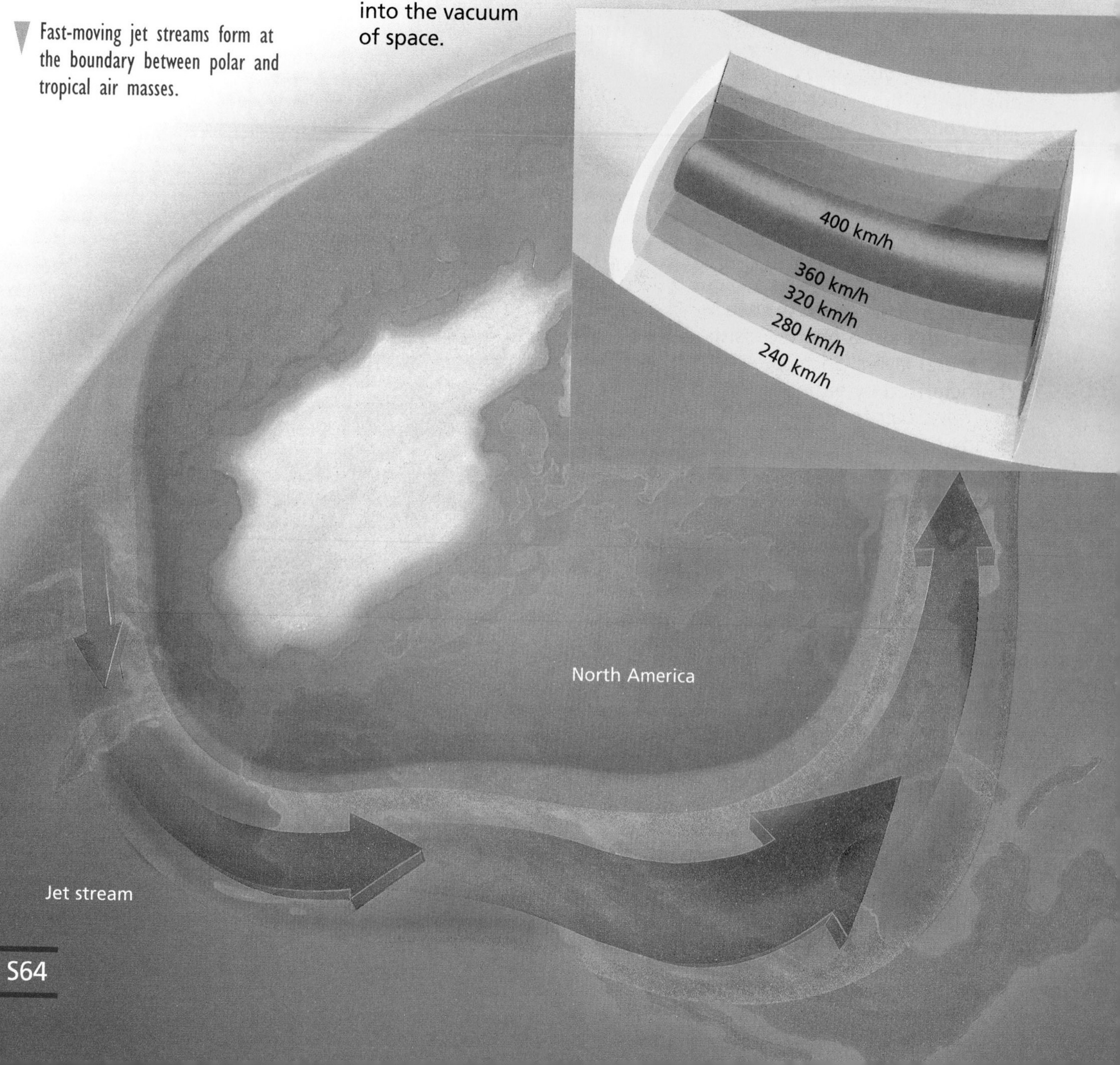

400 km/h

360 km/h

320 km/h

280 km/h

240 km/h

North America

Jet stream

The Ionosphere   Within the uppermost part of the mesosphere and the lower part of the thermosphere is a region called the *ionosphere.* It begins at a height of about 60 km and continues up to about 400 km. The ionosphere is so named because it contains electrically charged particles called *ions.* These ions are formed when atoms and molecules in the atmosphere absorb solar radiation and lose some of their electrons. Colorful light displays, known as *auroras,* result from the recapture of electrons by ions in the ionosphere. Auroras, however, are usually visible only near the poles. Ions, which are guided by Earth's magnetic field, are concentrated there because the magnetic field is much stronger near the poles.

Have you noticed that AM radio signals often travel greater distances at night than during the day? The ionosphere is responsible for this effect because it reflects these radio waves. At night, the lower layers of the ionosphere dissipate because ions cannot form in the absence of the sun's rays. Thus, radio waves are able to travel farther because they are reflected off a higher level at night.

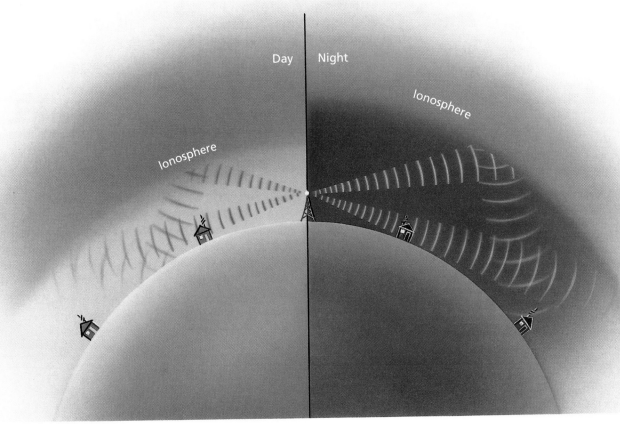

Day   Night

Ionosphere

Ionosphere

AM radio waves are able to travel around the curve of the Earth because they are reflected by the ionosphere. These radio waves travel long distances by bouncing back and forth between the Earth and the ionosphere. But every time a radio wave is reflected, some of its energy is lost, so it eventually fades out. At night, the lowest layer of the ionosphere dissipates, allowing the waves to travel farther before being reflected back to the Earth. This is why AM radio waves can travel farther at night.

S U M M A R Y

The atmosphere is one of Earth's most vital resources. The Earth's atmosphere most likely originated as gases released by volcanic action. The atmosphere evolved over millions of years through chemical reactions and the actions of living things. Living things were the primary producers of the oxygen in the atmosphere of today. The composition of the atmosphere is now fairly stable. The atmosphere is divided into several invisible, yet distinct, layers.

WEATHER AND CLIMATE

Weather

The condition of the atmosphere at a given time and place

Climate

The average weather for a particular place over many years

If you live along the Gulf Coast, in Hawaii, or in Florida, you know that it will be warm and humid for most of the year. If you live in the Northeast or the Midwest, your summers will be warm and humid, but winters will be cold and snowy. If you live in the Southwest, it will be warm and dry most of the year. If you live in the Pacific Northwest, it will be cool and rainy most of the year.

Weather is the condition of the atmosphere at a particular time and place. Specific weather conditions, such as temperature, humidity, clouds, winds, and precipitation, usually change from day to day and often from hour to hour. The average weather conditions over a long period of time constitute the **climate** of that region. Weather and climate both are created by the sun's uneven heating of Earth's atmosphere.

Heating the Atmosphere

Although the sun releases a huge amount of radiant energy into space, only about three-billionths of this energy is received by the Earth. Yet even this tiny fraction of the sun's radiation contains a very large amount of energy. In fact, the total amount of energy that reaches the Earth is thousands of times greater than all the energy used by every country on Earth combined. So what happens to all this energy? About 20 percent of the solar radiation that reaches Earth is absorbed by the gases of the atmosphere. Another 33 percent is reflected back into space by clouds, snow, and ice. The remaining 47 percent is absorbed at the Earth's surface. As you know, because of the greenhouse effect, some of the energy absorbed by the Earth and re-emitted as heat does not escape back into space but is absorbed by atmospheric gases. As a result, the atmosphere warms up.

▼ Only about half of the sun's energy reaches the Earth's surface. The rest is reflected away or is absorbed by the atmosphere.

S66

However, the solar radiation striking the Earth is not evenly distributed. Because Earth is a sphere, the sun's rays strike different places at different angles. This uneven distribution of solar radiation causes unequal heating of the surface. Since the troposphere is heated by the Earth's surface, it too is heated unevenly. For example, air over the equator is heated much more than air over the poles. This causes the air over the equator to be less dense and to have a lower pressure than the air over the poles.

DID YOU KNOW...

that the troposphere is about twice as thick at the equator as at the poles?
In the equatorial regions, strong updrafts caused by intense solar heating push the boundary between the troposphere and stratosphere upward. But in the calm polar regions the boundary drops down to within 5 km of the surface.

▼ The sun's rays strike the Earth more directly at the equator and more obliquely at the poles.

Air Circulation

In general, air flows from a region of high pressure toward a region of low pressure. Thus, cold dense air from the poles flows toward the equator along the Earth's surface. At the equator, the air becomes warmer, expands, and rises. The heated air flows away from the equator along the top of the troposphere. At the poles, the air cools, becomes denser, sinks, and starts a new cycle as it flows back toward the equator. The first illustration on this page shows this idealized pattern of air circulation.

However, the Earth's pattern of air flow actually consists of several smaller *wind belts.* As equatorial air flows toward the poles along the top of the troposphere, it gradually cools. About one-fourth of the way to the poles, most of this air sinks back to the surface. When it strikes the surface, the mass of sinking air divides. Some of it flows back toward the equator, while the rest continues on toward the poles.

Because the Earth rotates, the path taken by air moving between the equator and the poles is not a straight line. Rather, it appears to be deflected. This phenomenon, called the **Coriolis effect,** makes objects moving north or south in the Northern Hemisphere drift to the right when one faces the direction of their movement. In the Southern Hemisphere, the drift is to the left. In the Northern Hemisphere, the Coriolis effect causes air that is moving toward the equator to blow from northeast to southwest. Such is the case with the wind belts known as the *trade winds* and the *polar easterlies.* Air moving toward the poles creates winds that blow from southwest to northeast. These winds are called *prevailing westerlies.* Notice how the direction of the winds in the wind belts differs between the Northern and Southern Hemispheres.

▲ At the equator, air is warmed and rises. At the poles, air is cooled and sinks.

▶ Much of the warmed air loses energy and sinks at about 30 degrees latitude. This breaks up the equator-to-pole circulation into smaller convection cells.

Coriolis effect

The apparent bending of the path of a moving object due to Earth's rotation

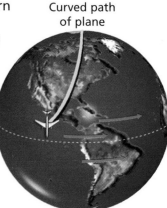

Curved path of plane

Earth spinning

▲ Because of the Coriolis effect, objects traveling in a southerly direction in the Northern Hemisphere appear to curve to the right.

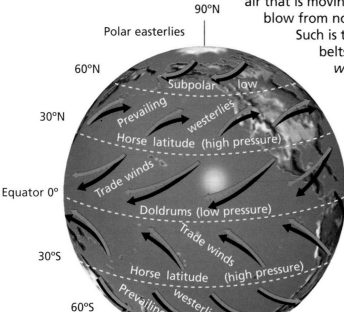

90°N
Polar easterlies
60°N
Subpolar low
30°N
Prevailing westerlies
Horse latitude (high pressure)
Trade winds
Equator 0°
Doldrums (low pressure)
Trade winds
30°S
Horse latitude (high pressure)
Prevailing westerlies
60°S
Polar easterlies
Subpolar low
90°S

◀ A combination of the Coriolis effect and the circulation of air to and from the poles causes the global wind belts.

Water Enters the Atmosphere

When water is heated by energy from the sun, some of it evaporates and enters the atmosphere as water vapor. All bodies of water contribute water vapor to the atmosphere. Plants also contribute water vapor to the air. The oceans, however, are by far the most important source of water vapor.

Water vapor in the air is called *humidity*. Air, however, can hold only a certain amount of water vapor. This amount is determined by the temperature of the air. Warm air can hold more water vapor than cold air. Air that is holding as much water vapor as possible is said to be *saturated*.

But air does not usually contain all the water vapor that it could possibly hold. The term **relative humidity** refers to the amount of water vapor that is currently in the air, as compared with the maximum amount the air could possibly hold. In other words, relative humidity is the percent to which air is saturated with water vapor. For example,

if a certain volume of air could hold 40 g of water but has only 20 g, the relative humidity would be 50 percent. If the same mass of air had 30 g of moisture, the relative humidity would be 75 percent. When this air mass contains 40 g of water, its relative humidity is 100 percent, and it is saturated.

Relative humidity changes as air temperature changes. The amount of water vapor itself does not change, but the amount of water vapor that air can hold does change. For example, as air temperature drops, the air can hold less water vapor. Therefore, the relative humidity rises. When air is warmed, it can hold more water vapor. Therefore, the relative humidity drops.

This graph shows the maximum amount of water vapor that air can hold over a range of air temperatures. As you can see, the ability of air to hold moisture increases sharply with increases in temperature.

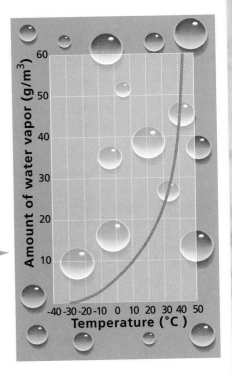

This graph shows how the relative humidity might vary throughout a typical sunny day. As the temperature changes, so does the relative humidity, even though the actual amount of moisture in the air stays constant.

Temperature is highest; therefore, relative humidity is lowest.

Temperature

Relative humidity

Temperature is lowest; therefore, relative humidity is highest.

$$\text{Relative humidity} = \frac{\text{Amount of moisture air currently holds}}{\text{Amount of moisture air can hold}}$$

6 A.M.   12 noon   6 P.M.   12 midnight   6 A.M.

Water Leaves the Atmosphere

Each day about 1100 km³ of water move from the Earth's surface into the atmosphere. And each day, about the same amount of water returns to the surface as precipitation. Yet the moisture does not return evenly to the Earth. Some regions are drenched while others receive little moisture for years on end. Why does this happen? Where will it rain today? Weather forecasters attempt to answer these challenging questions all the time, often unsuccessfully. A discussion of some of the factors involved in the return of water to Earth's surface follows.

Condensation   The temperature at which the relative humidity reaches 100 percent is called the *dew point.* If the temperature continues to drop, water vapor in the air will change back into a liquid. This process is called **condensation.** Water vapor may condense directly on cool objects, such as rocks, metal, windows, or plants. This moisture is called *dew.* If the dew point is below 0°C, *frost* will form. Frost is composed of solid ice crystals that form directly on an object when water vapor condenses below its freezing temperature. Frost is not frozen dew.

There are several ways that water can be cooled to the dew point. Air may be cooled as it is pushed upward to rise over mountains. It may also be pushed upward by denser air moving under it. Visible water droplets will form in the air when it is cooled to the dew point, provided there are tiny particles such as dust or salt crystals in the air. The particles give water vapor a surface on which to condense. The visible droplets that result form clouds.

Precipitation   Cloud droplets are so small that even slight air movements keep them floating. When cloud droplets collide and merge, however, larger, heavier drops form and may fall as *rain* or *drizzle.* Whenever the temperature in a cloud is below 0°C, ice crystals may form. Ice crystals grow as they fall, forming snowflakes. These may fall as *snow* if the air is cold enough all the way to the ground. However, if snowflakes pass through a layer of warm air, they will melt and become rain. Sometimes, rain passing through a cold layer of air freezes as it falls. It is then called *sleet.* In violent storms, strong updrafts may carry raindrops high enough to freeze. As they fall, a new layer of water collects on them and freezes as they rise again. Often, they are carried up and down several times, causing multiple layers of ice to accumulate. This forms *hail.* Sometimes hail can grow large enough to damage objects that it strikes on the ground.

Condensation

The changing of a gas into a liquid

At the dew point, the moisture in the air condenses. When conditions are right, fog—a cloud very near the Earth's surface—is formed.

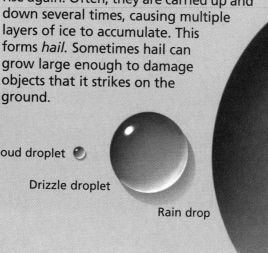

Cloud droplet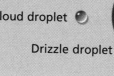

Drizzle droplet

Rain drop

Relative sizes of condensed water vapor

Clouds In 1803 an Englishman named Luke Howard developed a classification system to describe various types of clouds. Cloud formation depends on certain conditions in the atmosphere, such as air temperature, air pressure, humidity, and air currents (wind). Howard identified and named three basic cloud types based on their appearance: *cumulus*, from the Latin for "heap"; *cirrus*, from the Latin for "lock of hair"; and *stratus*, from the Latin for "layer."

If you have done much cloud gazing, you know that clouds do not always fit into one of the three categories above but appear to be combinations of the three basic types. Howard suggested names for these combinations as well. For example, sometimes cumulus clouds become so crowded together that they almost form a layer across the sky. Howard called these clouds *stratocumulus*.

The modern classification system for clouds also takes into account the altitude at which clouds form and whether they produce rain. Clouds that form at altitudes above 7 km are given the prefix *cirro,* and

those that form between 2 and 7 km above the ground are given the prefix *alto*. Clouds that produce rain or snow are called *nimbus clouds. Nimbus* comes from the Latin word meaning "rainy cloud." The illustration below shows some of the characteristics of these basic cloud types.

▼ Classification of clouds according to height and form

Powerful updrafts form towering cumulonimbus clouds, or thunderheads. These clouds can sometimes reach an altitude of almost 20 km. Cumulonimbus clouds produce lightning, rain, and wind.

Cirrus clouds form high in the troposphere, where the temperature is very cold. Water vapor there forms ice crystals, which are blown by the high winds of the upper troposphere into patterns resembling delicate strands of hair.

On a sunny day, warm air rising from the Earth's surface cools and forms puffy white cumulus clouds. Cumulus clouds are sometimes called fair-weather clouds.

Nimbostratus clouds produce light to heavy rain, sleet, or snow, but no lightning.

Stratus clouds form when a layer of air is cooled past its dew point. These broad, flat clouds often form a solid blanket over the entire sky.

Air Masses

When air remains over one part of the Earth's surface for a long period of time, it takes on the characteristic temperature and humidity of that region. Such a body of air is called an **air mass**. In general, air masses that form over land are dry, and those that form over oceans are humid. Air masses that form in polar regions are cold, and those that form in regions close to the equator are warm.

Only regions that have light winds and consistent surface features over a large area can form air masses. For example, the Earth's surface at the equator is mostly covered by oceans. At the poles, vast areas are covered by the polar icecaps. Regions such as these are called *source areas*.

Fronts

As air masses move, they encounter each other. When air masses meet, little mixing takes place. Instead, one air mass pushes the other air mass out of the way, replacing it. The boundary between air masses is usually fairly sharp. Such a boundary is called a **front**. Temperature, pressure, and wind direction usually change significantly with the passage of a front. Most fronts form when one air mass moves into a region occupied by another air mass.

Air mass

A large body of air with uniform temperature and humidity

Front

The boundary between two air masses

Air masses that affect the weather in North America ▶

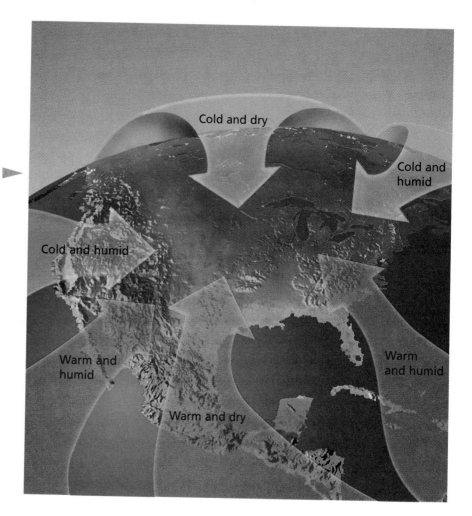

Cold Fronts   If a cold air mass invades warmer air, a *cold front* is created. Since colder air is denser, it pushes under warmer air like a wedge. The lifting of a warm air mass usually causes cumulus clouds to form along the cold front. These clouds often bring stormy weather and heavy rain. Cold fronts tend to move quickly because dense, cold air easily pushes into warmer air. The turbulent weather associated with a cold front usually passes quickly. The air mass behind a cold front normally brings cool or cold, dry weather.

Warm Fronts   When a warm air mass pushes into a cold air mass, a warm front is formed. Because it is less dense, the invading warm air rides up and over the colder air. As the warm air is lifted, cirrus clouds are formed high in the troposphere. As time passes, the cloud cover becomes lower and thicker until a solid sheet of stratus clouds covers the sky. A long period of gentle rain may occur, followed by slow clearing and rising temperatures. The weather changes that accompany a warm front are not as dramatic as those that accompany a cold front.

Severe Weather   Sometimes the interaction of air masses causes very turbulent weather to occur. For example, thunderstorms are common along cold fronts. The thunderstorms occur when warm, humid air is lifted very rapidly by the cold air mass. Towering *cumulonimbus* clouds, or thunderheads, develop as a result. Rain, lightning, and wind accompany the thunderstorms. Sometimes, cold fronts are preceded by a solid line of thunderstorms, called a *squall line.* While severe weather is common along cold fronts, it is rare along warm fronts. This is because warm fronts, unlike cold fronts, are a zone of gradual transition between air masses.

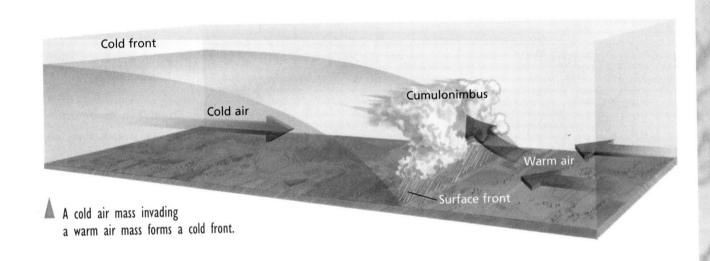

▲ A cold air mass invading a warm air mass forms a cold front.

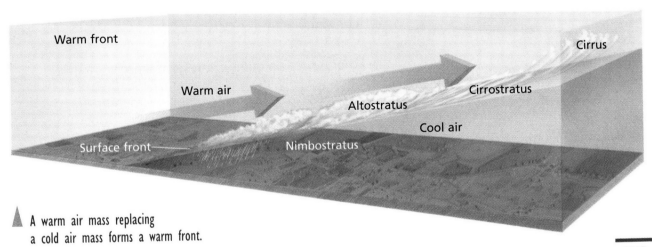

▲ A warm air mass replacing a cold air mass forms a warm front.

About 5 percent of all thunderstorms develop to the point that they become classified as severe. Severe thunderstorms produce heavy rain, dangerous lightning, and damaging winds. They may also produce hail. Occasionally, small but extremely violent phenomena called *tornadoes* accompany severe thunderstorms. A tornado is a violently whirling column of air, usually seen as a funnel-shaped cloud extending from the base of a thunderhead. Tornadoes can be anywhere from 3 m to 3 km across. Scientists are not completely sure how tornadoes form, but they do know the kinds of weather conditions that are likely to produce them. These conditions are typically found along cold fronts, where strongly contrasting air masses—one cool and dry, the other warm and moist—collide. Such conditions are most common in the spring.

Tornadoes are common in only a few places on Earth. The vast majority occur in the United States, which averages more than 800 every year. Most of these occur in "Tornado Alley," a corridor running from Texas to Michigan. Although tornadoes usually last only a few minutes, they can be extremely destructive.

▲ A towering thunderhead

▼ Tornadoes are the most concentrated form of wind energy. Their fierce winds, sometimes more than 400 km/h, can destroy almost everything in their path.

SUMMARY

Weather is the condition of the atmosphere at a given time and place. Climate is the average weather of a region over a long period of time. Weather is created by the uneven heating of the atmosphere by the sun. There is a general pattern of circulation between the equator and the poles. This pattern is broken up into several zones. The rotation of the Earth deflects the winds from a straight-line path. Water vapor in the atmosphere condenses when it cools, forming clouds and precipitation. The motion and interaction of air masses cause many weather phenomena.

OCEAN RESOURCES

S een from space, perhaps the Earth's most compelling feature is the enormous, deep blue oceans that seem to dominate the planet. The oceans are ancient; life most likely began there some 3.5 billion years ago. For more than 3 billion years, living things were found only in the oceans; the land was barren. The ocean is and has been home to a vast number of living things. Today, the oceans provide a wealth of important living and nonliving natural resources to the Earth's dominant life-form—humans.

Food From the Sea

Oceans cover almost two-thirds of the Earth's surface. Living things can be found in every part of the oceans. In theory, oceans could produce as much food as the land. Then why does less than 10 percent of the world's food supply come from the ocean? Part of the answer is the type of food we collect there.

This image shows the concentration of living things in the ocean. Most of the ocean's living things are microscopic plankton (inset).

Plankton   Most of the mass of the organisms (biomass) in the ocean is in the form of **plankton**. Most plankton are so small that they can be seen clearly only with a microscope. Microscopic plants, called *phytoplankton,* are found mostly in the upper few meters of the sea, where there is enough light for photosynthesis. Tiny animals, called *zooplankton,* feed on the phytoplankton. Both kinds of plankton are eaten by larger animals that, in turn, become food for even larger fish and other marine animals. Thus, phytoplankton are the beginning of a food chain that supports the ocean's animal life. Plankton are not evenly dispersed throughout the ocean. Because of its remoteness from land, much of the ocean is poor in nutrients and is therefore a biological desert. Such regions are poor even in plankton.

Fish   Fish are the main food we take from the ocean. They are caught primarily in areas of the ocean that lie over continental shelves. There the water is rich in the phytoplankton that are the basis of most ocean food chains. Some areas near the coasts act like natural fish farms. This is partly because of *upwelling*, which is the rising of cold water from below to replace surface waters that are removed by steady winds and evaporation. Upwelling brings up nutrients from the ocean bottom that are needed for the growth of phytoplankton. Upwelling also occurs in parts of the open ocean.

Plankton

The small plants and animals floating near the surface of the ocean

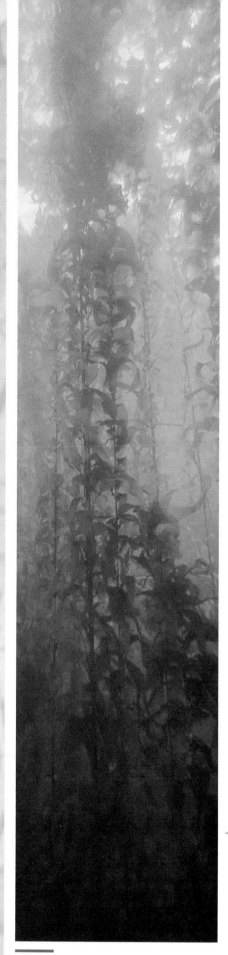

Today, humans catch nearly one-third of all the fish large enough to be taken in nets. Unfortunately, careless overharvesting has taken a toll on the population of certain fishes used as food. Some kinds of fish that were once plentiful are now scarce. Laws have been enacted to protect some endangered fish populations. However, in the absence of strong enforcement, these laws are often broken. Scientists are trying to determine the conditions in the ocean that affect fish populations. This information may help researchers to learn how many fish can be harvested without doing irreparable damage. Using that information could help ensure a constant supply of fish.

Other Ocean Food Sources

Like fish, shrimp and shellfish are also important sources of food. Seaweeds, especially certain red and brown algae that grow anchored to the ocean floor, are another food resource. In many countries, the seaweed *kelp* is a popular food. Substances extracted from algae, such as algin, agar, and carrageenan, are used as stabilizers in foods such as ice cream, cheese spreads, and salad dressings. One potential source of food is *krill.* This small, shrimplike animal is found in large numbers in the cold waters of the Southern Hemisphere. Krill can be used directly for food or as part of other foods, just as shrimp are.

Kelp, which forms "forests" up to 30 m tall, grows in cold and rocky coastal waters. It is both a food source and a source of raw materials for humans. It is also a key player in aquatic ecosystems, providing food and shelter for dozens of species of marine animals.

Overuse of a resource can sometimes destroy it. Redfish were almost eliminated from the Gulf of Mexico because of their popularity as food fish. Because of strict limits on the allowable catch, redfish have made a dramatic comeback.

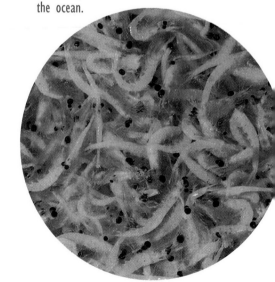

Krill are small shrimplike animals that are found in huge numbers in some parts of the ocean.

Other Ocean Resources

For years, proposals to mine the oceans for their mineral wealth have been discussed. Because rivers, streams, and undersea hot springs dump huge amounts of dissolved material into the ocean, the ocean floor and even the water itself are rich in minerals. However, any ocean mining operation would face considerable difficulties. For example, to recover useful amounts of minerals from the water, large quantities of sea water would have to be processed, at great expense. It is far too costly to extract even such a valuable mineral as gold from sea water. Only a few highly concentrated minerals—sodium, bromine, and magnesium salts—can now be economically extracted from sea water.

Another option is to mine the sea floor for its mineral wealth. The mid-ocean ridges are honeycombed with mineral lodes deposited by hot springs. In addition, vast areas of the ocean floor are littered with *manganese nodules*. Trillions of tons of these egg- to potato-sized lumps of high-grade ore are thought to exist. Nodules contain high concentrations of manganese, iron, nickel, copper, and cobalt. They also contain smaller amounts of zinc, silver, gold, and lead. As rich a mineral source as manganese nodules are, though, mining them is not currently economical. Because these nodules are found only at great depths and far from shore, mining them would be difficult and very expensive.

Vast areas of the ocean floor are carpeted with deposits of manganese nodules. Manganese, nickel, cobalt, and other valuable metals are concentrated in nodules such as these.

Fuel From the Sea   One of the most important sources of energy from the sea is *petroleum*. When marine organisms die, their remains fall to the ocean floor along with other debris. Since the water at the ocean bottom does not contain much oxygen, organic matter does not readily decay. Instead, it accumulates along with other, inorganic sediments. As more sediments accumulate, the buried organic matter is eventually chemically converted into petroleum. The largest known petroleum deposits on Earth formed when the African continent drifted northward and pushed organic-rich sediments against Asia.

Artist's conception of an undersea mining operation

S77

These sediments ultimately became the large petroleum reserves of the Middle East. Other large reserves are found on the north shore of Alaska, beneath the North Sea, and along the coast of the Gulf of Mexico.

Protecting Ocean Resources

The technical difficulty of obtaining ocean resources is not the only problem to be overcome in utilizing the ocean's resources. Deciding who owns these resources may ultimately be an even bigger problem, one that could lead to serious conflict if not solved. Many questions remain to be settled: How far into the ocean do the borders of a country extend? Do nations have the right to collect these resources from the ocean beyond their borders? Do landlocked countries have any rights to ocean resources? Until there is agreement among nations, there cannot be full or fair use of the ocean's resources.

Some people would say that we should be less concerned about how to use the oceans and more concerned about how to preserve them. For example, pollution is a serious problem that threatens the vitality of the ocean. The chief sources of ocean pollution are sewage and garbage, industrial waste, and agricultural chemicals and wastes. These pollutants not only are detrimental to ocean life in the areas where they are dumped, but also can be spread by currents to affect the entire ocean. Oil spills from tankers and drilling platforms can also cause major harm. One accident can damage a coastal ecosystem for many years.

Because the oceans belong to no single nation, preserving them requires the cooperation of all nations. This will not be an easy task, but humanity must act soon to safeguard this valuable resource or face the consequences.

Huge, costly drilling platforms are used to extract the oil trapped beneath the ocean floor.

Pollution threatens the vitality of the oceans.

S U M M A R Y

The Earth's oceans contain many valuable natural resources. Food is taken from the ocean, mostly in the form of fish. Other sources of food, such as plankton, krill, and algae, may one day be of greater importance. The ocean also contains many valuable minerals, either dissolved in the water or in undersea deposits. The vitality of the ocean is threatened by pollution and overuse of resources. A major challenge facing humanity is to establish a way to share ocean resources fairly.

Concept Mapping

The concept map below illustrates major ideas in this unit. Complete the map by supplying the missing terms. Then extend your map by answering the additional question below. Write your answers in your ScienceLog. **Do not write in this textbook.**

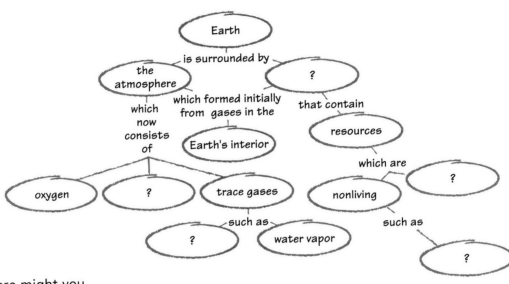

How and where might you connect the terms *argon, petroleum,* and *food*?

Checking Your Understanding

Select the choice that most completely and correctly answers each of the following questions.

1. The Earth's early atmosphere
   a. was probably much like the atmosphere today.
   b. contained little oxygen.
   c. enabled living things to evolve rapidly.
   d. consisted largely of ozone.

2. Which layer of the atmosphere is capable of supporting life?
   a. troposphere
   b. stratosphere
   c. mesosphere
   d. thermosphere

3. Which is an example of weather?
   a. a squall line
   b. a short growing season
   c. scant year-round rainfall
   d. hot, humid summers

4. Outside, the temperature cools to the dew point, causing
   a. rain to fall.
   b. condensation to occur.
   c. dew to evaporate.
   d. the relative humidity to exceed 100 percent.

5. Mineral resources from the ocean
   a. are off-limits to all nations by international agreement.
   b. belong to all nations by international agreement.
   c. are retrieved only with great difficulty.
   d. have largely replaced land-based mineral stores.

Interpreting Illustrations

The photo here shows a home weather station. According to the readings on the station, can you tell what season of the year it is? What is the current air temperature? What is the relative humidity? How much rain has fallen since the rain gauge was last emptied? In which direction is the wind blowing, and how fast is it blowing?

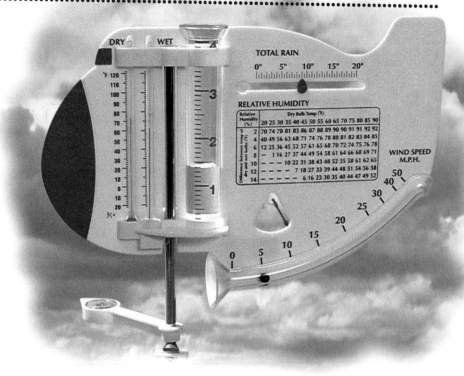

Critical Thinking

Carefully consider the following questions, and write a response in your ScienceLog that indicates your understanding of science.

1. How are volcanoes connected to the formation of the atmosphere and oceans?

2. How did the evolution of living things affect the evolution of the atmosphere?

3. Large thunderstorms sometimes form anvil-shaped clouds like the one shown below. How might the jet stream contribute to this type of cloud's characteristic shape?

4. How do you think the Coriolis effect would change (if at all) if the Earth were to rotate twice as fast? Explain.

5. Without the atmosphere, the Earth's surface would be very different. Describe a few major ways in which the atmosphere and the surface interact.

Portfolio Idea

Use your knowledge of air masses to put together a weather forecast for your area. Start by examining weather maps for the last 3 days. (Your local newspaper may have these.) Then use your knowledge of weather systems to predict the changes that will take place in your area over the next 48 hours. Do NOT use existing forecasts; make your own. At the end of the 48-hour period, compare your predictions, and those of your classmates, with the actual results.

Electromagnetic Systems

IN THIS UNIT

Now *that you have been introduced to electromagnetic systems, take a closer look at the basics of these systems by answering the following questions.*

1. What is static electricity? How is it different from current electricity?
2. How do the properties of certain materials cause magnetism?
3. What is the relationship between magnetic effects and electric effects?
4. What kinds of devices make and use electricity? How do they work?

In this unit, you will examine electricity and magnetism, the relationship between them, and the methods and machines used to make, distribute, and harness electricity.

TWO KINDS OF ELECTRICITY

When you think of electricity, you probably think of something that comes from a wall outlet and operates machines and appliances. That certainly is electricity—electricity in motion. But have you ever walked across a carpet and then felt a shock when you touched a metal doorknob?

Do your clothes sometimes stick together after being in the dryer? Have you ever rubbed a balloon on your shirt or hair and then stuck the balloon to a wall? Does your hair ever crackle and cling when you comb it? Such experiences are the result of *static electricity*.

The effects of static electricity can be demonstrated by this simple activity.

Static Electricity

Since ancient times people have noticed something curious about a natural substance called *amber*. Amber is ancient tree sap that has hardened over the centuries into a solid. Amber ranges in color from yellow to brown and sometimes contains the remains of insects. The ancient Greeks noticed that when amber is rubbed with cloth, it attracts objects, such as bits of straw and paper. The Greek word for amber is ηλεκτρον (*elektron*), and amber's attraction for small objects is due to what we now call *electric charges*.

◀ When amber is rubbed with a cloth, bits of cloth and paper are attracted to it.

Electric Charges Benjamin Franklin, who was a scientist as well as a statesman, studied electric charges. He is famous for having demonstrated that lightning is a form of electricity. In his day, it was known that there were different types of electricity. It was also known that electric charges could be created by rubbing two materials together. Franklin hypothesized that there are only two kinds of electric charges—positive (+) and negative (–). Franklin defined a *negative charge* as the type of charge given to a rubber rod when it is rubbed with fur or wool. He defined a *positive charge* as the charge given to a glass rod that has been rubbed with a silk cloth. An object that has neither a positive nor a negative charge is considered *neutral*. But how do objects become electrically charged?

Franklin's experiment, while very dangerous, demonstrated that lightning is a form of electricity.

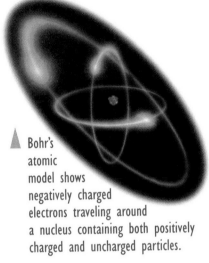

Bohr's atomic model shows negatively charged electrons traveling around a nucleus containing both positively charged and uncharged particles.

You have learned that all substances are made of atoms. According to current scientific models, atoms are composed of three basic particles—negatively charged electrons, positively charged protons, and neutrons that have no charge. When atoms have the same number of electrons and protons, they are electrically neutral. Normally, you are neutral because the negative electrons in your body are balanced by an equal number of positive protons.

Producing Charges   Rubbing two materials together causes electrons to be torn away from one material and added to the other. For example, clothes taken from a dryer often cling together because they are electrically charged. This charge results from the different kinds of cloth rubbing together as the clothes are tumbled in the dryer. Clothes that have picked up electrons take on a negative charge. Those that have lost electrons take on a positive charge. Likewise, running a comb through your hair may cause electrons to move from your hair to the comb. In this case, the comb takes on a negative charge because it has acquired extra electrons. Your hair is left with a positive charge because it has lost electrons. When you walk across a carpet, the friction of your shoes on the carpet may cause you to pick up extra electrons, giving your body an overall negative electric charge.

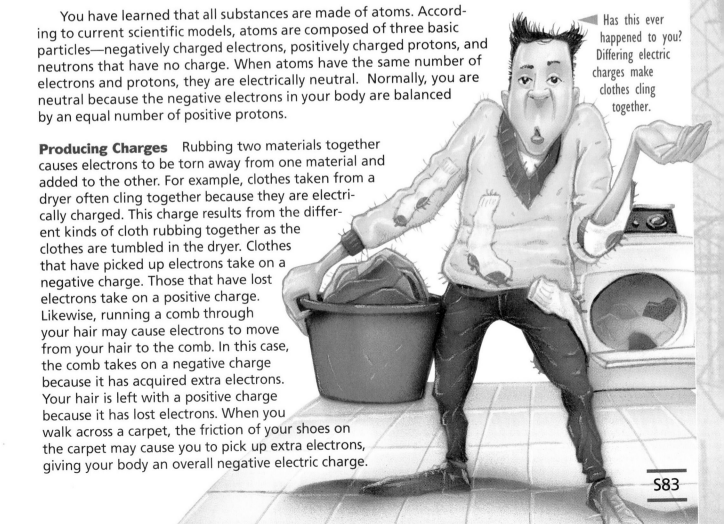

Has this ever happened to you? Differing electric charges make clothes cling together.

Static electricity

Electric charges that accumulate on an object

Electric charges ▶ can make your hair stand on end. The electric charge here was produced by a Van de Graaff generator.

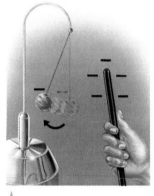
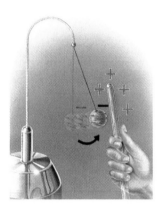

▲ A rubber rod that has been rubbed repels a negatively charged ball (left). A glass rod that has been rubbed attracts the ball (right).

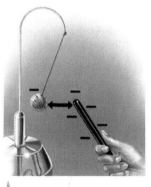
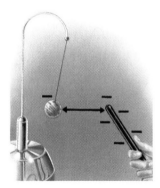

▲ The electric force between objects depends on their charges and the distance between the objects. The precise relationship is given by Coulomb's law.

Electric force

The force that causes two like-charged objects to repel each other or two unlike-charged objects to attract each other

An electric charge results when an object either gains or loses electrons. Because this type of electric charge builds up on objects and does not move, as does electricity through a wire, it is called **static electricity**. The word *static* means "staying in one place." Any substance can be given a static electric charge.

Electric Forces and Fields When two charged objects are brought close together, they either attract or repel each other. The behavior of objects with electric charges can be described by a simple rule: *Like charges repel, and unlike charges attract.* This is called the *law of electric charges*.

Recall that force is always needed to cause an object to move. Thus, when two electrically charged objects cause each other to move, a force must be involved. This force is called **electric force**. Electric forces cause charged objects to move apart or come together depending on whether their charges are alike or different.

Two things seem to affect the strength of an electric force. First, the more an object is rubbed to give it an electric charge, the stronger the electric force that is produced. In other words, the strength of the electric force between two objects increases as the amount of electric charge on those objects increases.

The distance between two charged objects also affects the strength of an electric force. Charged objects have a greater effect on each other as they come closer together. For example, if the distance between two charged objects is halved, the strength of the force becomes four times greater. In the same way, doubling the distance between two charged objects reduces the force to one-fourth of what it was. These two relationships are expressed in **Coulomb's law**: *The force between two charged objects is directly proportional to the product of their charges and inversely proportional to the square of the distance between them.* How is this law similar to Newton's law of gravitation?

Around every charged object there is a space in which the effects of the electric force may be observed. Such a region of space is called an **electric field**, and it surrounds the charged object in all directions. As the distance from a charged object increases, the strength of its field becomes weaker. At a great enough distance, the electric field is too weak to be noticeable.

Electric field

The region of space around an electrically charged object in which the effects of the electric force may be observed

 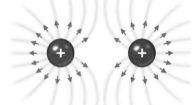

The lines indicate the field existing between two opposite charges (left) and two similar charges (right).

Electricity on the Move

A bolt of lightning has tremendous power. In fact, a single flash of lightning can heat the air to a temperature higher than the surface of the sun. Lightning has melted holes in church bells and welded chains into iron bars. Potatoes in a field struck by lightning have been cooked in the ground. But for all its great power, lightning is basically a spark—very much like the spark that jumps from your hand to a doorknob, only much larger.

Lightning begins as static electricity that is caused by charges building up in storm clouds. These excess charges are rapidly transferred to the ground and result in the spark we see as a bolt of lightning.

Scuffing your feet across a carpet may give your body an overall electric charge. But you don't notice this static charge unless you touch a metal object such as a doorknob. Then you suddenly feel a small jolt as extra electrons jump from your hand to the metal object. When you touch a doorknob, you complete a path over which the excess electrons can travel. As electrons travel through a conducting path, they become electricity in motion. Although you see a spark when this happens, you cannot actually see the electric flow itself. As with lightning, a spark results from the rapid movement of electric charges, which heats the air and causes it to glow.

Electric current

Electrons moving from one place to another

Electric Current

The electron model can be used not only to explain how objects become electrically charged, but also to help explain how electrons move from one place to another. In some ways, electrons behave like water, which will flow through a pipe when pressure of a pump is applied. In a similar way, a form of "electric pressure" can cause electrons to "flow" from an area where they are in excess to an area where they are lacking. For example, electrons will travel from the negative pole to the positive pole of a battery. However, in order for electrons to flow, there must be a path. A *conductor*, such as a wire, forms that path. If the conductor touches only the negative pole of the battery, the electrons will pile up on the conductor—static electricity. But if the negative and positive poles of a battery are connected by a conductor, electrons will flow through the conductor from one terminal to the next. This flow of electrons from one place to another is called an **electric current**.

Electric charges can build up on your body when you walk across a rug. When you touch a doorknob, the charges can jump from your hand to the metal knob.

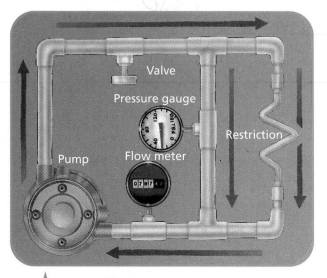

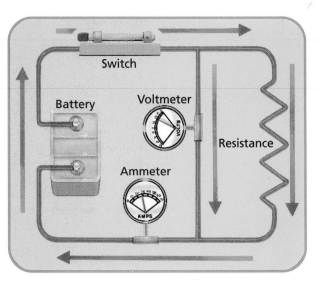

The flow of water in a system of pipes (left) can be compared to the flow of electricity through a circuit (right).

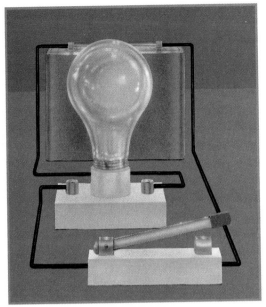

An electric circuit must form a continuous path in order for electrons to travel.

Describing Electric Current

Water will not flow in a pipe unless a force pushes or pulls it along. That force can be gravity or it can be pressure supplied by a pump. For example, a pump is often used to lift water to the top of a water tower. The water at the top of the tower has *potential* energy. When the water is allowed to flow from the tower, its potential energy is converted into the kinetic energy of motion, which can do work.

In the same way, potential energy can be stored by electric charges. Due to the electrical force of attraction between opposite charges, it takes work to pull them apart. Once these charges are separated, they have potential energy. When the separated charges are allowed to recombine, this potential energy is converted into other forms of energy—heat, light, and sound, for example.

Water stored in a water tower is similar to electric charges stored in a battery.

Potential Difference   The difference in charge from one point in a circuit to another is called **potential difference**. The greater the difference, the greater the pressure forcing electrons through the circuit. This pressure is measured in *volts* (V) and is referred to as *voltage*. The greater the voltage, the more work that can be done.

Potential difference

The difference in charge between two points in a circuit

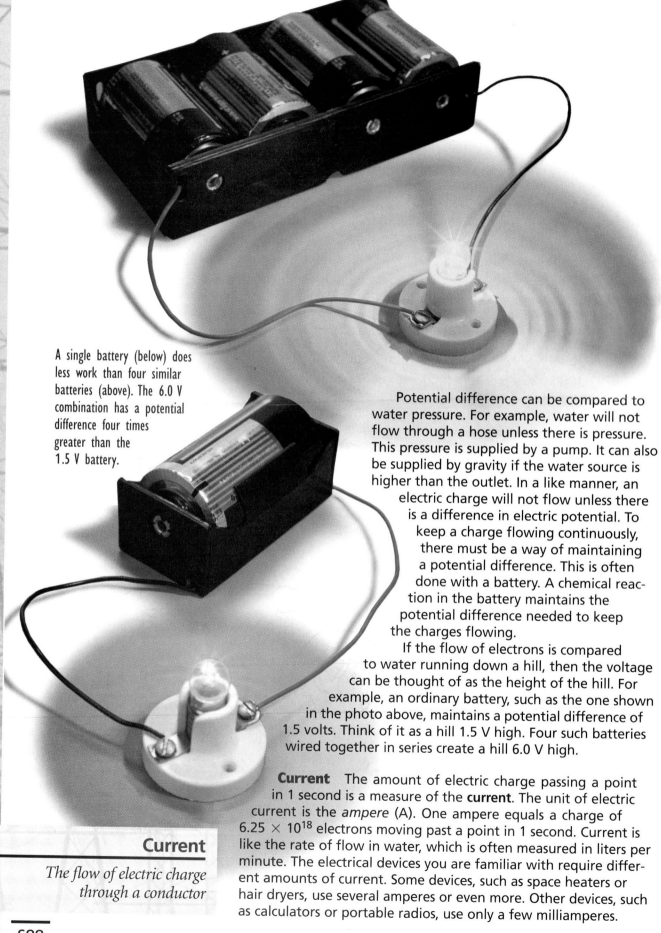

A single battery (below) does less work than four similar batteries (above). The 6.0 V combination has a potential difference four times greater than the 1.5 V battery.

Potential difference can be compared to water pressure. For example, water will not flow through a hose unless there is pressure. This pressure is supplied by a pump. It can also be supplied by gravity if the water source is higher than the outlet. In a like manner, an electric charge will not flow unless there is a difference in electric potential. To keep a charge flowing continuously, there must be a way of maintaining a potential difference. This is often done with a battery. A chemical reaction in the battery maintains the potential difference needed to keep the charges flowing.

If the flow of electrons is compared to water running down a hill, then the voltage can be thought of as the height of the hill. For example, an ordinary battery, such as the one shown in the photo above, maintains a potential difference of 1.5 volts. Think of it as a hill 1.5 V high. Four such batteries wired together in series create a hill 6.0 V high.

Current The amount of electric charge passing a point in 1 second is a measure of the **current**. The unit of electric current is the *ampere* (A). One ampere equals a charge of 6.25×10^{18} electrons moving past a point in 1 second. Current is like the rate of flow in water, which is often measured in liters per minute. The electrical devices you are familiar with require different amounts of current. Some devices, such as space heaters or hair dryers, use several amperes or even more. Other devices, such as calculators or portable radios, use only a few milliamperes.

Current

The flow of electric charge through a conductor

Resistance   Water moving through a hose encounters friction with the walls of the hose. The longer and narrower the hose, and the rougher its inner wall, the greater the friction will be. The greater the friction, the more the flow is reduced. Similarly, when electrons move through a material, they meet resistance. **Resistance** is an opposition to electric current that limits the flow of electrons in a circuit. Resistance is measured in *ohms* (Ω). A circuit has a resistance of 1 Ω when a potential difference of 1 V produces a current of 1 A.

You may recall that conductors have low resistances, but insulators have high resistances. For example, copper wires, such as those in your home, offer very little resistance to electric current, while rubber and glass offer a very large resistance. The amount of current in any electric circuit is determined by *both* the voltage and the resistance in the circuit. Think again of water flowing down a hill through a pipe. The amount of water that will pass through the pipe is determined by both the height of the hill *and* the size of the pipe. A narrow pipe offers more resistance to the flow of water than a large pipe, and it will therefore carry less water at a given pressure. A narrow pipe is similar to an electric circuit with high resistance. When an electric circuit has high resistance, the current in the circuit is reduced for any given voltage.

Resistance can be thought of as electrical friction. When electric current meets resistance, heat is produced. Consider the wires in a toaster oven. These wires are made of metal that has a relatively high resistance. As electricity flows through the wires, it encounters resistance, or electrical friction. The wires heat up as a result. This is similar to what happens when you rub your hands together. The harder you press them together or the faster you rub them, the warmer they get.

Resistance

The opposition to current in an electric circuit

▼ The resistance to flow in these pipes is mainly a function of their diameter.

◄ The wires in a toaster oven provide a lot of resistance to electric current.

Ohm's Law The voltage, current, and resistance in an electric circuit are related to each other by a rule known as *Ohm's law*. This relationship was discovered by a German schoolteacher, Georg Ohm, in the early 1800s. Ohm experimented with electric circuits made of wires having different amounts of resistance. By doing so, he discovered a general rule that describes the relationship among voltage (*E*), current (*I*), and resistance (*R*) in a circuit. Ohm's law can be written in equation form as follows:

$$\text{current} = \frac{\text{voltage}}{\text{resistance}}$$

or

$$I = \frac{E}{R}$$

For example, suppose an automobile with a 12 V battery has headlights with a resistance of 4 Ω. When the lights are on, the current needed to run the lights is

$$I = \frac{E}{R} = \frac{12\,\text{V}}{4\,\Omega} = 3\,\text{A}$$

By mathematically rearranging the terms, Ohm's law can also be written in other ways, for example:

$$E = IR \qquad \text{or} \qquad R = \frac{E}{I}$$

In short, if you know two of the values for a circuit, you can always find the third value by using Ohm's law.

▲ Using this illustration may help you recall the alternative forms of Ohm's law.

S U M M A R Y

Electric phenomena can be explained in terms of positive and negative electric charges. An electric charge results when something either gains or loses electrons. Like charges repel each other, and unlike charges attract each other. Static electricity refers to stationary electric charges. The force between two charged objects depends on the amount of charge and the distance between the objects. An electric current is the flow of electrons as they move from a place where there are more of them to a place where there are fewer. Resistance is an opposition to electric current. Ohm's law states that the electric current increases with increasing voltage but decreases with increasing resistance.

MAGNETS AND MAGNETIC EFFECTS

Y ou can see the effects of electricity all around you, and you have learned that electricity is related to magnetism. But what is magnetism, and what produces magnetic forces? Over 2000 years ago, the Greeks discovered a naturally occurring material with unusual properties. This material, called lodestone, attracted pieces of iron. In honor of the region of Magnesia, where the lodestone was found, the term *magnet* was given to any material that attracted iron and other metals. Common magnets are specially treated pieces of steel that can attract iron. They also attract steel (which is largely iron), nickel, and cobalt.

▲ Lodestone is a naturally occurring magnetic material.

▲ Magnets come in many different shapes and sizes.

Describing Magnetism

Magnetism is not evenly distributed throughout a magnet. If you dip a bar magnet into a box of nails, you will find that the nails cling mostly to the ends of the bar. The ends of a bar magnet are called *poles*. Poles are the points on a magnet at which its magnetic attraction is the strongest.

▲ Attractive force is strongest at the poles of a magnet.

Magnetic Poles   A simple demonstration illustrates an important fact about magnetic poles. If you allow a bar-shaped magnet to swing freely from a string, it will always point in a north-south direction. One pole points north; the other pole points south. If you disturb the magnet, after a moment or two it will return to this same orientation. Because magnets always act this way, the pole of the magnet that points north is called its north-seeking pole or *north pole*. The opposite end is called the *south pole* of the magnet. A compass is simply a magnet that is free to turn so that its north pole always points north.

▲ Because a magnet always aligns itself in a north-south direction, its two ends are called its north and south poles.

Magnetic force

The attraction or repulsion between the poles of magnets

Magnetic field

A region of space around a magnet in which magnetic forces are noticeable

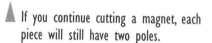

▲ If you continue cutting a magnet, each piece will still have two poles.

▲ Similar poles repel each other, whereas opposite poles attract, even at a distance.

▼ This horseshoe magnet has a circular magnetic field. Like all magnets, the field is strongest at the magnetic poles.

What do you suppose would happen if you cut a bar magnet in half? You might think that the north pole would be in one piece and the south pole would be in the other piece. The actual result is shown above. Each piece becomes a complete magnet with both north and south poles. If the pieces are cut into even smaller ones, the smallest piece will still be a complete magnet with a north and a south pole.

Magnetic Forces   You probably know what happens when you try to put two magnets together—like poles repel and unlike poles attract. This attraction and repulsion is called **magnetic force**. If the north pole of one magnet is brought near the north pole of another magnet, a repulsive force develops between them. If the north pole of one magnet is brought near the south pole of a second magnet, an attractive force develops. In this way, magnetic force is similar to electric force. Magnetic force is similar to electric force in other ways as well. First, both kinds of force work at a distance. In other words, two magnets do not need to touch in order to attract or repel each other. Furthermore, both magnetic force and electric force become weaker as the distance between two magnets or two electrically charged objects increases. As two magnets are brought closer together, the push or pull between them becomes much greater.

Magnetic Fields   You can feel the force between two magnets as they either attract or repel one another. The region of space in which magnetic forces are noticeable around a magnet is called the **magnetic field**. You can see the shape of a magnetic field by sprinkling iron filings around a magnet. The filings line up along curving lines running from one end of the magnet to the other. These lines are known as *lines of force*.

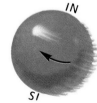

The needle of each compass shows the orientation of the magnetic field around the magnet.

As you move a compass around a magnet, the compass needle turns. This happens because the magnetic compass needle is affected by the magnetic field produced by the magnet. At any location in the field, the compass needle comes to rest in line with the magnetic lines of force. These lines, however, are only imaginary. Like the lines of latitude and longitude on a map, magnetic lines of force are "inventions" that help us understand how things work. They can be used to predict which way a compass needle will point.

Magnetic lines of force never cross each other. When two or more magnets produce fields that overlap, the result is a combined field. If opposite poles are aligned, most of the lines run from the north pole of one magnet to the south pole of the other.

The Source of Magnetism

Although people have known about magnets for centuries, it was not until the early nineteenth century that theories were developed to explain magnetism. The currently accepted theory suggests that magnetism is actually a property of electric charges in motion. You already know that the electrons of an atom are in constant motion. They not only orbit the nucleus but also "spin" about an axis. If the electron spins one way, it causes one type of magnetic force. If the electron spins the other way, it causes the opposite magnetic force. A pair of electrons spinning in opposite directions has no net magnetism because the forces cancel each other. Therefore, the magnetic properties of an atom are related to how many unpaired electrons it has. The motion of unpaired electrons can give an atom north and south poles, making it act like a tiny magnet.

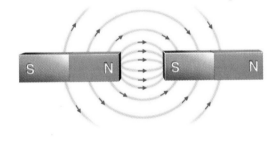

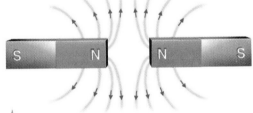

Magnetic lines of force never overlap, even when the poles of two magnets are brought close to one another, as shown here.

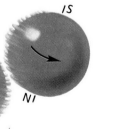

Spinning electrons act like tiny magnets. While paired electrons cancel each other out, unpaired electrons cause the atom to be magnetic.

Within materials that have magnetic properties, such as an iron bar, atoms are grouped together in clusters called **domains**. The atoms within each domain have about the same magnetic alignment. In other words, all of their individual north and south poles point in the same direction. In an unmagnetized piece of iron, these small domains are arranged in a random manner. Because of this, their magnetic properties cancel each other. But when you bring the south pole of a magnet near the iron bar, the magnet attracts the north poles of the atoms in the iron bar and causes most of the domains to line up in the same general direction. When most of the domains are aligned, it makes the whole iron bar into a single large magnet.

Domains

In magnetic materials, clusters of atoms in which the magnetic fields of most atoms are aligned in the same direction

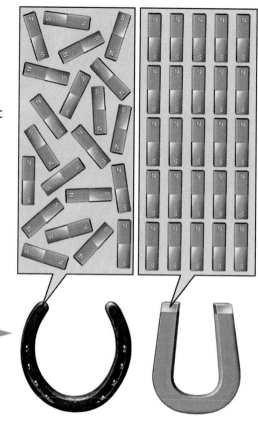

In most materials, such as this horseshoe, magnetic domains are randomly arranged. In magnets, however, the domains are aligned and produce a magnetic effect.

▲ Some alnico magnets are strong enough to support your weight.

Types of Magnets

The ancient Chinese discovered that a piece of lodestone will always point in the same direction when allowed to move freely. As a result, lodestone was used as the first compass. The Chinese also found that iron can be magnetized by rubbing it with lodestone. Lodestone is actually a naturally magnetic form of iron oxide called *magnetite*. Artificial magnets, on the other hand, are made of metals such as iron, cobalt, and nickel. Materials that are easily magnetized but quickly lose their magnetism are called *temporary magnets*. Some harder metals, such as steel, are harder to magnetize but tend to retain their magnetism better. A magnet made of such material is called a *permanent magnet*.

Artificial Magnets There are only five elements that can be made into magnets—iron, cobalt, nickel, gadolinium, and dysprosium. In their pure forms, however, these metals can be magnetized only temporarily. Excess heat or a sudden blow may cause their domains to become disorganized. To make a permanent magnet, you need an *alloy*. An alloy is a mixture of two or more metals, or a mixture of metals and nonmetals. The most common material used for making permanent magnets is steel, an alloy of iron and carbon. Strong permanent magnets can also be made of an alloy called *alnico*, which contains iron, aluminum, nickel, cobalt, and copper.

The best material known for permanent magnets is an alloy called *magnequench*. First fabricated in 1985, magnequench consists of the elements iron, neodymium, and boron. The magnetic field produced by this material is 10 times stronger than that produced by the best alnico magnets.

Electromagnets As you know, electromagnets are temporary magnets made by using an electric current. Due to the close relationship between electricity and magnetism, magnets can be used to produce electricity (as in a generator), and electricity can be used to make temporary magnets. These magnets can be turned on and off simply by flipping a switch. Powerful electromagnets, such as the one shown to the right, have been constructed for a variety of uses.

The Earth as a Magnet

In 1600, William Gilbert, who was the physician of Queen Elizabeth of England, proposed that the reason a suspended lodestone lines up in a north-south direction is that the Earth itself is a giant lodestone! To test his theory, Gilbert performed an experiment in which he carved a natural lodestone into the shape of a sphere. He found that a compass placed on the surface of the round lodestone acts just as a compass does at different places on the Earth's surface.

In fact, the Earth behaves as if a giant bar magnet were running through its center. Because the north pole of a magnet (by definition) is the pole that points northward, the magnetic pole at that location on Earth must in fact be a magnetic south pole. This follows from the behavior of magnetic poles—unlike poles attract. So when you use an ordinary compass, you are actually making use of two magnets. One of these magnets is small—the needle of the compass. The other magnet is very large. That magnet is the Earth itself.

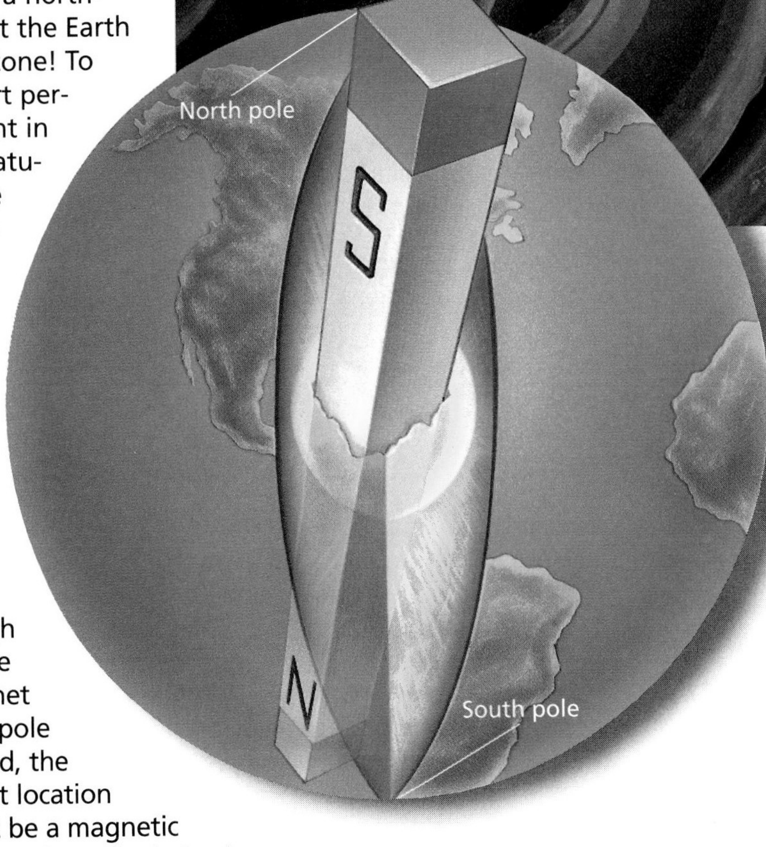

North pole

South pole

▲ An electromagnet attached to this crane can lift hundreds of kilograms at a time.

◄ The Earth itself is a magnet, with north and south poles. The north magnetic pole of the Earth is actually a magnetic south pole.

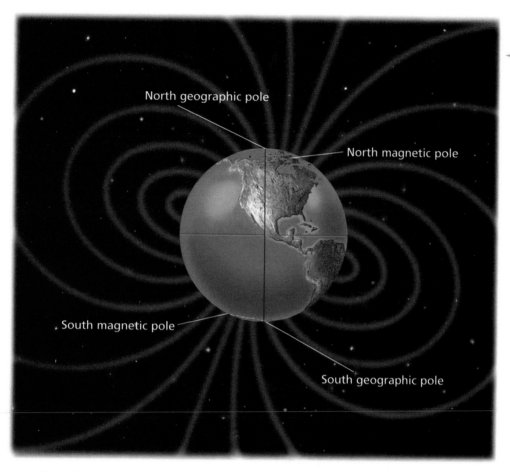

North geographic pole

North magnetic pole

South magnetic pole

South geographic pole

◀ The Earth has a magnetic field that is strongest at the poles. Notice that the Earth's geographic and magnetic poles are not in the same location.

Like all magnets, the Earth is surrounded by a magnetic field. Recall that you can see evidence of a magnetic field by sprinkling small pieces of iron on a piece of paper over a magnet. If there were some way to do this for the Earth, you would have a picture of this magnetic field. It is a little-known fact that the magnetic poles drift as much as several kilometers every year. Since it was first discovered in 1831, the north magnetic pole has drifted about 1000 km to the northeast.

SUMMARY

A magnet is any material that attracts iron and certain other metals. A magnet always has both a south pole and a north pole. As with electric charges, like magnetic poles repel and unlike poles attract. Magnetic fields surround magnets much like electric fields surround electrically charged objects. The magnetic properties of an atom are related to the movement of its electrons and to the number of unpaired electrons it has. In a magnetized bar of iron, the magnetic domains line up with like poles pointing in the same direction. Artificial magnets are made from alloys, and electromagnets are produced by the action of an electric current. The Earth also has a magnetic field, and it behaves as if a giant bar magnet were running through its center.

Using Electricity

To use electricity, you need four basic things: a source of electricity, a way to manipulate voltage, a device for converting the electricity into useful work, and a pathway for the current. We will now take a look at each of these requirements.

Starting the Flow of Current

Electricity must be generated before it can be used. To start an electric current flowing, an electric generator can be used to change mechanical energy into electrical energy. Generators come in a variety of sizes. Some, such as the generator that operates a bicycle lamp, could fit in the palm of your hand. Others, such as those that supply cities with electricity, are enormous devices weighing tons. We know what a generator is, but how does it work?

Basically, an electric generator consists of a coil of wire, called an *armature*, and a permanent magnet. A simple generator is shown above.

In most working generators, the field structure rotates and the armature stays fixed. Furthermore, the armature consists of hundreds of turns of wire, rather than the single coil shown here. In all generators, the part that rotates (whether it is the armature or the magnet) is called the *rotor*. The part that remains fixed is called the *stator*.

Alternators   As you know, household current is alternating current, called simply AC. A device that generates alternating current is called an **alternator**, or AC generator. You may have heard this term before, probably in reference to the small electric generator found in automobiles. Automotive alternators convert a small fraction of the engine's mechanical energy into electrical energy to operate the car's lights, radio, and gauges. The operation of an alternator is summarized in the illustration below.

Slip rings · Coil · Brushes · Permanent magnet

▲ In a simple generator, the rotating armature sweeps through lines of magnetic force produced by the magnet. The changing magnetic field sets electrons within the armature in motion, causing an electric current to flow.

Alternator

A mechanical device for generating alternating current

◄ How an alternator works

1 Each loop of the armature sweeps one way across magnetic lines of force. This causes current to flow one way through the wire.

Slip rings · Field structure

S · N · Brushes

2 When the armature reaches a point where its loops are no longer crossing lines of force, the current drops to zero.

Armature

3 The loops then cut across magnetic lines of force from the other way, causing current to flow in the opposite direction through the circuit.

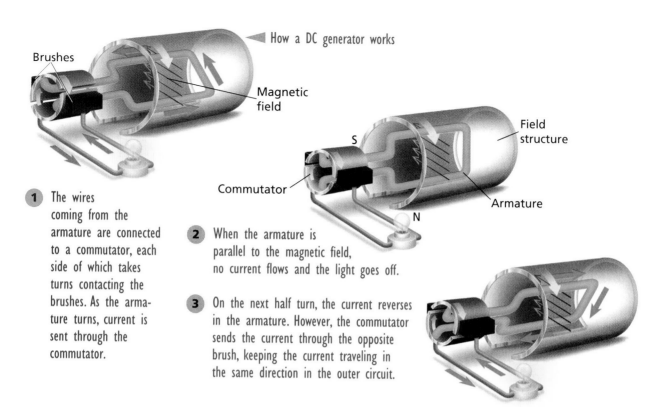

◀ How a DC generator works

Brushes

Magnetic field

Field structure

Commutator

S

N

Armature

1 The wires coming from the armature are connected to a commutator, each side of which takes turns contacting the brushes. As the armature turns, current is sent through the commutator.

2 When the armature is parallel to the magnetic field, no current flows and the light goes off.

3 On the next half turn, the current reverses in the armature. However, the commutator sends the current through the opposite brush, keeping the current traveling in the same direction in the outer circuit.

Being Direct Not all generators produce alternating current. A DC generator produces direct current. The diagram above summarizes its operation. You might notice a similarity between the AC and DC generators: in both, the current in the coil drops to zero and then reverses direction with every half turn. To compensate, the DC generator has a mechanical device called a *commutator*, which switches the connections to the circuit to keep the current flowing in the same direction.

Transforming Electricity

In getting electricity to do what you want it to do, it is often necessary to *transform* it—raise or lower its voltage. Since voltage is a measure of the amount of "push" that an electric current has, the higher the voltage, the stronger the push.

As you already know, most of our electricity comes in the form of alternating current (AC). You may wonder why this type of current is used instead of direct current (DC), which flows only in one direction. After all, wouldn't the constantly changing current cause problems?

The primary reason AC is used is that it is easy to change its voltage. The device used to do this is called a *transformer*, shown below. In its simplest form, a transformer consists of coils of wire wound around an O-shaped iron core. Alternating current from the source flows through the *primary coil*. With each surge of current, the iron core becomes an electromagnet. Every time the current alternates, or switches direction, the electromagnet changes polarity. This causes the other coil of the transformer to respond as though it were rotating between the poles of a magnet. Current is therefore produced in the *secondary coil*.

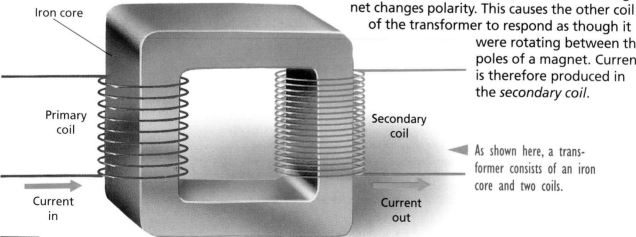

Iron core

Primary coil

Secondary coil

Current in

Current out

◀ As shown here, a transformer consists of an iron core and two coils.

Wind-Up Voltage The relative amount of voltage produced in the secondary coil of a transformer depends on the number of turns of wire in each of the two coils. If the primary coil has more turns than the secondary coil, the voltage output will be lowered. This type of transformer is called a *step-down* transformer. If, on the other hand, the secondary coil has more turns than the primary coil, the voltage will increase. This is called a *step-up* transformer.

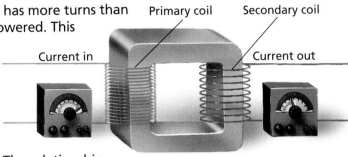

You can figure out the voltage in either of the two coils if you know the voltage in one coil and the number of turns in both coils. The relationship between voltage and turns in a transformer is given by the following equation:

$$\frac{\text{Voltage in the primary}}{\text{Voltage in the secondary}} = \frac{\text{Turns in the primary}}{\text{Turns in the secondary}}$$

$$\text{or} \quad \frac{E_P}{E_S} = \frac{T_P}{T_S}$$

If you know three values, you can find the fourth. For example, if the primary coil had 100 turns and the secondary coil had only 50 turns, the voltage would be transformed from high to low voltage. If this transformer were plugged into a standard wall socket at 120 V, the output from the secondary coil would be 60 V.

▲ A step-down transformer (top) and a step-up transformer (bottom)

Effects on Current If you transform the electricity by increasing or decreasing the voltage, there is a corresponding effect on current. A transformer can change the voltage (force) in a circuit, but the total electrical energy (work) must remain the same. In other words, when voltage is stepped up, current is stepped down. Remember that electrical energy is a function of both voltage and current. The change in current can be calculated as follows:

$$\frac{\text{Voltage in the primary}}{\text{Voltage in the secondary}} = \frac{\text{Current in the secondary}}{\text{Current in the primary}}$$

$$\text{or} \quad \frac{E_P}{E_S} = \frac{I_S}{I_P}$$

If, for example, the voltage is boosted from 100 V to 1000 V and the current in the primary coil is 2 A, the current in the secondary coil will be 0.2 A.

From the generating plant, electricity is usually transmitted over special lines at very high voltages, up to 765,000 V. This high voltage produces very low currents and therefore reduces losses caused by resistance in the power lines. Electrical substations reduce this high voltage to around 100,000 V for distribution to local power stations. A substation reduces the voltage to around 15,000 V for distribution within neighborhoods. And near your house, a small transformer further reduces the voltage to about 110 V, which is what most of your appliances use.

Electricity Into Motion

Motor

A device for converting electrical energy into mechanical energy

A **motor** is a device that changes electrical energy into mechanical energy that can be used to do work. There are three basic types of motors: AC motors, DC motors, and universal motors. As their names suggest, AC motors work with alternating current, and DC motors work with direct current. Universal motors will work with either AC or DC.

How Motors Work   Motors operate on the same general principle as generators, only in reverse. In fact, some motors can be used as generators if they are turned by an outside force. Like a generator, an electric motor consists of an armature and a permanent magnet. The magnet creates a magnetic field in which the armature turns. The rotating magnetic field of the armature interacts with the magnetic field of the permanent magnet to cause the rotational motion of the motor.

1 In a motor, the armature becomes an electromagnet when it is supplied with electricity from an outside source. Because like poles of magnets repel and unlike poles of magnets attract, the armature assembly turns in response to the magnetic field of the permanent magnet.

Like a generator, a motor consists of a permanent magnet and an armature. In a real motor, the armature would be attached to a drive shaft that turns and does work.

2 As the armature turns, however, the commutator causes the current passing through the armature to change direction. The polarity of the electromagnet is thus reversed, and the armature is forced to turn again. These continuously changing magnetic fields keep the armature turning for as long as the motor is in operation.

Brushes

N

S

Permanent magnet

Commutator

Armature

Direction of rotation

N

S

Source of electricity

POWER STATION

Circuit

A continuous path through which electricity flows

BUZZZZZ!

FUSEBOX

Pathways for Current

As you know, in order for electricity to flow, there must be an unbroken pathway that electrons can move through. Such a pathway is called a **circuit**. Circuits have three parts: a source of electric current, an output device, and a connection.

The source is typically a generator or a battery. When you plug in a blow-dryer, the source is not the wall socket but the distant generator that supplies your house with current. Output devices, such as lamps and motors, use the energy carried by the circuit to perform useful functions. A lamp, for example, converts electricity into light (and heat); a motor converts electrical energy into rotary motion. The connection, such as a wire, joins the components of a circuit together, providing a pathway for the electrons. As long as the pathway is intact, electricity will flow. If the connection is broken, however, electricity will cease to flow and the output device will quit working.

Controlling Circuits For electricity to be useful, the current must be controlled in some way. For example, there should be a way to turn an electrical device off. The simplest device for controlling electric current is a *switch*. A switch stops the flow of electric current by introducing a gap into the circuit. Another common current-control device is the *rheostat*. A rheostat allows the amount of current flowing through a circuit to be varied at will by adding resistance to the current. The dimmer switch on a lamp and the volume knob on a stereo are examples of rheostats.

Circuits may be simple or quite complex. Some circuits, such as those in a computer, may contain hundreds or thousands of individual components.

As you know, there are two basic types of circuits—*series* and *parallel*. A series circuit has a single current pathway. The components of a series circuit are wired "in series," meaning that each component is located on the same current pathway. The entire electric current passes through every component in the circuit, one after another.

A parallel circuit, on the other hand, has at least two current pathways. Typically, a parallel circuit consists of branches coming off of a main current pathway. Each of the branches goes through an electrical component or output device. An electric current passing through a parallel current divides. Some current continues along the main circuit and some diverts into the branching circuits. After passing through the branches, the separate currents rejoin the common current pathway.

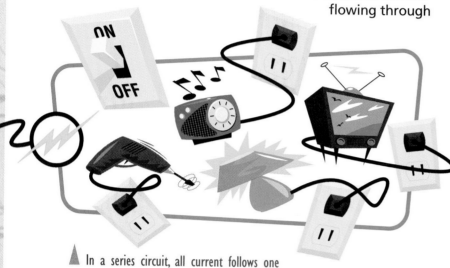

In a series circuit, all current follows one path. If the circuit is broken (or the switch is thrown), all of the appliances will go off.

In a parallel circuit, the current follows several branching paths. If the circuit is broken at any one branch, only the appliances on that branch will go off. The current will follow the alternate paths.

S U M M A R Y

Generators convert mechanical energy into electrical energy. Alternators generate alternating current, and DC generators generate direct current. Transformers allow for regulating the voltage in AC circuits. Motors convert electrical energy into mechanical energy. Circuits are the pathways through which electric current flows. Series circuits consist of a single current pathway, whereas parallel circuits consist of two or more current pathways.

Concept Mapping

The concept map shown here illustrates major ideas in this unit. Complete the map by supplying the missing terms. Then extend your map by answering the additional question below. Use your ScienceLog. **Do not write in this textbook.**

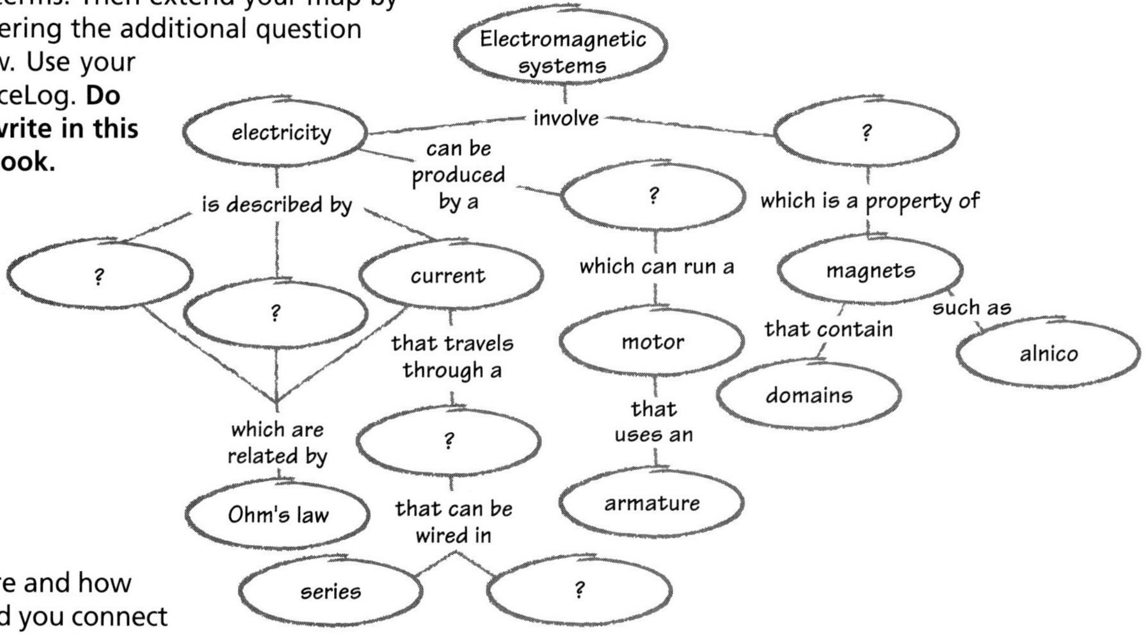

Electromagnetic systems

involve

electricity

can be produced by a

?

which is a property of

is described by

?

current

which can run a

magnets

such as

?

?

that contain

alnico

that travels through a

motor

domains

which are related by

?

that uses an

Ohm's law

that can be wired in

armature

series

?

Where and how would you connect the terms *electrons, poles,* and *transformer*?

Checking Your Understanding

Select the choice that most completely and correctly answers each of the following questions.

1. An electric current results from
   a. the flow of positively charged particles through a conductor.
   b. the flow of electrons between areas of greater and lesser charge.
   c. the movement of ions through wires.
   d. the flipping of a switch.

2. Which of the following correctly describes the behavior of charged particles?
   a. Negatively charged particles attract positively charged particles.
   b. Negatively charged particles attract other negatively charged particles.
   c. Positively charged particles attract other positively charged particles.
   d. Negatively charged particles attract both positively charged and negatively charged particles.

3. Voltage can be compared to
   a. flow rate.
   b. current velocity.
   c. pressure.
   d. friction.

4. "The greater the distance between two objects, the weaker their interaction." This statement applies to
   a. only an electric force.
   b. only a magnetic force.
   c. both an electric and a magnetic force.
   d. neither an electric nor a magnetic force.

5. A generator converts
   a. electrical potential into electrons.
   b. electrical energy into mechanical energy.
   c. direct current into alternating current.
   d. mechanical energy into electrical energy.

Critical Thinking

Carefully consider the following questions, and write a response in your ScienceLog that indicates your understanding of science.

1. Suppose that you have an iron nail and a compass. How could you demonstrate whether the nail was magnetized or not?

2. A certain transformer has a primary coil with 30 turns of wire and a secondary coil with 10 turns; the voltage in the secondary coil is 30 V. What is the voltage in the primary coil? Show your work.

3. Suppose you coupled a generator to an electric motor so that the motor drove the generator and the generator supplied the motor with electricity. Would this system run itself indefinitely? Why or why not?

4. Why is AC, but not DC, easily transformed? What would happen if you passed DC through a transformer? Explain.

5. All magnets have a *Curie point*, the temperature at which the materials cease to be magnetic. Above that temperature, no magnetism is present. As you know, there is a relationship between temperature and molecular behavior. How does this relationship help to explain the Curie point?

Interpreting Photos

Some of the bulbs in this string of lights are not working. What type of circuit—series or parallel—would account for the on-and-off pattern of bulbs shown here? Explain.

Portfolio Idea

Imagine that you are a scientist in prehistoric times, and you have just discovered magnetism. You are describing its properties to a colleague in a letter. Since you have discovered it well before the modern age of science, most of the words associated with magnetism do not yet exist. Describe your discoveries about magnets and magnetism to your colleague without using the terms *magnet*, *magnetism*, *pole*, *magnetic field*, or *lines of force*.

S104

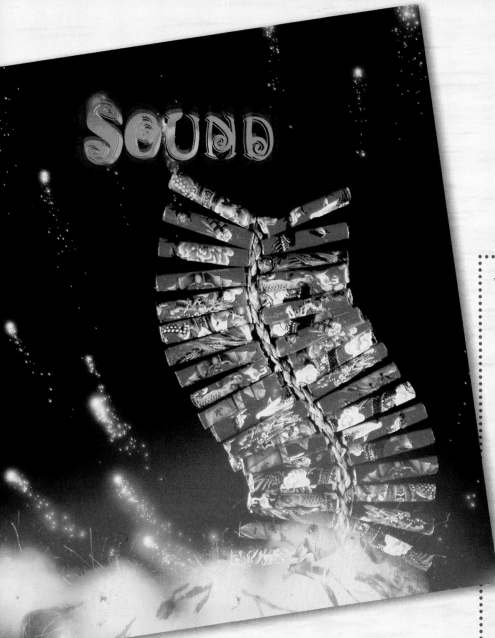

SOUND

IN THIS UNIT

Now *that you have been introduced to sound and the nature of its qualities, consider the following questions.*

● **1.** How do waves transmit energy from one place to another? How can you observe this with sound waves?

2. Does the fact that sound travels as waves affect how we hear it?

3. In what ways do we use sound, and how can we protect ourselves from sound levels that are too high?

In this unit, you will take a closer look at the physical basis for sound—waves—and at how modern technology uses sound.

WAVES

Y ou have been learning about sound waves. But what, exactly, are waves? There are many kinds of waves in addition to sound waves. You have probably heard of water waves, radio waves, microwaves, earthquake waves, and electromagnetic waves, as well as other types of waves. Because sound is made of waves, it has certain properties that are characteristic of all waves. So let's take a step back and investigate the phenomenon of waves in general.

If you have ever dropped a stone into a quiet pond, you produced many waves, or disturbances, in the surface of the water. A **wave** is a disturbance that transmits energy as it travels through matter or empty space. The material through which a wave travels is called a *medium*. The waves on the pond are produced when kinetic energy from the stone is trans-ferred to the medium, which is the water. The waves then carry the energy outward through the medium, from the point where the stone hit the water.

It is important to understand that the energy of the wave moves *through* the medium; the medium itself does not travel with the wave's energy. It is like a long line of people standing close together. Imagine that the person at the end of the line pushes the next person in line, and that person pushes the next, and so on. Can you see how the disturbance can pass to the head of the line without any person taking a step forward?

Wave

A disturbance that transmits energy as it travels through matter or space

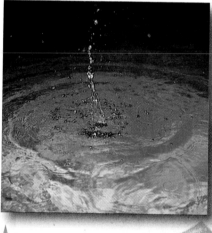

Energy from a stone radiates outward in the form of waves.

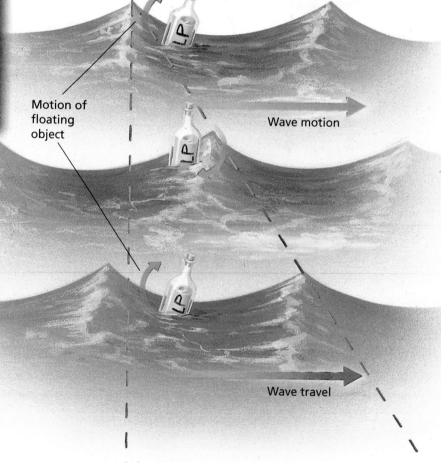

Motion of floating object

Wave motion

Wave travel

Notice that although the wave continues traveling to the right, the floating object (as well as the water itself) remains in one place, making small bobbing motions.

Types of Waves

A liquid, such as water, is one medium for waves. But gases, like those in the atmosphere, and solids, such as iron, can also be mediums for waves. We can study the action of waves without getting wet by considering waves in a rope. The rope acts as the medium for the waves. To study waves in more detail, look at the illustration below.

Transverse Waves The waves produced by the up-and-down motion of a medium are called **transverse waves**. *Transverse* means "moving across." In a transverse wave, the material of the medium moves across (at right angles to) the direction in which the wave travels. Look at the illustration again. Notice that although the wave travels to the right, no part of the rope moves to the right. Look at the colored markers on the rope. If you follow the motion of any one of them from one drawing to the next, you will see that each marker moves up and down. Every point on the rope follows the motion of the student's hand.

The top of a transverse wave is called a *crest*. Each crest is followed by a depression called a *trough*. Transverse waves consist of a series of crests and troughs that follow each other through a medium. The motion of the particles of the rope is perpendicular to the direction in which the wave is traveling. If the rope were shaken from side to side, would the wave still be transverse?

Transverse wave

A wave in which the motion of the particles of the medium is perpendicular to the path of the wave

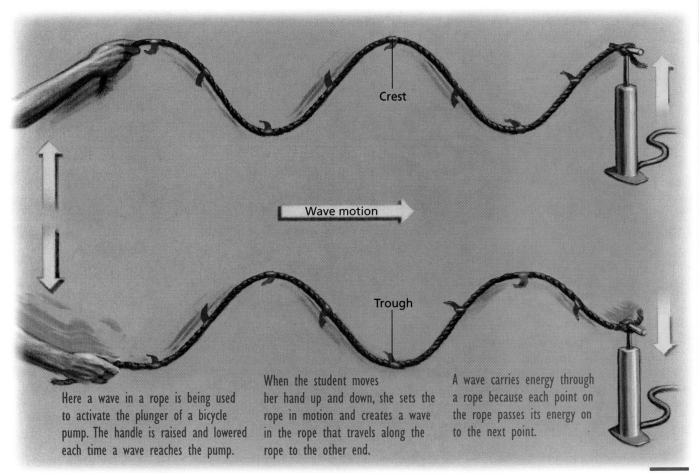

Crest

Wave motion

Trough

Here a wave in a rope is being used to activate the plunger of a bicycle pump. The handle is raised and lowered each time a wave reaches the pump.

When the student moves her hand up and down, she sets the rope in motion and creates a wave in the rope that travels along the rope to the other end.

A wave carries energy through a rope because each point on the rope passes its energy on to the next point.

Longitudinal wave

A wave in which the particles of the medium move back and forth, parallel to the direction of motion of the wave

Longitudinal Wave A different kind of wave motion can be demonstrated by using a coiled spring like the one shown in the illustration below. In this case, a student produces a wave by moving her hand toward and away from a bicycle pump with a motion that is parallel to the length of the spring. The particles of the spring also move toward and away from the pump. Waves in which the particles of the medium move back and forth parallel to the path of the wave are called **longitudinal waves**.

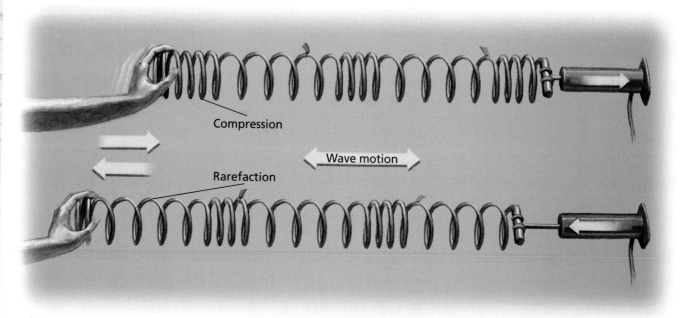

Compression

Rarefaction

Wave motion

▲ Here a spring is being used to activate the plunger of a bicycle pump. Each time a complete wave reaches the pump, it moves the pump handle in and out.

Notice that longitudinal waves do not have crests and troughs. Every time the student pushes the spring forward, she pushes some of the coils of the spring closer together. In other words, the student causes a *compression* in the spring. In each compression of a longitudinal wave, the particles of the medium are more dense than they are in the surrounding material. When the student pulls her hand back, she separates the coils. This causes an expansion in the spring, where the particles of the medium become less dense. This part of a longitudinal wave is called a *rarefaction*. Longitudinal waves consist of alternating compressions and rarefactions moving through a medium.

The crest of a transverse wave can be compared to the compression of a longitudinal wave, while a trough can be compared to a rarefaction.

Transverse wave

Crest

Trough

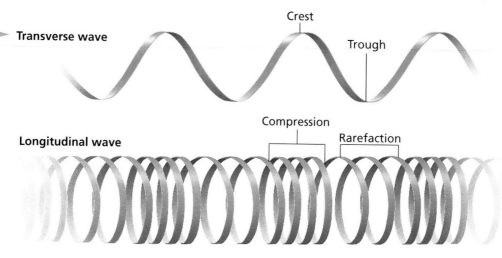

Longitudinal wave

Compression

Rarefaction

Properties of Waves

Just as you can describe matter by various physical properties, such as color, size, mass, and density, waves also have certain properties by which they can be described. Although they are produced in many different ways and cause many different effects, all waves can be described by four basic properties.

Amplitude   You can make a small wave in a rope by flicking your wrist while holding one end of the rope. You could make a much larger wave by moving your whole arm to shake the rope. Scientists use the term *amplitude* when referring to the size of a wave. **Amplitude** is the maximum distance that a certain point on a wave moves from a rest position. The amplitude of any given wave depends on the amount of energy in the wave. The greater the energy that the wave carries, the greater the wave's amplitude is.

Amplitude

The maximum distance that a wave rises or falls from a normal rest position

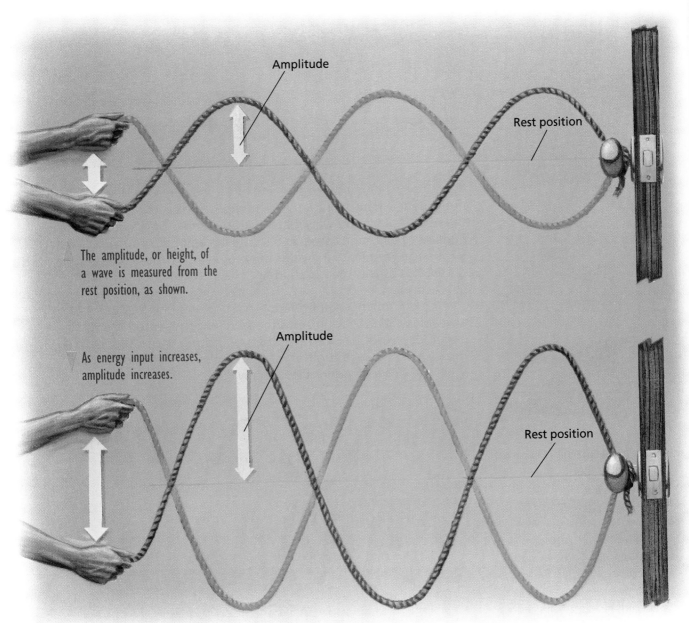

Amplitude

Rest position

The amplitude, or height, of a wave is measured from the rest position, as shown.

As energy input increases, amplitude increases.

Amplitude

Rest position

Wavelength   Waves are characterized not only by their amplitude, but also by their *wavelength*. **Wavelength** is the distance between a particular point on one wave and the identical point on the next wave. The easiest way to find the wavelength of a transverse wave is to measure the distance from one crest to the next. The easiest way to find the wavelength of a longitudinal wave is to measure the distance between two compressions.

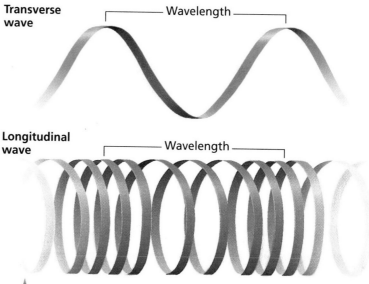

Transverse wave

⎯⎯ Wavelength ⎯⎯

Longitudinal wave

⎯⎯ Wavelength ⎯⎯

▲ The wavelengths of transverse and longitudinal waves are illustrated here.

Frequency   If you shake one end of a rope very slowly, you will not produce many vibrations in the rope. However, if you shake the rope rapidly, you will create a large number of vibrations. The number of vibrations produced in a given amount of time is called **frequency**. Frequency indicates the rate at which vibrations are produced. It is usually measured in hertz (Hz), the number of vibrations per second.

There is an important relationship between the frequency of a wave and its wavelength. As the frequency increases, the wavelength decreases. You can see this by watching the vibrations in a rope as you shake it at different rates. The faster you shake the rope, the closer together the crests will be.

▲ Objects that vibrate with constant frequencies can be used to measure time. The pendulum of a grandfather clock, for example, might swing back and forth once each second—a frequency of 1 Hz—while the quartz crystals used in most digital watches vibrate at a frequency of 32,768 Hz. Frequencies also determine phenomena such as the pitch of musical tones and the color of light.

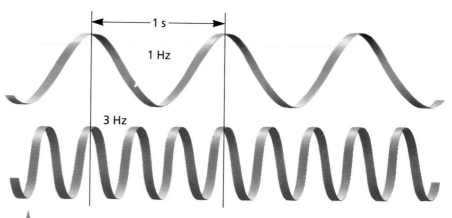

1 s

1 Hz

3 Hz

▲ The wave on the bottom has a wavelength one-third that of the wave on the top, while it has a frequency three times as great.

Wave Speed Another basic property of waves is the speed at which they travel. The speed of a wave can be determined by observing a certain point on the wave as it moves. In longitudinal waves, you could observe a single compression or rarefaction as it travels through the medium. In transverse waves, you could observe a single crest or trough. For example, the speed of a wave through water can be found by observing one crest and measuring the distance it moves in a certain amount of time. Wave speed is usually given in meters per second (m/s).

Wave speed can be calculated if you know the wavelength and frequency of a wave. The relationship between speed, frequency, and wavelength is expressed in the following formula:

$$\text{speed} = \text{frequency} \times \text{wavelength}$$

or

$$v = f\lambda$$

For example, suppose that you are fishing from an anchored boat. Several waves pass the boat. You could find the speed of those waves by calculating the number of waves that pass in 1 second (frequency) and measuring the length of 1 wave. If the frequency is 2 waves/s (2 Hz) and the wavelength is 0.5 m, the speed is

$$v = f\lambda$$
$$v = 2 \text{ waves/s} \times 0.5 \text{ m/wave}$$
$$v = 1 \text{ m/s}$$

The same formula can be used for sound waves. The speed of sound (v) is known to be about 346 m/s. So if you can determine either the frequency (f) or the wavelength (λ) of a sound, you can calculate the other value.

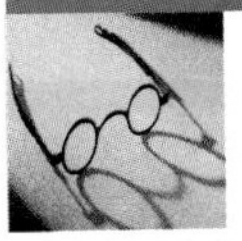

Wavelength

The distance between two identical points on neighboring waves

Frequency

The number of waves produced in a given amount of time

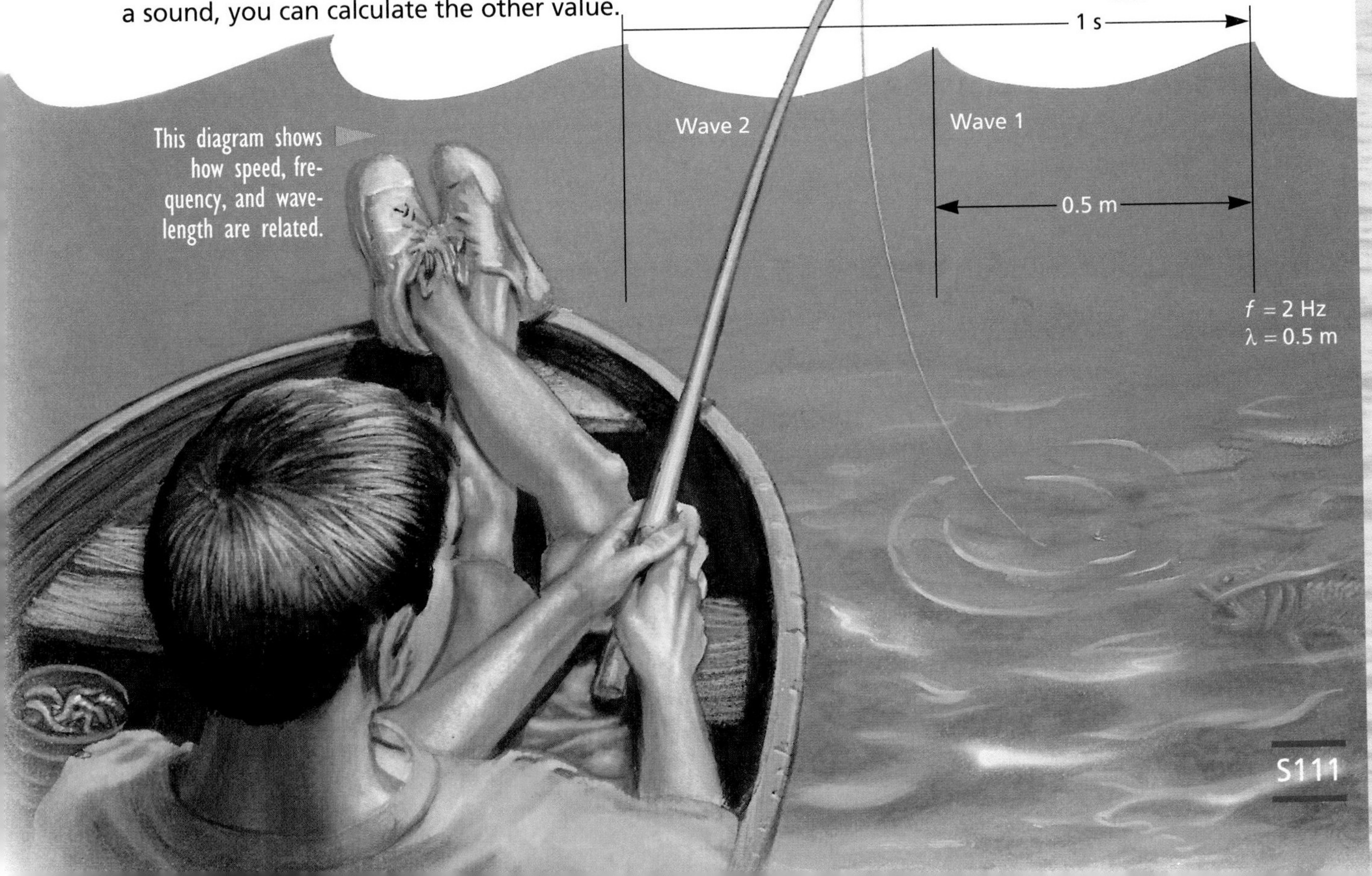

1 s

Wave 2

Wave 1

0.5 m

$f = 2$ Hz
$\lambda = 0.5$ m

This diagram shows how speed, frequency, and wavelength are related.

Characteristics of Waves

There are certain characteristics that all waves have in common. *Reflection*, for example, is displayed by waves when they encounter obstacles. Other characteristics include *refraction*, *diffraction*, and *interference*.

Reflection

The process in which a wave bounces back after striking a barrier that does not absorb all of the wave's energy

Reflection   Have you ever watched or played a game of pool? A skillful player knows how to use the side rail of the table to his or her advantage. Success often depends on getting the balls to bounce at the desired angles. Normally, a ball will bounce off the side rail of the table at the same angle at which it strikes it.

▲ When playing pool, you must consider the laws of reflection.

Waves bounce off barriers in the same manner that a pool ball does. This is called **reflection**. To demonstrate that waves are reflected, try these experiments. Tie one end of a rope to a doorknob. Hold the other end of the rope and send a transverse wave pulse through the rope. You will see that when the wave reaches the doorknob, it bounces back. Longitudinal waves are also reflected by barriers. Hold a coiled spring at one end. If you send a longitudinal wave down the spring, the wave will come back when it reaches the end of the spring. To demonstrate the reflection of water waves, make a wave in a sink or tub full of water by tapping your finger on the surface of the water. When the wave reaches the side of the sink or tub, you will see it bounce back. For an example of the reflection of sound waves, recall how sound is reflected from walls or the sides of buildings to produce echoes.

◀ This circular wave was reflected from a straight barrier.

If waves approach the shore at an angle, the ends closest to the shore slow down first, and the waves are refracted.

Refraction   The bending of a wave when it passes from one medium to another is called **refraction**. Refraction is due to a change in the speed of a wave. You can observe the refraction of waves at the ocean shore. As waves in the ocean approach the shore, they enter shallow water and slow down. This is because waves travel faster in deep water than in shallow water. If a wave approaches the shore at an angle, the end closest to the shore slows down first. As a result, the wave bends toward the beach.

Refraction can be compared to the column of a marching band that marches at an angle from hard pavement into a patch of soft mud. The band members at one side of the front row will encounter the mud first and, finding it harder to march, will slow down and march with shorter strides. The others in the front row will continue at their initial speed until each, in turn, enters the mud. In effect, each row in the column of marchers will turn toward the mud. The opposite effect will occur when the column returns from the mud to the pavement.

Diffraction   Waves can also go around corners. Water waves clearly illustrate this behavior. The image to the right shows ripples from a vibrating source touching the water surface in a "ripple tank." Straight water waves (coming from the top of the picture) bend around the edge of the barrier. This is called **diffraction**. All waves, including sound waves, exhibit diffraction. For example, you can hear a sound from around a corner, even if there is nothing to reflect the sound to you.

Refraction

The process in which a wave changes direction because its speed changes

Diffraction

The bending of waves as they pass the edges of objects

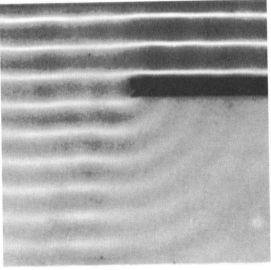

Straight waves can be bent around the edges of a barrier.

Interference

The result of two or more waves overlapping

Interference Have you ever thrown two stones into a still pond at the same time and seen the ripples from the two splashes pass through each other?

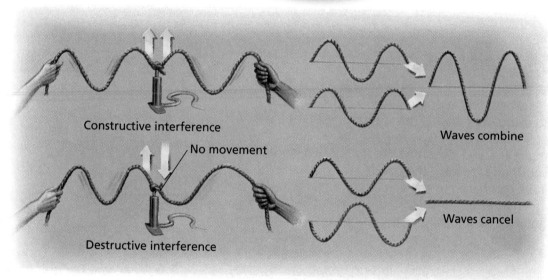

When two waves cross each other's path, they affect each other's amplitude.

Waves that meet in this manner affect each other by a process called **interference**. When two troughs meet, they combine to make a single, larger trough. Likewise, when two crests meet, they combine to make a single, larger crest. The result is deeper troughs and higher crests. This phenomenon is called *constructive interference*. On the other hand, when a crest and a trough of equal amplitude meet, they cancel each other out. This is called *destructive interference*. Interference is a characteristic of all waves, including sound waves.

Constructive interference

No movement

Destructive interference

Waves combine

Waves cancel

Waves that undergo constructive interference (top) produce a single, stronger wave. Waves that undergo destructive interference (bottom) tend to cancel each other out.

SUMMARY

A wave transmits energy through a medium or through empty space. Both transverse waves and longitudinal waves exhibit the properties of amplitude, wavelength, frequency, and speed. Reflection, refraction, diffraction, and interference are characteristics common to all waves. Reflection is the bouncing back of a wave when it meets a barrier. Refraction is the bending of a wave when it passes at an angle from one medium to another, either slowing down or speeding up. Diffraction is the bending of straight waves around the edge of a barrier. When two or more waves meet, they interfere with each other.

SOUND WAVES

W ithout air, there would be no familiar sounds. In fact, we would hear no sounds at all. This is because air is a medium for sound. Sound moves through air in the same way that ripples move through water. A sound begins when a vibrating object, such as a guitar string, pushes on the air particles around it. These particles then push on other particles, causing a disturbance that is transmitted outward from the source. As sound waves move through the air, the air particles are alternately crowded together and spread apart. This back-and-forth motion of the particles of air is similar to the way longitudinal waves move through a spring. In other words, sound consists of longitudinal waves.

When air is removed from the bell jar, the alarm clock will not be heard. No air molecules are there to transfer the energy as a wave.

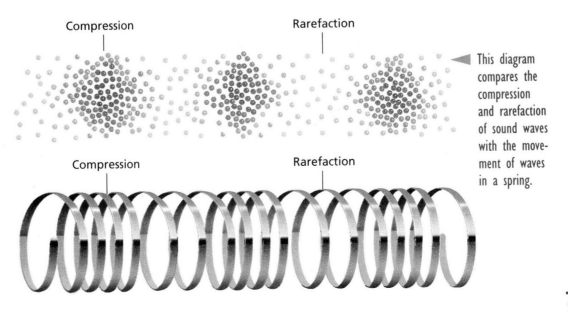

Compression

Rarefaction

Compression

Rarefaction

This diagram compares the compression and rarefaction of sound waves with the movement of waves in a spring.

Vibrating Strings

Pythagoras was a Greek philosopher and mathematician who lived in the sixth century B.C. He is famous for his proof of the relationship among the three sides of a right triangle. However, Pythagoras also pursued many other areas of study. For example, he was very interested in the sounds produced by vibrating strings. He discovered that pleasant-sounding chords could be produced if the lengths of the strings were in the ratio of two small numbers. For example, consider two similar strings that are under the same tension. If one string is two times the length of the other, they will produce two notes that are one octave apart (for example, from C to C). If the lengths of the strings are in the ratio of two to three, then a fifth chord is produced (for example, from C to G). These chords are considered to be very pleasing to the ear.

Pythagoras was so impressed by his discovery that he made it the basis for a whole school of thought. He believed that the beauty of the universe was related to the beauty of these chords. For example, he thought that the movements of the planets and stars were based on the same ratios that he discovered in vibrating strings. The phrase "music of the spheres" comes from this idea. Later scientists disproved Pythagoras's theory concerning the movements of stars and planets. However, his law of vibrating strings, which came from direct observation, has withstood the test of time.

DID YOU KNOW...

that Pythagoras was the leader of a secret society? The investigations he and his colleagues did with numbers and ratios were jealously guarded as wonders that the common people would not understand.

▲ Pythagoras thought that the sun and the planets moved in perfect harmony, one that could be expressed with simple numbers.

Standing Waves

An important characteristic of vibrating strings is the patterns of waves that they form. Suppose you send a wave through a rope and the wave is reflected by a barrier. If you send only one crest, it will travel to the barrier and then be reflected back toward you. The crest will encounter no interference. However, if you continue to produce waves, the waves traveling toward the barrier will soon encounter waves reflected from the barrier. In this situation, the waves will interfere with one another.

Because the strings on a musical instrument like the guitar are fixed at both ends, they reflect from both ends. Waves traveling in opposite directions that have matching characteristics—amplitude, frequency, and speed—combine to form **standing waves**.

The points on standing waves that have no vibration are called *nodes*. Nodes are caused by destructive interference between waves moving in opposite directions. The points on standing waves that vibrate with the greatest amplitude are called *antinodes*. Antinodes are the result of constructive interference between waves moving in opposite directions.

Standing waves

Waves that form a pattern in which portions of the waves do not move and other portions move with increased amplitude

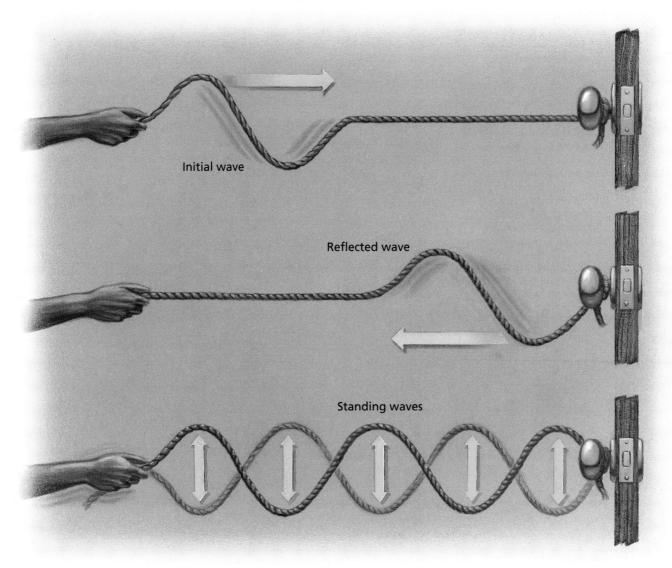

Initial wave

Reflected wave

Standing waves

▲ When reflected waves meet waves moving in the opposite direction, standing waves are formed.

Standing waves look very different from traveling waves. Standing waves do not appear to move through the medium. Instead, the waves cause the medium to vibrate in a series of stationary *loops*. Each loop is separated from the next by a node, or point of no movement. The distance from one node to the next is always one-half of a wavelength.

The following diagram shows the vibrations of a string. In each example, a standing wave has been produced. Notice that both ends of the string are fastened tightly and cannot move. Therefore, there must be a node at each end of the string.

Standing waves can also be formed by longitudinal waves. Standing waves in a column of air cause the sound you hear when you blow across the top of a soda bottle. Columns of standing waves are also responsible for the sounds of flutes, clarinets, and other wind instruments.

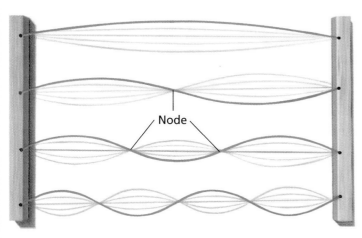

Node

1 The frequency of the first vibration has one loop in the wave. Since the loop is one-half wavelength long, the wavelength is twice the length of the string.

2 In this wave, the string has a total of three nodes. The entire string is equal in length to one wavelength.

3 There are four nodes in this wave. Here the string is equal to the length of one and one-half wavelengths.

4 Five nodes are present in this wave. The entire length of the string includes two wavelengths.

Each of these wavelengths occurs at a different frequency. When you pluck a string, it produces all of these frequencies at the same time. The lowest frequency is called the *fundamental*; higher frequencies are called *overtones*. The number and relative intensity of the overtones give the characteristic sound to the instrument.

▼ When guitar strings are plucked, a variety of standing waves are produced, blending to form a guitar's distinctive sound quality.

The Doppler Effect

Have you ever heard the blast of the warning horns as a train passed through a railroad crossing? Did you notice how the sound of the horns changed as the engine passed by? You probably heard a sudden drop in the pitch of the sound. But why?

The pair of photographs below shows the pattern of sound waves produced by a moving sound source. Notice that as the horns on the train move forward, the sound waves in front of the train are closer together, giving these waves a shorter wave-length and thus a higher frequency. The sound waves in back of the train are spread farther apart, so they have a longer wavelength and a lower frequency. This change in the apparent frequency of the waves from in front of to behind a moving wave source is called the **Doppler effect**.

As sound waves are crowded toward you, the frequency at which they reach your ears increases. Remember that higher frequency sound waves have a higher pitch. Therefore, as a train approaches you, you hear a higher pitch than you would if the train were standing still. Once the train has passed by, the apparent pitch of each horn decreases. However, the actual pitch does not change. The engineer on board the train always hears the same sound.

Doppler effect

A change in the apparent frequency of waves caused by the motion of either the observer or the source of the waves

▲ As the train approaches, the sound waves are "bunched up."

▲ The pitch of the horn seems to get lower after the train passes by because the sound waves are spread out.

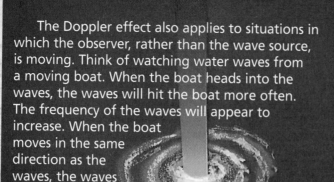

The Doppler effect also applies to situations in which the observer, rather than the wave source, is moving. Think of watching water waves from a moving boat. When the boat heads into the waves, the waves will hit the boat more often. The frequency of the waves will appear to increase. When the boat moves in the same direction as the waves, the waves will hit the boat less often. The frequency of the waves will appear to decrease. To someone watching the waves from the shore, the actual frequency of the waves will not change at all.

The Doppler effect is a property of all waves, not just sound and water waves. Light waves from distant galaxies also display the Doppler effect, in the form of a shift toward the red end (longer wavelengths) of the spectrum of light. This *red shift* indicates that these galaxies are moving away from us at high speed.

As indicated by this artist's conception, light from faster-moving galaxies appears shifted more toward the red end of the spectrum.

SUMMARY

There would be no sound without air or some other transmitting medium. Sounds in air begin when vibrating objects push on air particles, causing back-and-forth motions of these particles, or longitudinal waves. Standing waves are formed when similar waves traveling in opposite directions interfere with each other. They form a pattern in which the medium vibrates with nodes and antinodes. The Doppler effect—a property of all waves—is a change in the apparent frequency of waves caused by the movement of the wave source, its observer, or both.

SOUND EFFECTS

In addition to the sounds we can hear, there are sounds we cannot hear. These sounds can be used in a variety of ways, from locating earthquakes and measuring the depth of the ocean to ensuring good health. Whether produced naturally or artificially, audible and inaudible sounds are an important aspect of our daily lives.

Infrasonic waves

Sound waves at a frequency below that which humans can hear

Infrasonic Waves

Sounds with frequencies below 20 Hz consist of **infrasonic waves**. These frequencies are too low for human ears to detect. Earthquakes produce infrasonic waves that move through the solid material of the Earth. Two types of waves originate from the focus of an earthquake—primary and secondary waves. *Primary waves*, or P waves, are longitudinal waves that travel by compressing rock and soil in front of them and stretching rock and soil behind them. These waves move quickly and are the first ones to reach earthquake-recording stations. It is for this reason that they are called primary waves. *Secondary waves*, or S waves, are transverse waves. These waves travel more slowly than P waves and cause the ground to move up and down or from side to side.

Primary and secondary earthquake waves come from the focus of an earthquake, which is often far below the epicenter where surface waves originate.

The interaction of P and S waves with the Earth's surface produces a third type of wave, called a *surface wave*. These waves cause the surface of the Earth to shake and roll. Surface waves are unique in that they originate from the epicenter of an earthquake, not from its focus. Their rolling action, which is like the rising and falling of ocean waves, causes many buildings to collapse.

The instrument used to record earthquake waves is called a *seismograph*. A seismograph consists of a rotating drum wrapped with paper and a pen attached to a suspended weight. The pen presses gently against the paper-wrapped drum. The recording from a seismograph is called a seismogram.

Since P waves travel fastest, they are the first waves recorded on the seismogram. The difference in travel time between the arrival of P and S waves is used to determine the distance between the recording station and the epicenter of the earthquake.

Seismologists keep detailed records of earthquake activity.

Volcanic eruptions and tornadoes also produce infrasonic waves that travel for thousands of kilometers through the atmosphere. When Krakatau erupted in one of the largest volcanic events in recorded history, the infrasonic waves it produced could still be detected after they had circled the Earth several times.

Infrasonic waves are not always connected with disasters. Scientists use infrasonic waves to investigate the inner structure of the sun. *Helioseismology*, the study of solar "earthquakes," is based on the assumption that sound waves generated within the sun cause a pattern of peaks and valleys on the sun's surface. Because the movement of waves through a medium depends on factors such as elasticity and density, helioseismologists can learn about processes inside the sun by analyzing the surface wave patterns.

Ultrasonic Waves

Insects and other animals can detect sounds that are much higher than those detected by the human ear. Sounds in this upper range are called ultra-high-frequency sounds, or **ultrasonic waves**. Ultrasonic waves have been put to use in a variety of ways, from industrial cleaners to medical diagnostic and treatment instruments.

Many ships have a navigation system that uses echoes of ultrasonic waves to find the depth of the water. This system is called *sonar*, which stands for *so*und *na*vigation *ra*nging. This system works in very much the same way that echolocation works in bats. A sonar device sends short pulses of ultrasonic waves through the water. When the sound waves hit the sea floor or other underwater objects, some of the waves are reflected back to the ship as an echo. A receiver detects the echo, and the time delay indicates the distance. Some autofocus cameras operate in the same way.

Ultrasonic waves can also be used to clean jewelry, machine parts, and electronic components. A device called an *ultrasonic cleaner* consists of a container that holds a bath of water and a mild detergent. Sound waves are sent through the bath, causing intense vibrations in the water that remove dirt from the items placed in the bath.

Ultrasonic waves

Sound waves at frequencies too high for humans to hear

▲ Bats use ultrasonic waves to locate prey in the dark.

Submarine pilots rely completely on sonar in order to navigate underwater. ▶

DID YOU KNOW...

that elephants communicate with infrasonic sound waves?
Their large ears enable them to pick up very-low-frequency sounds that can travel many kilometers from one elephant to another.

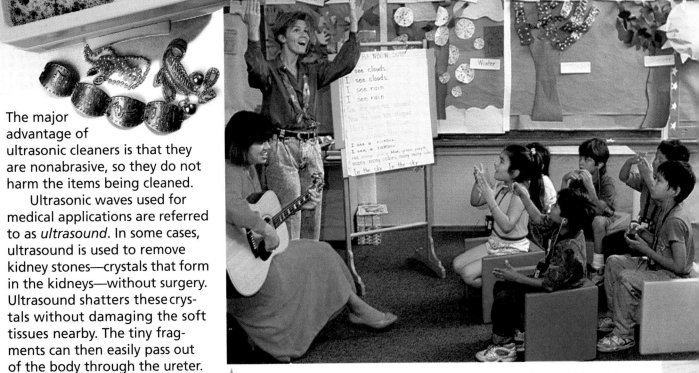

An ultrasonic cleaner, such as the one shown here, uses high-frequency sound waves to clean jewelry.

The major advantage of ultrasonic cleaners is that they are nonabrasive, so they do not harm the items being cleaned.

Ultrasonic waves used for medical applications are referred to as *ultrasound*. In some cases, ultrasound is used to remove kidney stones—crystals that form in the kidneys—without surgery. Ultrasound shatters these crystals without damaging the soft tissues nearby. The tiny fragments can then easily pass out of the body through the ureter.

Ultrasound also provides a way to "see" inside the body. Ultrasonic waves bounce off high-density tissues and are converted into electrical signals that are fed into a computer. The computer uses these signals to form a type of picture called a *sonogram*. Using this technology, doctors can locate tumors and gallstones. They can also examine a developing baby inside its mother to determine whether it is forming normally and if it is in the proper position.

By using electronic technology, people with hearing loss can interact with sound.

Sounds of Silence

Most people can hear a wide range of sounds. This is not the case for individuals who are hard of hearing. However, even a person who is completely deaf can detect certain sound waves in the environment. Since sound waves cause vibrations in solid materials, these vibrations can be felt through the sense of touch. Low-frequency sounds, especially, can be felt this way. Deaf people can dance to the beat of a loud band by sensing the strong beat of the bass guitar and drums. For people with impaired hearing, hearing aids amplify frequencies that are hard to hear. Modern electronic technology has produced hearing aids so small that they can be inserted completely into the ear canal.

DID YOU KNOW...

that sound can be used to produce light? Ultrasonic waves can be used to form tiny bubbles of air within a beaker of water. When the bubbles collapse, they emit intense flashes of light.

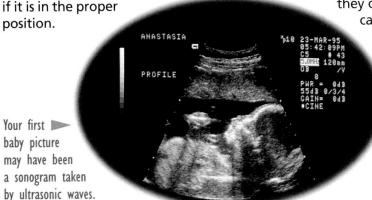

Your first baby picture may have been a sonogram taken by ultrasonic waves.

Hearing loss can result from exposure to music played at high volumes or from the extended use of personal stereo headphones.

Loud sounds and even loud music can permanently damage your hearing. Sound intensity is measured in *decibels* (dB). The higher the decibel level, the more intense the sound. Generally, the more intense a sound, the louder it is. Above a certain level of intensity, noise can harm your health. A quiet conversation, for example, is only about 40 dB, a harmless level of sound. Hearing damage may begin, however, when a person is exposed to noise levels of about 75 dB for 8 hours a day. A noise of more than 115 dB causes immediate pain. Loud music and crowd noise at rock concerts can reach 100 to 130 dB. Repeated exposure to high noise levels at rock concerts and the long-term use of personal stereos may cause permanent hearing impairment. Do not let the world of sound fade away due to carelessness!

Sound Intensity Levels		
Type of sound	Intensity (dB)	Hearing damage
Whisper	10–20	None
Soft music	30	None
Conversation	60–70	
Vacuum cleaner	70	After long exposure
Heavy street traffic	70–80	
Construction equipment	100	Progressive
Thunder	110–120	
Loud rock concert	110–130	Immediate and irreversible
Jet engine at takeoff	120–150	

S U M M A R Y

Sound waves, both audible and inaudible, can be used in many ways. Infrasonic waves are very-low-frequency sound waves. They are produced by several natural phenomena, such as earthquakes, and are even used to study the sun's interior. Ultrasonic waves are used by sonar devices, ultrasonic cleaners, and some medical instruments. People who are hearing-impaired cannot hear the sounds that are constantly around us, but they can sense strong vibrations. Loud sounds or music can cause permanent hearing loss.

Concept Mapping

Starting with the terms supplied here, construct a concept map that illustrates major ideas from this unit. Arrange the terms in an appropriate manner and connect them with linking words. Then extend your map by adding as many additional terms from the unit as you can. Use your ScienceLog. **Do not write in this textbook.**

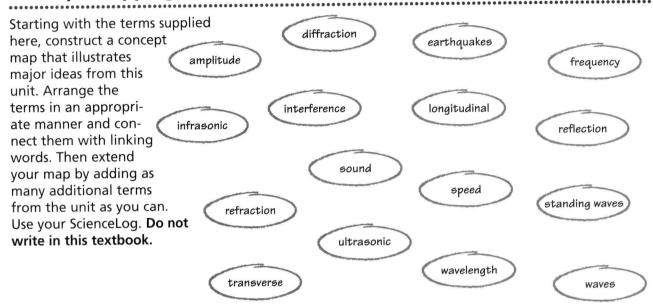

diffraction earthquakes frequency

amplitude

interference longitudinal reflection

infrasonic

sound speed

refraction standing waves

ultrasonic

transverse wavelength waves

Checking Your Understanding

Select the choice that most completely and correctly answers each of the following questions.

1. When a wave moves the particles of its medium at right angles to the direction of the wave itself, the wave is a
   a. transverse wave.
   b. longitudinal wave.
   c. sound wave.
   d. wavelength.

2. The relationship between frequency and wavelength is expressed by which of the following?
   a. As frequency increases, wavelength increases.
   b. As frequency decreases, wavelength decreases.
   c. As frequency increases, wavelength decreases.
   d. As frequency decreases, wavelength stays the same.

3. The action of waves bending around a corner is called
   a. reflection.
   b. refraction.
   c. diffraction.
   d. interference.

4. A train sounding its horn moves toward you at a constant speed. As it passes the location where you are standing, the pitch of the sound gets lower because
   a. the frequency of the sound produced decreases.
   b. the wavelength of the sound produced increases.
   c. the frequency of the sound heard increases.
   d. the wavelength of the sound heard increases.

5. Which of the following would pose the most immediate danger to your hearing?
   a. a singing trio that couldn't stay in tune
   b. a supersonic jet flying over your house
   c. sitting in the third row of a loud rock concert
   d. operating a vacuum cleaner for 30 minutes a day

Interpreting Graphs

According to the graph shown here, which animal's range of hearing corresponds to the longest bar on the graph? Which animal actually has the greatest range of hearing? How can you explain this apparent discrepancy?

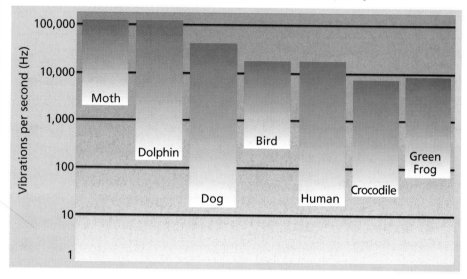

Critical Thinking

Carefully consider the following questions, and write a response in your ScienceLog that indicates your understanding of science.

1. In what ways are sound waves like other waves? In what ways are they different?

2. A popular science-fiction film used the following statement in its advertisements: "In space, no one can hear you scream . . ." Explain this statement in terms of what you have learned about sound waves.

3. Assume that sound travels through air at 346 m/s. If you were able to produce a sound wave with a frequency of 440 Hz, what would the wavelength of the sound wave be?

4. Suppose that you were at an auto-racing track. If you could listen only to the sound of the cars as they passed, would there be a way to estimate which car was moving the fastest? Explain.

5. If a tree falls in a forest and no one is there to hear it, does the tree make a sound when it hits the ground? Explain your answer.

Portfolio Idea

When new highways are constructed through residential areas, noise-abatement walls are often built along the roadway. These walls help reduce the noise of traffic entering neighborhoods, noise that can often reach levels exceeding 70 dB. Do some research to find out how these walls work and in what cities they have been used. Also investigate some of the problems associated with using or not using noise-abatement walls along busy highways.

LIGHT

IN THIS UNIT

Now that you have been introduced to light, its relationship to heat and color, and how images form, consider the following questions.

● **1.** How does light behave—as a stream of particles, a series of waves, or both?

2. What causes us to see the different colors of the objects around us?

3. What is "invisible" light?

4. How does modern technology make use of the properties of light?

In this unit, you will take a closer look at the composition of light and at some of the high-tech ways in which light is used.

LIGHT: A SPLIT PERSONALITY?

Ever since scientists first began investigating light, there have been conflicting ideas about it, not only about what gives light color, but also about the very makeup of light. Light seems to act in different ways under different circumstances. Various theories of light have been advanced to explain some of its properties. But, as you will see, none of these theories explains all the known properties of light.

The Particle Theory

Isaac Newton, whose laws of motion and gravity are the basis of traditional physics, wrote a scientific paper on light in 1672. Newton performed many experiments with light at his home in England. Besides using prisms to study color, he also studied how light is affected by objects in its path. He wanted to learn how light creates shadows, how mirrors reflect light, and how various materials refract light. Newton explained the results of his experiments by assuming that rays of light consist of streams of tiny particles. He called these particles "corpuscles."

Newton developed a theory that explained all the phenomena of light known in his day. For example, Newton argued that light particles are reflected in the same way that a ball bounces off a wall. In both cases, the angle of incidence equals the angle of reflection. Also, light changes its direction when it enters a new medium. This behavior of light is known as *refraction*. Newton compared the refraction of light to the behavior of a ball rolling down an incline.

Newton's most compelling evidence in favor of a particle theory of light, however, was that light seems to travel in a straight line. For example, an object placed in the path of light casts a shadow; in other words, the object stops the light. In the same way, a rolling ball can be stopped by a barrier placed in its path.

Newton thought that light consists of streams of tiny particles.

The ball rolls faster down the incline, and its direction changes.

The ball slows down when it leaves the incline, and its direction changes again.

The path of a rolling ball bends as the ball rolls down an incline, as seen by the changes in angles. Because light bends in a similar way, Newton saw this as evidence that light consists of particles. However, while the ball in this model speeds up, refracted light slows down.

Meanwhile, the Dutch physicist Christian Huygens theorized that light consists of waves. To support this idea, he pointed out that one beam of light passes through another without either beam being disturbed. Waves can do this; streams of particles presumably cannot. A wave theory could also explain reflection and refraction of light because both phenomena are common to waves such as water waves. But Huygens could not explain why light seems to travel only in straight lines, whereas sound waves can travel around barriers. The bending of waves around barriers is called *diffraction*. If both sound and light consist of

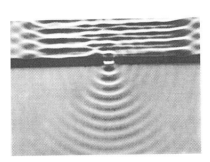

◄ Huygens thought that light, like sound, consists of waves.

waves, why can you hear, but not see, around corners? If light consists of waves, it should also be diffracted around barriers. Because there was not enough evidence supporting Huygens's wave theory, Newton's particle theory of light became the accepted theory for the next 100 years.

The Wave Theory

In 1801, another English scientist, Thomas Young, discovered that light, like sound, *can* be diffracted. This was a powerful argument in favor of the wave theory

of light because diffraction is a property of waves but not of particles. An example of the diffraction of water waves is seen in the photos to the left, below.

To see whether light behaves as particles or as waves, Young did a simple experiment. He made two narrow, parallel slits in a card. He then placed a light source on one side of the card and a screen on the other. Young reasoned that if light consists of particles, two bright bands of light should appear on the screen. But if light consists of waves, the light should be diffracted as it passes through the slits. Because each slit acts as a separate circular wave source, the waves from these two wave sources would *interfere* with each other, resulting in a pattern of bright and dark bands on the screen. This is exactly what happened in Young's experiment, as the diagram below shows.

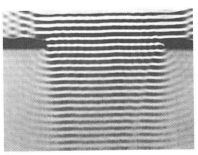

▲ When water waves pass through a sufficiently small opening, they are diffracted, or bent around the edges of the opening. When water waves pass through a large opening, little or no diffraction takes place.

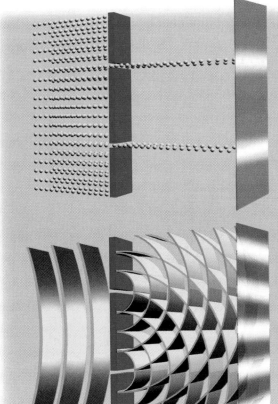

If light consists of particles (left), no interference will occur. If light consists of waves (below), diffracted waves will interfere with one another.

Bright bands appear where the crests and troughs of waves from two wave sources coincide.

Dark bands appear where the crests of waves from one wave source coincide with the troughs of waves from the other wave source.

Huygens may not have detected any diffraction of light because, in order for a wave to be diffracted, the opening through which it passes must not be much larger than its wavelength. Because the wavelengths of visible light are very small (about 0.0005 mm), an opening must be very narrow in order to diffract light. The slits Young used were narrow enough to cause the diffraction of light and the interference of light waves. As a result of Young's experiments, the wave theory eventually replaced Newton's particle theory as the accepted model of light.

Broadening the Spectrum

In 1864, the wave theory of light received a major boost from the Scottish physicist James Clerk Maxwell. While studying the relationship between electricity and magnetism, Maxwell showed that light consists of **electromagnetic waves**. Using mathematics, he showed that electromagnetic waves, which have an electric field and a magnetic field at right angles to each other, would travel at a constant velocity. Maxwell was surprised to find that this calculated velocity is the same as the velocity of light! Although he was not looking for a theory of light, the coincidence was striking. Thus, Maxwell suggested that light is an electromagnetic wave.

The German scientist Heinrich Hertz performed an experiment that verified Maxwell's theory. Hertz generated electromagnetic waves with a wavelength of approximately 1 m. Later experiments showed that a wide range of electromagnetic waves could be generated. Electromagnetic waves can have wavelengths that range from thousands of kilometers to billionths of a centimeter. These waves are a type of energy called *electromagnetic radiation*. Visible light, whose wavelength (and frequency) is approximately in the middle of the range of all electromagnetic waves, is only part of a much broader spectrum of electromagnetic radiation.

Electromagnetic waves

Waves that carry both electric and magnetic energy and move through a vacuum at the speed of light

DID YOU KNOW...

that the speed of electromagnetic waves is not the same in all materials? Electromagnetic waves move more slowly in denser materials. The degree to which a wave is slowed depends on its wavelength.

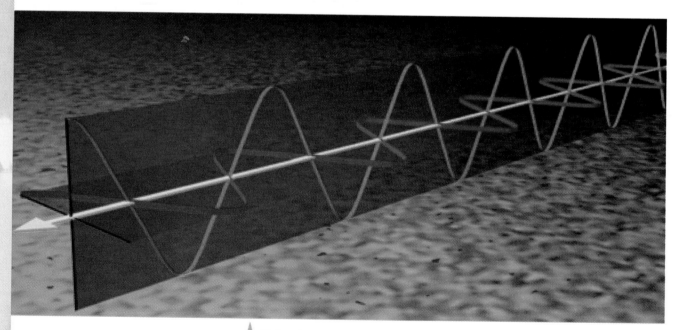

Notice that the electric and magnetic fields of an electromagnetic wave are at right angles to each other and to the direction of travel.

The Particle Theory Revisited

After the discovery of electromagnetic waves, many scientists were satisfied that they knew all there was to know about light. However, in 1887, Heinrich Hertz discovered the **photoelectric effect**. Hertz found that some metals emit electrons when light hits a thin plate of metal. However, the light must be above a certain *threshold frequency* for each kind of metal. If the frequency of the light hitting a particular metal plate is slightly less than the threshold frequency, no electrons are emitted, even if the light intensity is great. On the other hand, if the frequency of the light is slightly greater than the threshold frequency, electrons are emitted no matter how weak the light is.

This discovery presented a problem—the wave theory of light could not explain the photoelectric effect. If light consists of waves, then any frequency of light should give electrons enough energy to escape from a metal plate, as long as the intensity of the light is great enough. According to the wave theory, the energy of light would be spread out over the front edge of the light wave. Thus, as the intensity of the light increased, so would the amount of energy spread over the wave front. But the characteristic threshold frequency of each material showed that this is not the case.

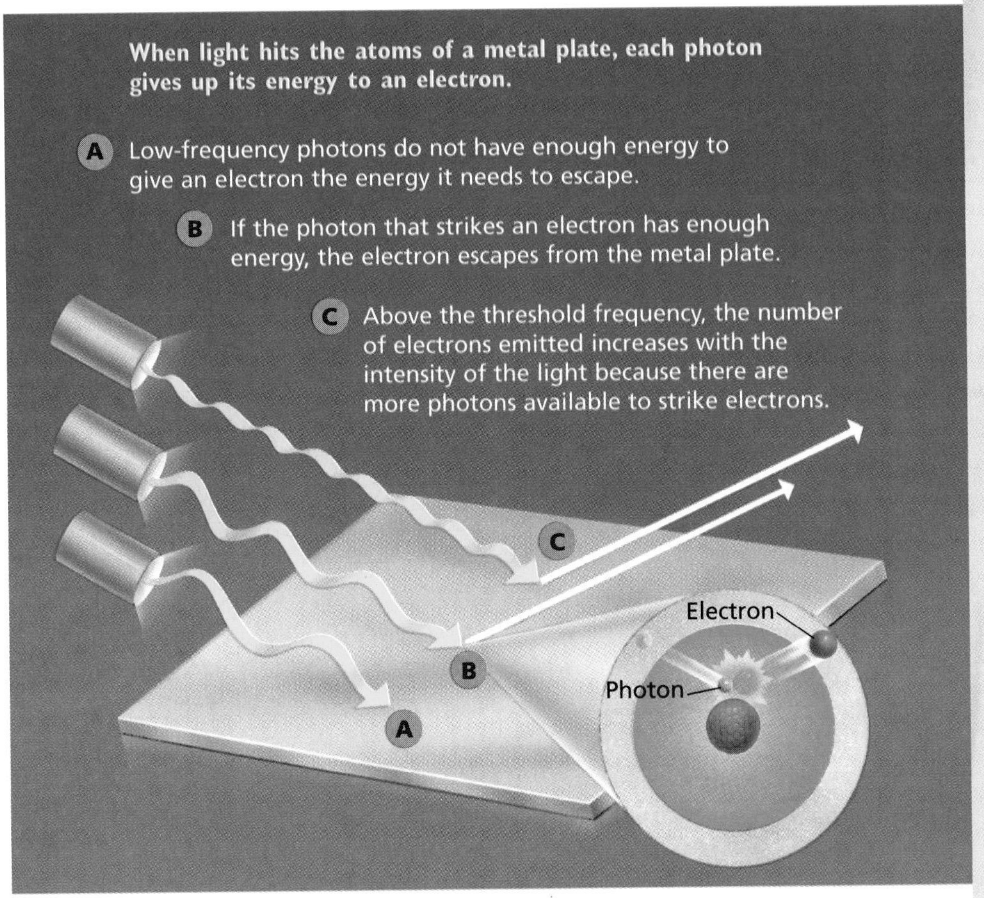

Then, in 1905, Albert Einstein offered an explanation of the photoelectric effect. Using the ideas of the German physicist Max Planck, Einstein proposed that light is composed of tiny packets (particles) of light energy. These packets of energy were later called **photons**. According to Einstein, each photon has a definite amount of energy that depends on the frequency of the light. To see how the photoelectric effect works, look at the diagram above.

The photoelectric effect has many practical applications. When many photons strike a light-sensitive metal plate, the many electrons emitted can initiate an electric current. Electric currents produced by the photoelectric effect operate photographic light meters, produce the sound in a movie soundtrack, and turn street lights on and off.

Photoelectric effect

The emission of electrons by a substance when illuminated by light of a sufficient frequency

When light hits the atoms of a metal plate, each photon gives up its energy to an electron.

A Low-frequency photons do not have enough energy to give an electron the energy it needs to escape.

B If the photon that strikes an electron has enough energy, the electron escapes from the metal plate.

C Above the threshold frequency, the number of electrons emitted increases with the intensity of the light because there are more photons available to strike electrons.

Electron

Photon

The photoelectric effect occurs when some metals emit electrons as light shines on them.

Photon

A tiny package of electro-magnetic energy

The Particle-Wave Theory

Newton was the first to suggest that light consists of particles because it travels in a straight line. But in 1801, Young showed that light behaves as waves because it is diffracted and shows interference. In 1864, Maxwell discovered that light waves and other electromagnetic waves behave in the same way. However, to explain the photoelectric effect, Albert Einstein proposed in 1905 that light *does* sometimes behave as particles. Thus, it seems that neither waves nor particles alone can explain the behavior of light. What is light then—a wave or a stream of photons?

Today scientists realize that the wave model is the best explanation for some behaviors of light, while the particle model is the best explanation for other behaviors of light. This has led to what is called the *particle-wave theory* of light. How can light exhibit the qualities of both particles and waves, yet not be purely one or the other? Perhaps scientists have yet to determine the true nature of light. In other words, a third explanation may await discovery.

When light passes through a large opening (left), it behaves as the particle model predicts. When light passes through a small opening (right), it behaves as the wave model predicts.

Light has both particle and wave characteristics. Individual photons can be detected as they hit the screen, but a wavelike interference pattern is also evident.

SUMMARY

The photoelectric effect and the fact that light travels in a straight line suggest that light consists of a stream of particles. However, behaviors such as diffraction and interference suggest that light is a wave. Other behaviors of light, such as reflection and refraction, can be explained by either a wave theory or a particle theory of light. Light waves are electromagnetic waves, which can travel through a vacuum. Individual packets of light energy are called photons. The amount of energy in a photon of light increases as the frequency of the light increases. Today, physicists explain the properties of light with a combined particle-wave theory.

VISIBLE LIGHT

Visible light consists of all the electromagnetic waves that humans can see. Within the entire range of electromagnetic waves, visible light consists of only a very narrow range of wavelengths. These wavelengths have a range of about 400–760 nanometers (nm), or billionths of a meter. Because our eyes are able to detect visible light waves and our brains are able to interpret them, we can see the objects that surround us.

Sources of Light

Light comes from many different sources. The sun and stars are natural sources of light. Candles and electric lamps are artificial sources of light. However, most light originates from its source in the same way—by the excitement of electrons. Electrons orbit atoms in regions called energy levels. The electrons in the energy level closest to an atom's nucleus have less energy than the electrons in energy levels that are farther away from the nucleus. If an electron absorbs energy, it jumps to a higher energy level, as the diagram above shows. When the electron drops back to its normal energy level, it emits a photon of light. The amount of energy in the photon emitted is determined by how far the electron jumps. Many types of energy can excite electrons. For example, energy that comes from nuclear reactions inside the sun produces light by exciting electrons in the sun's outer layers. Electricity

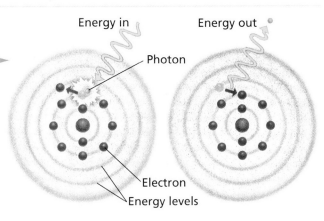

When an electron ▶ absorbs the energy of a photon (left), it rises to a higher energy level. As it falls (right), it emits a photon of light.

Energy in

Energy out

Photon

Electron

Energy levels

excites the electrons in a light bulb. Light excites electrons in luminescent materials, causing them to glow in the dark.

▲ Electrons in the luminescent paint on the dial of this timer are excited by light and then slowly give off their own light, enabling you to see how much time remains.

Color

As you can see by looking at a rainbow or at the objects around you, you interpret visible light in many different colors. Nearly 300 years ago, the English physicist Isaac Newton observed all the colors of a rainbow by passing sunlight through a prism. The colors that we see are now known as the **visible spectrum**. Newton also noted that when all the wavelengths of light in the visible spectrum are recombined, white light results. Thus, he concluded that white light contains all the wavelengths of visible light.

▼ A rainbow reveals that visible light consists of many colors.

Visible spectrum

All the colors of visible light

Each color of light has a different ► wavelength. The colors of light separate when passing through a prism because the waves with shorter wavelengths are bent more than those with longer wavelengths.

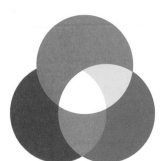

▲ Red, green, and blue are the primary colors of light. When two of these colors overlap, one of the secondary colors (magenta, yellow, or cyan) is produced.

But why does a prism separate the colors of light? As you know, light slows down when it passes from air into a denser medium, such as the glass of a prism. As a beam of light slows down, its path of travel is bent (refracted). However, waves with long wavelengths are not bent as much as waves with shorter wavelengths. Therefore, each color of light is bent at a different angle when it passes through a prism, as the illustration above shows.

Now imagine a bowl of fruit sitting on a table. The bowl contains an apple, a banana, and a plum. When white light shines on the fruit, you see that the apple looks red, the banana looks yellow, and the plum looks purple. Since you know that white light is a mixture of all colors, what do you think happened to the other colors of light that fell on each fruit? The fruits contain substances called *pigments*, which absorb some wavelengths of light and reflect others. The color of each fruit results from the wavelengths of light that it reflects. In fact, the wavelengths of reflected light determine the perceived color of each object that we see.

White light has three *primary colors*: red, green, and blue. These colors, seen in the diagram to the left, are called the primary colors because all the colors we see can be made from them. You are able to see the colors of light reflected by an object because your eyes have color receptors called *cones*. There are three different types of cones, each type sensitive to one of the primary colors of light. When two primary colors of light mix, a new color that is a combination of these two colors is produced.

The colorful picture of the balloon above is produced by overlaying three colors of film—magenta, yellow, and cyan—and a piece of black film, which provides contrast.

Rays and the Behavior of Light Waves

The wave theory can explain much about the behavior of light. As you know, light waves appear to travel out from a source as a series of expanding circles. But every point on each circle is traveling directly away from the source along a straight line called a *ray*. A ray is, in effect, the radius of the circle. Light rays are not real, however. They are simply imaginary lines that represent the path that light waves take in a particular direction.

The *ray model* of light is a very useful way to describe the behavior of light waves—how they are refracted, how they are reflected, and how they form images (likenesses) of objects.

Lenses and Refraction As you know, when light travels from one medium to another, it is refracted, or bent. *Lenses* are used to refract light rays so that images of an object can be formed. A lens is a curved piece of glass, plastic, or some other transparent material. Lenses refract light because the light moves slower through the material of the lens than it does through the air.

There are two basic types of lenses—convex and concave, as shown in the photos to the right. A *convex lens*, which is thicker in the center than at the edges, causes light rays to converge. A *concave lens*, which is thinner in the center than at the edges, causes light rays to diverge.

By using the ray model, the way that refracted light forms images of objects can be

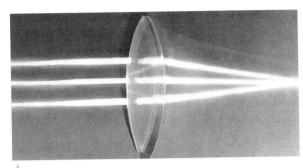

▲ Convex lenses cause light rays to converge, or come together.

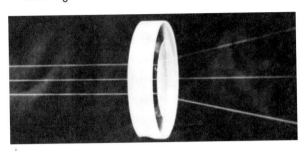

▲ Concave lenses cause light rays to diverge, or spread apart.

explained. A **real image** forms where light rays converge, or are *focused*. A **virtual image** forms not where light rays converge, but where they appear to originate. Because concave lenses cannot make light rays converge, they cannot form real images. However, a virtual image forms in front of a concave lens, where light rays appear to originate.

◄ A convex lens forms a real image where light passing through the lens converges. A convex lens can also form a virtual image at the location where light passing through the lens appears to originate.

◄ A concave lens forms a virtual image where light rays appear to originate. A real image cannot be made because light rays from the object never converge.

Real image

An image that forms where light rays coming from an object converge; it can be projected on a screen

Virtual image

An image that is not formed by the convergence of light rays; it cannot be projected on a screen

Mirrors and Reflection When light strikes a surface and bounces off, we say that it has been reflected. Most of the light you see has been reflected from some object. Most surfaces are rough and reflect light in all directions. This type of reflection is called *diffuse reflection*.

Light is reflected from a surface in the same way that a ball bounces off a wall—the angle of incidence equals the angle of reflection. You can see this behavior of light when light is reflected from a smooth surface, such as a mirror. As the diagrams on this page indicate, there are three types of mirrors—*plane mirrors*, *convex mirrors*, and *concave mirrors*. All three types of mirrors are used to produce images of objects.

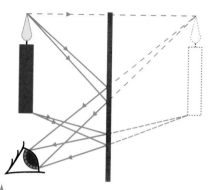

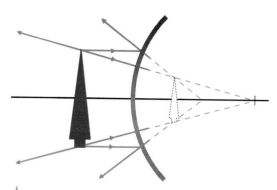

Plane mirrors are flat and therefore cannot focus light. A plane mirror produces a virtual image that appears to be the same size as the object. Notice how the light rays appear to diverge from a point behind the mirror. This is where the image appears to be located.

Convex mirrors curve outward, causing light rays to diverge. A convex mirror produces a virtual image that appears to be smaller than the object. The image appears to form at a point behind the mirror, at the location from which the light rays appear to originate.

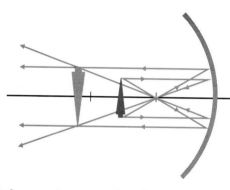

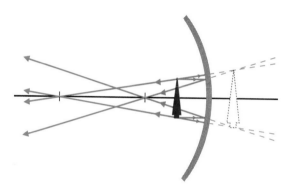

Concave mirrors curve inward and therefore can focus light. A concave mirror produces both real images and virtual images, depending on how close the object is to the mirror. Real images form in front of the mirror, and virtual images appear behind the mirror.

S U M M A R Y

Visible light is only a small portion of the entire range of electromagnetic radiation. Light is produced by excited electrons that emit photons as they fall back from their excited state to lower energy levels. White light consists of many different wavelengths, each of which is perceived as a particular color. A prism separates light because each wavelength of light is bent at a different angle when it passes through the prism. The color of an object is produced by the various wavelenghs of light that are reflected from the object to your eyes. The ray model of light can be used to describe the reflection and refraction of light and the images formed by light. Real images form where light rays from a mirror or lens converge. Virtual images form where light rays appear to originate.

INVISIBLE LIGHT

As you know, visible light is not the only form of electromagnetic radiation. We usually think of light as waves we can see, but there are many more types of electromagnetic radiation that we cannot see. The electromagnetic waves that we cannot see may be thought of as invisible light.

The Electromagnetic Spectrum

Electromagnetic waves have frequencies from below 10^1 Hz to above 10^{21} Hz. (Remember, frequency is the number of complete waves that pass a point in 1 second.) The total range of electromagnetic wave frequencies (or wavelengths) is called the **electromagnetic spectrum**. Visible light, which has a frequency of about 10^{15} Hz, represents only a small portion of the electromagnetic spectrum. Just as the color of visible light varies according to its frequency, the properties of all electromagnetic waves vary according to their frequency. Therefore, the electromagnetic spectrum is divided into several regions based on the characteristics of the waves within each range of frequencies. The major regions of the electromagnetic spectrum are shown in the diagram below.

Electromagnetic spectrum

The entire range of electromagnetic waves, of which visible light is a small part

▼ The electromagnetic spectrum. Notice that visible light represents only a small portion of the electromagnetic spectrum.

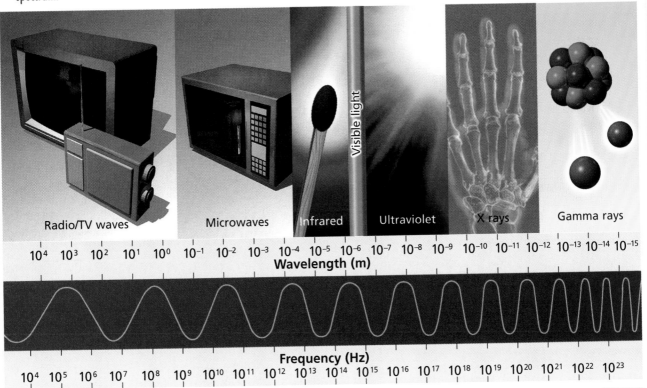

Radio/TV waves　Microwaves　Infrared　Visible light　Ultraviolet　X rays　Gamma rays

Wavelength (m): 10^4 10^3 10^2 10^1 10^0 10^{-1} 10^{-2} 10^{-3} 10^{-4} 10^{-5} 10^{-6} 10^{-7} 10^{-8} 10^{-9} 10^{-10} 10^{-11} 10^{-12} 10^{-13} 10^{-14} 10^{-15}

Frequency (Hz): 10^4 10^5 10^6 10^7 10^8 10^9 10^{10} 10^{11} 10^{12} 10^{13} 10^{14} 10^{15} 10^{16} 10^{17} 10^{18} 10^{19} 10^{20} 10^{21} 10^{22} 10^{23}

Power Waves *Power waves* are very low frequency electromagnetic waves that are produced by electric generators. These waves are produced as electric current moves through transmission lines. They cause static on your car radio when you drive near them. Their frequencies range between 10 Hz and 100 Hz, with wavelengths of thousands of kilometers.

Radio Waves When you turn on a radio, you hear the effects of another type of electromagnetic wave. *Radio waves* consist of a wide range of electromagnetic waves with frequencies between 10^4 and 10^{12} Hz. Ordinary AM radio broadcasts use the lower-frequency waves between 535 kilohertz (kHz) and 1605 kHz. (One kilohertz is 1000 waves per second.) AM radio waves are reflected back to the Earth's surface by the layer of the atmosphere called the *ionosphere*. Thus, AM radio broadcasts can be received far away from the radio station. At night, you might be able to pick up AM radio stations from halfway across the country. FM radio waves normally use higher frequencies, between 88.1 megahertz (MHz) and 107.9 MHz. (One megahertz is 1,000,000

waves per second.) FM radio waves are not reflected by the ionosphere, so they cannot be received past the curve of the Earth like AM waves. Television (TV) broadcasts use high-frequency radio waves that also travel only in straight lines. Satellites are used to relay radio and television waves around the world.

Radar waves are radio waves of even higher frequencies than TV waves. Waves at these frequencies are reflected by many materials, especially metals. Reflected radar waves function like reflected sound waves, which are heard as an echo. A radar echo can be used to "see" objects in the dark, through

These air traffic controllers use radar to scan the sky for aircraft in the vicinity.

fog, or at a great distance. For this reason, radar is used for navigation on ships and planes.

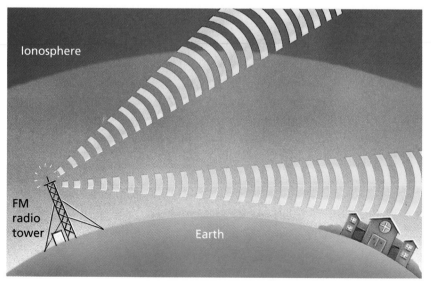

AM radio waves (above) have longer wavelengths and are reflected by the ionosphere. FM waves (left) have shorter wavelengths, which allow them to pass through the ionosphere.

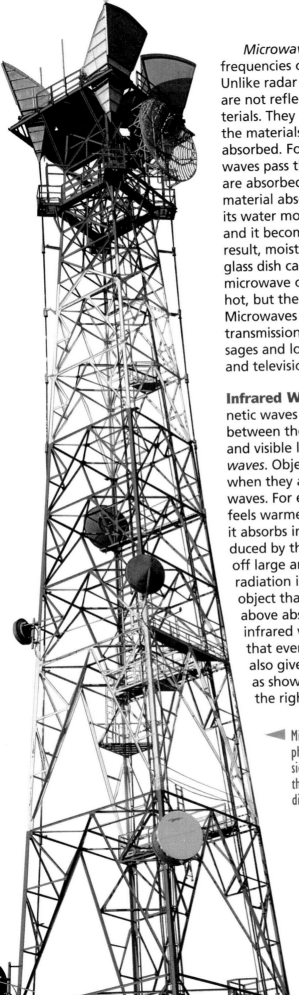

Microwaves have the highest frequencies of any radio waves. Unlike radar waves, microwaves are not reflected by most materials. They either pass through the materials or are easily absorbed. For example, microwaves pass through glass but are absorbed by water. When a material absorbs microwaves, its water molecules move faster and it becomes hotter. As a result, moist food or water in a glass dish can be heated in a microwave oven. The food gets hot, but the glass dish does not. Microwaves are also used in the transmission of telephone messages and long-distance radio and television broadcasts.

Infrared Waves Electromagnetic waves with frequencies between those of microwaves and visible light are *infrared waves*. Objects become warmer when they absorb infrared waves. For example, your skin feels warmer in sunlight because it absorbs infrared waves produced by the sun. The sun gives off large amounts of infrared radiation into space. But any object that has a temperature above absolute zero radiates infrared waves. This means that everything around you also gives off infrared waves, as shown in the image to the right.

◄ Microwave towers relay telephone conversations and television programs to areas where the installation of cables is difficult or impossible.

▲ The lighter colors in this infrared image indicate higher temperatures.

Visible Light Waves *Visible light* is the very narrow band of frequencies that can be seen with the human eye. We see different frequencies within this narrow band as different colors. As you know, the lowest frequencies of visible light are seen as red; the highest frequencies are seen as blue or violet. White light contains all the frequencies of visible light.

S139

Ultraviolet light is used to treat jaundice in newborn babies.

Ultraviolet Light If you spend a lot of time outdoors in the sun, you should be aware of *ultraviolet light*. The waves in this part of the electromagnetic spectrum have frequencies just above those of visible light. The ultraviolet light in sunlight causes sunburn. Too much ultraviolet light can be dangerous because it may cause harmful mutations and skin cancer. In limited amounts, ultraviolet light can be beneficial. For example, when skin cells are exposed to ultraviolet light, they begin to produce vitamin D, a substance necessary for healthy teeth and bones. Ultraviolet light also kills germs and is often used for this purpose in hospitals.

X Rays *X rays* are high-frequency electromagnetic waves that are very useful in the field of medicine. They easily pass through skin and other tissues, but not through bone. X rays can produce an image on film or a picture on a screen. This means that X rays can be used to see inside your body. A doctor or dentist has probably made X-ray photographs of parts of your body. However, too much exposure to X rays can kill living cells. Therefore, machines that produce X rays must be used only by trained personnel.

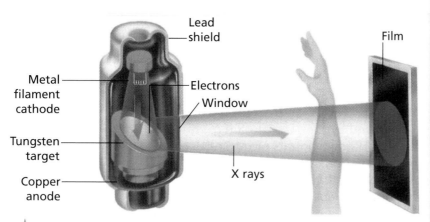

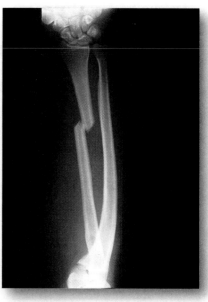

X rays are produced in a special tube when electrons strike a tungsten target. The reflected X rays pass through the object to form an image on film.

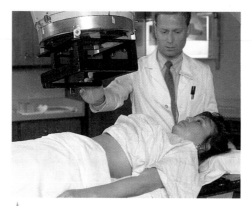

Gamma rays are used to treat patients with cancer.

Gamma Rays *Gamma rays* are very high frequency waves similar to X rays. They are much more dangerous, however, because they pass through matter more easily. Nevertheless, gamma rays are useful in fighting cancer. Beams of gamma rays can be aimed at the location of the cancer. The rays pass deep into the body to reach and kill the cancer cells.

Quicker Than the Eye

Because visible light is the easiest type of electromagnetic radiation to detect, most early investigations were done with light. However, many of the characteristics of light apply to other forms of electromagnetic radiation. One of the most impressive characteristics of light, and of all other electromagnetic radiation, is its incredible speed.

Speed of Light   How long does it take light from a lamp to get to your eyes? The first scientists who tried to measure the speed of light found that light travels at a speed that is far too fast to measure by ordinary means. In the sixteenth century, the Italian scientist Galileo attempted to measure the time it took light to travel about 2 km. However, light travels this distance almost instantly. Galileo had no way to measure such a short period of time. In about 1676, however, the Danish astronomer Ole Rømer succeeded in calculating the speed of light. By studying the eclipses of the moons of Jupiter, he was able to figure out how long it takes for light to travel from Jupiter to Earth. His calculations turned out to be quite accurate.

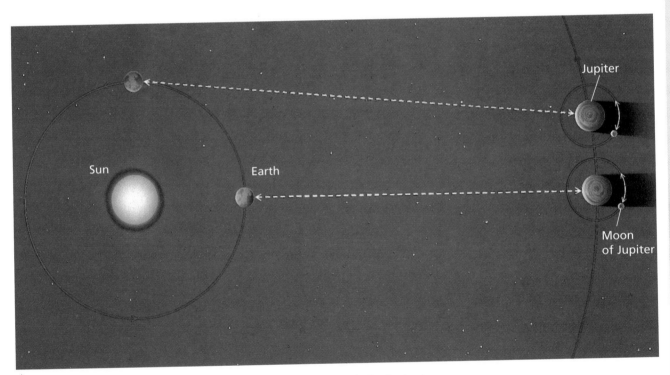

Light from an eclipse of one of Jupiter's moons takes longer to reach the Earth when the Earth is farther away from Jupiter. By comparing this difference in time, Rømer estimated the speed of light.

Modern measurements, of course, are much more accurate than Rømer's calculations. Today, the value used for the speed of light has been determined to eight decimal places. In a vacuum, light travels at a speed of 2.99792458×10^8 m/s, or about 3×10^8 m/s. At this speed, a beam of light could travel from Los Angeles to Atlanta in less than one-hundredth of a second. All forms of electromagnetic radiation travel through space at the speed of light.

Light is refracted when it travels from one medium to another because it has a different speed in each medium.

Index of refraction

The ratio of the speed of light in a vacuum to its speed in another medium

Speed Limits The speed of light noted on the previous page is only valid for light in a *vacuum*. In air, the speed of light is slightly slower. Also, when light moves from one material into another, its speed changes. For example, when light moves from air into water, it slows down. Light waves travel about 25 percent slower in water than they do in air. On the other hand, light moving from water into air speeds up. This change of speed causes light entering or leaving water to be refracted along a new path.

Some materials slow light down more than water does. The more that light slows down when it passes from one medium to another, the more it is bent. The degree to which a material can bend light is given by its **index of refraction**. To find the index of refraction for any material, simply divide the speed of light in a vacuum by the speed of light in that material. For example, you can find the index of refraction for a diamond by doing the following calculation:

$$\text{index of refraction} = \frac{\text{speed of light in a vacuum}}{\text{speed of light in diamond}}$$

$$= \frac{3.00 \times 10^8 \text{ m/s}}{1.24 \times 10^8 \text{ m/s}}$$

$$= 2.42$$

The index of refraction of some common materials is given in the table to the right. The larger the index of refraction, the greater the change in the path of the light rays.

Infrared binoculars demonstrate that electromagnetic waves other than visible light are also affected by the materials through which they travel. Images form in infrared binoculars because their lenses refract infrared waves in much the same way that the lenses of a normal set of binoculars refract visible light.

Radio waves can also undergo refraction. Because the density of the atmosphere changes at different altitudes, radio waves are gradually refracted. This refraction sometimes carries FM radio waves past the horizon.

Indices of Refraction	
Substance	**Index of refraction**
air	1.00
ice	1.31
water	1.33
quartz	1.46
glass	1.52
amber	1.54
ruby	1.76
diamond	2.42

By using infrared binoculars, objects in complete darkness become visible.

S U M M A R Y

Visible light is only one part of a larger range of electromagnetic radiation called the electromagnetic spectrum. Electromagnetic waves are classified according to frequency into power waves, radio waves, infrared waves, visible light, ultraviolet light, X rays, and gamma rays. The speed of light (and other electromagnetic radiation) in a vacuum is about 3.00×10^8 m/s. As light enters another medium, its speed decreases. This change of speed causes light to be refracted by an amount determined by each material's index of refraction.

High-Tech Light

T he unique properties of light have led to many innovations that affect our daily lives. From the sunglasses you wear at the beach to the lasers that scan prices at the grocery store to the fiber-optic cables that carry your phone conversations with your friends, the technology of light is all around you. Both the wave and particle models of light have inspired important advances in technology and improved our understanding of the universe. In this section, you will look at a few of these technological advances and discover how we use some of the properties of light.

Polarized light

Light consisting of light waves that vibrate in only one plane

Polarization of Light

Picture yourself on the shore of a lake on a beautifully clear day. Sunlight is reflected off the sparkling lake, making you squint. You put on a pair of polarizing sunglasses, and the glare disappears. Now you can even see into the water. How can a pair of sunglasses make such a difference? Sunglasses that eliminate glare contain polarizing filters.

Like all electromagnetic waves, light waves are transverse waves, which vibrate at right angles to their direction of travel. In a beam of light, however, some of the waves vibrate up and down, some vibrate from side to side, and others vibrate at all the angles in between. As light waves pass through a polarizing filter, all waves are blocked except those that vibrate in the direction allowed

by the filter. The diagram below shows how polarizing filters work.

Light that vibrates in only one plane or orientation is called **polarized light**. The light causing the glare at the surface of the lake is polarized horizontally when it is reflected from the surface of the water. Your sunglasses have a vertical polarizer that blocks the glare from the lake so that it does not reach your eyes. Because the rest of the light in your environment is not polarized, there are plenty of light waves that vibrate vertically and can pass unobstructed through your sunglasses. You can also see the effect of a polarizing filter on glare by looking into a store window, like the one in the photos on the next page.

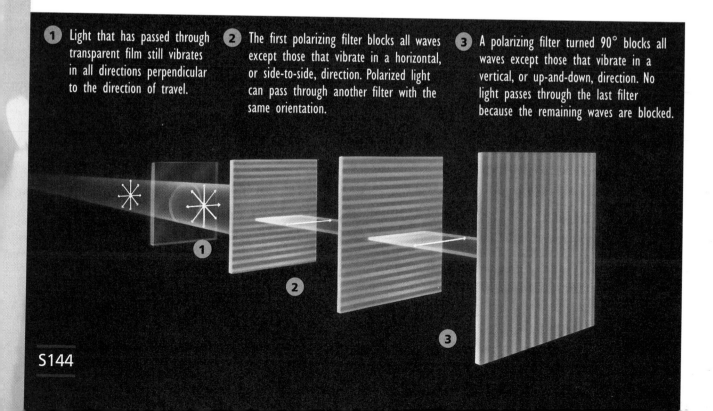

1 Light that has passed through transparent film still vibrates in all directions perpendicular to the direction of travel.

2 The first polarizing filter blocks all waves except those that vibrate in a horizontal, or side-to-side, direction. Polarized light can pass through another filter with the same orientation.

3 A polarizing filter turned 90° blocks all waves except those that vibrate in a vertical, or up-and-down, direction. No light passes through the last filter because the remaining waves are blocked.

While you cannot detect polarized light with your eyes alone, you can find out whether light is polarized by using a polarizing filter. To do so, look at the light through a polarizing filter and then rotate the filter 90 degrees. If the light is polarized, it will get dimmer or brighter as you rotate the filter.

Lasers

American industries spend billions of dollars every year making and using laser light for hundreds of purposes. From medical applications and space research to home entertainment, lasers have made a large impact on modern technology. But how is laser light different from ordinary light?

▲ The glare from this window makes it difficult to see the inside of the store.

▲ Because this photograph was taken through a polarizing filter, much of the glare has been eliminated.

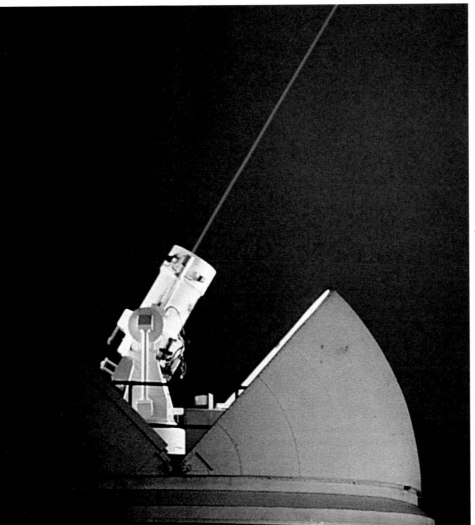

◀ Lasers can be used to measure the distance from locations on Earth to satellites in orbit around the Earth.

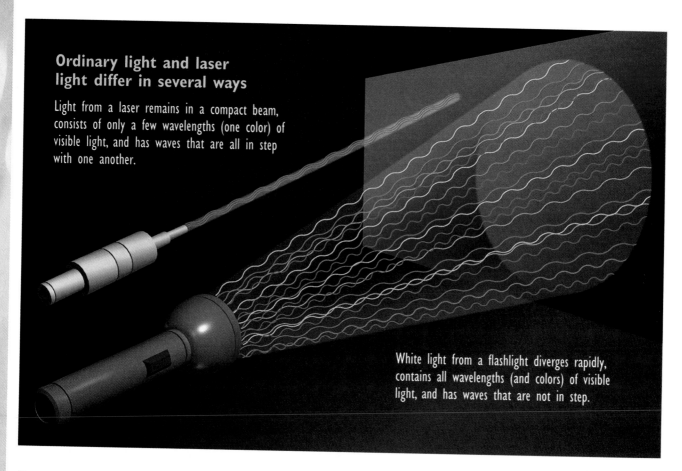

Ordinary light and laser light differ in several ways

Light from a laser remains in a compact beam, consists of only a few wavelengths (one color) of visible light, and has waves that are all in step with one another.

White light from a flashlight diverges rapidly, contains all wavelengths (and colors) of visible light, and has waves that are not in step.

Properties of Laser Light

The beams of light from an ordinary flashlight and a low-power laser can have about the same amount of total energy. But there is a big difference between them. You can see some of these differences in the diagram above.

First, a flashlight beam that is about 3 cm across when it leaves the flashlight spreads to a width of about 40 cm by the time it reaches the other end of a room. By contrast, a beam of laser light that is 5 mm wide when it leaves the laser makes the same size spot when it

reaches the other end of the room—5 mm. Laser light spreads out so little that a laser beam from Earth can be reflected from mirrors left on the moon by American astronauts.

Second, a flashlight's white light contains most wavelengths of visible light. The light in a laser beam, however, contains only a few, similar wavelengths. For instance, a helium-neon laser produces visible light with long wavelengths, and therefore the laser beam looks red. Other types of lasers produce light with different wavelengths (and colors).

Third, the crests and troughs of light waves in ordinary light overlap. However, the light waves in a laser beam are *in step*. That is, crests travel with crests and troughs travel with troughs. Light consisting of waves that are in step is called **coherent light**.

How Lasers Work   A laser is a device that produces coherent light. The most commonly used laser is the low-power, helium-neon laser. It produces a thin, straight beam of red light. This light has the same wavelength as the light from red neon signs (like those you see in store windows) and originates in much the same way.

The tubing used in a neon sign is filled with neon gas at low pressure. As an electric current passes through the tube, electrons in the neon atoms absorb energy. When an electron has extra energy, it is said to be in an *excited state*. However, electrons cannot stay in an excited state, and they quickly release energy by emitting a photon of light. This causes the neon gas in the tube to glow. When atoms give off photons of light energy, their electrons are said to return to the *ground state*.

Coherent light

Light consisting of waves that are in step

Every photon released by excited neon atoms has the same wavelength. However, in a neon tube, the photons are released in all directions and at different times.

Because all of the photons released by neon atoms have the same wavelength, a neon tube can be converted into a laser. When a photon of red light strikes an already excited neon atom, the atom immediately releases another, identical photon of red light. This process, seen in the diagram to the right, is called **stimulated emission**. If each of the two photons produced by stimulated emission from an excited neon atom strikes another excited neon atom, four photons result, then eight, then sixteen, and so on, as the diagram below shows. In a microsecond, so many photons are produced that the neon tube glows brightly. In a laser, the two ends of the tube are coated with a reflecting surface

that causes any photons traveling down the tube to bounce back and forth between the ends. The bouncing photons stimulate emission from other atoms that they strike. Soon an enormous number of photons flow back and forth in the tube, all in step with one another. One of the mirrored ends has a thinner coating, which allows some of the light to pass through. From this end, 5 to 10 percent of the coherent light escapes. This light is the laser beam.

Stimulated emission

A process of forcing identical photons into step in order to produce coherent light

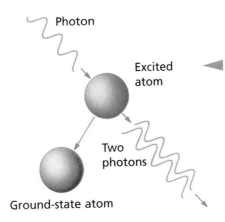

Photon

Excited atom

Two photons

Ground-state atom

When a photon strikes an already excited atom, the excited atom will emit an additional photon. The two photons will then leave the atom together, in step with one another.

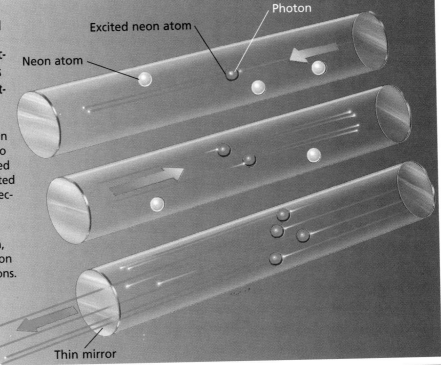

How a laser beam is produced

A laser beam is initiated when excited electrons within the neon atoms inside a glass tube drop back to their ground state, emitting photons of red light.

1. If a photon of red light strikes an already excited neon atom, two photons of red light are emitted by the atom. These newly emitted photons travel in the same direction, in step.

2. The mirrored ends of the tube reflect photons back and forth, continually causing excited neon atoms to emit additional photons.

3. Light that escapes from the thinly coated end forms the laser beam.

Neon atom

Excited neon atom

Photon

Thin mirror

Holography Laser light is used to make *holograms*. A hologram is a picture that often looks three-dimensional. Sometimes the images in a hologram seem to protrude toward the observer. Looking at a hologram can be like looking through an open window. By looking through different parts of the window, you can see the same scene from different angles. Some holograms can be viewed from all sides.

In a holographic camera, a beam splitter divides a laser beam in two. Both beams are aimed by a mirror and spread out by a lens. The object beam strikes the object and is reflected to the film. The reference beam hits the film without ever striking the object. Notice that light waves from every spot on the object intersect light waves in the reference beam.

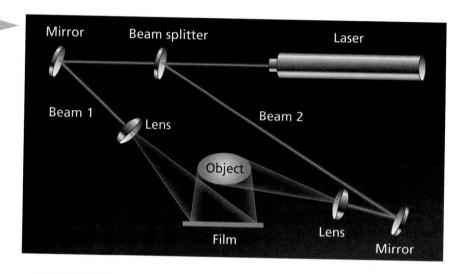

In order to make a hologram of an object, the light from a laser is split into two beams, an *object beam* and a *reference beam*. Mirrors and lenses then redirect the two beams in different ways, as seen in the diagram above. The two beams of light interact to produce a microscopic interference pattern that is recorded on a piece of film. With proper lighting, this interference pattern "reconstructs" a three-dimensional image of the object.

Holograms have many uses. Engineers use holograms to study how pressure deforms objects. Biologists use holograms to study the structure of bacteria. Artists use holograms to create unusual images. The largest user of holograms, however, is the banking industry. In the future, it may be possible to make holographic television and movie films that make it seem as if you are actually in the picture.

The image in this hologram can be viewed from all sides.

Some banks use holograms as security devices on their credit cards. This hologram is difficult to reproduce, making it hard to counterfeit the card.

Even though the fiber-optic cable (right) is a lot smaller, it can carry much more information than the copper-wire cable (left).

Fiber Optics

Have you ever seen telephone-company workers installing new cables along the highway? The cables are probably not made of copper wire, which carries electricity, but of fine glass fibers that carry light. As amazing as it seems, new long-distance telephone lines transmit conversations in the form of tiny pulses of light.

The principle that explains how glass fibers carry light has been known for a long time. When light travels through a medium in which the light travels slower than it would travel through air, the light may not be able to leave the medium. When light passes through the boundary between two mediums, it is usually refracted. However, if the light's angle of incidence at the boundary of the medium is too great, the light is reflected inward, as the photo to the right shows. The reflection of light from the inside surface of the medium through which it travels is called *total internal reflection*.

Total internal reflection is used in many kinds of optical instruments, including binoculars, telescopes, and the viewfinders of cameras. Total internal reflection is also used in the operation of a *light pipe*. As long as the light pipe's curvature is not too great, the light always strikes the inner surface at a high angle of incidence and is reflected inward. Images can be transmitted by the white light traveling through the light pipe, as shown below.

Two of the beams of light passing from the water in this tank into the air are refracted. If the angle of the light beam is great enough, it is reflected at the surface of the water. This type of reflection is called total internal reflection.

DID YOU KNOW...

that a beam of light can carry more information than an electrical signal? More information can be encoded in light waves because of their high frequency, which is about 100,000 times greater than the highest radio-wave frequency.

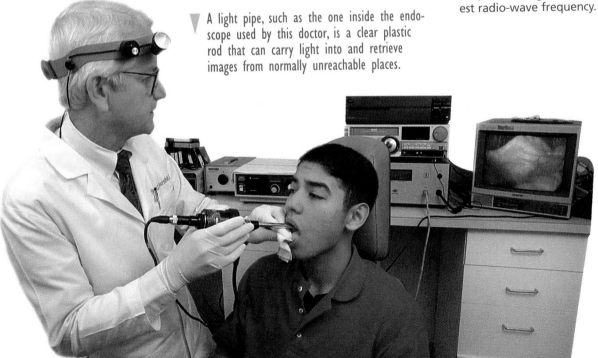

A light pipe, such as the one inside the endoscope used by this doctor, is a clear plastic rod that can carry light into and retrieve images from normally unreachable places.

An **optical fiber** is a long, thin light pipe. Optical fibers are made of glass and encased in plastic. Light travels more slowly in glass than it does in plastic, making total internal reflection possible. The fiber can be bent and twisted into many shapes without losing the light.

Fiber-optic telephone cables also consist of optical fibers encased in plastic. When you talk on the telephone, your voice is converted into an electrical signal. At a local telephone station, this signal is changed into a digital code. The electric bits of the code are used to trigger a tiny laser. The laser translates this code into a series of flashes of infrared light, which travel

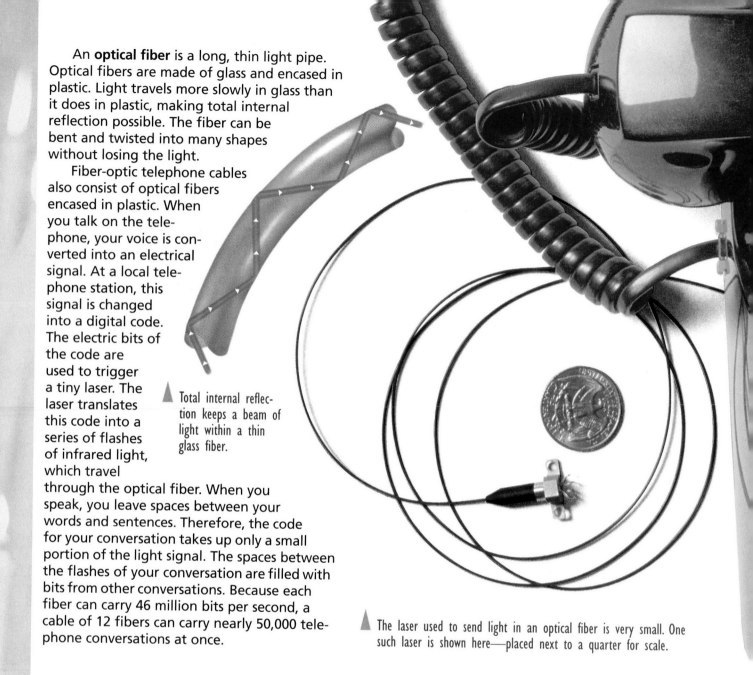

▲ Total internal reflection keeps a beam of light within a thin glass fiber.

through the optical fiber. When you speak, you leave spaces between your words and sentences. Therefore, the code for your conversation takes up only a small portion of the light signal. The spaces between the flashes of your conversation are filled with bits from other conversations. Because each fiber can carry 46 million bits per second, a cable of 12 fibers can carry nearly 50,000 telephone conversations at once.

▲ The laser used to send light in an optical fiber is very small. One such laser is shown here—placed next to a quarter for scale.

Optical fiber

A long, thin light pipe that uses total internal reflection to carry light

S U M M A R Y

The unique properties of light have led to innovations that affect our daily lives. Polarizing sunglasses eliminate glare because they contain filters that allow only light waves vibrating in one plane to pass. A laser is a device that produces coherent light by forcing identical photons into step through a process called stimulated emission. Laser light can be used to make holograms. Optical fibers carry information in beams of light that are guided by total internal reflection. Telephone systems use both optical fibers and lasers to greatly enhance our ability to communicate.

Concept Mapping

The concept map shown here illustrates major ideas in this unit. Complete the map by supplying the missing terms. Then extend your map by answering the additional question below. Use your ScienceLog. **Do not write in this textbook.**

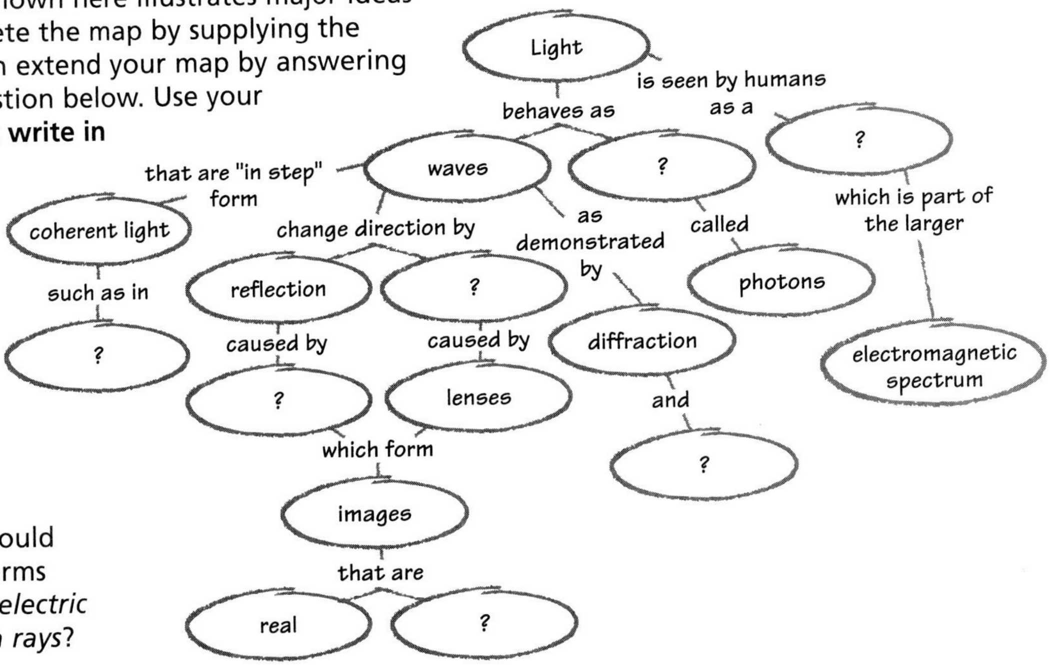

Light
behaves as
is seen by humans as a
?
which is part of the larger
waves
?
called
that are "in step" form
coherent light
change direction by
as demonstrated by
photons
electromagnetic spectrum
such as in
reflection
?
diffraction
?
caused by
caused by
and
?
lenses
?
which form
images
that are
real
?

Where and how would you connect the terms *fiber optics, photoelectric effect,* and *gamma rays*?

Checking Your Understanding

Select the choice that most completely and correctly answers each of the following questions.

1. Which of the following phenomena of light cannot be explained by the wave theory of light?
   a. reflection
   b. photoelectric effect
   c. virtual images
   d. diffraction

2. Light is produced when
   a. photons are absorbed by an atom.
   b. electrons jump to a higher energy level and stay there.
   c. photons jump to a higher energy level.
   d. photons are emitted as excited electrons fall to lower energy levels.

3. Which colors of light would combine to produce white light?
   a. red, blue, and green
   b. red, yellow, and blue
   c. magenta, yellow, and green
   d. magenta, cyan, and blue

4. Which of the following types of electromagnetic waves are NOT currently used in communications?
   a. X rays
   b. radio waves
   c. visible light
   d. microwaves

5. Laser technology is used in all but which of the following?
   a. holograms
   b. fiber optics
   c. polarizing lenses
   d. measuring long distances

Interpreting Illustrations

Show how the light rays in these diagrams would be either reflected or refracted. Copy the diagrams into your ScienceLog, and draw the resulting reflected or refracted rays and the images produced.

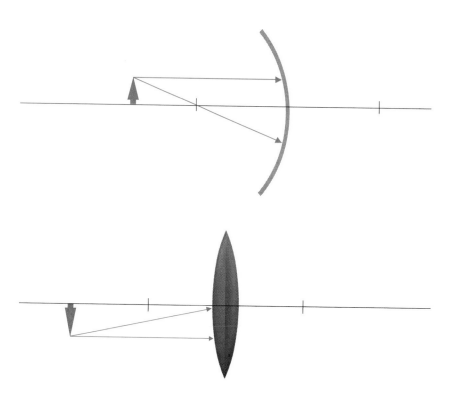

Critical Thinking

Carefully consider the following questions, and write a response in your ScienceLog that indicates your understanding of science.

1. How can light be explained both as streams of particles and as waves?

2. If a beam of red light and a beam of green light were focused on the same spot on a white wall, what color would the spot appear to be? Explain.

3. Both sound waves and electromagnetic waves transmit energy. Sound waves require a medium in order to be transmitted, while electromagnetic waves do not require a medium. Therefore, when a radio station uses electromagnetic waves, such as radio waves, to broadcast music and news, are these waves transmitting sound? Explain.

4. How could you test a pair of sunglasses to determine whether they have polarizing lenses? Explain how the method you propose works. Use diagrams to support your explanation.

5. Suppose that your eight-year-old cousin asks you to explain the difference between the beam of light from a flashlight and the beam of light from a laser. Develop one or more analogies that you can use to explain the difference between ordinary light and laser light so that your cousin will understand.

Portfolio Idea

Today, laser technology plays a very important role in medicine and in the entertainment industry. Do research on one of these fields and write a report to describe how lasers are used. Be sure to include the types of lasers that are used, how the light from the lasers is produced and controlled, and what possible dangers are involved in the use of lasers.

CONTINUITY OF LIFE

IN THIS UNIT

Now that you have been introduced to heredity and the processes of reproduction and development, consider these questions.

● **1.** What evidence did the discoveries of mitosis and meiosis provide in support of Mendel's laws of heredity?

2. How does our modern understanding of genetics differ from Mendel's concept?

3. How does the structure of DNA make it an ideal molecule for the transmission of inherited traits?

4. What is asexual reproduction and how does it occur?

In this unit, you'll take a closer look at some patterns of inheritance and DNA, the molecule of heredity. You will also learn about the different ways that organisms reproduce.

FOUNDATIONS OF HEREDITY

▲ What principle of reproduction does this litter seem to violate?

Heredity

The passing of traits from parents to offspring

The process of *reproduction* ensures the continuation of a species by forming new individuals of that species. The organisms that result from reproduction resemble their parents—at least in the most basic ways. Live-oak trees produce offspring that are also live-oak trees, and human beings produce other human beings. In other words, each species produces offspring that belong to that species.

Although offspring generally resemble their parents, they also may have differences. External differences are easiest to see. For example, there can be differences in height, coloration, and shape of body parts. But there may be internal differences as well. For instance, most people have type O or type A blood, while others have type B or type AB blood. Such differences, external and internal, are called *variations*.

For thousands of years, people knew that many characteristics are passed from parents to offspring. With this knowledge, people selected and bred certain plants and animals to bring out desirable characteristics. Yet it was not until the twentieth century that we discovered *how* those characteristics, or *traits*, pass from parents to offspring.

Mendel's Discoveries

Gregor Mendel was an Austrian botanist who lived from 1822 to 1884. He is famous because he discovered the basic principles of **heredity**. Mendel became interested in plants while growing up on his family's farm. At the age of 21, he entered a monastery, where he studied to become a priest and a teacher. Later, the monastery sent Mendel to the University of Vienna, where he studied science and mathematics. After his return to the monastery, Mendel taught natural science at a nearby high school.

◀ Sometimes organisms are selectively bred "just for show." These fancy goldfishes are called lionhead goldfishes. You may have seen them in a pet store, or perhaps you have some in your aquarium at home.

Mendel also applied what he had learned at the university to his interest in plants. In the monastery garden, he did research with pea plants that led to many important discoveries about heredity. Much of Mendel's success was due to his understanding of mathematics. Mendel analyzed his results by using the mathematical principles he had learned at the University of Vienna. He was among the first to analyze biological experiments using mathematics.

Crossing Pea Plants Mendel studied seven pea-plant traits, each with two contrasting forms. No matter which trait Mendel was studying (stem length, flower position, seed shape, etc.), he always began his experiments by *cross-pollinating* purebred plants that exhibited contrasting forms of the trait. A **purebred** plant produces offspring that all have one form of a particular trait. For example, a purebred tall pea plant produces only tall pea plants, generation after generation. And a purebred short pea plant produces only short pea plants.

In one experiment, Mendel cross-pollinated tall plants and short plants that were purebred for the trait.

Parent

Parent

Mendel found that this type of cross produced only *tall* plants. The contrasting form, shortness, had disappeared. The offspring of such crosses are called *hybrids*. In crosses involving the six other traits that he studied, the same thing happened. Only one form of each trait appeared among the hybrid offspring.

2nd generation

2nd generation

2nd generation

In the next stage of the experiment, Mendel allowed the hybrid offspring to *self-pollinate*. When he grew the seeds that resulted from these self-pollinations, both forms of each trait appeared among the resulting plants. Tall hybrid pea plants produced both tall and short offspring. The form of the trait that had disappeared in the hybrid offspring reappeared in the next generation.

Mendel also noticed that among the offspring of the hybrids, the contrasting forms of the trait appeared in the same ratio. *Three-fourths* of the plants showed the form of the trait that appeared in the hybrid offspring. Only *one-fourth* showed the form that had disappeared and then reappeared.

3rd generation

3rd generation

3rd generation

3rd generation

Mendel's Factors Seeing that one form of each trait had disappeared in his hybrids but reappeared in the next generation, Mendel inferred that some "distinct element" must be responsible for each of the traits he observed. He called these distinct elements *factors*. He also came to several important conclusions about how traits are inherited, as illustrated below.

Mendel's hybrids displayed only the dominant form of each trait because each had received one dominant factor and one recessive factor. The recessive factors reappeared in the offspring of the hybrids because some of them received the factor for shortness from both the male and female reproductive cells. Only when two recessive factors were paired together was the recessive factor expressed.

Mendel published the results of his work in 1866. However, no one paid much attention to them at the time. But over the next 35 years, important discoveries about cells were made. Those discoveries prepared the scientific community for the rediscovery of Mendel's work in 1900. Unfortunately, Mendel had already died.

2
The factors for a trait separate when reproductive cells are formed. As a result, reproductive cells receive and carry only one factor for each trait. When two reproductive cells combine to produce a new individual, there is again a pair of factors for each trait. This principle can be called the *law of segregation.*

3
There are different forms of the factor for each expression of a trait. In peas, for example, there are two kinds of factors for the trait of plant height—one for tallness and the other for shortness.

1
Each parent contributes one factor for each trait to its offspring. Thus, individuals have a *pair* of factors for each trait.

4
One of the two factors for a trait prevents the expression of the other. For example, the factor for tallness prevents the expression of the factor for shortness. The factor that prevents the expression of the other is the **dominant** factor. The factor whose expression is prevented is the **recessive** factor.

Cell Division

During the 1860s, several scientists began to examine living cells that were in the process of dividing. The scientists noticed certain structures in the nucleus of each cell undergoing the division process. These structures were named *chromosomes* because they absorbed a colored dye that made them more visible under the microscope. *Chroma* means "color" in Greek. The scientists also noticed that when a cell divided, equal numbers of chromosomes were distributed to each new cell. They also observed that each new cell ended up with the same number of chromosomes as the original cell.

Later, scientists who were studying the formation of reproductive cells observed a second kind of cell division. During the process of reproductive-cell formation, two successive divisions occur. The result is four new cells, each having only one-half of the original cell's number of chromosomes. Let's take a closer look at these two types of cell division.

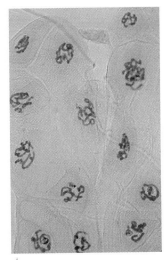

The darkly stained objects in this photograph are chromosomes in the cells of a fruit fly.

Mitosis   Look at your hand. Each square centimeter of skin is made of more than 150,000 cells. Yet many of these cells will be gone by tomorrow. These surface skin cells are actually dead, and normally they are shed or get washed away. But don't worry, you'll never run out of skin, because brand-new skin cells are continually being produced. These new cells will replace the skin cells you lose. In other parts of your body, cells are also reproducing rapidly. For example, as you grow taller, your bones grow in length by adding new bone cells at the ends of the bones.

The chromosomes in the nucleus of a cell are the "blueprints" for that cell. They contain all the information needed to build new cell materials and to control the cell's activities. Suppose you want to have two identical houses built at the same time. You would not cut the blueprints in two and give one-half to each builder. Instead, you would make an exact copy, or duplicate, of the blueprints so that each builder could have a full set of instructions. The same is true for the blueprints of a cell. Each new cell must carry a full set of chromosomes. Therefore, chromosomes must be duplicated before the nucleus of a cell divides. This duplication process occurs during a certain stage in the life of a cell.

A *parent* cell reproduces by dividing into two new cells.

The new cells are called *daughter* cells.

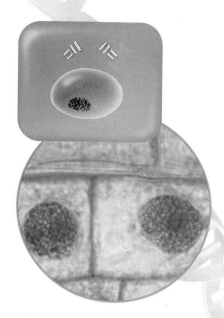

A cell spends most of its life in a stage called *interphase*. This is the time when the cell *seems* to be at rest. Actually, the cell is not resting at all. It's really very busy—growing and carrying out all of its normal life activities. Toward the middle of interphase, each chromosome in the cell duplicates. This doubling of the number of chromosomes in a cell is a sign that cell division is about to begin.

The first part of a cell that divides is the nucleus. This division of the nucleus is called **mitosis**. During mitosis, the division of the nucleus occurs in four consecutive stages that enable each daughter cell to receive one copy of each chromosome. Scientists call these stages—from first to last—*prophase, metaphase, anaphase,* and *telophase*.

Late Interphase Each chromosome has been duplicated. Once chromosomes have been duplicated, they exist as two strands joined together by a structure called a *centromere*.

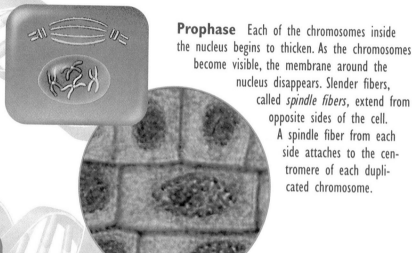

Prophase Each of the chromosomes inside the nucleus begins to thicken. As the chromosomes become visible, the membrane around the nucleus disappears. Slender fibers, called *spindle fibers*, extend from opposite sides of the cell. A spindle fiber from each side attaches to the centromere of each duplicated chromosome.

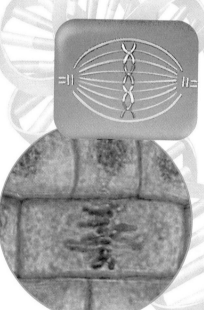

Metaphase The duplicated chromosomes line up across the middle of the cell. Then the centromeres divide, separating the duplicate chromosomes.

Mitosis

The process by which a cell's nucleus divides, leading to the formation of two identical cells

Anaphase The spindle fibers shorten and pull the duplicate chromosomes through the cytoplasm toward opposite sides of the cell.

Meiosis Many organisms reproduce by uniting two sex cells—one male and one female—during a process called *fertilization*. As a result, a fertilized cell, or *zygote*, forms. The zygote is the first cell of a new organism. It will divide by mitosis many times and develop into a new organism.

If sex cells contained the same number of chromosomes as other body cells, the number of chromosomes in a zygote would end up being double the normal amount. For example, every human body cell, such as a skin cell or a liver cell, has 46 chromosomes in its nucleus. If each human sex cell (egg or sperm) also contained 46 chromosomes, then a human zygote would contain 92 chromosomes. This means that every body cell of the new human being that developed from that zygote would also contain 92 chromosomes. This simply does not happen. Each species has a characteristic number of chromosomes in each of its cells; that number remains stable from one generation to the next. Scientists realized that for this to be the case, the number of chromosomes in sex cells would have to be exactly half the number of chromosomes in body cells.

While studying the formation of sex cells, scientists discovered a cell-division process that

This tiny sperm cell is attempting to fertilize a much larger egg cell. This particular sperm cell and egg cell are from a sea urchin.

results in daughter cells with half as many chromosomes as the parent cell. This process is called **meiosis**. In meiosis, the nucleus divides in stages that are very similar to those of mitosis, with one major exception. In meiosis, the duplicated chromosomes do not line up randomly during metaphase of the first division, as they do in mitosis. Instead, each duplicated chromosome lines up across from another duplicated chromosome of the same size and shape. Seeing this, researchers realized that chromosomes occur *in pairs*.

Meiosis

The process by which a parent cell divides twice, leading to the formation of four sex cells, with half the original number of chromosomes

Telophase The separated chromosomes cluster at the opposite sides of the cell. A nuclear membrane forms around each group of chromosomes. At the same time, the chromosomes untwist, becoming longer and thinner.

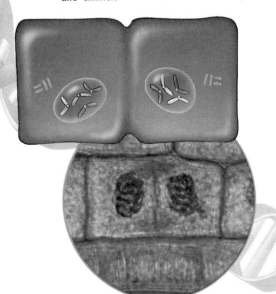

Cytokinesis At the end of telophase, the cytoplasm is divided between the two new cells, and they separate.

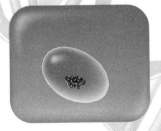

Cells that contain pairs of chromosomes are described as being *diploid*. Diploid cells are represented by the symbol 2*n*. The pairs of chromosomes separate during the first division of meiosis, and then the duplicates separate during a second division. The result is that each of the four daughter cells carries only one chromosome from each pair. Cells with a half-set of chromosomes are called *haploid* cells and are represented by the symbol *n*. Although he had never observed meiosis itself, Mendel had described this same pattern in explaining the inheritance of factors. Traits are determined by a pair of factors, the factors of each pair separate during formation of sex cells, and only one factor for each trait is transmitted by each parent.

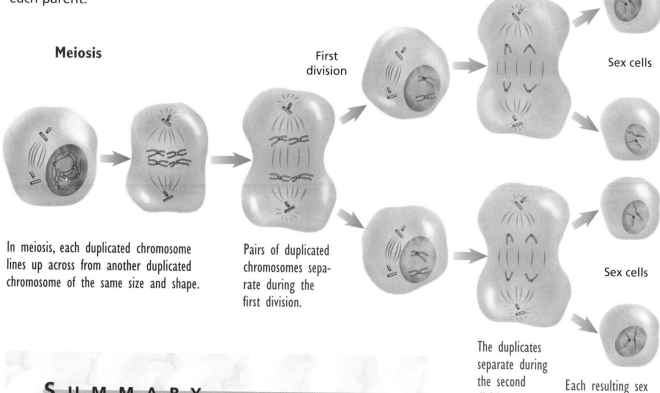

Meiosis

First division

Second division

Sex cells

Sex cells

In meiosis, each duplicated chromosome lines up across from another duplicated chromosome of the same size and shape.

Pairs of duplicated chromosomes separate during the first division.

The duplicates separate during the second division.

Each resulting sex cell has only one of each kind of chromosome.

S U M M A R Y

Reproduction is the means by which a species continues. During reproduction, characteristics are passed from parents to their offspring. Gregor Mendel was the first to describe the basic principles by which traits are inherited. He theorized that a trait is determined by a pair of factors. These factors separate during the formation of sex cells. Each parent transmits one factor for a trait to its offspring. Factors occur in different forms. A dominant factor prevents the expression of a recessive factor. Cells reproduce themselves through a process called cell division. During mitosis, the chromosomes in the nucleus of a parent cell are doubled and then divided so that two daughter cells receive the same kind and number of chromosomes as the parent cell. During meiosis, pairs of like chromosomes separate so that four daughter cells are produced, each having half as many chromosomes as the parent cell.

CHROMOSOMES, GENES, AND HEREDITY

Without knowing what they were or where they were located, Mendel reasoned that "factors" inherited from parents are responsible for the characteristics of an organism. Many years later, the behavior of chromosomes was observed during the formation of reproductive cells. However, the significance of this behavior was not immediately recognized. When Mendel's work was rediscovered in 1900, a new science, called **genetics**, was born. This name, which comes from a Greek word meaning "beginning," was first suggested at a scientific meeting on heredity in 1906.

The Chromosome Theory

In 1903, Walter S. Sutton, a researcher at Columbia University, was conducting a microscopic examination of meiosis in grasshoppers. Having read Mendel's paper about the inheritance of traits in pea plants, Sutton realized that the behavior of Mendel's factors could be explained by the behavior of chromosomes during meiosis. He inferred that, because of this relationship, these factors must somehow be related to chromosomes. He also realized that for every organism, there are many more traits than there are chromosomes. Sutton reasoned that more than one hereditary factor must be located on each chromosome. Sutton's ideas became known as the *Chromosome theory*, which simply states that *hereditary factors are found on chromosomes*. Today, scientists call these hereditary factors *genes*.

The term *gene* was first used to refer to a hereditary factor in 1909. Over the next 50 years, *geneticists* studied the way in which many genes are inherited. They even identified the exact locations of specific genes on the chromosomes of many organisms. However, the chemical nature of a gene and how it works remained a mystery until the 1950s. Even today, there is still much more to be learned about genes and chromosomes.

Hair color

Eye color

Skin color

Height

Shoe size

◄ Humans have 23 pairs of chromosomes, each one containing hundreds of genes that together determine your inherited traits.

Genetics

The science in which heredity is studied

Working With Genes

You can represent individual genes by using letters. Capital letters are used to indicate dominant genes. For example, *T,* is used to represent the gene for tallness in pea plants, and *R,* is used to represent the gene for round seed shape in pea plants. Remember that both of these traits are dominant in pea plants. The lowercase version of the *same* letter indicates the recessive gene for the same trait. The gene for shortness, then, is *t,* and the gene for wrinkled seed shape is *r.*

Since traits are produced by *pairs* of genes, a pair of letters can be used to represent the genes that a particular individual has for a trait. For example, pea plant height is determined by one of three possible gene pairs—*TT, Tt,* or *tt.* Pea-seed shape can be determined by *RR, Rr,* or *rr.* The specific pair of genes that an organism has for a trait is called its **genotype**. The physical description of a genotype, or what the organism actually looks like, is called a **phenotype**. Round seeds, wrinkled seeds, tall plants, or short plants are terms that describe phenotype. If both genes for a trait are identical (*TT* or *tt*), the genotype is said to be *homozygous*, or pure, for that trait (*homo* means "same"). If the two genes are different (*Tt*), the genotype is said to be *heterozygous*, or hybrid (*hetero* means "different").

Today, you can make the same genetic crosses that Mendel made about 150 years ago. But you don't need to find garden space to do it. You just have to know how to use a special chart called a *Punnett square.*

Genotype

The genetic makeup of an organism according to its genes

Phenotype

The external appearance of an organism as determined by its genotype

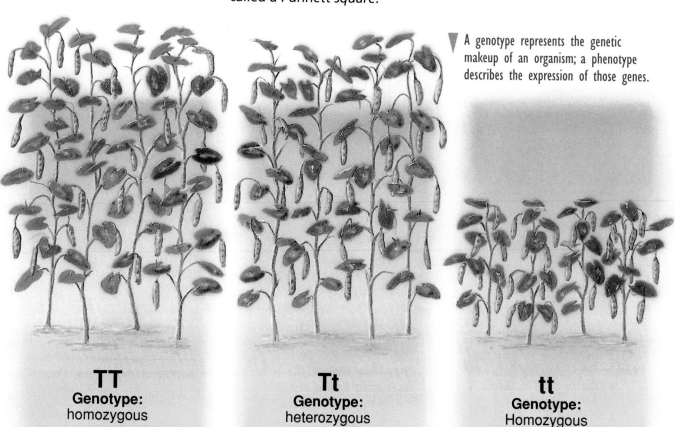

A genotype represents the genetic makeup of an organism; a phenotype describes the expression of those genes.

TT
Genotype:
homozygous

Phenotype:
tall plants

Tt
Genotype:
heterozygous

Phenotype:
tall plants

tt
Genotype:
Homozygous

Phenotype:
short plants

Constructing a Punnett Square The Punnett-square method for displaying the possible results of a genetic cross was developed by the British scientist Reginald Punnett in the early twentieth century. Follow the steps below to see how a Punnett square is used to predict the outcome of a cross between homozygous tall pea plants and homozygous short pea plants.

1 First draw a square and divide it into two rows and two columns.

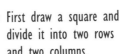

2 Now write symbols for the two genes of one parent's genotype at the top of the square, one above each column. Then write the symbols for the two genes of the other parent's genotype down the left side of the square, one beside each row. These letters represent the gene content of the parents' reproductive cells.

3 Finally, copy the gene symbol from the top of each column into the boxes below, and copy the gene symbol from the left side of each row into the boxes of that row. Typically, the dominant gene is written first, followed by the recessive gene. All of the possible gene combinations from the cross now appear in the boxes. These are the possible genotypes of the offspring.

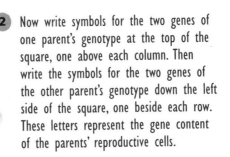

4 Notice that all of the genotypes in the square are heterozygous. Since tallness *(T)* is dominant to shortness *(t)*, what will the phenotype of all the offspring be?

5 Now take a couple of the offspring and cross them in another Punnett square. What is the phenotype of each member of the second generation? What is the phenotype ratio?

Pedigree

A chart that traces the inheritance of a genetic trait through several generations of a family

Using a Pedigree Suppose you decide to raise rabbits with black fur (B), which is dominant to brown fur (b). You go to a pet store to buy several pairs of black rabbits. But how will you know if the rabbits you buy will produce only offspring with black fur? You cannot tell just by looking at them, because a black rabbit can have either one of two different genotypes—homozygous BB or heterozygous Bb. If some of the rabbits you buy have the Bb genotype, your rabbit warren will eventually contain some brown rabbits. To be sure that you buy only homozygous black rabbits, you must obtain a **pedigree** for each rabbit.

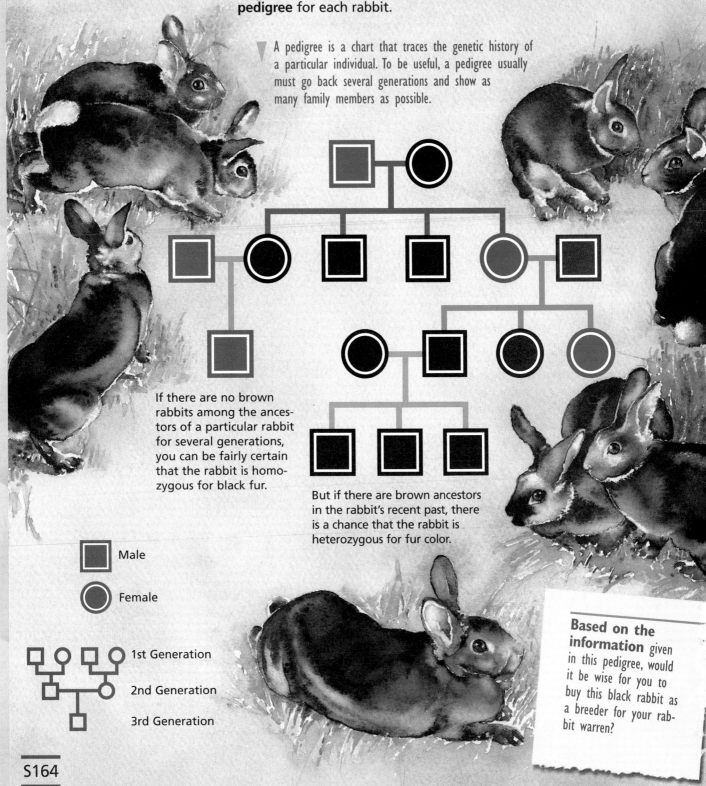

▼ A pedigree is a chart that traces the genetic history of a particular individual. To be useful, a pedigree usually must go back several generations and show as many family members as possible.

If there are no brown rabbits among the ancestors of a particular rabbit for several generations, you can be fairly certain that the rabbit is homozygous for black fur.

But if there are brown ancestors in the rabbit's recent past, there is a chance that the rabbit is heterozygous for fur color.

☐ Male

⬤ Female

☐○☐○ 1st Generation

2nd Generation

3rd Generation

Based on the information given in this pedigree, would it be wise for you to buy this black rabbit as a breeder for your rabbit warren?

Sex Determination

What determines whether an individual is male or female? In the early 1900s, biologist Thomas Hunt Morgan found an answer to this question. He made his discovery while conducting genetic experiments with fruit flies, the tiny insects that fly around overripe bananas.

Morgan noticed that the cells of female flies have four pairs of chromosomes that are alike in shape and size. In male flies, however, only three of the pairs match; the fourth pair does not. One chromosome of this pair is of normal size, while the other is much shorter. Morgan believed that this mismatched fourth pair was responsible for the sex of the male fruit fly. Morgan called the nonmatching chromosome (the short one) in the male the *Y chromosome*. The other chromosome in the pair he called the *X chromosome*.

Further research has revealed that sex in most animals, including humans, is determined by the inheritance of X and Y chromosomes. Females have two X chromosomes, but males have one X and one Y chromosome. Some conditions in humans, such as hemophilia and colorblindness, are caused by abnormal genes that are located on the X chromosome.

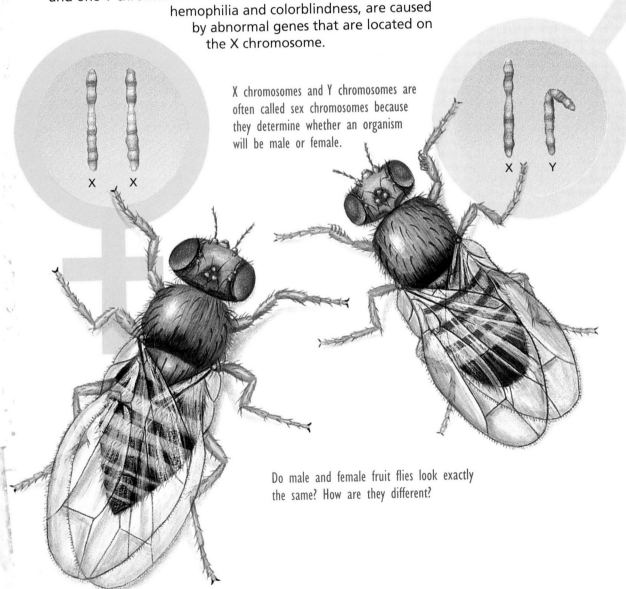

X chromosomes and Y chromosomes are often called sex chromosomes because they determine whether an organism will be male or female.

X   X

X   Y

Do male and female fruit flies look exactly the same? How are they different?

Other Patterns of Inheritance

Mendel described only one pattern of inheritance. The traits he studied in peas are each controlled by one pair of genes. For each of these traits there are two forms of the gene—one dominant and one recessive. Modern geneticists refer to this pattern of inheritance as *simple dominance*. However, simple dominance is only one way that genetic traits are inherited.

For some traits, there are no dominant or recessive genes. Instead, *both* genes in a heterozygous pair are expressed. This pattern of inheritance is called *incomplete dominance*. Flower color in snapdragons, for example, shows incomplete dominance. Crosses made between a red-flowered plant and a white-flowered plant produce all pink-flowered plants. They would not produce red-flowered plants or white-flowered plants, as you would expect if one color were completely dominant over the other.

Many traits in humans are determined by several pairs of genes. This pattern of inheritance is called *polygenic inheritance*. The effect of each pair of genes is added together to produce the phenotype of the individual. Skin color, for example, is a trait produced by the action of several pairs of genes.

When plants with red flowers are crossed with white-flowered plants, the resulting offspring have pink flowers.

These students exhibit a wide range of skin colors.

S U M M A R Y

Hereditary factors called genes are located on chromosomes. Dominant genes are indicated by capital letters, while recessive genes are indicated by lowercase letters. A Punnett square is used to determine the genotypes and phenotypes of the offspring from genetic crosses. A pedigree is used to trace an individual's genetic history. In many animals, sex is determined by the inheritance of X and Y chromosomes. In addition to simple dominance, there are other patterns of inheritance, including incomplete dominance and polygenic inheritance.

DNA—THE MATERIAL OF HEREDITY

DNA (deoxyribonucleic acid) was discovered in 1869. Because scientists found it in the nucleus of cells, they dubbed it a "nucleic acid." By the 1940s, scientists knew that chromosomes were made of both DNA and protein. But they didn't know whether it was the DNA or the protein that was the genetic material of cells. It was not until the 1950s that scientists were able to demonstrate that DNA is the material responsible for heredity. Yet, how does DNA work? And what is it made of?

Structure of DNA

Have you ever put together a model of an airplane or an automobile? Would you be able to put all the tiny pieces of the model together correctly if you lost the instructions? You might if you knew what the model was supposed to do. Back in 1953, the American biologist James Watson and the English physicist Francis Crick knew what DNA was supposed to be able to do—duplicate itself and act as the genetic material of a cell. Armed with this knowledge, and with the information collected by many other scientists, they were able to build a three-dimensional model of DNA.

The DNA molecule is twisted into a spiral, or *helix*. This twisted ladder shape is called a *double helix*.

Phosphate group

Nitrogen bases

Sugar molecule

Each DNA molecule consists of two very long chains of smaller units called *nucleotides*. DNA's two chains are connected by crosspieces, or "rungs," that give the molecule a ladderlike appearance.

Four different nitrogen bases are found in the nucleotides of DNA. They are *adenine*, *guanine*, *cytosine*, and *thymine*. The nitrogen bases are often abbreviated as A, G, C, and T, respectively. They can be thought of as the letters of the genetic alphabet.

C

A

G

T

Phosphate group

Nitrogen base

Sugar molecule

Each nucleotide in a DNA molecule is made of three parts—a sugar molecule, a phosphate group, and a molecule called a *nitrogen base*.

Functions of DNA

DNA has two major functions. It undergoes duplication, also called *replication*, and it directs *protein synthesis*. In order to understand how DNA accomplishes either of these processes, you must first know the way that the nucleotides are arranged to form the "sides" and the "rungs" of the DNA "ladder."

Nucleotides are bonded together in a specific way to form the double helix. The sides are formed by bonding the sugar of one nucleotide to the phosphate of the next nucleotide in a continuous chain. The nitrogen bases stick out from the sugar and phosphate sides. Each of the nitrogen bases of one side is bonded to a base from the other side. Look again at the DNA model on the previous page. The resulting nitrogen-base pairs form the rungs of the ladder. But each base can be paired with only one other base. Adenine (A) always pairs with thymine (T), and guanine (G) always pairs with cytosine (C).

DNA Replication Replication is the making of an exact copy of a DNA molecule. Replication occurs during the inter-phase portion of a cell's life cycle. The process of replication begins when the two chains of a DNA molecule begin to separate.

Original strands

Enzyme

1

New strand

2

3

Free-floating nucleotides

New strand

1 The nitrogen-base pairs are pulled apart by a type of protein called an *enzyme*. The separation of the two sides of the DNA molecule is often compared to the unzipping of a zipper. When the chains have separated, individual nucleotides float-ing freely in the nucleus line up across from the nitrogen bases of the separated chains.

2 New base pairs form as the nitrogen bases of the new nucleotides bond to the nitrogen bases on each half of the old molecule. Remember, there are only two kinds of nitrogen-base pairs that can be made—C always pairs with G, and A always pairs with T. As a result, the base-pair sequence of the original DNA reappears.

3 Once replication is complete, each side of the original DNA has produced a new DNA molecule. The two new DNA molecules are identical to each other and to the original molecule.

Protein Synthesis The DNA in each of the chromosomes of an organism has a different sequence of nitrogen-base pairs. Genes are no more than sections of a DNA molecule. One gene may consist of thousands of nitrogen-base pairs. The individual genes of a DNA molecule contain instructions for making specific proteins. This is an extremely vital role of DNA, since the primary structural and regulatory chemicals of living things are proteins. Some of the many different proteins made by an organism are used to build its various parts. Other proteins, called enzymes, assist in making the variety of other substances that an organism produces. You just read about how one enzyme functions to unzip DNA during replication.

▼ Each "word" in the genetic code, as found in RNA, is composed of three "letters," or nitrogen bases. Sixty-four such three-letter "words," or triplets, can be made.

The Genetic Code

The instructions that DNA contains for making proteins are not written in words that are familiar to you. Instead, they are written in a "language" that uses the nitrogen bases A, T, G, and C as "letters." This language is called the **genetic code**.

Recall that amino acids are the molecules that make up proteins. The string of code words in a gene causes many amino acids to come together in a certain order to make a particular protein. But this cannot be a direct process because proteins are made in the cytoplasm of a cell, while DNA is found only in the nucleus. Instead, proteins are made in a more indirect way, by way of another nucleic acid, called RNA. RNA copies the information for making a protein from DNA. The RNA then brings the protein "recipe" to the cytoplasm.

RNA, or *ribonucleic acid*, is similar to DNA. Both kinds of nucleic acids are made of nucleotides. RNA, however, is only a single chain of nucleotides. It also has two chemical parts that are different from DNA. RNA contains a sugar called *ribose*,

while DNA contains a similar sugar called *deoxyribose*. As you can see, the molecules' names are based on their sugar parts. Instead of thymine, RNA has a nitrogen base called *uracil* (U). As with DNA, the nitrogen bases of RNA also form pairs, but there is one difference. The U of RNA pairs with A, since U takes the place of T in RNA.

Genetic code

The language in which the instructions for proteins are written in DNA

Some RNA Codes and Their Associated Amino Acids	
Triplet	**Amino acid**
AAA	Lysine
GUG	Valine
UUA	Leucine
AGA	Arginine
GUU	Valine
CAG	Glutamine
UGU	Cysteine
CAC	Histidine
AUU	Isoleucine
UUU	Phenylalanine
CCC	Proline
AUG	Methionine

▲ Use the chart to determine which amino acids are coded for in the strand of RNA being produced below.

Each triplet "codes" for a particular amino acid.

From DNA to RNA to Protein Prior to protein synthesis, a special RNA molecule, called *messenger RNA* (mRNA), is synthesized from the DNA of a gene and carries the protein-making instructions from the DNA in the nucleus to a ribosome in the cytoplasm. Ribosomes are the structures on which proteins are assembled. In the cytoplasm, there is another kind of RNA, known as *transfer RNA* (tRNA). Each tRNA molecule (which has a three-looped, or cloverleaf, shape) carries a particular amino acid. For example, one kind of tRNA carries the amino acid *leucine*, while another carries *valine*. The amino acids carried by the tRNA molecules line up end to end at the ribosome. Then the amino acids are joined together by chemical bonds to form a protein.

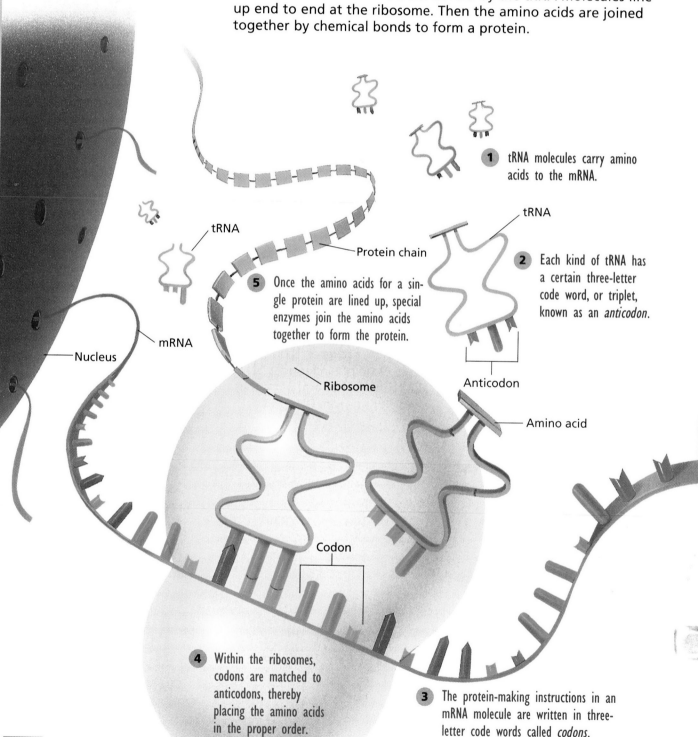

1 tRNA molecules carry amino acids to the mRNA.

tRNA

2 Each kind of tRNA has a certain three-letter code word, or triplet, known as an *anticodon*.

Anticodon

Amino acid

tRNA

Protein chain

5 Once the amino acids for a single protein are lined up, special enzymes join the amino acids together to form the protein.

mRNA

Nucleus

Ribosome

Codon

4 Within the ribosomes, codons are matched to anticodons, thereby placing the amino acids in the proper order.

3 The protein-making instructions in an mRNA molecule are written in three-letter code words called *codons*.

Changes in DNA

Right before a cell divides, its DNA is replicated so that each new cell receives a copy of the complete set of genetic instructions for an organism. Since the nitrogen bases of DNA pair only in certain ways, the copies are normally exact. However, mistakes during copying are sometimes made. Also, factors from the environment, such as chemicals and radiation, can cause accidents that destroy parts of a DNA molecule. In both cases, the result is a change in the nitrogen-base sequence of a DNA molecule. Thus, its genetic instructions may change as well. Such a change is called a *mutation*.

Mutations are important because they are the source of new genetic variations. As such, they have played a major role in the evolution of life on Earth. Although mutations can be beneficial, they are more frequently harmful. Fortunately, a mutation usually affects only one cell, and so there is essentially no effect on the entire organism. Only mutations that occur in reproductive cells can affect an entire organism. Remember that during fertilization, reproductive cells come together to form a new organism. In this way, these mutations can be introduced

Various types of mutations can occur in chromosomes.

into offspring. There are two basic types of mutations—*gene mutations* and *chromosome mutations*.

Gene Mutations   A gene mutation is a change in the structure of a single gene and is therefore also called a point mutation. Most often, this kind of mutation affects only one nucleotide, or maybe several nucleotides, in the gene. But still, a change as small as even one base in the sequence of bases that make up a gene can have a significant effect. It can

cause the gene to produce an altered protein or, sometimes, no protein at all.

Chromosome Mutations
Chromosome mutations are changes caused by the breaking of chromosomes. Normally, chromosome breaks are repaired. But if the parts of a broken chromosome are improperly rejoined or if some parts are lost, major changes in the genetic instructions of a cell can occur. Chromosome mutations affect many genes. As a result, they usually have a far more serious effect on a developing embryo than gene mutations. In fact, most are fatal and the offspring is never born.

◀ This black rat snake is white because a gene mutation has resulted in the inability to produce normal pigments.

Genetic Engineering

Humans have learned how to engineer changes in DNA. Early in the 1970s, scientists discovered several bacterial enzymes that could be used to cut DNA molecules into pieces. By using these enzymes, scientists are able to remove a gene out of a cell from one organism and place it into a cell from another organism. The gene that was removed combines with the DNA in the new cell. That cell then has some new hereditary instructions. The new DNA formed by adding DNA from one organism to the DNA of another organism is called **recombinant DNA**.

Bacteria containing recombinant DNA are now used to make substances that are vital to human beings. One such substance is insulin, which is needed by many people who suffer from a disease called diabetes.

Most recently, recombinant DNA has been used to replace incomplete or undesirable DNA instructions in plants and animals. Scientists have been able to improve certain food crops and farm animals by adding genes for higher nutritional value and disease resistance. In a new kind of medical procedure called *gene therapy*, doctors use recombinant DNA techniques in attempts to cure certain human genetic diseases.

Recombinant DNA

A molecule made from the DNA of different organisms

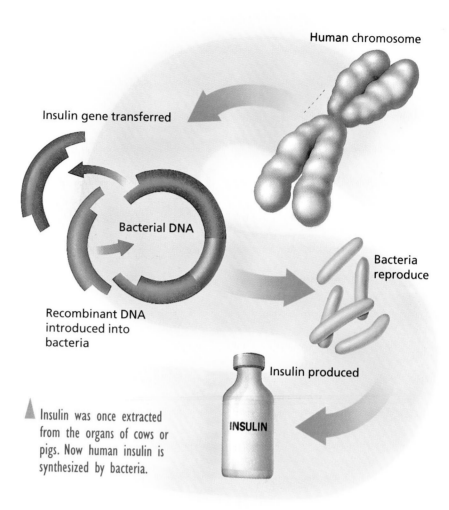

Human chromosome

Insulin gene transferred

Bacterial DNA

Recombinant DNA introduced into bacteria

Bacteria reproduce

Insulin produced

INSULIN

Insulin was once extracted from the organs of cows or pigs. Now human insulin is synthesized by bacteria.

S U M M A R Y

DNA is the material responsible for heredity. A molecule of DNA is a double strand of nucleotides that is twisted into a spiral shape called a double helix. The hereditary information carried by DNA is in a code that is contained in the sequence of the four nitrogen bases adenine, thymine, cytosine, and guanine. Because only certain bases can bond together in pairs, a DNA molecule can be copied by the process of replication. The instructions of the genetic code are used to put amino acids together to make proteins. Messenger RNA molecules take the instructions for making proteins from the nucleus into the cytoplasm, where proteins are made. Transfer RNA molecules carry amino acids to the messenger RNA. A change in the sequence of bases in a DNA molecule is called a mutation. A single gene or an entire chromosome may be changed by a mutation. Humans can alter DNA by taking DNA pieces from one cell and placing them into another cell, thus making recombinant DNA.

REPRODUCTION AND DEVELOPMENT

Ou now know about chromosomes and genes and the role that meiosis plays in passing traits from one generation to the next. The sperm cells and egg cells that result from meiosis combine during a form of reproduction called **sexual reproduction.** But not all organisms reproduce by using sex cells, also called *gametes.* Some organisms, such as bacteria and sponges, produce offspring by **asexual reproduction,** which doesn't rely on the formation of sperm and eggs.

Asexual Reproduction

If you watch live bacteria under a microscope long enough, you should eventually see some of them divide in two. These dividing bacteria are reproducing asexually—reproducing without using gametes. Many single-celled organisms, such as bacteria, reproduce by simply splitting. This form of asexual reproduction is called fission. *Fission* means "splitting in half." In fission, one parent cell produces two identical daughter cells. Similarly, asexual reproduction in some protists occurs by mitosis. But fission and mitosis are not the only ways that organisms can reproduce asexually. Nor does asexual reproduction occur only in bacteria and protists.

This bacterium is undergoing fission, a type of asexual reproduction.

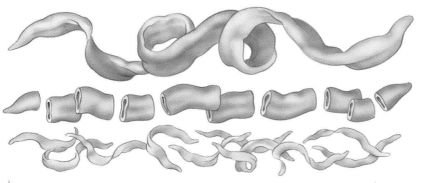

Some simple worms reproduce by *fragmentation.* In fragmentation, the worm's body breaks up into many pieces. Each piece grows into a complete new worm.

Asexual reproduction

Producing new individuals without using gametes

Sexual reproduction

Producing new individuals by uniting gametes

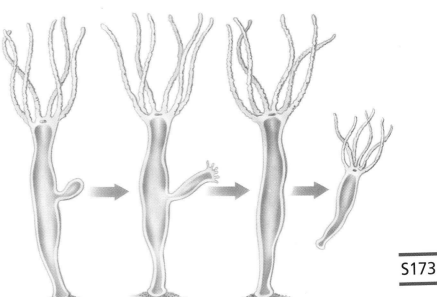

The hydra, a relative of the jellyfish, reproduces by *budding.* In budding, a small knob of tissue grows from the body wall of the hydra. Before long, this outgrowth develops into a new hydra. Usually the offspring detaches from the parent and begins living on its own.

S173

Sexual Reproduction

Many protists, fungi, plants, and animals reproduce sexually, using gametes. During the life cycle of a sexually reproducing organism, there are two different cellular phases. In one phase, cells are diploid (contain pairs of chromosomes). In the other phase, they are haploid (contain a half-set of chromosomes). Remember that haploid cells result from meiosis.

The Plant Life Cycle   Plants have a life cycle that is described as an *alternation of generations*. This description comes from the fact that plants alternate between a multicellular diploid stage and a multicellular haploid stage. The diploid plant is called a *sporophyte*. Sporophytes are so named because they produce haploid cells, called *spores,* by meiosis. Some of the spores that are produced by a sporophyte undergo mitosis and develop into multicellular *gametophytes*, which are haploid. The gametophytes produce haploid sperm and egg cells, which unite to form another sporophyte. Let's take a look at a fern and see how it alternates between a sporophyte stage and a gametophyte stage during its life cycle.

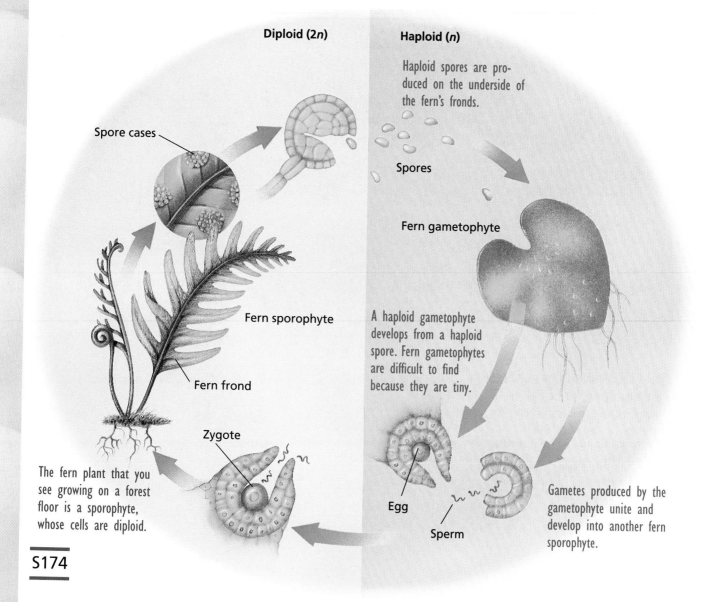

Diploid (2*n*)

Haploid (*n*)

Haploid spores are produced on the underside of the fern's fronds.

Spore cases

Spores

Fern gametophyte

Fern sporophyte

Fern frond

A haploid gametophyte develops from a haploid spore. Fern gametophytes are difficult to find because they are tiny.

Zygote

The fern plant that you see growing on a forest floor is a sporophyte, whose cells are diploid.

Egg

Sperm

Gametes produced by the gametophyte unite and develop into another fern sporophyte.

The Animal Life Cycle The life cycle of an animal is different from a plant's life cycle. In animals, the only haploid cells that exist are gametes. All the other cells in an animal's life cycle are diploid. The diploid zygote and the diploid individual that develops from the zygote are the most obvious parts of the animal life cycle.

These male and female salmon are breeding in a stream in northern California. The salmon are diploid and the gametes that they shed are haploid.

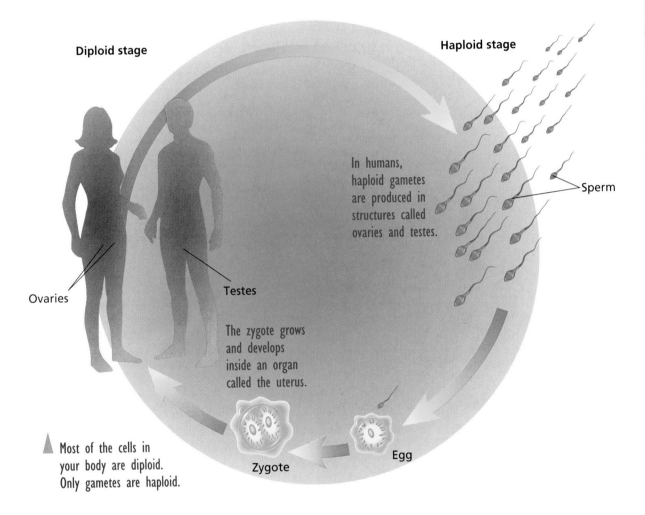

Diploid stage

Haploid stage

Ovaries

Testes

In humans, haploid gametes are produced in structures called ovaries and testes.

Sperm

The zygote grows and develops inside an organ called the uterus.

Most of the cells in your body are diploid. Only gametes are haploid.

Zygote

Egg

Human Reproduction

You are born with all the parts of your reproductive system, but it doesn't fully develop until you reach puberty. **Puberty** is the stage of life during which a person matures sexually. It usually begins sometime between the ages of 10 and 14. During puberty, males begin producing sperm cells, and females begin producing egg cells. The production of these gametes is mostly under the control of chemicals called *hormones*. Hormones are made by structures called *endocrine glands*. Hormones enter the bloodstream and cause changes in the body by affecting specific tissues and organs.

Puberty

The stage of physical development when sexual reproduction first becomes possible

Male Reproductive System   The male reproductive system has two functions in the process of reproduction. It produces sperm, and it delivers the sperm to the female. Both nerves and endocrine glands control the male reproductive system. The testes are the reproductive organs that produce sperm cells. They also act as endocrine glands that produce hormones.

One hormone produced by the testes is responsible for the development of a male's *secondary sexual characteristics*, which include male features such as body hair, strong muscles, and a deep voice. The *pituitary gland*, located on the underside of the brain, produces other hormones that cause cells inside the testes to develop into sperm.

The testes are located outside the body cavity in a sac called the scrotum. Sperm are sensitive to high temperatures and develop normally only at the slightly cooler temperatures found outside the body cavity.

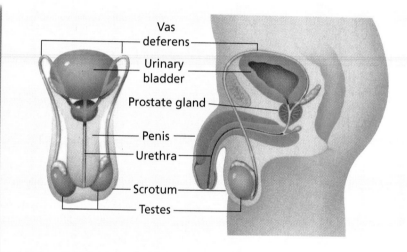

Vas deferens

Urinary bladder

Prostate gland

Penis

Urethra

Scrotum

Testes

- Millions of sperm are produced inside the tiny, tightly coiled tubes that are found within the testes.
- The prostate gland produces fluids that mix with the sperm as they exit the body. The mixture of sperm cells and fluid is called *semen*.
- The vas deferens and the urethra are tubes that carry sperm out of the male's body.

▼ A sperm cell consists of three regions: the head, the midsection, and the tail.

The head contains DNA, the hereditary material.

Head

Midsection

Mitochondria

The midsection contains many mitochondria. Mitochondria release energy.

Tail

The tail moves back and forth, propelling the sperm cell forward.

Female Reproductive System

The female reproductive system produces hormones and eggs, or *ova*. It also receives sperm from a male and provides a place for the development of offspring. Ova develop near the surface of two *ovaries*, which are located in the abdomen. Generally, one ovary releases an ovum (singular of ova) once every month.

Hormones produced by the ovaries are responsible for a female's secondary sexual characteristics, such as breasts and widened hips. These same hormones, along with hormones from the pituitary gland, also control the formation and release of ova and cause a thickening of the inner lining of the uterus. These changes in the female reproductive system occur each month during the *menstrual* (MEN struhl) *cycle*.

If the ovum released during the menstrual cycle has not been fertilized inside the fallopian tube, the tissues of the thickened lining of the uterus leave the body through the vagina. This process is called *menstruation* (men STRAY shuhn). If the ovum has been fertilized by a male's sperm, menstruation does not occur. Instead, a zygote begins to develop into a new human being.

The development and release of an egg cell and the thickening of the lining of the uterus are controlled by changing levels of hormones in the bloodstream.

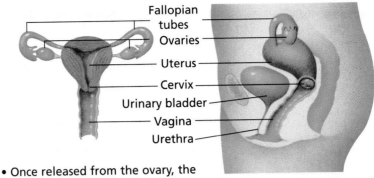

Fallopian tubes
Ovaries
Uterus
Cervix
Urinary bladder
Vagina
Urethra

- Once released from the ovary, the ovum enters the fallopian tube, which leads from the ovary to the uterus.
- The uterus is a muscular, pear-shaped organ. It is used to house a developing embryo.
- Sperm, which enter the female's body through the vagina, must swim a long way to fertilize an egg in the fallopian tube.

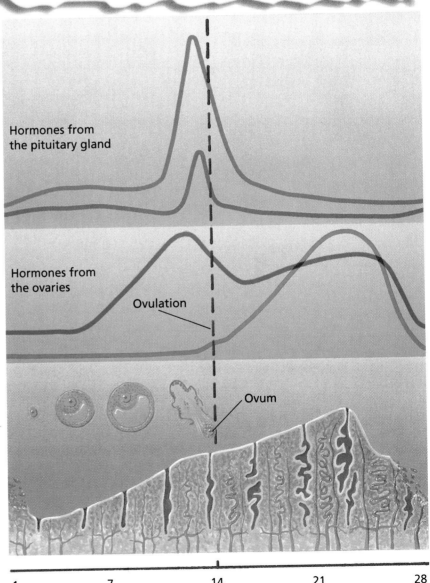

Hormones from the pituitary gland

Hormones from the ovaries

Ovulation

Ovum

1   7   14   21   28
Day of cycle

Stages of Human Development The process of development that takes place between fertilization and birth lasts about 265 days. After fertilization, the zygote passes down the fallopian tube and into the uterus. During this 5 to 7 day trip, cell division occurs several times. By the time it reaches the uterus, the zygote has become a ball of tightly packed cells called an **embryo**. For development to continue, the embryo must attach itself to the wall of the uterus. Enzymes, given off by the embryo, break down a tiny spot in the thick wall of the uterus. The embryo attaches itself to this tiny spot. After 1 week of development, the embryo is about the size of the period at the end of this sentence. The new organism is called an embryo from the second to the eighth week. From the ninth week until birth, it is called a **fetus**.

Embryo

Describes a human from the second to the eighth week of development

Fetus

Describes a human from the ninth week of development until birth

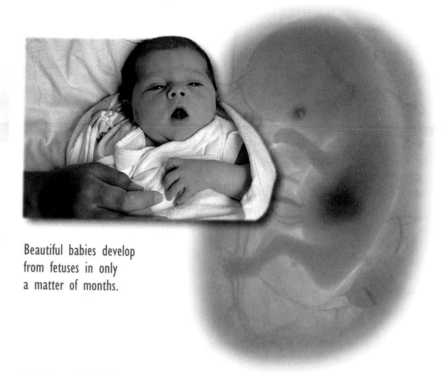

Beautiful babies develop from fetuses in only a matter of months.

S U M M A R Y

Some organisms reproduce asexually, without sex cells. Types of asexual reproduction are binary fission, mitosis, fragmentation, and budding. Other organisms reproduce sexually, with sperm and eggs. These organisms have life cycles that include a diploid phase and a haploid phase. Humans become sexually mature at puberty. The male reproductive system produces sperm and delivers sperm to the female. It also produces hormones. The female reproductive system produces hormones and ova, and receives sperm. An ovum is released and the lining of the uterus thickens during the menstrual cycle. The development of a zygote into an embryo, and then into a fetus, occurs inside the uterus.

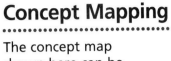

Unit CheckUp

Concept Mapping

The concept map shown here can be used to illustrate major ideas in this unit. Complete the map by placing terms from the list in the appropriate position. Then extend your map by answering the additional question below. Use your ScienceLog. **Do not write in this textbook.**

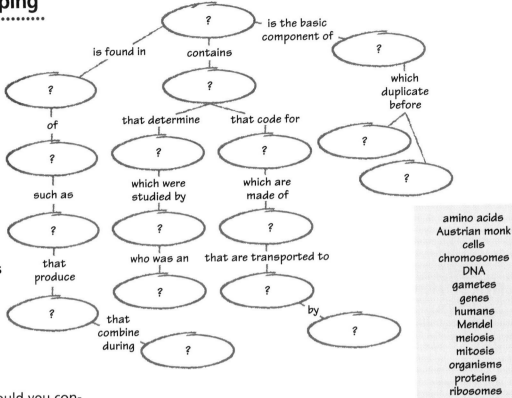

amino acids
Austrian monk
cells
chromosomes
DNA
gametes
genes
humans
Mendel
meiosis
mitosis
organisms
proteins
ribosomes
sexual reproduction
traits
tRNA

Where and how would you connect the terms *pedigree*, *codons*, and *mutation*?

Checking Your Understanding

Select the choice that most completely and correctly answers each of the following questions.

1. If you crossed homozygous tall pea plants with homozygous short pea plants, you would expect the offspring to be
   a. all short.
   b. all tall.
   c. mostly short.
   d. of medium height.

2. The "factors" that Mendel described are today called
   a. chromosomes.
   b. daughter cells.
   c. genes.
   d. spindle fibers.

3. Meiosis in humans occurs in the
   a. testes.   c. ovaries.
   b. uterus.   d. Both *a* and *c*

4. Which event results from meiosis but NOT from mitosis?
   a. Four daughter cells form.
   b. A spindle develops.
   c. Chromosomes duplicate.
   d. Two daughter cells form.

5. If the series of nitrogen bases in one strand of a DNA molecule is ATGCATG, what are the bases in the opposite strand?
   a. TCCCATG
   b. TACGTAC
   c. ATGCATG
   d. GTACCCT

S179

Interpreting Photos

Most fruit flies have red eyes. The white eyes of the fruit fly shown on the left are caused by a gene mutation in one of its sex chromosomes. Which sex chromosome do you think contains the mutation? If you had to guess the sex of this fly, what would your guess be, male or female? Why did you make the guess that you did?

Critical Thinking

Carefully consider the following questions, and write a response in your ScienceLog that indicates your understanding of science.

1. In pea plants, yellow peas are dominant to green peas. Use a Punnett square to show the possible outcome of a cross between two pea plants that are both hybrid for color. Write down the ratios of the phenotypes and the genotypes of the offspring that result from this cross.

2. A scientist removes a cell from an organism and observes it undergoing meiosis. At the end of meiosis, the scientist counts six chromosomes in each of the four sex cells that result. Based on this information, how many chromosomes would there be in a skin cell from the organism? Explain how you determined your answer.

3. Part of a protein contains the following sequence of amino acids: arginine-isoleucine-lysine. What is the nitrogen-base sequence of the DNA that coded for these amino acids? What are the anticodon sequences of the transfer RNA molecules that carried these amino acids to the messenger RNA? (Hint: the mRNA codons are AGA, AUU, and AAA.)

4. As a group, organisms that reproduce sexually adapt better to changes in their environments than do organisms that reproduce asexually. Explain why. (Hint: think about the genetic makeup of the offspring that result from both groups.)

Portfolio Idea

Imagine that Walter Sutton is suddenly zapped back in time and lands right in the middle of Gregor Mendel's pea garden. After talking for just a short while, Sutton and Mendel discover their mutual interest in science. From that moment on, they decide to work side by side to answer the questions they have about genetics and heredity. In 1866, they publish their findings. Write a one-page summary of the paper that Sutton and Mendel coauthored.

A

Absorption the action of light being taken in by a substance rather than being reflected or transmitted **(443)**

Abyssal plains flat, featureless, and largely lifeless areas that make up much of the deep-ocean bottom **(240)**

Active transport the movement of materials into and out of cells against a concentration gradient, requiring the use of energy **(S15)**

Adhesion the force of attraction between unlike particles **(50, S28)**

Air the mixture of gases in the atmosphere **(S62)**

Air mass a large body of air with uniform temperature and humidity **(252, S72)**

Albinism an abnormal condition in plants and animals in which some or all natural pigments are absent **(71)**

Alternating current an electric current that reverses direction at regular intervals **(320)**

Alternator a mechanical device for generating alternating current **(S97)**

Ampere (amp) the basic unit of measure of electric current, equal to 1 coulomb of charge passing a given point in 1 second **(343)**

Amplitude the distance that a wave rises or falls from its normal rest position **(371, S109)**

Asexual reproduction reproduction that does not involve the union of sex cells, as when a single-celled organism divides into two new cells **(505, S173)**

Atmosphere the layer of gases that surrounds the Earth **(251, S60)**; a unit of measure of pressure, equal to approximately 101 kilopascals **(268, S32)**

Atom the smallest particle into which an element may be divided and still be the same substance **(92)**

ATP (adenosine triphosphate) an organic molecule used to deliver energy for life processes **(S11)**

Avogadro's number the number of particles in 1 mole of a substance $(6.022137 \times 10^{23})$ **(S24)**

B

Barometer an instrument for measuring atmospheric pressure **(266)**

Battery a group of connected chemical cells that convert chemical energy into electrical energy to produce an electric current **(310)**

Biomass any organic material, such as plant or animal wastes, that can be used as an energy source **(67)**

Blade the flattened green part of a leaf that varies in shape from plant to plant **(18)**

C

Caesarean section a surgical incision made in the lower abdomen and uterus of a pregnant woman so that the fetus and placenta can be removed **(528)**

Calibrate to establish a reference scale for an instrument in order to standardize measurements made with the instrument **(250)**

Capillary action the rising of a fluid in a thin tube caused by the force of attraction between the liquid and the walls of the tube **(48)**

Carbohydrates organic molecules composed of one or more monosaccharides; often used for energy by cells **(S3)**

Cell the smallest unit of life that is capable of functioning independently **(498)**

Cell division the process by which a single cell splits into two identical cells **(500)**

Chemical cell a device, consisting of an electrolyte and two electrodes, that converts chemical energy into electrical energy to produce an electric current **(305)**

Chemical energy a form of potential energy that is stored in fuels, food, and other chemicals and that is released by chemical reactions **(160)**

Chlorophyll a green chemical that absorbs energy from the sun during photosynthesis **(23, S9)**

Chromosomes bodies in the nucleus of a cell which contain the genes that pass hereditary traits from parent to offspring; can be seen microscopically when a cell is dividing **(514)**

Circuit a continuous path for the flow of electrons **(296, S101)**

Circulatory system the body's internal transportation system that carries materials to and from the cells; includes the heart, blood vessels, and blood **(65)**

Circumstantial evidence indirect evidence, gathered through observation, that is used to make an inference about something unknown **(81)**

Climate the average weather for a particular place over many years **(256, S66)**

Clone an organism produced by asexual reproduction that is genetically identical to its parent **(547)**

Coherent light light consisting of waves that are in step with one another **(S146)**

Cohesion the force of attraction between like particles **(49, S28)**

Cold front the leading edge of an advancing cold air mass **(253)**

Complementary colors of light specific pairs of colored lights, consisting of one primary and one secondary color, that produce white light when mixed **(438)**

Compound a substance consisting of two or more elements that are chemically combined **(95)**

Concave mirror a mirror in which the reflective surface curves inward **(473)**

Concentration gradient the difference between the concentrations of a substance in two areas **(S14)**

Condensation the process by which a gas or vapor changes to a liquid **(34, S70)**

Conductor a material through which electric current flows easily **(301)**

Conservation of energy the law stating that energy can be transferred or changed from one form to another but cannot be created or destroyed **(181)**

Continental climate a climate typical of large land-masses, characterized by relatively large daily and seasonal temperature changes **(256)**

Control an experimental setup or subject used as a standard of comparison with another setup or subject that is identical except for one variable **(216)**

Controlled experiment an experiment performed to evaluate another experiment that is identical except for one variable **(216)**

Converging lens a lens that is thicker at its center than at its edge. Light rays bend toward each other as they pass through a converging lens. **(469)**

Convex mirror a mirror in which the reflective surface curves outward **(467)**

Coriolis effect the effect, caused by the rotation of the Earth, whereby moving objects (including winds and ocean currents) veer to the right in the Northern Hemisphere and to the left in the Southern Hemisphere **(273, S68)**

Coulomb a unit of measure of electrical charge, equal to the charge of 6.24×10^{18} (6.24 billion billion) electrons **(343)**

Cross-pollination the transfer of pollen from a flower of one plant to the flower of another plant **(508)**

Current a flow of electric charge through a conductor. Current is measured in amperes. **(343, S88)**

Cytologist a person who studies the structure, function, behavior, growth, and reproduction of cells **(505)**

D

DNA (deoxyribonucleic acid) the chemical in chromosomes which carries the genetic information that instructs the cells of living organisms **(537)**

Daughter cells the two cells that result from the division of a single parent cell **(501)**

Decibel (dB) a unit of measure of the loudness of sound **(405)**

Density the mass per unit volume of a given substance **(125, 245)**

Density current currents caused by density differences in liquids or gases. The denser substance flows downward, beneath the less dense substance. **(248)**

Dew point the temperature at which water vapor in the air begins to condense **(231)**

Diffraction the bending of waves as they pass the edges of objects **(S113)**

Diffuse reflection scattered reflection; that is, light reflected in more or less random directions **(455)**

Diffusion the uniform intermingling of particles of one substance with particles of another substance because of the motion of both types of particles. As a result of diffusion, particles move from regions of greater concentration to regions of lesser concentration. **(36, S30)**

Digestion the process by which an organism breaks down food into substances that can be used by individual cells **(55)**

Digestive system the organs associated with the intake, digestion, and absorption of food **(65)**

Direct current an electric current that flows continuously in one direction only **(321)**

Doldrums a region of the ocean near the equator characterized by calm weather and very light winds **(275)**

Domains in magnetic materials, clusters of atoms in which the magnetic fields of most atoms are aligned in the same direction **(S94)**

Dominant describing a genetic factor, or trait, that is always expressed. The presence of a dominant trait prevents a recessive trait from being expressed in offspring. **(508, S156)**

Doppler effect a change in the apparent frequency of waves caused by the motion of either the observer or the source of the waves **(S119)**

Dry cell a type of chemical cell in which the electrolyte is a paste rather than a liquid **(310)**

E

Echolocation the determination of the positions of objects by emitting sound that is reflected back to the sender as echoes. Bats use echolocation to navigate. **(389)**

Efficiency the ratio of the work done by a machine to the work put into it **(176, S55)**

El Niño effect a variation in worldwide weather patterns that recurs every 3–5 years and that is caused by changes in the wind conditions over the eastern Pacific Ocean **(276)**

Electric current the flow of electrons from one place to another **(300, 303, S86)**

Electric field the region of space around an electrically charged object in which the effects of the electric force may be observed **(S85)**

Electric force the force that causes two like-charged objects to repel each other or two unlike-charged objects to attract each other **(S84)**

Electricity the energy of negative particles (electrons) flowing in a conductor **(293)**

Electrode a terminal that conducts electrons into or away from the electrolyte in a battery **(305)**

Electrolyte a substance that conducts electricity; usually dissolved in water or some other solvent **(305)**

Electromagnet a magnet produced by an electric current flowing through a coil of insulated wire wrapped around a core of steel or iron **(338)**

Electromagnetic spectrum the entire range of electromagnetic waves, of which visible light is a small part **(S137)**

Electromagnetic waves waves that carry both electric and magnetic energy and that move through a vacuum at the speed of light **(S130)**

Electron a negatively charged particle found in atoms **(130, 302)**

Element a substance that consists of only one kind of atom and that cannot be separated into other substances by ordinary chemical changes **(95)**

Embryo a developing organism in the early stages of growth; a human from the second to the eighth week of development **(516, S178)**

Endothermic describes processes in which heat is absorbed **(115)**

Energy the capacity to do work **(153)**

Energy converter a physical system in which energy is changed from one form into another **(166, 427)**

Evaporation the process by which a liquid becomes a gas **(34, 104)**

Evaporative cooling the removal of heat from the surroundings of a liquid that is undergoing evaporation **(116)**

Exothermic describes processes in which heat is given off **(115)**

Excretion the removal of metabolic wastes **(S17)**

Excretory system the organs associated with collecting and eliminating metabolic wastes **(65)**

Exponent a small number placed at the upper right of another number that tells how many times the lower number is to be multiplied by itself. For example, $10^3 = 10 \times 10 \times 10$; 3 is the exponent. **(80)**

F

Family tree a diagram for a family that shows parents and offspring for a number of generations **(492)**

Fertilization the union of a sperm cell with an egg cell, or ovum, resulting in the development of a new organism **(515)**

Fetus a developing organism that is in the later stages of development but is still in the womb or egg; a human from the ninth week of development until birth **(524, S178)**

Filter a process or device used for screening out something. An example of a filter is a colored piece of glass that absorbs certain colors of light while permitting other colors of light to pass through. **(431)**

Flywheel a heavy rotating wheel that stabilizes the speed of the device to which it is attached **(187)**

Focal length the distance from the focal point of a lens or mirror to the lens or mirror itself **(471)**

Focal point the point at which light beams passing through a lens or reflecting off a mirror converge **(471)**

Forced vibration the vibration produced in an object when it comes in contact with another object that is already vibrating **(403)**

Fraternal twins two individuals who are born at the same time from the same mother but who developed from different fertilized eggs. Such individuals have different genetic makeups. **(521)**

Frequency the number of repetitions in a given interval of time, such as the number of vibrations of an object per second or the number of complete waves passing a point per second **(371, S111)**

Front the boundary between two air masses **(253, S72)**

Fuse a safety device containing a strip of metal that melts when too much current passes through a circuit containing the device **(349)**

G

Galvanometer an instrument that measures small amounts of electric current **(306)**

Gear a toothed wheel that meshes with another toothed wheel in order to transmit force **(187)**

Generation a stage in a family's line of descent that includes all the individuals born within one life cycle **(492)**

Genes individual components of a chromosome which carry the factors that pass hereditary traits from parents to offspring **(514)**

Gene therapy the process of replacing defective genes in cells by injecting healthy genes into the cells **(546)**

Genetic code the language in which the instructions for proteins are written in DNA **(S169)**

Genetic engineering the manipulation of DNA by splicing and recombining it in order to produce new characteristics in species **(544)**

Geneticist a person who specializes in the study of genetics **(544)**

Genetics the science in which heredity is studied **(509, S161)**

Genotype the genetic makeup of an organism according to its genes **(540, S162)**

Global warming the apparent overall warming trend of the Earth's atmosphere **(218)**

Greenhouse effect the warming of the Earth's surface and lower atmosphere caused by heat rising from the surface of the Earth and being re-radiated back toward Earth by gases in the atmosphere **(220)**

Guard cells the kidney-shaped cells found on either side of each stoma on a plant leaf. Guard cells control the passage of gases into and out of the plant by causing the stomata to open and close. **(20)**

Gulf Stream a warm ocean current that flows from the tropical Atlantic Ocean northward along the eastern coast of North America **(270)**

H

Harmonic a component of a musical tone, the frequency of which is a multiple of the fundamental frequency **(413)**

Heredity the passing of traits from parents to offspring **(512, S154)**

Hertz (Hz) the basic unit of measure of frequency, equal to 1 cycle per second **(321, 396)**

Horse latitudes regions of the ocean at about 35°N and S that are known for their lack of winds **(275)**

Hurricane a severe tropical cyclonic storm with sustained winds in excess of 118 kilometers per hour **(277)**

Hybrid an offspring that contains one dominant gene and one recessive gene for a given trait **(509)**

Hydrometer an instrument that measures the density of liquids **(249)**

Hypertonic solution a solution with a higher concentration of a dissolved substance than another solution (52)

Hypotonic solution a solution with a lower concentration of a dissolved substance than another solution (52)

I

Ideal machine an imaginary machine in which there is no friction (S52)

Identical twins two individuals produced from the same fertilized egg. Such individuals have an identical genetic makeup. (521)

Image the visual impression of an object produced by reflection in a mirror or refraction by a lens (463)

Impermeable not allowing substances to pass through (41)

Incident beam a beam of light that falls on or strikes something, such as a beam of light falling on a mirror (452)

Inclined plane a simple machine consisting of a plane set at an angle to a horizontal surface, forming a ramp (174)

Index of refraction the ratio of the speed of light in a vacuum to its speed in another medium (S142)

Inference a hypothesis, drawn from observations, that attempts to explain or to make sense of the observations (81)

Infrasonic waves sound waves at frequencies below those which humans can hear (S121)

Insulator a material through which electric current passes with difficulty or not at all (301)

Insulin the chemical substance released into the blood by the pancreas that enables the body to use sugar as a fuel in the process of respiration (64)

Interference the result of two or more waves overlapping (S114)

In vitro fertilization fertilization that takes place outside a living organism. The fertilized egg is maintained in an artificial environment until it can be placed into the uterus of a female for normal embryonic development. (520)

Isobar a line on a weather map that connects points of equal pressure (279)

Isotonic solution a solution with the same concentration of a dissolved substance as another solution. The term often refers to a solution of salt and water with the same salt concentration as blood. (45)

J

Jet stream a belt of high-speed wind circling the Earth between the troposphere and the stratosphere (S64)

Joule (J) a unit of measure of work, equal to the amount of work done when a force of 1 newton is exerted over a distance of 1 meter (155)

K

Kinetic energy the energy of motion (162)

L

Larynx the upper part of the trachea that contains the vocal cords and produces vocal sounds; the "voice box" (368)

Lens a piece of glass or other transparent material that is curved on one or both sides and that is used to refract light (469)

Lever a simple machine consisting of a bar that pivots about a fixed point; used to transmit or increase force or motion (178)

Light electromagnetic radiation in the wavelength range including infrared, visible, ultraviolet, and X rays; often used to refer specifically to the range that is visible to humans (425)

Lipids organic molecules composed of fatty acids and glycerol that store energy and make up cell membranes in living things (S4)

Longitudinal wave a wave, such as a sound wave, in which the particles of the medium move back and forth, parallel to the direction of motion of the wave (S108)

M

Machine a device that helps to do work by multiplying force or distance (149)

Magnetic field a region of space around a magnet in which magnetic forces are noticeable (S92)

Magnetic force the attraction or repulsion between the poles of magnets; the force that electric currents exert on each other (317, S92)

Magnetic lines of force invisible curved paths, between the poles of a magnet, along which a magnet exerts its force (317)

Maritime climate a climate typical of the ocean, characterized by relatively small daily and seasonal temperature changes (256)

Mechanical advantage the factor by which a machine multiplies force (177, S53)

Mechanical energy the kinetic energy of moving objects (164)

Mechanical system a machine composed of more than one simple machine (150)

Medium a substance, such as air or water, through which a wave travels (380)

Megaphone a cone-shaped device used to direct or amplify the voice (403)

Meiosis a process, occurring within the nucleus during cell division, that leads to the formation of sex cells with half of the organism's usual number of chromosomes (515, S159)

Metabolism all of the chemical reactions of an organism (S16)

Mitosis the process by which a cell's nucleus divides, leading to the formation of two identical cells (504, S158)

Model a representation of a phenomenon that simulates the structure, function, or effect of the phenomenon and that allows predictions to be made about the phenomenon (83)

Mole the SI unit for the amount of a substance (S23)

Molecular biology the branch of biology that deals with the organization of living matter and genetic inheritance at the molecular level (544)

Molecule the smallest unit of a substance that has all of the physical and chemical properties of the substance and that is composed of two or more atoms (95)

Motor a device for converting electrical energy into mechanical energy (S100)

Mutation a change in the DNA of a gene (543)

N

Negative electrode the terminal that conducts electrons into the electrolyte in a chemical cell (313)

Nervous system the network of structures, including the brain and spinal cord, that control the actions and reactions of the body (65)

Neutron a particle with no charge that is found in the nucleus of an atom (135)

Newton (N) the SI unit of measure for force. One newton (1 N) is approximately equal to the weight (gravitational force) of a 100 g mass. (260)

Noise undesired sound, especially nonmusical sound that includes a random mix of frequencies (360)

Nucleic acids large organic molecules composed of nucleotides that store the information for building proteins (S6)

Nucleus the central core of an atom that contains protons and neutrons and that accounts for most of the mass of the atom (133); the central mass of protoplasm that is found in most plant and animal cells, controls the activities of the cell, and contains genes (498)

O

Observation direct evidence obtained through use of the senses; also, the act of obtaining direct evidence through use of the senses (81)

Octave the musical distance, or interval, between a musical tone and one of twice the frequency, such as from middle C to the C just above it (400)

Offspring the child or children of a parent or parents; in the case of one-celled organisms, the new cells are the offspring of the original cell (505)

Optical fiber a long, thin light pipe that uses total internal reflection to carry light (S150)

Organ part of an animal or plant that performs a specialized function (17)

Organic compound a compound that contains carbon and usually is produced by living things (S2)

Oscilloscope an electronic instrument that shows waves as curves on a screen (409)

Osmosis the movement of water across a semipermeable membrane from areas where water particles are more concentrated to areas where they are less concentrated (43)

Ovum a female sex cell produced as a result of meiosis; also known as an egg (515)

P

Pancreas the gland between the stomach and small intestine that enables the body to use sugar as a fuel in the process of respiration by releasing insulin into the blood (64)

Parallel circuit an electrical circuit arranged in such a way that the current passes through more than one pathway simultaneously (331)

Particle any small unit of matter that, with others, forms a larger whole (88, S22)

Pascal a unit of measure of pressure, equal to 1 newton per square meter (266)

Pedigree a chart that traces the inheritance of a genetic trait through several generations of a family (514, S164)

Permeable allowing substances to pass through (41)

Petiole the stalk that fastens a leaf to the stem of a plant (18)

Phenotype the external appearance of an organism as determined by its genotype (540, S162)

Photoelectric effect the emission of electrons by a substance when illuminated by light of a sufficient frequency (S131)

Photon a tiny package of electromagnetic energy (S131)

Photosynthesis the process by which green plants and plantlike organisms use energy from sunlight to convert water and carbon dioxide into sugars that can be used for food (14)

Piezoelectric effect the production of electric currents by certain crystals when they are squeezed or stretched (323)

Pigment a material that gives a substance its color by absorbing some colors of light and reflecting others (24)

Pitch the highness or lowness of a sound, determined by the frequency of the sound wave (361)

Placenta an organ that develops around an embryo and attaches the embryo to the wall of the mother's uterus. The placenta exchanges nutrients, wastes, and gases between the blood of the mother and the blood of the embryo. (517)

Plane mirror a flat mirror; reflects light to produce virtual images (463)

Plankton the small plants and animals floating near the surface of the ocean (S75)

Polar molecule a molecule that carries unevenly distributed electrical charges (S7)

Polarized light light consisting of light waves that vibrate in only one plane (S144)

Positive electrode the terminal that conducts electrons away from the electrolyte in a chemical cell (313)

Potential difference the difference in charge between two points in a circuit (S87)

Potential energy stored energy, such as the energy in a stretched or compressed spring or in an object raised above the surface of the Earth (159)

Power the rate at which work is done (157, S44)

Primary colors of light the three colors of light—red, green, and blue—that can be mixed in various combinations to form different colors or that can be mixed together equally to form white light (**438**)

Primary colors of paint the three colors of paint—red, blue, and yellow—from which all other paint colors can be created (**458**)

Pressure the amount of force exerted per unit of area (**261**)

Prism a piece of glass or other transparent material used to separate beams of light into a spectrum. A prism used in optics has three rectangular sides and triangular ends. (**431**)

Proteins large organic molecules composed of amino acids that either act as structural materials in an organism or regulate the chemical activities of an organism (**S5**)

Proton a positively charged particle found in the nucleus of an atom (**130, 302**)

Puberty the stage of physical development in which sexual reproduction first becomes possible (**S175**)

Pulley a simple machine consisting of a wheel attached to a shaft and a rope or cord that passes over the wheel (**171**)

Punnett square a diagram showing all possible genetic combinations that can be passed to the offspring of mating parents (**512**)

Q

Quark a theoretical particle that is thought to make up protons and neutrons (**S35**)

R

Radiometer an instrument that detects and measures the intensity of light or other radiant energy (**426**)

Real image an image that forms where light rays coming from an object converge and that can be projected onto a screen (**466, S135**)

Recessive describing a genetic factor, or trait, that is not always expressed. A recessive trait is prevented from showing up in offspring when a dominant trait is present. (**508, S156**)

Recombinant DNA a molecule made from the DNA of different organisms (**S172**)

Reflected beam a beam of light after it bounces off a reflecting surface, such as a mirror (**452**)

Reflection the process by which a wave bounces back after striking a barrier that does not absorb all of the wave's energy (**S112**)

Refraction the bending of a beam of light or other waveform as it passes from one medium into another (**479, S113**)

Regulation the coordination of the internal activities of an organism (**S17**)

Relative humidity the amount of water vapor in air compared with the maximum amount of water vapor the air could hold at that temperature (**S69**)

Resistance the opposition to the flow of current in an electric circuit (**329, S89**)

Resistor a device placed in an electric circuit to provide resistance (**330**)

Resonance a phenomenon in which a source that is vibrating at the natural frequency of a nearby object or medium causes the object or medium to vibrate with a relatively large amplitude at this natural frequency (**404**)

Respiration the process in which organisms combine oxygen from air or water with digested food in order to release chemical energy for use by their cells (**57**)

Respiratory system the system of organs and air passages that accomplish the exchange of oxygen and carbon dioxide between the body of an organism and its surroundings (**65**)

Root hair a hairlike extension of the epidermal, or outer, cells of a root that helps a plant obtain moisture from the soil (**35**)

Root pressure a phenomenon caused by the higher concentration of water around the root hairs than inside the root hairs. The osmosis of water into the root hairs forces water upward through the plant. (**47**)

S

Scattering the reflection of light in random directions by a surface or by the particles of a medium through which the light is passing (**443**)

Secondary colors of light the three colors of light—cyan, yellow, and magenta—produced by mixing pairs of the primary colors of light in equal quantities (**438**)

Semipermeable allowing only certain substances to pass through (**41**)

Series circuit a circuit arranged so that the components are joined end to end, with only one path through which the current can flow (**331**)

Sex cells egg or sperm cells, formed through meiosis. Sex cells contain half of an organism's usual number of chromosomes. (**515**)

Sexual reproduction reproduction that involves the union of an egg cell (ovum) and a sperm cell, usually from two separate parents (**505, S173**)

Simple machine one of six elementary devices (lever, pulley, wheel and axle, wedge, screw, and inclined plane) that change the size or direction of a force. Simple machines can be combined to form complex machines. (**178, S47**)

Solidification the process by which a gaseous or liquid substance becomes solid (**104**)

Sonar short for **s**ound **n**avigation **a**nd **r**anging, a nautical navigation system that uses echoes to determine the depth of the water, to locate objects beneath the water, etc. (**390**)

Sonic boom a loud noise produced by the shock wave of an object traveling faster than the speed of sound (**384**)

Spectrum (plural, spectra) a continuous range or sequence, such as the pattern of colored bands produced by passing white light through a prism (**431**)

Specular reflection reflection of light off a smooth surface such as a mirror (**454**)

Sperm a male sex cell produced as a result of meiosis **(515)**

Spontaneous generation the now discredited theory that living organisms can come to life spontaneously from a nonliving source within a short time **(496)**

Starch a white, tasteless, odorless food substance produced and stored by plants **(7)**

Standing waves waves that form a pattern in which portions of the wave do not move and other portions move with increased amplitude **(S117)**

Static electricity potential energy in the form of a stationary electric charge **(303, S84)**

Stationary front a nearly motionless boundary between two air masses **(254)**

Stimulated emission a process of forcing identical photons into step in order to produce coherent light **(S147)**

Stomata (singular, stoma) slitlike openings located mainly on the underside of a leaf that allow water vapor and other gases to pass into and out of the leaf **(20)**

Sublimation the process by which a solid becomes a gas or a gas becomes a solid without passing through the liquid state **(124)**

Subsystem a system that is a part of a larger system **(150)**

Switch a device for opening and closing a circuit in order to start or stop the flow of electric current **(336)**

Syrinx the sound-producing organ of birds **(368)**

T

Total internal reflection the complete reflection of a beam of light by the inside surface of a medium so that none of the light leaves the medium **(478)**

Trade winds steady winds that blow toward the equator from about 35°N and S **(252)**

Trait a genetically determined characteristic **(492)**

Transgenic describing an organism into which new genetic material has been transferred **(553)**

Translucent describing materials that allow some light to pass through but scatter some of it, so objects are not clearly visible when viewed through the material **(448)**

Transmission the passage of energy through space or a medium **(443)**

Transparent describing materials that allow light to pass through without scattering, so objects are clearly visible when viewed through the material **(447)**

Transpiration the movement of water out of a plant by evaporation from the leaves **(32)**

Transverse wave a wave in which the motion of the particles of the medium is perpendicular to the path of the wave **(S107)**

Trimester one of the three 3-month periods into which a human pregnancy is divided **(525)**

Tuning fork a metal device consisting of two prongs that vibrate at a single frequency when struck and that can be used to tune musical instruments **(362)**

U

Ultrasonic waves sound waves at frequencies too high for humans to hear **(S122)**

V

Variable resistor a device in which the amount of resistance can be changed to control the flow of current in an electrical circuit **(330)**

Veins in plants, the channels through which materials move into and out of a leaf and that together help to strengthen the leaf **(18)**

Vibration one complete back-and-forth movement of an object, from one side to the other and back again **(370)**

Virtual image an image that is not formed by the convergence of light rays and that cannot be projected onto a screen **(466, S135)**

Visible spectrum all of the colors of visible light **(S133)**

Volt (V) the basic unit of measure of voltage **(344)**

Voltage a measure of the force that pushes current through a conductor, expressed in volts **(312, 344)**

W

Warm front the leading edge of an advancing warm air mass **(253)**

Water cycle the continuous circulation of water between the atmosphere and the Earth **(34)**

Watt a unit of measure of power, equal to 1 joule per second **(345)**

Wave a disturbance that transmits energy as it travels through matter or space **(S106)**

Wavelength the distance between two identical points on neighboring waves **(379, S111)**

Weather the condition of the atmosphere at a given time and place **(256, S66)**

Westerlies steady winds that blow in a general west-to-east direction over the middle latitudes of both the Northern and Southern Hemispheres **(269)**

Wet-bulb depression the difference, in degrees, between the wet- and dry-bulb air temperatures **(231)**

Wet-bulb thermometer a type of thermometer used to measure atmospheric humidity **(117, 229)**

Wet cell a type of chemical cell that uses a liquid electrolyte **(311)**

Wheel and axle a simple machine consisting of a wheel attached to a smaller wheel or shaft to increase force or speed **(182)**

Work the result of using force to move an object over a distance **(153, S42)**

Work input the work put into or done on a machine **(173)**

Work output the work done by a machine **(173)**

INDEX

Boldface numbers refer to an illustration on that page.

A

Absorption spectrum, 26, **26**

Abyssal plains, 240

Acids
amino, S5
nucleic, S6

Adhesion, 50, S7, S28

Aeronautical engineering
noise control and, 419
Richard Linn and, 419, **419**

AIDS (acquired immune deficiency syndrome), 142

Air
circulation, S68
defined, S62

Air mass
defined, 252, S72
motion of, 253, S72
types of, 252, **252**, S72, **S72**

Alternating current, 320

Alternator, S97, **S97**

Amino acids, S5

Amniocentesis, 545, **545**

Ampere
defined, 343
human body and, 347
understanding of, 345

Amphibians
effect of pollution on, 72
ozone and, 72
shrinking population of, 72

Amplitude (of waves), 371, S109, **S109**

Anaphase, S158, **S158**

Animals
carnivorous, 9
life cycle of, S175

Anvil (of ear), 393

Archimedes, S49

Art
machines as, 209, **209**

Artificial magnets, S94

Asexual reproduction, 505, S173, **S173**

Atmosphere, S60–S65
air in, S62
defined, 251, S60, **S60**
density currents in, 251–252
evolution of, S60–S61, **S60–S61**
heating of, S66, **S66**
layers of, **260**, S63, **S63**
oxygen levels in, S61–S62, **S61**
pressure unit, 268, S32
temperature changes in, S63

Atmospheric pressure, 261, S63

Atoms, S2, **S2**
Albert Einstein and, 140

defined, 92
John Dalton and, 127–128
models of, 127–135

ATP (adenosine triphosphate), S11

Auditory nerve, 394

Avogadro's number, S24

Axle
wheel and, 182, S50, **S50**

B

Barometer, mercury, 266

Battery
defined, 310
examples of, **313**

Beam
incident, 452–454
object, S148
reference, S148
reflected, 452–454

Belts
gears and, 190
wheels and, 190

Bicycles, 208

Biochemicals, S3
types of, S3–S6

Biomass, 67

Blade (of leaf), 18, **18**

Body systems
circulatory, 59, **59,** 65
digestive, 65
excretory, 65
nervous, 65
respiratory, 65

Bohr, Niels, 133, **133**

Bohr model, 133–135, **133–135**

Boyle, Robert, 267, S32

Boyle's law, S32–S33

Breathing
rate, 59–60
respiration and, 58
structures, 58–59, **59**

C

Caesarean section, 528

Capillary action, 48, **48**

Carbohydrates, S3, **S3**

Carbon dioxide
as component of air, 13
limewater test for, 13, **13**
sources of, 224
starch and, 14
use of by plants, 13–14

Carnivorous animals, 9

Cars
electric, 353, **353**

Cell division, 500, **501–502, 504**

Cells
blood, **59**
chemical, 305, 310–314
division, 500–504, **501–502, 504,** S157–S160, **S157–S160**
dry, 310–313
examples of, 7, **22, 498,** 515, **515, S11, S157, S159**
membranes of, 41
of leaves, 20–21, **20–21**
parts of, 498, **498,** 500, **500**
solar, 322, **322**
wet, 311

Cellular respiration, 57, S11

Charged particles, 301–303
discovery of, 302
J. J. Thomson and, 302
negatively charged, 301–303
positively charged, 301–303
static electricity and, 303
uncharged, 303

Chemical cell
defined, 305
technology, 311–314

Chemical energy, 160
is a form of potential energy, 160

Chemical role of water, S8

Chlorophyll, 23, **24**
defined, S9
function of, 23
photosynthesis and, 24, S9–S11

Chloroplasts, S9, **S9**

Chromosomes, S157, **S157**
defined, 514
human, 514, **514**
mutations in, S171
sexual reproduction and, 515, S159–S160, **S159–S160**

Chromosome theory
Walter S. Sutton and, S161

Chu, Paul, **352**
superconductors and, 352

Circuits (electric), 296
components, **327**
controlling, 335–337, S102
defined, S101
examples of, **336–337, S86–S87**
symbols, **328**
types of, 331, S102, **S102**

Circulatory system, 59, **59**

Circumstantial evidence, 81–82

Cirrus clouds, S71, **S71**

Climate, 256
defined, S66

types of, 256

Clones, 547, **547**

Clouds
formation of, 34, **34,** S71
types of, S71, **S71**

Cochlea (of ear), **392,** 394

Coherent light, S146

Cohesion, 49, 50, S7, S28

Cold front, 253, **253**, S73, **S73**

Colors, S133–S134
complementary (light), 438
light and, 436–438
primary (light), 438, **438,** S134, **S134**
primary (paint), 458–459
secondary (light), 438, **438,** S134, **S134**
visible spectrum and, S133

Complementary colors of light, 438

Compounds, 92
defined, 95
organic, S2
inorganic, S2

Compression (of a wave), S108, **S108, S115**

Concave lenses, S135, **S135**

Concave mirrors, 473–476, S136, **S136**

Condensation, 34, S70

Condensation reaction, S8

Conductors, electrical, 301

Coniferous trees, 25

Constructive interference, S114, **S114**

Continental climate, 256

Convex lenses, S135, **S135**

Convex mirrors, 467–468, S136, **S136**

Coriolis effect, 272
defined, 273, S68

Coriolis, Gustave-Gaspard de, **271**
wind and, 271

Coulomb
defined, 343
understanding of, 345

Cross-pollination, 508, S155

Cumulonimbus clouds, S71, **S71**

Cumulus clouds, S71, **S71**

Current
alternating, 320
deep-ocean, 248, **248**
defined, 343, S88
direct, 321
effects of voltage on, S99
electric, 300, 303, S86
electricity and, S97

Cytokinesis, S159, **S159**

PHOTO CREDITS

Abbreviated as follows: (t) top; (c) center; (b) bottom; (l) left; (r) right; (bckgd) background, (bdr) border.

All HRW photos by Sam Dudgeon except where noted.

FRONT COVER: (t), Richard Kaylin/Tony Stone Images; (br), Carl Roessler/Animals Animals

BACK COVER: (bc), Jeff Smith/FotoSmith/Reptile Solutions of Tucson

TITLE PAGE: (bl), Jeff Smith/FotoSmith/Reptile Solutions of Tucson

TABLE OF CONTENTS: Page iv(tr), John Langford/HRW; (cr), Renee Lynn/Photo Researchers; (bl), John Langford/HRW; v(tc), John Langford/HRW; (bl), (br), Michelle Bridwell/HRW; vi(tl), John Langford/HRW; (tr), HRW; vii(tc), Stephen Dalton/Photo Researchers; (tr), John Langford/HRW; (cr), Scott Van Osdol/HRW; (bl), John Langford/HRW; viii(t), John Langford/HRW; (tr), Jennifer Dix/HRW; (br), John Langford/HRW; ix(tl), Peter Van Steen/HRW; (tc), Comstock; (cr), Peter Van Steen/HRW; (bl), John Langford/HRW; (br), Howard Sochurek/Stock Market; x(t), Lennart Nilsson/From *The Incredible Machine*, National Geographic Society; (bc), Runk/Schoenberger/Grant Heilman Photography; (br), C & M Denis-Huot/Peter Arnold; xi(t), Bill Ross/Westlight; (cr), Bill Beatty/Visuals Unlimited; (br), Philippe Plailly/Science Photo Library/Photo Researchers; xii(cl), Patrick R. Dunn/HRW; (br), HRW; xiii(tl), Carlos Austin; (c), Michelle Bridwell/HRW; xiv(l), (br), HRW; xv(tl), Michelle Bridwell/HRW; (cr), (bl), HRW; xvi(cr), Jana Birchum/HRW; (bl), Michelle Bridwell/HRW; xvii(bc), Daniel Schaefer/HRW.

UNIT 1: Page 2-3(bkgd), Gary Braasch/Tony Stone Images; 2(tl), Grant Heilman/Grant Heilman; 4(c), Renee Lynn/Photo Researchers; (bl), (br), HRW; (bkgd), Index Stock Photography; 5(tl), John Langford/HRW; (bl), (bc), (b), Michelle Bridwell/HRW; 6(tl), Lance Nelson/Stock Market; 6(tc), (tr), (br), John Langford/HRW; 7(l), Manfred Kage/Peter Arnold; 9(tl), (r), (bl), John Langford/HRW; 11(bl), Runk/Schoenberger/ Grant Heilman; (bc), Grant Heilman/Grant Heilman; 12(tl), (tr), (cr), HRW; (bc), Cindy Bland Verheyden/HRW; 18(bl), (bc), Runk/Schoenberger/Grant Heilman; 19, HRW (sycamore,lily); Runk/Schoenberger/Grant Heilman(elm, grass,maple,oak); Jerome Wexler/Photo Researchers(carrot); Fritz Pölking/Peter Arnold(beech); Ed Reschke/Peter Arnold(horse chestnut); 20(bl), Spike Walker/Tony Stone Images; 21(c), (bl), John Langford/HRW; (bkgd), Diamar Portfolios Landscapes & Scenery CD ROM by Diamar Interactive Corp.; 23(b), HRW; 25, HRW(sycamore,pine,orange), (tr); Runk/ Schoenberger/ Grant Heilman(maple); (bl), Cindy Bland Verheyden/HRW; 26(t), Michelle Bridwell/HRW; 27(bl), Jim Frank/Photonica; 28(l), John Langford/HRW; (tr), David Nunuk/Science Photo Library/Photo Researchers; (br), Andrew Syred/Science Photo Library/Photo Researchers; 29(bkgd), Kevin Kelley/Tony Stone Images; 30(tl), Manfred Kage/Peter Arnold; (tc), Charles Philip/Westlight; (cl), Alan Carey/Photo Researchers; (c), Stephen Dalton/Photo Researchers; (tcr), Renee Lynn/Photo Researchers; (bcr), Index Stock; (bl), Rod Planck/Photo Researchers; (br), R. W. Jones/Westlight; (bkgd), Ralph Clevenger/Westlight; 31(bkgd), John Langford/HRW; 32(tr), Michael Fogden/DRK Photo; 36(tr), (cl), (bl), (bc), (br), John Langford/HRW; (b), Superstock; 37(cr), Michelle Bridwell/HRW; (bl), (bc), (br), John Langford/HRW; 45(tr), John Langford/HRW; (c), Robert Wolf; (cr), Phil A. Harrington/Peter Arnold; (br), Michael Mitchell; 46(t), Queen's Printer for Ontario, 1993, Reproduced with permission/Ontario Ministry of Agriculture and Food; (b), David Phillips/HRW; 47(r), Tom & Pat Leeson; 50(b), Michelle Bridwell/HRW; 51(tr), John Langford/HRW; 53(t), HRW; (bl), John Langford/HRW; 54(tr), Manoj Shah/Tony Stone Images; (br), NASA/Science Source/Photo Researchers; 57(tc), Nawrocki Stock Photo; (cl), Craig Lovell/Viesti Associates; (cr), Lori Adamski Peek/Tony Stone Images; (bc), HRW; (br), Paula Gaetani/Stock Shop; 58(bl), Stuart Westmorland/ Tony Stone Images; (bc), Rod Planck/Tony Stone Images; (br), Leonard Lee Rue III/Tony Stone Images; 59(br), CNRI/Science Photo Library/Photo Researchers; 60(c), (bl), (bc), (br), John Langford/HRW; 61(r), (cl), (br), John Langford/HRW; 62(tr), John Langford/HRW; 64(t), John Langford/HRW; 65(br), John Langford/HRW; 66(tr), NASA/Science Source/ Photo Researchers; 67(bl), Ron Sherman/Tony Stone Images; (br), Peter May/Peter Arnold; 69(tl), (tc), (tr), (cr), HRW; 70(cl), Ken Eward/Science Source/Photo Researchers; (tr), Gary Braasch/Tony Stone Images; 72(cr), John Swedberg/Bruce Coleman; (bl), Zefa Germany/Stock Market; (tr), Michael Fogden/Bruce Coleman; 73(tl-ICON), Rogge/Stock Market; (tr), NASA; 74 (tl-ICON), Tony Stone Images; (tr), Charles C. Place/Image Bank; (bl), Dr. Eric Pianka; 75(tl-ICON), Howard Sochurek/Stock Market; (tc), Ken Karp/HRW; (tr), (cl), HRW; (bl), Superstock; (br), Stock Editions/HRW.

UNIT 2: Page 76-77(bkgd), Lawrence Berkeley Laboratory/Science Source/Photo Researchers; 76(tl), John Raffo/Science Source/Photo Researchers; 78(tl), Kenneth Love/ National Geographic Society; (bl), Ken Lax/Stock Shop; (bkgd), Patrice Loiez, Cern/ Science Photo Library/Science Source/Photo Researchers; 79(cl), Michelle Bridwell/ HRW; 81(tr), Christopher Springmann/Stock Market; 82(tl), David Phillips/HRW; (br), John Langford/HRW; 83(tr), David Phillips/HRW; (bl), John Langford/HRW; 84(cr), John Langford/HRW; 85(tc), John Langford/HRW; (b), S. J. Krasemann/Peter Arnold; 86(tl), Neal Graham/Stock Shop; (tr), Steve Vidler/Nawrocki Stock Photo; (cl), B. Rubel/FPG Int'l.; (bc), Steven Ferry; (bkgd), HRW; 87(tr), Michelle Bridwell/HRW; 88(br), HRW; 90(c), HRW; 92(tr), (cr), Culver Pictures; (bc), Vaughn Fleming/Science Photo Library/ Photo Researchers; (br), Frank Lawrence Stevens/Nawrocki Stock Photo; (bl), Tom McHugh/Photo Researchers; 93(cl), Culver Pictures; 94(tl), F. & A. Michler/Peter Arnold; (tr), Bob Pizaro/Comstock; 95(bl), (bc), David Phillips/HRW; 96(tr), (tc), David Phillips/ HRW; 100(cr), Rossane Olson/Tony Stone Images; 101(tr), Garry Hunter/Tony Stone Images; (other), Specular Replicas CD ROM by Specular Int'l. (helicopters); Jeff Greenberg/Peter Arnold (city); 102(bc), David Phillips/HRW; (bkgd), Tom Bean/Stock Market; 103(bl), Michelle Bridwell/HRW; 107(cr), (bl), John Langford/HRW; 114(br), Simon Fraser/Science Photo Library/Photo Researchers; 116(bc), Tom Lyle/Stock Shop; (bkgd), David Phillips/HRW; 118(tr), John Langford/HRW; (other), Superstock (cloud); Clyde H. Smith/Peter Arnold(dark sky); Peter Arnold (lake); Robert MacKinlay/Peter Arnold(hills); 122(tc), (tr), (br), John Langford/HRW; 123(cl), (c), (tr), John Langford/ HRW; (cr), Michelle Bridwell/HRW; 124(tr), Specular Replicas CD ROM by Specular Int'l.; 125(bc), Specular Replicas CD ROM by Specular Int'l.; 126(bc), John Langford/HRW; 129(tr), Bettmann Archive; (b), Michelle Bridwell/HRW; 130(bl), Science Photo Library/ Photo Researchers; 133(bl), Bettmann Archive; (br), Michelle Bridwell/HRW; 135(tr), A. Barrington Brown/Science Photo Library/Photo Researchers; 137(cr), Michelle Bridwell/ HRW; 138(cl), Michelle Bridwell/HRW; (tr), Lawrence Berkeley Laboratory/Science Source/Photo Researchers; 139(cl), Michelle Bridwell/HRW; 140(tl), UPI/Bettmann Archive; (bc), Bettmann Archive; 141(tl-ICON), Tony Stone Images; (c), D. Dzurisin/U.S. Geological Survey/Westlight; (bl), Paul Chauncey/Stock Market; 142(tl-ICON), Howard Sochurek/Stock Market; (tc), Richard J. Green/Photo Researchers; 143(tr), (cr), Georges Seurat, French, 1859-1891, *A Sunday on La Grande Jatte* 1884, oil on canvas, 1884-86, 207.5X308 cm, Helen Birch Bartlett Memorial Collection, 1926.224/©1994, The Art Institute of Chicago, All Rights Reserved.

UNIT 3: Page 144-145(bkgd), Vicki Miller/Featherstone Kite courtesy of Mid-America Museum, Hot Springs, Ark.; 144(tl), Steven Gottlieb/FPG Int'l.; (l), Stuart Cohen/ Comstock; 146(tr), (bl), John Langford/HRW; 147(r), Superstock; (cr), Don Morley/Tony Stone Images; (bl), Peter Le Grand/Tony Stone Images; (cl), David Phillips/HRW; 148(tc), HRW; (c), Randy Duchaine/Stock Market; (cr), (b), David Phillips/HRW; (bkgd), Superstock; 149(tc), Tony Stone Images; (br), John Langford/HRW; (cr), Peter Grumann/ Image Bank; 150(tc), (tl), David Phillips/HRW; (cr), Steven Ferry; (bl), NASA; 151(tr), HRW; (bl), Steven Ferry; (br), John Langford/HRW; 152(tr), Patti McConville/Image Bank; (cr), Guy Gillette/Photo Researchers; (bl), Jose L. Pelaez/Stock Market; (br), Superstock; 153(t), Comstock; (br), Michelle Bridwell/HRW; 154(tr),(cr), (br), (bl), (bc), John Langford/HRW; 155(tl), (cl), (bl), (br), (bc), John Langford/HRW; 156(b), David Phillips/HRW; 157(r), David Phillips/HRW; (bl), John Langford/HRW; 159(tr), (bl), David Phillips/HRW; 160(t), (cr), David Phillips/HRW; (bl), Eugen Gebjardt/FPG Int'l.; (br), Superstock; (bkgd), The Harold E. Edgerton 1992 Trust/courtesy of Palm Press, Inc.; 161(tl), David Madison; (cl), Michelle Bridwell/HRW; (tr), (cr), John Langford/HRW; (bl), David Phillips/HRW; (br), Superstock; 163(t), (c), (b), John Langford/HRW; 164(cr), John Langford/HRW; (b), David Phillips/HRW; 165(b), The Harold E. Edgerton 1992 Trust/ courtesy of Palm Press, Inc.; 166(r), Barbara Van Cleve/Tony Stone Images; (br), Scott Van Osdol/HRW; (bl), David Phillips/HRW; (cr), Steven Ferry; (bkgd), Tony Stone Images; 167 (t), David Phillips/HRW; (br), HRW; 169(tl), David Phillips/HRW; (tr), (cr), Ron Kimball; (bl), Cindy Bland Verheyden/HRW; 172(l), Michelle Bridwell/HRW; 173(br), Michelle Bridwell/HRW; 175(br), John Langford/HRW; 178(tc), Michelle Bridwell/HRW; (tr), Mike & Carol Werner/Comstock; (cr), David Phillips/HRW; (b), J. Barry O'Rourke/ Stock Market; 179(tl), David Phillips/HRW; (tr), Steven Ferry; (cr), (br), Michelle Bridwell/HRW; (b), Superstock; 181(br), HRW; (tr), (cr), Culver Pictures; 185(br), David Phillips/HRW; 186(tr), David Phillips/HRW; (br), Norco Products, Ltd.; 187(bl), David Phillips/HRW; (bkgd), Russ Kinne/Comstock; 190(br), Michael Melford/Image Bank; 193(tr), (br), HRW; 194(tr), (bc), HRW; (cr), Michelle Bridwell/HRW; 195(t), (b), HRW; (c), Specular Replicas CD ROM by Specular Int'l.; 197(cl), Michelle Bridwell/HRW; 198(bl), Brian Smith/Black Star; (cl), Steve Chenn/Westlight; (cr), Comstock; (br), Superstock; 199(br), David Phillips/HRW; 201(tr), (cr), Michelle Bridwell/HRW; 204(cl), John Langford/HRW; (tr), Vicki Miller/Featherstone Kite courtesy of Mid-America Museum, Hot Springs, Ark.; 206(tl-ICON), Rogge/Stock Market; (bl), Peter Menzel; (br), IBM Corporation, Research Division Almaden Research Center; 207(tl-ICON), Tony Stone Images; (tr), A.W. Stegmeyer/Upstream; (bl), Kathy Raddatz/*People Weekly* 1986; 208(tl-ICON), Ron Kimball; (tr), Loren Callahan; (cl), Michelle Bridwell/HRW; (bl), Barry Gregg/Advanced Transportation Products; 209(tr), Museum of Modern Art, New York "Tizio Lamp" by artist Richard Sapper, 1972, black metal; manufactured by Artemide; (bl), R. Beaudry/Waring Products Division.

UNIT 4: Page 210-211(bkgd), NASA; 210(l), Superstock; (tl), Warren Faidley/International Stock; 211(t), Superstock; 212(tl), Photri/Stock Market; (tc), Superstock; (t), US Geological Survey/Science Photo Library/Photo Researchers; (bl), Rod Planck/Photo Researchers; (bc), Superstock; 213(c), Keith Gunnar/FPG Int'l.; (bl), Zefa-Hummel/Stock Market; (br), Dallas & John Heaton/Westlight; (b), Superstock; 214(tc), NASA/Science Source/Photo Researchers; (bc), Photri/Stock Market; 215(tc), Jack Zehrt/FPG Int'l.; (bc), US Geological Survey/Science Library/Photo Researchers; 217(br), David Phillips/HRW; 218(b), Mountain High Maps CD ROM by Digital Wisdom and Quarto Publishing; 219(bl), HRW; (bkgd), Superstock; (bkgd), City Builder CD ROM by Dedicated Digital Imagery; 220(tc), Specular Replicas CD ROM by Specular Int'l.; (br), Science Photo Library/Photo Researchers; 221(cl), Ron Watts/Westlight; (b), E. Nagele/FPG Int'l.; 224(t), Mide Dobel/ MasterFile; (cr), E.R. Degginger/Photo Researchers; (bc), Stephen J. Krasemann/ Photo

Researchers; (b), Clyde Smith/FPG Int'l.; 225(tr), Bill Brooks/ MasterFile; 226(tr), John Langford/HRW; (b), NASA's Goddard Space Flight Center; 231(bkgd), Wesley Bocxe-Upon/Photo Researchers; 232(bl), John Langford/HRW; 233(b), Superstock; (cb), Marcel Isy-Schwart/Image Bank; (ct), Marc Romanelli/Image Bank; (t), Mike Schneps/Image Bank; 237(tr), Superstock; (tc), John Langford/HRW; (cl), Superstock; (tcr), Steven Ferry/ HRW; (bcr), Gary Bell/MasterFile; (bc), Planet Art Antique Maps CD ROM; (bl), David Phillips/HRW; 238-239(other), Bruce C. Heezen and Marie Tharp; 241(tr), Global Relief Data CD ROM from the National Geophysical Data Center, NOAA, U.S. Department of Commerce; (bl), Dudley Foster/Woods Hole Oceanographic Institution; (br), Al Giddings/ Images Unlimited; 243(tr), Superstock; (b), E. R. Degginger; 248(tr), Souricat/Animals Animals; (b), Mountain High Maps CD ROM by Digital Wisdom and Quarto Publishing; 249(bc), (br), Scott Van Osdol/HRW; 251(br), NASA; 252(tr), Mountain High Maps CD ROM by Digital Wisdom and Quarto Publishing; (br), John Langford/ HRW; 254(cr), (bl), Mountain High Maps CD ROM by Digital Wisdom and Quarto Publishing; 256(tr), Weather Graphics courtesy of Accu-Weather; 257(tr), Janet Foster/ MasterFile; (bl), T. Tracy/FPG Int'l.; 258(tr), Larry Williams/MasterFile; (bl), John Langford/HRW; 259(tl), Mark M. Lawrence/Stock Market; (tr), John Langford/HRW; (bl), R. Ian Lloyd/Westlight; 260(br), Specular Replicas CD ROM by Specular Int'l.; 261(bc), Stephen Dalton/Photo Researchers; (br), John Langford/HRW; 262(tr), (br), John Langford/HRW; 265(cr), Bettmann Archive; (bc), (br), Scott Van Osdol/HRW; 266(tr), Science Photo Library/Photo Researchers; (cr), Granger Collection, New York, N.Y.; 268(other), Bettmann Archive (Trieste); M. C. Chamberlain/DRK Photo (diver); Photoworld/FPG Int'l. (Titanic); Norbert Wu/DRK Photo (squid); Doug Perrine/DRK Photo (mini-sub); Bassot/Photo Researchers (viper fish); 269(tr), ARCHIV/ Photo Researchers, Anonymous painting, 16th Century, oil on canvas, Museo de America, Madrid, Spain; (b), Mountain High Maps CD ROM by Digital Wisdom and Quarto Publishing; 270(tr), Mountain High Maps CD ROM by Digital Wisdom and Quarto Publishing; (b), NOAA; 271(t), Tom Van Sant/Photo Researchers; (br), Science Photo Library/Photo Researchers; 272(bl), Scott Van Osdol/HRW; 273(br), Runk/Schoenberger/ Grant Heilman; 277(tr), Telegraph Colour Library/FPG Int'l.; 280(bl), Mountain High Maps CD ROM by Digital Wisdom and Quarto Publishing; 281(tr), David Phillips/HRW; (bl), Mountain High Maps Globe Shots CD ROM by Digital Wisdom; 282(cl), David Phillips/HRW; (tr), NASA; 284(tl-ICON), Ron Kimball; (t), Chuck O'Rear/Westlight; 285(tl-ICON), Rogge/Stock Market;(tr), Michael Sexton Photography; (c), Caroline Parsons/Aria Pictures; 286(tl-ICON), HRW; (c), Warren Bolster/Tony Stone Images; (cr), Bob Paz/Caltech; 287(tl-ICON), Tony Stone Images; (tr), NASA/Stock Market; (tc), Michael Lyon.

UNIT 5: Page 288-289(bkgd), Telegraph Colour Library/FPG Int'l.; 290(cr), Dave Reede/ First Light/Westlight; (bl), John Langford/HRW; (bkgd), Zefa-London/Stock Market; (bkgd), Barry Seidman/Stock Market; 291(tr), (bc), John Langford/HRW; (cr), George Haling/Photo Researchers; 294(bl), Michelle Bridwell/HRW; (br), David Phillips/HRW; 295(tr), Scott Van Osdol/HRW; (br), John Langford/HRW/Racing car courtesy Tradd Racing; 296(b), John Langford/HRW; 297(tr), Scott Van Osdol/HRW; 298(b), Scott Van Osdol/HRW; 299(b), Scott Van Osdol/HRW; 300(bl), John Langford/HRW; 302(br), Superstock; 306(tl), (tc), (tr), (bl), (bc), (br), John Langford/HRW; 307(cl), (cr), Scott Van Osdol/HRW; 308(bl), Michelle Bridwell/HRW; 309(cr), Scott Van Osdol/HRW; (bl), Michelle Bridwell/HRW; 312(b), HRW; 317(tr), E.R. Degginger/Animals Animals/Earth Scenes; 318(tr), Superstock; 321(b), HRW; 322(bl), Telegraph Colour Library/FPG Int'l.; (br), NASA Photo/Research by Grant Heilman; 326(tr), (cl), (bl), John Langford/HRW; 329(bl), John Langford/HRW; 330(c), (bl), Scott Van Osdol/HRW; 331(cl), John Langford/ HRW; 338(br), David Frazier/Photo Researchers; 341(tr), HRW; (cl), Sam Dudgeon/HRW Photo; (cr), Scott Van Osdol/ HRW; 345(t), (tr), HRW; (br), Reddy Kilowatt® appears with permission of The Reddy Corporation Int'l., Albuquerque, New Mexico; 346(tl), HRW; (bl), John Langford/ HRW; 350(tr), Telegraph Colour Library/FPG Int'l.; (cl), John Langford/HRW; 351(br), Superstock; 352(tl-ICON), University of Houston/Texas Center Superconductivity; (tc), Larry Hamill; (b), dpa/Photoreporters; 353(tl-ICON), Tony Stone Images; (c), Bob Abraham/ Stock Market; 354(tl-ICON), Howard Sochurek/Stock Market; (tc), (cr), Howard Sochurek; 355(tl-ICON), HRW; (bl), Duane Dick Photography/Westlight; (tr), Bettmann Archive.

UNIT 6: Page 356-357(bkgd), Erich Schrempp/Photo Researchers; 358(tr), HRW; (bl), John Langford/HRW; 359(tr), John Langford/HRW; (cr), HRW; (br), Tom Copi/Michael Ochs Archives; 360(tr), John Langford/HRW; (b), Dick Luria/Science Source/Photo Researchers; 362(cl), (cr), Scott Van Osdol/HRW; 363(l), (cl), HRW; (tr), Scott Van Osdol/HRW; (br), John Langford/HRW; 365(br), John Langford/HRW; 367(br), (tr), John Langford/HRW; 368(l), F. Stuart Westmorland/Photo Researchers; (tr), Christoph Burki/Tony Stone Images; (br), Ron Austing/Photo Researchers; 369(tr), Robert and Linda Mitchell; 370(t), Jennifer Dix/HRW; 371(tr), John Langford/HRW; 373(c), Scott Van Osdol/HRW; 374(tr), John Langford/HRW; 375(tr), NASA; (bl), John Langford/HRW; 376(tc), Scott Van Osdol/HRW; (c), (br), HRW; 378(t), John Langford/HRW; 379(r), John Langford/HRW; (c), Scott Van Osdol/HRW; 381(tl), HRW; (tc), John Langford/HRW; 382(tr), John Langford/HRW; (bc), Scott Van Osdol/HRW; 383(other), Dennis O'Clair/ Tony Stone Images (Motorcyclist); (t), Michelle Bridwell/HRW; (other), Denver Bryan/ Comstock (Cheetah); Kenneth R. Morgan/Animals Animals (Falcon); Lori Adamski Peek/Tony Stone Images (Runner); 383-384(other), Bob Thomas/Tony Stone Images (Crowd); 384(other), Tony Stone Images (Race Car); 386(br), John Langford/HRW; (other), Tony Stone Images (SST); 387(br), John Langford/HRW; (other), Globus Brothers/Stock Market (Bullet); 390(tr), Stephen Dalton/Animals Animals; (bc), Steve Bloom - TCL/MasterFile; 391(tl), John Langford/HRW; (tr), Scott Van Osdol/HRW; (br), HRW; 395(tr), Peter Van Steen/HRW; (bc), John Langford/HRW; 396(cr), Julian Baum/ Science Photo Library/Photo Researchers; (other), NASA/Science Photo Library/ Photo Researchers (Shuttle); 398(bl), Scott Van Osdol/HRW; 401(cl), Scott Van Osdol/HRW; (br), Michelle Bridwell/HRW; 402(cl), John Langford/HRW; (c), HRW; 403(tr), (cl), (bc), John Langford/HRW; 404(tl), (bc), John Langford/HRW; 405(c), John Langford/HRW; (br), Robert Wolf; 408(bl), Lennart Nilsson/*From The Incredible Machine*/National Geographic Society; 409(tr), Scott Van Osdol/HRW; 410(c), Douglas Mesney/Leo De Wys; 412(t), Scott Van Osdol/HRW; 413(bl), (tr), (cr), (br), HRW; 414(br), Scott Van Osdol/ HRW; 415(b), John Langford/HRW; 416(cl), Scott Van Osdol/HRW; (tr), Eric Schrempp/ Photo Researchers; 418(c), Courtesy of Robert Moog; 419(tl-ICON), Tony Stone Images; (tr), Peter Vadnai/Stock Market; (cl), Michael Lyon; 420(tl-ICON), HRW; (cl), Gary Benson/Tony Stone Images; 421(tl-ICON), Ron Kimball; (tr), Dr. Curtis R. Smith/Auburn University.

UNIT 7: Page 422-423(bkgd), West Stock; 422(tl), Kathleen A. Culbert/Tony Stone Images; (tr), Phototone Alphabets CD ROM by Esselte Letraset Ltd.; 424(t), Ronald Royer/Science Photo Library/Photo Researchers; (cl), Tim Beddow/Tony Stone Images; (b), Rafael Macia/Photo Researchers; 425(b), Peter Van Steen/HRW; 426(c), (br), Scott Van Osdol/HRW; 427(tr), Ed Pritchard/Tony Stone Images; (cr), Christian Grzimek/ Okapia/Photo Researchers; (b), Scott Van Osdol/HRW; (bc), Robert Wolf; (br), Martin Dohrn/Science Photo Library/Photo Researchers; (bl), Michelle Bridwell/HRW; 428(cr), Comstock; 429(r), Michelle Bridwell/HRW; 430(tr), Donald Smetzer/Tony Stone Images; (tc), Comstock; (tl), Mark C. Burnett/Photo Researchers; (br), Hale Observatories/ Science Source/Photo Researchers; 434(tr), Bettmann Archive; 435(cr), Scott Van Osdol/HRW; (bl), John Langford/HRW; 436(bl), Howard Sochurek/Stock Market; 438(tr), Scott Van Osdol/HRW; 439(c), Scott Van Osdol/HRW; 440(tr), (c), Michelle Bridwell/ HRW; 442(tr), J. A. Kraulis/MasterFile; (bl), HRW; (bkgd), J. A. Kraulis /MasterFile; 443(r), John Langford/HRW; 447(bl), (br), Michelle Bridwell/HRW; (bc), Gregory M. Nelson; 448(cl), Scott Van Osdol/HRW; (br), John Langford/HRW; 449(tr), Comstock; (cr), Manoj Shah/Tony Stone Images; (br), Paul Steel/Stock Market; 457(l), (bc), Sam Dudgeon/HRW Photo; (tc), Scott Van Osdol/HRW; (br), Peter Van Steen/HRW; 460(tr), (c), John Langford/HRW; (bl), John Langford/HRW/Original costume by Susan Branch, courtesy of Friends of the Summer Musical; 461(bl), Steve Ferry/HRW; 462(tl), Charles D. Winters/Photo Researchers; (cl), (bl), Scott Van Osdol/HRW; (cr), Richard Hutchings/ Photo Researchers; 463(tr), Peter Van Steen/HRW; 465(tr), Bettmann Archive; (br), Peter Van Steen/HRW; 466(tr), (bl), John Langford/HRW; (cr), HRW; 467(tr), (cr), John Langford/HRW; 468(cr), Michelle Bridwell/HRW; 469(tr), Russ Kinner/Comstock; (c), (bl), (br), John Langford/HRW; 470(br), Michelle Bridwell/HRW; 473(tc), (tr), (br), John Langford/HRW; 476(tr), HRW; (b), Runk/Schoenberger/Grant Heilman; 478(br), John Langford/HRW; 481(tr), (cr), John Langford/HRW; 482(tr), West Stock; (cl), Michelle Bridwell/HRW; 483(tr), Courtesy The Boston Red Sox; 484(tl-ICON), Rogge/Stock Market; (tc), (tr), Sovfoto/Eastfoto; (cl), Chad Ehlers/Tony Stone Images; 485(tr), Sharon Gaisford; (bl), Michelle Bridwell/HRW; 486 (tl-ICON), Tony Stone Images; (tc), NASA; (tr), Superstock; 487(tl-ICON), Private Collection of Garrett A. Morgan Family; (tc), US Patent and Trademark Office; (bc), Superstock.

UNIT 8: Page 488-489(bkgd), BIOS (R. Leguen)/Peter Arnold; 490(t), Michelle Bridwell/HRW; (cr), Steven Ferry; 491(t), Stephen Trimble; (cr), Steven Ferry; (b), Erika Stone/Peter Arnold; 492(r), Michelle Bridwell/HRW; (br), Dennie Cody/FPG Int'l.; 493(bl), Science VU/Visuals Unlimited; 494(tl), Tom Lyle/Stock Shop; (tlc), Wiesehahn/ Stock Shop; (cl), (c), (bc), (tc), HRW; (bl), Steven Ferry; 495(bl), (bc), (br), Michelle Bridwell/HRW; 498(tl), Manfred Kage/Peter Arnold; (cl), Runk/Schoenberger/Grant Heilman; (bl), Michael Abbey/Photo Researchers; (br), Bruce Iverson; 500(tl), Biophoto Associates/Science Source/Photo Researchers; (b), Runk/ Schoenberger/Grant Heilman; 502(r), Michael Abbey/Science Source/Photo Researchers; 504(b), Biology Media/Photo Researchers; 505(c), T. R. Broker/Phototake; 506(br), Bettmann Archive; 509(br), John Langford/HRW; 510(tr), HRW; 512(t), (tc), HRW; 513(cr), The Kobal Collection; (br), Retna Ltd.; 514(br), Custom Medical Stock Photo; 515(tc), Philip Matson/Tony Stone Images; (bkgd), Lennart Nilsson/From *A Child Is Born*, Dell Publishing; 518(b), Lennart Nilsson/ From A Child Is Born, Dell Publishing; 520(br), De Keerle/Gamma-Liaison; 522(tc), Lennart Nilsson/From *Behold Man*, Little, Brown and Co.; (cr), (bc), Lennart Nilsson/From *The Incredible Machine*, National Geographic Society; 523(r), Custom Medical Stock Photo; 524(r), Lennart Nilsson/From *A Child Is Born*, Dell Publishing Co.; 525(tl), Lennart Nilsson/From *A Child Is Born*, Dell Publishing Co.; 526(l), Lennart Nilsson/From *Behold Man*, Little Brown and Co.; (c), Lennart Nilsson/From *A Child Is Born*, Dell Publishing Co.; 527(tr), Lennart Nilsson/From *A Child Is Born*, Dell Publishing Co.; 528(tr), Keith/Custom Medical Stock Photo; (br), Ariel Skelley/Stock Market; 529(tr), John Colwell/Grant Heilman; 531(bl), David York/Stock Shop; 532(tl), Leonard Lessin/Peter Arnold; (tr), Millard H. Sharp/Photo Researchers; (cl), C & M Denis-Huot/Peter Arnold; (c), Adam Hart-Davis/Science Photo Library/Photo Researchers; 533(r), HRW; 534(tr), (cr), Steven Ferry; (cl), HRW; (br), Michael Tamborrino/Stock Shop; 535(tr), So. Ill. University/Photo Researchers; (br), Biology Media/Photo Researchers; 536(tr), Runk/Schoenberger/Grant Heilman; (c), Rosalind Franklin/From *The Double Helix* by James D. Watson, 1968 Atheneum Press, NY; (cr), CSHL Archives/Peter Arnold; 537(tr), A. C. Barrington Brown/ Photo Researchers; (r), Dan Richardson/Stock Shop; 538(cl), CNRI/Science Photo Library/ Photo Researchers; 539(t), John Langford/HRW; 540(c), Steve Prezant/Stock Market; 541(bl), (br), Lanpher Productions; 542(tr), Enrico Ferorelli; (br), Erika Stone/Peter Arnold; 544(bl), M. Baret/Science Source/Photo Researchers; (l), Peter J. Kaplan/Stock Shop; (br), Scott Spiker/Stock Shop; 545(bl), Michelle Bridwell/ HRW; 546(tl), Human Genome Management Information Systems; (br), Jeffrey Reed/Stock Shop; 547(t), Runk/ Schoenberger/Grant Heilman; (br), John Langford/HRW; 548(tl), Specular Replica CD ROM/Specular International; (tc), (tr), Michelle Bridwell/HRW; 549(t), (cr), HRW; (tr), Michael S. Thompson/Comstock; 550(cl), Michelle Bridwell/HRW; (tr), BIOS (R. Leguen)/ Peter Arnold; 551(b), Dan Richardson/Stock Shop; 552(tl-ICON), Ron Kimball; (l), (tc), (tr), Ambergene; 553(tl-ICON), Howard Sochurek/Stock Market; (tc), (tr), Jim Gipe Photo/ Courtesy Tufts University Dept. of Anatomy and Cellular Biology; (bc), Tom Smith/ Courtesy Tufts University Dept. of Anatomy and Cellular Biology; 554(tl-ICON), (bl), Diego

Goldberg/Sygma; (r), HRW; 555(tl-ICON), (tc), HRW; (bl), Cornell University New York State Center for Advanced Technology.

Art Credits

Abbreviated as follows: (t) top; (b) bottom; (l) left; (r) right; (c) center.

Illustration; 344(t), Mark Persyn Illustration; 347, Blake Thornton/Rita Marie & Friends; 348, Blake Thornton/Rita Marie & Friends.

UNIT 6: Page 361, Blake Thornton/Rita Marie & Friends; 364-365, Boston Graphics, Inc.; 366, Howard Fullmer/Cary & Company; 369, Jack Graham/Carol Chivlovsky Design, Inc.; 370-371, Boston Graphics, Inc.; 377(r), Patti Bonham/Washington-Artists' Represents, Inc.; 381, Darrel Tank/Sweet Represents; 385, Boston Graphics, Inc.; 386, Frank Demes; 388-389, Tim Ladwig/Suzanne Craig Represents; 392, Boston Graphics, Inc.; 393, Boston Graphics, Inc.; 393-394, Martens & Keiffer/Carol Chivlovsky Design, Inc.; 397(t), Michael Koester/Woody Coleman Presents; 397(c), Jack Graham/Carol Chivlovsky Design, Inc.; 397(b), David Chen/James Conrad Artist Representative; 399, Tim Ladwig/Suzanne Craig Represents; 402(t), Michael Koester/Woody Coleman Presents; 406, Valerie Marsella; 407, Martens & Keiffer/Carol Chivlovsky Design, Inc.; 409, Stephen Durke/Washington-Artists' Represents, Inc.; 412(c), Stephen Durke/Washington-Artists' Represents, Inc.; 414, Aletha Reppel/Suzanne Craig Represents.

UNIT 7: Page 428, Darrel Tank/Sweet Represents; 431-435, Patti Bonham/Washington-Artists' Represents, Inc.; 437, Jim Pfeffer; 443-445, Darrel Tank/Sweet Represents; 446(t), Darrel Tank/Sweet Represents; 446(b), Tim Ladwig/Suzanne Craig Represents; 450(t), Patti Bonham/Washington-Artists' Represents, Inc.; 450(b), Stephen Durke/Washington-Artists' Represents, Inc.; 451, Stephen Durke/Washington-Artists' Represents, Inc.; 453, Stephen Durke/Washington-Artists' Represents, Inc.; 454, Tim Ladwig/Suzanne Craig Represents; 456, Blake Thornton/Rita Marie & Friends; 468(t), Darrel Tank/Sweet Represents; 468(b), Stephen Durke/Washington-Artists' Represents, Inc.; 471(t), Darrel Tank/Sweet Represents; 471(b), Lori Anzalone/Jeff Lavaty Artist Agent; 472(l), David Fischer; 472(r), Martens & Keiffer/Carol Chivlovsky Design, Inc.; 474, Darrel Tank/Sweet Represents; 475, Patti Bonham/Washington-Artists' Represents, Inc.; 477, Stephen Durke/Washington-Artists' Represents, Inc.; 478, Uhl Studio; 479, Stephen Durke/Washington-Artists' Represents, Inc.

UNIT 8: Page 490, Martens & Keiffer/Carol Chivlovsky Design, Inc.; 496(t), Barbara Kiwak/Barbara Gordon Associates Ltd.; 497, Joani Pakula; 499, Jack Graham/Carol Chivlovsky Design, Inc.; 502, Aletha Reppel/Suzanne Craig Represents; 503(t), Aletha Reppel/Suzanne Craig Represents; 503(b), Bill Geisler; 504(t), Bill Geisler; 504(b), Aletha Reppel/Suzanne Craig Represents; 507-509, Joani Pakula; 516-517, David Fischer; 519, Bill Geisler; 521, Bill Geisler; 527, Bill Geisler; 530(t), Keith Locke/Suzanne Craig Represents; 530(b), Jack Graham/Carol Chivlovsky Design; 531, Don Sullivan; 537, Bill Geisler; 545, Bill Geisler.

SOURCEBOOK ILLUSTRATION CREDITS

UNIT 1: Page S2(br), Academy Artworks, Inc.; S3(tr), Academy Artworks, Inc.; S4(tr), Academy Artworks, Inc.; S5(tc), Academy Artworks, Inc.; S6(cl), Academy Artworks Inc.; S7(c), Academy Artworks Inc.; (br), Mark Persyn; S8(c), Academy Artworks Inc.; S9(cr), Martens & Keiffer/ Carol Chislovsky Design Inc. Design Inc.; S10, Wendy-Smith Griswold/ Melissa Turk & The Artist Network & The Artist Network; S11(tr), Martens & Keiffer / Carol Chislovsky Design Inc. Design Inc.; S12(ct) Greg Harris / Cornell & McCarthy; S14(tl), Uhl Studio; S14(cb), Academy Artworks, Inc.; S16(tc), Martens & Keiffer / Carol Chislovsky Design Inc. Design Inc.; S17(tl), Martens & Keiffer / Carol Chislovsky Design Inc. Design Inc.; S17(cr), Martens & Keiffer / Carol Chislovsky Design Inc. Design Inc.; S18(c), Martens & Keiffer / Carol Chislovsky Design Inc. Design Inc.; S18(tr), Martens & Keiffer / Carol Chislovsky Design Inc. Design Inc.; S20(tc), Glasgow & Associates.

UNIT 2: Page S22(bl), Academy Artworks, Inc.; S23(cr), Academy Artworks, Inc.; S24(l), Richard Wehrman; S26(tl), John Francis; S26(cr), Mark J. Persyn; S28(cl), John Francis; S29(tr), Academy Arts Inc., S29(cr), Academy Arts Inc.; S29(bl), Academy Arts Inc.; S31(b), Liaison Production Services; S32(b), Uhl Studio; S33(tr), John Francis; S33(b), John Francis; S34(tl), Uhl Studio; S35(tc), Uhl Studio; S36(b), Uhl Studio; S38(tl), Academy Arts, Inc.

UNIT 3: Page S43(t), Bob Dorsey; Page S43(b), Bob Dorsey; S45(t), David Merrell / Suzanne Craig Represents; S48(c), Joe McDermott / Koralik Associates; S49(b), Joe McDermott / Koralik Associates; S50(t),Patti Bonham / Washington; S51(t), Darrel Tank; S52(b), Gary Locke / Suzanne Craig; S53(tr) Joe McDermott / Koralik Associates; S54(t), Darrel Tank.

UNIT 4: Page S60(b), Steven Adler / Jeff Lavaty Artist Agent; S61, Steven Adler / Jeff Lavaty Artist Agent; S62(b), Greg Harris / Cornell & McCarthy; S63, Stephen Durke / Washington-Artists' Represents, Inc.; S64(all), Craig Attebury / Jeff Lavaty Artist Agent; S65(c), John Francis; S66, Richard Wehrman; S67, Stephen Durke / Washington-Artists' Represents,Inc.; S68(all), Stephen Durke / Washington-Artists' Represents, Inc.; S69(cr), Glasgow & Associates; S69(b), Liaison Production Services; S70(b), Stephen Durke / Washington-Ariitst' Represents, Inc.; S71, Todd Lockwood / Carol Guenzi Agents; S72, Glasgow & Associates; S73, Craig Attebury / Jeff Lavaty Artist Agent.

UNIT 5: Page S83(br), Keith Locke / Suzanne Craig Represents; S84(lc), Joani Pakula; S85(tr), Mark Persyn; S86(b), Don Brautigam / Bill Erlacher, Artists Associates; S87(t), Don Brautigam / Bill Erlacher, Artists Associates; S90(tl), Blake Thornton / Rita Marie & Friends; S92(t), Darrel Tank; S93(rc)(rb), Mark Persyn; S94(tr), Mark Persyn; S95(cb), Brian Harrold / American Artists Rep., Inc.; S96, Mark Persyn; S97(all), Jim Pfeffer; S98(t), Jim Pfeffer; S98(b), Tony Randazzo / American Artists Rep., Inc.; S99, Tony Randazzo / American Artists Rep., Inc.; S100, Tony Randazzo / American Artists Rep., Inc.; S101, Blake Thornton / Rita Marie & Friends; S102, Blake Thornton / Rita Marie & Friends; S104, Digital Art.

UNIT 6: Page S106(rc), Uhl Studio; S107, Todd Lockwood / Carol Guenzi Agents; S108(t), Todd Lockwood / Carol Guenzi Agents; S108(b), Uhl Studio; S109, Todd Lockwood / Carol Guenzi Agents; S110(t)(b), Uhl Studio; S111(b), Todd Lockwood / Carol Guenzi Agents; S114(c), Todd Lockwood / Carol Guenzi Agents; S115(b), Uhl Studio; S116, Paul Hess / Garden Studio Illustrators; S117, Todd Lockwood / Carol Guenzi Agents; S118(tl), Uhl Studio; S120, Uhl Studio; S121(tr),Wendy Smith-Griswold / Melissa Turk & The Artist Network.

UNIT 7: Page S128(tl), Joani Pakula; S128(bc), Keith Locke / Suzanne Craig Represents; S129(tl), Joani Pakula; S130, Liaison Production Services; S131, Brian Harrold / American Artists Rep., Inc.; S132(c), Brian Harrold / American Artists Rep., Inc.; S133(tr), Brian Harrold / American Artists Rep., Inc.; S135(bl), Liaison Production Services; S136, Liaison Production Services; S137, Digital Art; S138(c)(b), Brian Harrold / American Artists Rep., Inc.; S141, Uhl Studio; S144, John Francis; S146, Digital Art; S147(b), George Ladas / Melanie Kirsch Artist Represents; S148(t), Digital Art; S152, Liaison Production Services.

UNIT 8: Page S155, John Francis; S156, Joel Spector; S157(b), David Fischer; S158, Jean E. Calder; S159(b), Jean E. Calder; S160, Morgan Cain & Associates; S162, John Francis; S163, Michael Krone; S164, Joyce Kitchell / James Conrad; S165, Wendy Smith-Griswold / Melissa Turk & The Artist Network; S167(all), Martens & Keiffer / Carol Chislovsky Design Inc.; S168, Martens & Keiffer / Carol Chislovsky Design Inc.; S169, Martens & Keiffer / Carol Chislovsky Design Inc.; S170, Martens & Keiffer / Carol Chislovsky Design Inc.; S172, Jean E. Calder; S173(cl)(br), Morgan Cain & Associates; S174, Jean E. Calder; S175(c), Morgan Cain & Associates; S176(b), David Fischer; S177(br), Martens & Keiffer / Carol Chislovsky Design Inc. Design Inc.

Acknowledgment

For permission to reprint copyrighted material grateful acknowledgment is made to the following source:

The Humana Press, Inc.: From *The Life Within: Celebration of a Pregnancy* by Jean Hegland. Copyright © 1991 by the Humana Press, Inc.